INTRODUCTION TO
MOLECULAR BIOPHYSICS

CRC SERIES *in*
PURE *and* APPLIED PHYSICS

Dipak Basu
Editor-in-Chief

PUBLISHED TITLES

Handbook of Particle Physics
M. K. Sundaresan

High-Field Electrodynamics
Frederic V. Hartemann

Fundamentals and Applications of Ultrasonic Waves
J. David N. Cheeke

Introduction to Molecular Biophysics
Jack Tuszynski
Michal Kurzynski

INTRODUCTION TO
MOLECULAR
BIOPHYSICS

Jack A. Tuszynski
Michal Kurzynski

CRC Press
Taylor & Francis Group
Boca Raton London New York

CRC Press is an imprint of the
Taylor & Francis Group, an **informa** business

CRC Press
Taylor & Francis Group
6000 Broken Sound Parkway NW, Suite 300
Boca Raton, FL 33487-2742

First issued in paperback 2020

Library of Congress Card Number 2002031592

ISBN 13: 978-0-367-57854-1 (pbk)
ISBN 13: 978-0-8493-0039-4 (hbk)

Library of Congress Cataloging-in-Publication Data

Tuszynski, J.A.
   Introduction to molecular biophysics / Jack Tuszynski, Michal Kurzynski.
      p. cm. — (CRC series in pure and applied physics)
   Includes bibliographical references and index.
   ISBN 0-8493-0039-8 (alk. paper)
      1. Molecular biology. 2. Biophysics. I. Kurzynski, Michal. II. Title. III. Series.

QH506 .T877 2003
572.8—dc21   2002031592

Preface

Biology has become an appealing field of study for growing numbers of physicists, mathematicians, and engineers. The reason is obvious. Extensive media coverage has made much of the world familiar with biology's critical role on the front lines of scientific research. Former U.S. President Clinton said that the last 50 years belonged to physics and the next 50 will belong to biology. His assessment requires a slight correction: the last 300 years focused on physics. Only the last 10 or 20 concentrated on biology, but the concentration will certainly continue as technology accelerates progress.

The connection between the physical and biomedical sciences developed rapidly over the past few decades, particularly after the ground-breaking discoveries in molecular genetics. The need clearly exists for continuing dialogues and cross-fertilization between these two groups of scientists. Ideally, neither group should attempt to "civilize" the other. As a result of the interdisciplinary nature of modern life sciences, new areas of endeavor such as mathematical biology, biophysics, computational biology, biostatistics, biological physics, theoretical biology, biological chemistry (and its older sister, biochemistry), and biomedical engineering are emerging rapidly and contributing important information to our understanding of life processes.

This new appeal of biology and our growing knowledge of physical concepts that play important roles in biological activities have not proceeded without significant friction among the disciplines. The representative quotes below reflect the mutual apprehension evident over decades (if not centuries) of co-existence. Sydney Brenner, a biologist and recent Noble Prize winner, says:

> Biology differs from physics in that organisms have risen by natural selection and not as the solutions to mathematical equations.
> Biological organisms are not made by condensation in a bag of elementary particles but by some very special processes that are, of course, consistent with the laws of physics but could not easily be directly derived from them.
> The trouble with physics is that its deepest pronouncements are totally incomprehensible to almost everybody except the deepest physicists, and while they may be absolutely true, they are pretty useless ... to understand *E. coli*.
> In biology it is the detail that counts, and it counts because it is that natural selection needed to accomplish for there to be anything at all.

Of course physicists have other views in this matter:

> There is a feeling that something is missing .... Molecular biology has revolutionized the understanding of how biological processes work, but ... not ... *why*.
> (J. Krumhansl, 1995)

Physics studies replicas of identical atoms or molecules while in biology a single group of atoms existing in only one copy produces orderly events tuned in with each other and the environment. There is difficulty in describing life using ordinary laws of physics. While a new principle is needed, it should not be alien to physics. (E. Schrödinger, 1967)

Molecular biology remains a largely descriptive science.

Even the best known systems in biology may not be as well understood as is generally believed, which means that understanding is incomplete, and may even be misplaced. (J. Maddox, 1999)

It is legitimate to ask whether the two sciences and their objects of study use operating principles at variance with each other. The principles applied in physics to mathematically describe inanimate matter focus on:

- Universality
- Conservation laws
- Minimum energy conditions
- Maximum entropy
- Reproducibility

On the other hand, the principles governing the biological study (seldom mathematical) of animate matter appear to emphasize:

- Uniqueness
- Diversity
- Complexity
- Heterogeneity
- Robustness and adaptability to the environment and competition
- Maximum stability
- Evolutionary achievement retention (history)
- Nonequilibrium (open) character
- Hierarchical organization and interlocking of segments
- Communication, signaling, information (perhaps even meaning and intelligence)
- Repetition of motifs (all proteins are formed from 20 amino acids; all DNAs are formed from 4 nucleotides)

In essence, the difference between living organisms and inanimate matter is the ability of living organisms to reproduce, adapt, and control key biological events with great precision. Cells that cannot coordinate these activities will not survive. Many molecules found in living organisms are large and complex. Proteins are the most varied and have the most diverse range of functions. Their molecular masses range from tens of thousands up to millions of hydrogen masses. Conversely, the chemical subunits comprising biological molecules are not nearly so varied; essentially 20 amino acids serve as the building blocks of all proteins.

Diversity is a simple result of a multitude of possible combinations of a finite number of structural elements. The functioning of biological systems must also be derived from this complexity. The specific organizations of complex molecular systems provide specific functions but continue to be governed by fundamental physical laws. The principle of complexity begetting function is familiar to physicists and has often been referred to as an emergent phenomenon. It is characteristic of atomic systems to display new properties as they become more complex (e.g., the emergence of structural rigidity when a crystal is grown from its constituent atoms).

This hierarchical, interconnected, and synchronous organization of systems that sustains life poses perhaps the greatest challenge to our intellects. However, it is hard to believe that the mysteries of life can be solved without physics. It is also doubtful that the use of standard physics rules alone will solve the mystery of life and establish its scientific basis. Our search has been greatly aided by the proliferation of sophisticated experimental techniques that physics has devised; they include (Parsegian, 1997):

- Light microscope (resolution: 400 to 600 nm)
- Electron microscope (resolution: 10 to 100 nm)
- Neutron scattering (resolution: 1 to 10 Å)
- X-ray crystallography (resolution: 1 Å)
- Scanning tunneling microscope
- Atomic force microscope
- Nuclear magnetic resonance; magnetic resonance imaging
- Fluorescence spectroscopy
- Positron emission spectroscopy
- Microwave absorption
- Laser light scattering
- Laser tweezers

New physical concepts developed principally by nonlinear physicists are being evaluated as possible theoretical frameworks within which living systems can be better understood. These concepts involve:

- Nonlinearity
- Self-organization
- Self-similarity
- Cooperation versus competition (e.g., prey–predator models)
- Collective behavior (e.g., synergetics)
- Emergence and complexity

The full significance of these factors will not be known until a sufficient number of test cases are closely investigated. Due to their promise, however, we have included two appendices that summarize the most important ideas and results involved in nonlinear physics and phase transitions found in many-body interacting systems.

It is now generally accepted that the laws of physics apply to living organisms as much as they apply to inanimate matter. Attempts at applying physical laws to living systems can be traced to the early creators of modern science. Galileo analyzed the structures of animal bones using physical principles, Newton applied his optics to color perception, Volta and Cavendish studied animal electricity, and Lavoisier showed that respiration is simply another oxidative chemical reaction.

Robert Mayer was inspired by physiological studies to formulate the first law of thermodynamics. A particularly fruitful area of application of physics to physiology is hydrodynamics. Poiseuille analyzed blood flow by using physics principles. Air flow in the lungs has been described consistently via the laws of aerodynamics.

An important figure in the history of biophysics is German physicist and physiologist Hermann von Helmholtz who laid the foundations for the fundamental theories of vision and hearing. A long list of physicists made large impacts on biology and physiology. We will only name a few who crossed the now-disappearing boundary between physics and biology. Delbrück, Kendrew, von Bekesy, Crick, Meselson, Hartline, Gamow, Schrödinger, Hodgkin, Huxley, Fröhlich, Davydov, Cooper, and Szent-György (1972) have undoubtedly pushed the frontiers of life sciences in the direction of exact quantitative analysis. We hope and expect that the work they started will accelerate in the 21st century.

Physics has proven helpful in physiology, biology, and medical research by providing deeper insights into the phenomena studied by these sciences. In some fields of investigation, physics studies produced major analytic and diagnostic tools in the area of electrophysiology. Membranes of nerve cells are characterized by a voltage gradient called the action potential. The propagation of action potentials along the axons of nerve cells is the key observation made in investigating brain physiology. The theory of action potential propagation was developed by Huxley and Hodgkin, who earned a Nobel Prize for their discovery.

Likewise, the discovery of the structure of DNA by Crick and Watson sparked creation of a new discipline called molecular biology, which would not have been possible without experimental and theoretical tools developed by physicists. In this case, x-ray crystallography revealed the double helix structure of DNA.

More recently, investigations of DNA sequences have been pursued in the hope of revealing molecular bases for inherited diseases. Gel electrophoresis and fluorescent labelling are the crucial techniques perfected by physicists and biochemists for the studies of DNA sequences. Techniques that originated in physical laboratories have become standard equipment for most molecular biologists and chemists. Such devices usually start as probes of physical phenomena; they are later adapted for molecular biology and eventually transformed into common diagnostic and therapeutic tools. X-ray machines are used to detect abnormalities. Nuclear magnetic resonance (NMR), now called magnetic resonance imaging (MRI), aids in detecting tumor growth; tumors in turn can be treated by radiation.

Cardiologists use electrocardiography (ECG) to monitor heart activity; neurosurgeons can study electrical impulses in the brain via electroencephalography (EEG). Ultrasound has applications in diagnostic (e.g., fetal development) and therapeutic (gall and kidney stone shattering) fields. Optical fibers are used for noninvasive examination of internal organs (Tuszynski and Dixon, 2002).

This book is intended as a broad overview of molecular biophysics — the science that combines mathematics, physics, chemistry, and biology techniques to determine how living organisms function. The questions posed by physicists are, for example, how does the brain process and store information? How does the heart pump the blood throughout the body? How do muscles contract? How do plants extract light energy in photosynthesis? While biologists, physiologists, and geneticists work toward answering the same questions, biophysicists focus on the physics and physical chemistry of the processes. The questions apply to various levels of complexity and structural organization. On a large scale, biophysicists study how organisms develop and function. At a smaller scale, they investigate individual organs or tissues, for example, the nervous system, the immune system, or the physics of vision. Other groups quantify processes such as cell division that take place within single cells. Finally, at the finest level of organization, molecular interactions are analyzed via sophisticated experimental and theoretical techniques that overlap the areas of genetics, cell biology, biochemistry, and molecular physics.

The hierarchies listed above are interlocked and it is not always well-advised or even possible to confine investigations to a certain level of organization. This book will serve as a guided tour through the interlocking hierarchies, starting from the smallest molecular building blocks of life and ending with a panoramic view of the evolving landscape of living forms. The objects of study belong to the realm of biology; the language of description will be physics with sprinklings of mathematics and chemistry as needed. Since life is a far-from-equilibrium process (or a complex nonlinear fabric of interdependent processes), some aspects of the book will require introduction to the key ingredients of nonlinear physics in order to convey ideas clearly.

Biophysics is the study of the physics of certain complex macromolecular systems — cells and organisms — that function under conditions of insignificant temperature and pressure changes. An organism can be thought of as an intelligent, self-controlled, chemical machine that is self-regulated by molecular signals, molecular receptors, and transducers of information. The basic biological functional subsystems are nucleic acids, biopolymers (peptide chains), proteins, and specialized proteins called enzymes.

Biophysicists seek to understand biophysical processes by accounting for intramolecular and intermolecular interactions, and their resulting electronic and structural conformational changes; and by studying the transfers of electrons, protons, metallic ions, and energy within biological systems. In solid state physics, such problems are solved by the methods of quantum mechanics, statistical physics, and equilibrium and nonequilibrium thermodynamics. However, since isolated biophysical systems are not found in nature, the description is complicated by the openness of living systems and their far-from-equilibrium natures.

Studies of biological systems have been advanced and clearly dominated in the past by biochemistry, molecular and structural biology, and genetics. The domination accrued tremendous benefits, the most obvious of which are the availability of precise information regarding the chemical compositions of cells, macromolecules and other structural components, the discovery of the reaction pathways of the production of the synthesized components, and finally, the elucidation of the genetic code mechanism.

We may be witnessing the dawn of a new era in which physics and mathematics may find a new fertile ground for the application of their exact scientific methods and theories. While biochemistry studies mainly atoms in direct contact with each other, many biological phenomena arise from subtler, weaker, short- and long-range forces. The solvation and desolvation problem, for example, has yet to be treated theoretically due to inherent computational limits, although it is essential to the understanding of ligand binding in physiological environments.

Biological function results from specific chemical reactions and reaction cascades. Some molecules derive their functions solely from quantum mechanical interactions; others depend on classical interactions with surrounding molecules and external fields (e.g., electromagnetic fields). The main task facing theoretical biophysicists today is the investigation of the physical characteristics of biological molecules and very simple biological systems such as enzymes, functional proteins, and cellular membranes, while accounting for the openness of biological systems to the environment. Biological systems routinely exchange energy and matter with their environment. Many components of biological systems, such as proteins, undergo continual restructuring and renewal. Life is only possible because the timescale of protein stability is much longer than the timescales of the biological functions of proteins (Frauenfelder et al., 1999).

Finally, we should briefly address the concept of modeling due to its importance in the advancement of science. A biological model is often understood to be simply a diagram illustrating the interrelationships of various subsystems in a process. A biochemical model is typically a diagram of several complex chemical reactions for molecular pathways and possibly a table of values for their kinetic rates. A computational model is usually a computer simulation of a process with more or less arbitrary transition rules (e.g., a Monte Carlo or cellular automation model). However, a physical model is expected to be a theoretical description of a process involving a number of equations of motion stemming from the first principles (if possible), testable against a range of tunable experimental conditions. It must lead to a quantitative prediction and not simply reproduce already known results.

Hierarchical organization of knowledge

Every branch of science is more than a collection of facts and relations. It is also a philosophy within which empirical facts and observations are organized into a unified conceptual framework providing a more or less coherent concept of reality. Since biology is the study of life and living systems, it is simultaneously the study of human beings and it can be biased by philosophical and religious beliefs.

Understandably, biophilosophy has been the battleground for the two most antagonistic and long-lived scientific controversies between mechanism and vitalism. Mechanism holds that life is basically no different from nonlife; both are subject to the same physical and chemical laws. Living matter is simply more complex than nonliving matter. The mechanists firmly believe that life is ultimately explicable in physical and chemical terms. The vitalists, on the other hand, fervently argue that life is much more than a complex ensemble of physically reducible parts and that some life processes are not subject to normal physical and chemical laws. Consequently, life will never be completely explained on a physiochemical basis alone. Central to the vitalists' doctrine is the concept of *life force*, *vis vitalis*, or *elan vital*, a nonmaterial entity that is not subject to the usual laws of physics and chemistry. This life force is seen to animate the complex assembly and give it life. The concept is ancient and virtually universal, having appeared in some form in all cultures and providing the basis for most religious beliefs.

The early Greek philosopher–physicians such as Hippocrates were the first to consider organized concepts of the nature of life. The concepts developed within the framework of the medicine of that time and were based on clinical observations and conjecture. All functions of living things were considered the results of "humors" or liquids that had mystical properties and circulated within the body. Several centuries later, Galen founded the sciences of anatomy and physiology virtually single-handedly. He devised a complete, complex system based on his anatomical observations and an expanded concept of Hippocrates' humors. Galen's ideas were readily accepted and rapidly assumed the status of dogma, remaining unchallenged for more than a thousand years.

In the mid-16th century, Andreas Vesalius questioned the validity of Galen's anatomical concepts, performed his own dissections of human bodies, and published his findings in a book in 1543. His *De Humanis Corporus Fabricus* was the first anatomical text based on human dissection. In 1628, William Harvey published the first studies that described blood circulation as a closed circuit and the heart as the pumping agent. Vitalism, however, was still the only acceptable concept and Harvey naturally cited a *vital spirit* in the blood.

At mid-century, René Descartes, the great French mathematician, attempted to unify biological concepts of structure, function, and mind within a framework of mathematical physics. In his view, all life was mechanical and all functions were directed by the brain and nerves. He developed the mechanistic concept of a living machine complex that was fully understandable in terms of physics and chemistry. Even Descartes did not break completely with tradition. He believed that a machine required an *animating force* to produce life "like a wind or a subtle flame" located within the nervous system. Around the same time Malphighi, an Italian physician and naturalist, used the new compound microscope to study living organisms and found a wealth of detail and complexity.

Continued progress in biological sciences pushed the vitalistic view to the fringes of reputable science. The universe is estimated to be 14 to 16 billion years old. The solar system is about 4.6 billion years old and life on Earth is believed to have emerged 3.5 to 4 billion years ago. The life process is described in terms of its properties and functions including self-organization, metabolism (energy utilization), adaptive behaviors, reproduction, and evolution. A new approach was developed to explain the nature of the living state, namely functionalism. Functionalism implies that life is independent of its material substrate.

For example, certain types of self-organizing computer programs (lattice animals, Conway's Game of Life, etc.) exhibit life-like functions or *artificial life*. All components of living matter are in turn composed of ordinary atoms and molecules. This apparently demystifies life — an emergent property of biochemical processes and functional activities. The failure of functionalism can be seen in its inability

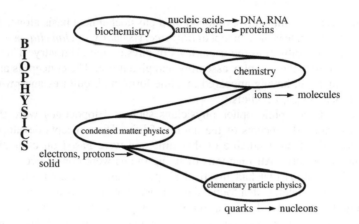

FIGURE 0.1
Hierarchical organization of science.

to consider the *unitary oneness* of all living systems. In 19th century biology, that characteristic of living systems, called the *life force*, or *life energy* was assumed to be of electromagnetic nature. Molecular biology systematically pushed vitalists (or animists) out of the spotlight by viewing the electromagnetic effects associated with life simply as effects, and not causes.

A.C. Scott (1995) describes an interesting aspect of the organization of science in his book titled *Stairway to the Mind*. Each branch of science exists almost autonomously on the broad scientific landscape and develops its own set of elements and rules governing interactions. This means that condensed matter physics would exist completely apart from elementary particle physics despite the fact that both disciplines use electrons and protons as the main building blocks for solid state systems. By extension, biology, as long as it does not violate the principles of physics, has little in common with physics. This hierarchical organization of knowledge allows only a tiny set of intersections between the hierarchies (see Figure 0.1) and involves:

- Human society
- Culture
- Consciousness
- Physiology of organisms
- Autonomous organs
- Neuronal assemblies (brains)
- Multiplex neurons (or other types of cells)
- Axon–dendrite-synapse systems
- Mitochondria–cell nuclei–cytoskeleton systems
- Protein–membrane–nucleic acid systems
- Phospholipid–ATP–amino acid systems
- Inorganic chemistry
- Atomic physics

- Molecular physics
- Nuclear physics
- Elementary particle physics

Mathematician W.M. Elsasser defined an *immense number* as $I = 10^{110}$. $I =$ atomic weight of the universe measured in proton masses (daltons) multiplied by the age of the universe in picoseconds (10^{-2} s). Since no conceivable computer, even one as big as the universe, could store a list of I objects and no person would live long enough to inspect it, an immense set of objects is a virtually inexhaustible arena for intellectual pursuits with no danger of running out of interesting relationships among its elements. Examples of immense sets are chess games, chemical molecules, proteins, nerve cells, musical compositions, and personality types. Thus, biology and physics are both legitimate areas of scientific exploration that can happily coexist with minimal overlap.

What makes this somewhat simplistic separation less applicable to the biology–physics divide is the existence of so-called emergent phenomena. We can better understand areas on a higher plane by knowing the organization rules for the elements whose roots are in the lower levels of the hierarchy. Knowing the interaction principles of electrons and protons certainly helped develop solid state physics. The knowledge of protein–protein interactions should give us a glimpse into the functioning of a dynamical cell. In general, the whole is more than the sum of its parts. Each level of the hierarchy adds new rules of behavior to the structure that emerges.

Scope of biophysics

Physics and biology have interacted for at least two centuries with a defining moment provided by Galvani's experiments on the electrical basis of the muscle contraction in frog legs. Another watershed event was Schrödinger's book titled *What is Life* which, while containing incorrect ideas and largely ignored by biologists, suddenly gave scientific legitimacy to physicists intrigued by living systems.

Physics has made many conceptual and experimental contributions to biology. The most important ones are x-ray structure determination of the double helix of DNA and crystallographic studies of proteins and macromolecules. The limitations of crystallography (poor availability of good crystals and inability to reveal positions of water molecules) have been overcome by the use of NMR and neutron diffraction. Neurophysiology gained significantly from the application of neural network models. Population biology and models of evolution were enhanced by sophisticated mathematical techniques such as Lotka-Volterra systems of differential equations.

The dialogue between physics and biology has been boosted by several key advocates. Szent-Gyorgi (1972) predicted the existence of conduction bands in proteins and the role of electrons in life processes. Per-Olov Loewdin (1989) postulated high organization levels of biological systems and the possibility of phase correlations. Herbert Froehlich (1968) proposed a theory of long-range phase correlations,

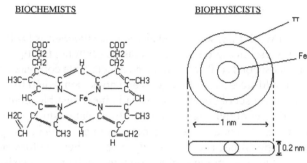

HEME AS SEEN BY

BIOCHEMISTS

BIOPHYSICISTS

EMPHASIS ON

INDIVIDUAL MOLECULES	COLLECTIVE PHENOMENA
CHEMICAL COMPOSITION	ENERGY BANDS
CONFIGURATION	SYMMETRIES
BINDING ENERGIES	COLLECTIVE EXCITATIONS
CHEMICAL REACTIONS	TRANSPORT PROPERTIES

FIGURE 0.2
A contrasting view of the heme as seen by biochemists and biophysicists.

i.e., biological coherence. Ilya Prigogine (1980) introduced pattern formation and nonlinear chemical kinetics into various branches of biological sciences including embryology. A.S. Davydov (1982) formulated a quantum mechanical model of transport in biological macromolecules that led to the prediction of solitonic models of excitation. Alwyn Scott (1995) promoted the ideas of nonlinearity in biology, in particular the concept of emergence.

It is noteworthy that biophysicists and biochemists perceive structures and functions differently (see Figure 0.2). Biophysicists have traditionally studied the following topics:

1. Thermodynamics of nonequilibrium processes (reaction–diffusion processes, entropy and other biological functions, energetics in mitochondria, actions of enzymes)
2. Membrane biophysics (thermodynamics of passive and active transport, structures and phases of membranes and lipid bilayers)
3. Nerve impulses (generation and propagation of nerve impulses, synaptic transmissions)
4. Mechanochemical processes (structures of muscle and muscle proteins, muscle contraction, movements of cilia and flagella, molecular motors, cell motility)
5. Mitochondrial processes (thermodynamics of oxidative phosphorylation, bioenergetics)
6. Photobiological processes (photosynthesis, fluorescence, chlorophyll and pigments, photoreception, membranes of photoreceptors)

7. Nonlinear dynamic processes (autocatalytic chemical reactions, nonlinear enzymatic reactions, periodic phenomena in membranes, cell cycle kinetics, heartbeat and arrhythmias, ECG and EEG processing, brain waves)
8. Developmental problems (evolution of living forms, ontogeny [cell division and growth], self-organization, epidemiology, gene mutations)

Biophysicists have just started the journey toward understanding biological structures and functions. Many biological phenomena are still largely unexplained, including:

1. Protein folding
2. Light-energy conversion
3. Muscle contraction
4. Infrared light detection of centrioles
5. Intracellular signaling via microtubules
6. Electronic and protonic conduction of biopolymers
7. Molecular level protein–protein interactions
8. Supramolecular protein assemblies
9. Cell division
10. Flagellar motors
11. Motor protein motion (especially their directionality)
12. Synaptic plasticity

And at a much grander scale, questions such as:

13. What is life?
14. What is a mind?

What is life?

Theories of life must include information about messages and reading devices at all levels of the hierarchy, not only at the DNA/RNA level. Just as letters spell words and words form sentences that have meanings, sentences can become chapters in books that are collected in libraries serving as repositories of human knowledge. Living systems are organized in similar fashion. The elements of a lower level hierarchy provide instructions, meanings, and signals to superior levels. The components are DNA, RNA, peptides, proteins, filaments, membranes, organelles, cells, cell assemblies, etc.

The physically interesting features of biological systems are dimensions (Table 0.1), organization of and cooperation between subsystems, sensitivities to external influences, stability, adaptability, dipolar characters of constituents, modes far from equilibrium, and nonlinear responses to external perturbations.

TABLE 0.1
Dimensions of Living System

Organism		10^{20} Atoms	Thermodynamics
Cell		10^{10} Atoms	Mesoscale
System		10^{5} Atoms	Mesoscale
Biomolecule		10^{3} Atoms	Mesoscale
Molecule		10^{1} Atoms	Quantum Chemistry
Atom		1 Atom	Quantum Physics

History and Physics

It is customary to think that unlike biology, physics is indifferent to history. Point masses in Newton's equations of motion move along periodic trajectories, then return near or to the same point in space. In the case of chaotic trajectories, as stated in Poincare's theorem in the late 19th century, point masses following chaotic motion return arbitrarily close to initial conditions. Equations of motion in classical mechanics, electrodynamics, and quantum mechanics are time-reversal invariant.

While Boltzmann introduced the arrow of time to describe a tendency to reach equilibrium, it applied to the future of a physical system, not its past. A typical sentence in a thermodynamic text is, "Consider 1 liter of hydrogen and 1 liter of oxygen in a closed container" to describe a system out of equilibrium. One might guess that the system will become a mixture of the two gases and they will form a puff of compressed steam. The process of reaching thermodynamic equilibrium is irreversible and a system may start from the same initial condition and follow a number of distinct pathways. Consequently, the knowledge of the final state is not sufficient to allow conclusions about the initial state of the system. However, one may still ask whether the puff of steam from this example is indeed in a state of thermodynamic equilibrium.

The atomic composition of a system before and after decompression is fixed. A water system has a certain number of oxygen atoms and twice as many hydrogen atoms. The main isotope of oxygen has the same number of protons and neutrons; hydrogen nuclei only contain protons. We now know the proton and neutron are simply two states of the same particle — a nucleon. Protons are in a low-energy state; neutrons occupy the excited high-energy state. The slightly higher mass of the neutron corresponds to a large energy difference, namely 1.3 MeV or in thermal energy units a temperature difference of 1.5×10^{10}K. If all the nucleons were in thermodynamic equilibrium with the electromagnetic radiation of the universe at 3K temperature, practically all would occupy their ground state and manifest themselves as protons. A universe in equilibrium would have to be composed exclusively of hydrogen. Since our universe is composed of oxygen, carbon, nitrogen, and other elements in addition to hydrogen, it is far from thermodynamic equilibrium.

Transformations of nucleons take place as results of very weak interactions with neutrinos. The atomic composition of the universe is a memory of this event that took place a second after the Big Bang, when the density of neutrinos was insufficient to force fast enough transitions between the neutron and proton states of the nucleons. These transitions could only take place spontaneously. The spontaneous transitions later ceased and nucleon composition became fixed.

Each physical system has structure, organization, and constraints imposed on its motion. These characteristics contain a memory of a past event when the structure, organization, or constraints became frozen, but not in the sense that we freeze or fix initial conditions when solving equations of motion. Freezing is a kinetic process that does not contradict the second law of thermodynamics guaranteeing a tendency to attain the state of thermodynamic equilibrium (thermal death of the universe). However, the time it takes to reach equilibrium is much longer than the time courses of the events and processes we try to explain. Kinetic freezing of a structure is believed to follow the laws of physics. Nonetheless, it has so much randomness (in the sense of deterministic chaos) that we are unable to deduce the nature of the underlying structure from the laws of physics. Hence, it must be introduced here as an externally supplied piece of information.

Our knowledge of physical structures allows us to try to reconstruct the history of the Earth, the solar system, the universe, and even time. The efforts of many physicists focus on these areas of investigation today. It is a general consensus that the laws of physics are well understood and it is time to apply them to systems and processes with high degrees of complexity.

Without a doubt, the greatest challenge for physicists today is understanding the phenomenon of life. Living systems are extremely complex and organized hierarchically as a result of evolutionary processes almost as old as the Earth. Not suprisingly, biology was the first branch of science that attempted to reconstruct past events from modern knowledge of the biosphere. The quest started with finding fossils of long-extinct species. A healthy dose of creativity and imagination applied to sets of more or less complete skeletal remains led to depictions of various extinct animal species. The dinosaurs are the most celebrated examples.

The history of life on Earth viewed from this perspective began in the Cambrian period at the dawn of the Paleozoic era (540 million years ago) when living creatures developed the ability to build solid skeletons based on calcium carbide. Contemporary techniques allow us to uncover fossils of simple organisms that do not possess skeletons. Fossils can be dated precisely and the emergence of life on Earth is set at 3.5 billion years ago.

Charles Darwin (1859) proposed a method that had great potential to reconstruct the history of life based on differences in selected characteristics of living animal species and extinct ones. In modern applications of his methodology, the most fundamental features are the nucleotide sequences in the genomes of selected individual organisms. Based on the mathematically defined differences, one tries to reconstruct the history of a genus or a species. Biochemistry and molecular biology, whose dynamic development flourished since the mid-20th century, provide several examples of living fossils. These are archaic metabolic pathways and more or less conserved domains in enzymes. The contemporary organization of animate matter reflects the history of its evolution and, conversely, the living structures that we encounter on Earth today are products of the evolution of life.

Biophysics attempts to describe the phenomenon of life using the conceptual framework of physics. It only partially explains the structures of the elements of living systems while treating other components as givens. Describing in this book the emergence of these given components as a historical process, we will strive to provide the most precise answer possible to Erwin Schrödinger's question: What is life?

References

Darwin, C., *On the Origin of Species by Means of Natural Selection, or the Preservation of the Favoured Races in the Struggle for Life*, 1859.

Davydov, A.S., *Quantum Mechanics and Biology*, Pergamon Press, London, 1982.

Elsasser, W.M., *Atom and Organism: A New Approach to Theoretical Biology*, Princeton University Press, Princeton, NJ, 1966.

Frauenfelder, H., in *Physics of Biological Systems: From Molecules to Species*, Springer, Berlin, 1997.

Frauenfelder, H., Wolynes, P.G., and Austin, R.H., *Rev. Mod. Phys.*, 71, S419, 1999.

Fröhlich, H., *Int. J. Quantum Chem.*, 2, 641, 1968.

Krumhansl, J., in *Nonlinear Excitations in Biomolecules*, Peyrard, M., Ed., EDP Sciences, Cambridge, MA, 1995.

Loewdin, P.O., Ed., *Proceedings of the International Symposium on Quantum Biology and Quantum Pharmacology*, John Wiley & Sons, New York, 1990.

Maddox, J., *What Remains to Be Discovered*, Touchstone Books, New York, 1999.

Parsegian, V.A. *Physics Today*, "Harnessing the Hubris: Useful Things Physicists Could Do in Biology," July issue, 23–27, 1997.

Prigogine, I., *From Being to Becoming: Time and Complexity in the Physical Sciences*, W.H. Freeman, San Francisco, 1980.

Schrödinger, E., *What is Life? The Physical Aspects of Living Cells*, Cambridge University Press, Cambridge, 1967.

Scott, A.C., *Stairway to the Mind: The Controversial Science of Consciousness*, Springer, New York, 1995.

Szent-Gyorgyi, A., *The Living State: With Remarks on Cancer*, Academic Press, New York, 1972.

Tuszynski, J.A. and Dixon, J.M., *Applications of Introductory Physics to Biology and Medicine*, John Wiley & Sons, New York, 2002.

Volkenstein, M.V., *General Biophysics*, Academic Press, San Diego, 1983.

Authors

Jack Tuszynski was born in Poznan, Poland in 1956. He earned an M.Sc. magna cum laude in physics from the University of Poznan and a Ph.D. in theoretical physics from the University of Calgary, Canada in 1983. After a brief post-doctoral fellowship in theoretical chemistry, he accepted a faculty position at Memorial University of Newfoundland. Since 1988, he has held a faculty position in physics at the University of Alberta in Edmonton and is now a full professor.

He has also held visiting professor appointments in Warwick, U.K., Lyon, France, Leuven, Belgium, Copenhagen, Denmark, and Dusseldorf and Giessen, Germany. He leads a research group whose focus is the development of mathematical models for pharmaceutical applications. He lives in Edmonton with his wife and two sons. For more information, visit his Web sites: http://www.phys.ualberta.ca/~jtus and http://www.phys.ualberta.ca/mmpd

Michal Kurzynski was born in Poznan, Poland in 1946. He earned a Ph.D. in solid state physics from the University of Poznan in 1973. Since 1980, he has held a professor position at the university's Institute of Physics. He has held a visiting professor appointment at the University Paris-Nord, France. He has served as a von Humboldt fellow at the University of Stuttgart, Germany, with Hermann Haken and at the Max Planck Institute, Goettingen, Germany, with Manfred Eigen. His main research interests are the theory of electronic and structural phase transitions in solids and nonequilibrium statistical physics. Recently, Dr. Kurzynski is involved in developing the foundations of the statistical theory of basic biochemical processes.

Contents

1

Origins and Evolution of Life

1.1 Initiation

Planet Earth was formed about 4.6 billion years ago as a result of accretions (inelastic collisions and agglomerations) of larger and larger rocky fragments formed gradually from the dust component of the gaseous dusty cloud that was the original matter of the Solar System. The Great Bombardment ended only 3.9 billion years ago when a stream of meteorites falling onto the surface of the newly formed planet reached a more or less constant intensity. The first well preserved petrified microstamps of relatively highly organized living organisms similar to today's cyanobacteria emerged about 3.5 billion years ago (Schopf, 1999), so life on Earth must have developed within the relatively short span of a few hundred million years.

Rejecting the hypothesis of an extraterrestrial origin of life, not so much for rational as for emotional reasons, we have to answer the question of the origins of the simplest elements of living organisms: amino acids, simple sugars (monosaccharides) and nitrogenous bases. Three equally probable hypotheses have been put forward to explain their appearance (Orgel, 1998). According to the first and the oldest theory, these compounds resulted from electric discharges and ultraviolet irradiation of the primary Earth atmosphere containing mostly CO_2 (as the atmospheres of Mars and Venus do today), H_2O, and strongly reducing gases (CH_4, NH_3, and H_2S). According to the second hypothesis, the basic components of living organisms were formed in space outside the orbits of large planets and transferred to the Earth's surface via collisions with comets and indirectly via carbon chondrites. The third hypothesis is that these compounds appeared at the oceanic rifts where the new Earth's crust was formed and where water overheated to $400°C$ containing strongly reducing FeS, H_2, and H_2S met cool water containing CO_2.

The origin issue is still open and all three hypotheses have been seriously criticized. First, the primary Earth atmosphere might not have been reducing strongly enough. Second, organic compounds from outer space may have deteriorated while passing through the Earth's atmosphere. Third, the reduction of CO_2 in oceanic rifts requires nontrivial catalysts.

The three most important characteristics of life that distinguish it from other natural phenomena were expressed by Charles Darwin, whose theory of evolution is so crucial to modern biology (Dawkins, 1986). Taking into account the achievements of post Darwinian genetics and biochemistry, we define life as a process characterized by continuous (1) *reproduction*, (2) *variability*, and (3) *selection* (survival of the fittest). An

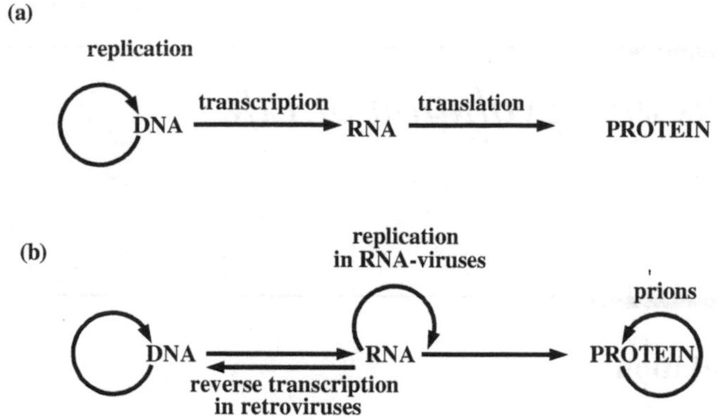

(a)

replication

(b)

replication
in RNA-viruses

prions

FIGURE 1.1
Processing genetic information. (a) The classical dogma. The information is carried by DNA which undergoes replication during biological reproduction and transcription into RNA when it is to be expressed; gene expression consists of translation of the information written in RNA onto a particular protein structure. (b) Modern version of the classical dogma. RNA can be replicated and transcribed in the opposite direction into DNA. Proteins also can carry information as is assumed to occur in prion diseases.

individual must have a replicable and modifiable *program*, proper *metabolism* (a mechanism of matter and energy conversion), and capability of *self-organization* to maintain life.

The emergence of molecular biology in the 1950s answered many questions about the structures and functioning of the three most important classes of biological macromolecules: DNA (deoxyribonucleic acid), RNA (ribonucleic acid), and proteins. However, in the attempts to develop a possible scenario of evolution from small organic particles to large biomolecules, a classical chicken-and-egg question was encountered: what appeared first? The DNA that carried the coded information on enzymatic proteins controlling the physiological processes that determined the fitness of an individual or the proteins that enabled the *replication* of DNA, its *transcription* into RNA, and the *translation* of certain sequences of amino acids into new proteins? See Figure 1.1a for illustration.

This question was resolved in the 1970s as a result of the evolutionary experimentation in Manfred Eigen's laboratory (Biebricher and Gardiner, 1997). The primary macromolecular system undergoing Darwin's evolution may have been RNA. Single-stranded RNA is not only the information carrier, program, or *genotype*. Because of a specific spatial structure, RNA is also an object of selection or a *phenotype*. Equipped with the concept of a hypercycle (Eigen and Schuster, 1977) and inspired by Sol Spigelman, Eigen used virial RNA replicase (Figure 1.1b), a protein, to produce new generations of RNA *in vitro*. The complementary RNA could polymerize spontaneously, without replicase, using the matrix of the already existing RNA as a template. Consequently, we can imagine a very early "RNA world" composed only of nucleotides, their phosphates, and their polymers — subject to Darwinian evolution, and thus alive based on the definition adopted (Gesteland et al., 1999).

A number of facts support the RNA world concept. Nucleotide triphosphates are highly effective sources of free energy. They fulfill this function as relicts in most chemical reactions of contemporary metabolism (Stryer, 1995). Dinucleotides act as cofactors in many protein enzymes. In fact, RNA molecules can serve as enzymes (Cech, 1986) and scientists now commonly talk about *ribozymes*. Contemporary ribosomes translating information from RNA onto a protein structure (see Figure 1.1a) fulfill their catalytic functions due to their ribosomal RNA content rather than their protein components (Ramakrishnan and White, 1998).

We have known for a number of years now about the *reverse transcriptase* that transcribes information from RNA onto DNA (Figure 1.1b). It also appears that RNA may be a primary structure and DNA a secondary one since modern organisms synthesize deoxyribonucleotides from ribonucleotides.

1.2   Machinery of prokaryotic cells

The smallest present-day system thought to possess the key function of a living organism, namely reproduction, is a cell. A sharp distinction exists between simple *prokaryotic* cells (that do not have nuclei) and far more complex *eukaryotic* cells (with well defined nuclei). Evidence points to an earlier evolution of prokaryotic cells. Eukaryotic cells are believed to have resulted from mergers of two or more specialized prokaryotic cells. Unfortunately, little is known about the origins of prokaryotic cells. The scenario below is only an attempt to describe some key functional elements of the apparatus possessed by all prokaryotic cells and is not a serious effort to reconstruct the history of life on Earth.

The world of competing RNA molecules must have eventually reached a point where a dearth of the only building materials, nucleotides triphosphate, was created. Molecules that could obtain adequate supplies of building materials gained an evolutionary advantage but they needed containers to carry their supplies and protect them from the environment. In the liquid phase, such containers were formed spontaneously from phospholipids.

The phospholipid molecules are amphiphilic — one part is hydrophilic (attracted to water) and the other is hydrophobic (repelled by water). See Section 2.4 for more details on this aspect. As a result of movement of the hydrophobic part away from water and movement of the hydrophilic part toward water, an unbounded lipid bilayer or a three-dimensional vesicle is formed (Figure 1.2). Since phospholipid vesicles can join to construct bigger structures from several small ones, they are important to the RNA molecules that can divide and compete for food. Merging into bigger vesicles can be advantageous in foraging for food. Division into small vesicles can be seen as a type of reproduction.

The phospholipid vesicle was not a complete answer to the problem because it required a way to selectively infuse nucleotides into its interior. Employing new types of biomolecules — amino acids, of which some were hydrophilic and some hydrophobic, solved that problem. Their linear polymers are called peptides and long peptides give rise to proteins. Proteins possess three-dimensional structures

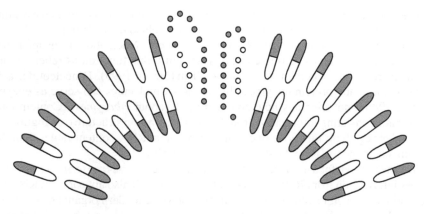

FIGURE 1.2
In a water environment, amphiphilic molecules composed of hydrophilic (shaded) and hydrophobic (white) parts organize spontaneously into bilayers closed into three-dimensional vesicles. Protein, a linear polymer of appropriately ordered hydrophilic (shaded circles) and hydrophobic (white circles) amino acids, forms a structure that spontaneously builds into the bilayer and allows selectively chosen molecules, e.g., nucleotide triphosphates, to pass into the lipid interior.

whose hydrophobicity depends on the order in which amino acid segments appear in a linear sequence. Such proteins may spontaneously embed themselves in a lipid bilayer and play the roles of selective ion channels (see Figure 1.2).

The first stage in the development of a prokaryotic cell was probably the enclosure of RNA molecules into phospholipid vesicles equipped with protein channels that enabled selective transfer of triphosphate nucleotides into the interior region (Figure 1.3a). The second stage must have been the perfection of these channels and a link between their structures and the information contained in the RNA molecules. Selective successes may have been scored by RNA molecules that could translate some of the information contained in the RNA base sequence into an amino acid sequence of an ion channel protein in order to synthesize it. This was the way to distinguish the so-called mRNA (*messenger* RNA) from tRNA (*transfer* RNA) and rRNA (*ribosomal* RNA). While mRNA carries information about the amino acid sequences in proteins, tRNA connects amino acids with their corresponding triple base sets. rRNA is a prototype of a ribosome, a catalytic RNA molecule that can synthesize amino acids transported to it by molecules of tRNA into proteins.

These amino acids had to be first recognized by triples of bases along the mRNA (Figure 1.3b). The analysis of the nucleotide sequences in tRNA and rRNA of various origins indicates that they are very similar and very archaic. The genetic code based on sequences of triples is equally universal and archaic. Contemporary investigations of prokaryotic and eukaryotic ribosomes provided solid evidence that the main catalytic role is played by rRNA and not the proteins contained within the ribosomes.

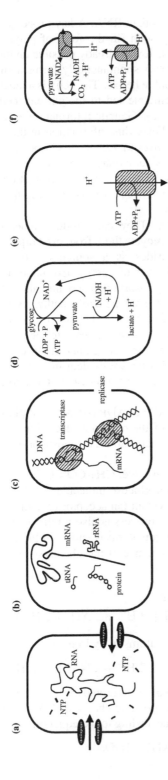

FIGURE 1.3

Development of the prokaryotic cell machinery. (a) The self-replicating RNA molecule with a supply of nucleotide triphosphates (NTP) is enclosed in a vesicle bounded by a lipid bilayer with built-in protein channels that allow selective passage of nucleotide triphosphates. (b) In an RNA chain, a distinction is made between mRNA and various types of tRNA and rRNA. rRNA is a prototype of a ribosome that can synthesize proteins based on the information encoded in mRNA. Proteins produced this way are more selective membrane channels and effective enzymes that can catalyze many useful biochemical processes. (c) Double-stranded DNA replaces RNA as an information carrier. Protein replicases double this information during division and protein transcriptases transfer it onto mRNA. (d) Protein enzymes appear to be able to catalyze lactose fermentation of sugars as a result of which the pool of high-energy nucleotide (mainly triphosphates ATP) can be replenished using low energy diphosphates (mainly ADP). The amount of oxidizer (hydrogen acceptor) NAD^+ remains constant; the cell interior becomes acidic. (e) Proton pumps can pump H^+ ions into the cell exterior via ATP hydrolysis. (f) Other proton pumps use hydrogen obtained from the decomposition of sugars through pyruvate as fuel. Due to the presence of a wall or a second cell membrane, pumped-out protons can return to the cell interior through the pumps of the first type that act in reverse to reconstruct ATP from ADP. Membrane phosphorylation becomes the basic mechanism of bioenergetics in all modern living organisms.

Proteins have much better catalytic properties than RNA. A key property is their high specificity *vis a vis* the substrate. They soon (in the form of polymerases) replaced RNA in the process of self-replication. It was already possible on the RNA template to replicate sister RNA and DNA. DNA spontaneously forms a structure composed of two complementary strands (a double helix). The helix is a much more stable information carrier than RNA. This principle led to the current method of transferring genetic information (see Figure 1.3c). Genetic information is stored in double-stranded DNA. Protein *replicases* duplicate this information in the process of cell division. If necessary, protein *transcriptases* transcribe this information onto mRNA, which is used during the process of translation (partly ribozymatic and partly enzymatic) as a template to produce proteins. The transfer of information in the *reverse direction* from RNA to DNA via *reverse transcriptases* is a fossil remnant that has been preserved in modern retroviruses.

Protein enzymes can perform useful tasks. They can produce much-needed triphosphate nucleotide building materials and recycle them from used diphosphates and inorganic orthophosphate, using saccharides as a source of free energy. Figure 1.4 illustrates the main metabolic pathways of energy and matter processing that are common to contemporary bacteria (prokaryotes) and animals and plants (eukaryotes). The central point at which many of these metabolic pathways converge is pyruvate. It is easy to see the vertical path of *glycolysis*, the reduction of the most common monosaccharide, glucose, to pyruvate. It is equally easy to see the circular *cycle* of the *citric acid* that is connected with pyruvate through one or more reactions. The archaic origins of the main metabolic pathways are evident in their universality (from bacteria to man) and in many of the reactions of nucleotide triphosphates, mainly ATP (adenosine triphosphate).

Reactions connected with the hydrolysis of ATP to ADP (adenosine diphosphate) are indicated in Figure 1.4 by P's at the starts of the reactions. Reactions linked to the synthesis of ADP and an orthophosphate group into ATP (phosphorylation) are indicated by P's at the ends of reactions.

The transition from glucose, $C_6H_{12}O_6$, to pyruvate, $CH_3\text{-}CO\text{-}COO^-$, is an oxidation reaction that takes hydrogen atoms from glucose molecules. NAD^+ (nicotinamide adenine dinucleotide) is a universal oxidant (an acceptor of hydrogen, i.e., simultaneously an electron and a proton). This process is also a relict of the RNA world. The acceptance by NAD^+ of two hydrogen atoms is shown in Figure 1.5 by the H at the end of each reaction. An overall balance of the glycolysis reaction or oxidation of glucose to a pyruvate takes the form:

$$C_6H_{12}O_6 + 2\ NAD^+ + 2\ ADP + 2P_i$$
$$\rightarrow 2CH_3\text{--}CO\text{--}COO^- + 2\ NADH + 2\ H^+ + 2\ ATP + 2\ H_2O. \quad (1.1)$$

Two molecules of NAD^+ are reduced by four atoms of hydrogen:

$$C_6H_{12}O_6 + 2\ NAD^+ \rightarrow 2\ CH_3\text{--}CO\text{--}COO^- + 2\ H^+ + 2\ NADH + 2\ H^+ \quad (1.2)$$

(Two protons are obtained from the dissociation of pyruvic acid into a pyruvate anion, whereas two other protons transfer the original positive charge of NAD^+) and two molecules of ADP are phosphorylated to ATP according to the equation:

$$ADP + P_i + H^+ \rightarrow ATP + H_2O. \quad (1.3)$$

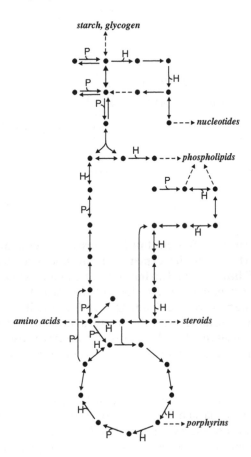

FIGURE 1.4
An outline of the main metabolic pathways. Substrates are represented by black dots; reversible or practically irreversible reactions catalyzed by specific enzymes are represented by arrows.

In a neutral water environment, ATP is present as an ion with four negative charges, ADP with three negative charges, and an orthophosphate P_i with two.

The primitive prokaryotic cells were properly equipped with the machinery of protein membrane channels able to select specific components from their environment. They also had protein enzymes to catalyze appropriate reactions. These cells became able to replenish their pools of nucleotide triphosphates at the expense of organic compounds of a fourth type — saccharides (see Figure 1.3d). The NAD^+ oxidant was recovered in the process of *fermentation* of a pyruvate into a lactate:

$$CH_3-CO-COO^- + NADH + H^+ \rightarrow CH_3-CHOH-COO^- + NAD^+. \quad (1.4)$$

This reaction is also used by modern eukaryotic organisms whenever they must rapidly obtain ATP under conditions of limited oxygen supply.

The lactic fermentation process that accompanies phosphorylation of ADP to ATP with the use of sugar as a substrate has several drawbacks. In addition to its low

(a)   (b)

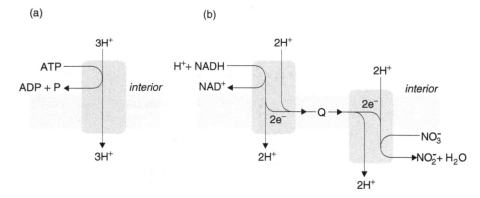

FIGURE 1.5
Proton pumps transport free proton H^+ across the membrane from the cell interior to its exterior at the expense of the following chemical reactions: hydrolysis of ATP into ADP and an inorganic orthophosphate (a) or oxygenation of hydrogen released in the decomposition of glucose to CO_2 and transportation by NAD^+. (b) A derivative of quinone Q is an intermediary in hydrogen transport. The molecule is soluble inside the membrane and oxidation is accomplished through NO_3^- reduced to NO_2^-. If pumps of both types are located in the same membrane, the first protons passing in the reverse direction can phosphorylate ADP to ATP.

efficiency (unused lactate), it leads to increased acidity of the cells. While sugars are neutral (pH near 7), lactate is a product of dissociation of lactic acid and in the process of breakdown of sugars, a free proton H^+ is released. The lowering of pH results in a significant slowdown or even stoppage of the glycolysis reaction.

For the decomposition of sugars to be effectively used in the production of ATP, a cell must find a different mechanism of fermentation whose product has a pH near 7 or whose proton H^+ can be expelled outside the cell. In yeast, a new type of fermentation consists of the reduction of pyruvate to ethanol with a release of carbon dioxide in the process:

$$CH_3-CO-COO^- + NADH + 2\,H^+ \rightarrow C_2H_5-OH + CO_2 + NAD^+. \quad (1.5)$$

Before this mechanism had been adopted, a *proton pump* was discovered utilizing the hydrolysis of ATP as a source of energy (Figure 1.3e). During the production of one molecule of ATP, one hydrated proton H^+ is released inside the cell. The hydrolysis of one molecule of ATP results in the pumping outside the cell membrane of three hydrated H^+ protons (see Figure 1.5a). The process is still energetically favorable.

However, from the viewpoint of ATP production, a more efficient process is further oxidation of a pyruvate to an acetate and a carbon dioxide:

$$CH_3-CO-COO^- + H_2O + NAD^+ \rightarrow CH_3-COO^- + CO_2 + NADH + H^+. \quad (1.6)$$

The equation above shows a reduction of one molecule of NAD^+ by two atoms of hydrogen. Subsequently, in the citric acid cycle of Krebs (see Figure 1.4),

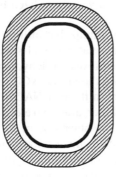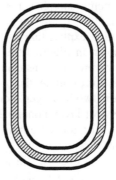

Gram - positive   Gram - negative

FIGURE 1.6
A bacterial cell is equipped with a cell wall composed of peptidoglycan, a complex protein-polysaccharide structure (shaded). It can also have a second, external membrane. The exposed thick peptidoglycan layer changes its color in the Gram dyeing procedure. The thin peptidoglycan layer covered by the external membrane does not change color. Hence bacteria are categorized as Gram-positive and Gram-negative.

acetate is oxidized to carbon dioxide and water. The net balance in the Krebs cycle is:

$$[CH_3-COO^- + H^+ + 2\ H_2O] + [3\ NAD^+ + FAD] + [GDP + P_i + H^+]$$
$$\rightarrow 2\ CO_2 + [3\ NADH + 3\ H^+ + FADH_2] + [GTP + H_2O]. \qquad (1.7)$$

Acetate enters the reaction bound to a so-called co-enzyme A (CoA) as acetyl CoA. During one turn of the Krebs cycle, a further reduction of three molecules of NAD^+ and one molecule of FAD (flavin adenine dinucleotide) involving eight atoms of hydrogen and phosphorylation of a molecule of GDP (guanosine diphosphate) to GTP (guanosine triphosphate) takes place. To enhance clarity, we used square brackets to indicate subprocesses.

Discussing the economy of the Krebs cycle makes sense only when a cell is able to utilize fuel in the form of hydrogen bound to the NAD^+ and FAD carriers for further phosphorylation of ADP to ATP. This became possible when a new generation of proton pumps was discovered. These pumps work as a result of the decomposition of hydrogen into a proton and an electron instead of ATP hydrolysis. These particles are further transported along a different pathway to the final hydrogen acceptor which, in the early stages of biogenesis, may have been an anion of an inorganic acid.

Primitive bacterial cells were endowed with cell membranes composed of peptidoglycan, a complex protein–saccharide structure, and later developed additional cell membranes (see Figure 1.6). This facilitated accumulation of protons in the spaces outside the original cell membrane from which they could return to the cell interiors using the proton pump of the first type (see Figure 1.3f). This pump, working in reverse, synthesizes ATP from ADP and an orthophosphate. This very efficient mechanism of membrane phosphorylation is universaly utilized by all present-day living organisms.

A more detailed explanation of the proton pump that utilizes the oxidation of hydrogen is shown in Figure 1.5b. In the original bacterial version, the pump is composed of two protein transmembrane complexes: a dehydrogenase of NADH and a reductase, for example, the one changing the nitrate NO_3^- to nitrite NO_2^-. In the first complex, two hydrogen atoms present in the pair NADH–hydronic ion are transferred to FMN (flavin mononucleotide). Later, after two electrons are detached, the hydrogens (as protons) are transferred to the other side of the membrane. The two electrons are accepted in turn by one and then another iron–sulfur center (iron is reduced from Fe^{3+} to Fe^{2+}). Subsequently, at a molecule of quinone derivative Q, they are bound to another pair of protons that reached the same site from the interior of the cell. An appropriate derivative of quinone Q is soluble inside the membrane and serves as an intermediary that ferries two hydrogen atoms between the two complexes. At the other complex, two hydrogen atoms are again split into protons and electrons. The released protons are transferred to the exterior of the membrane and the electrons and the protons from the interior of the membrane are relocated to a final acceptor site that may be a nitrate anion. Thus, the created nitrite can oxidate another reaction that can be used by another reductase:

$$NO_2^- \rightarrow N_2. \tag{1.8}$$

Alternatively, the nitrite can be involved with other inorganic anions such as an acid carbonate or sulfate in reactions leading to the formation of compounds with hydrogen: ammonia, methane, or sulfurated hydrogen:

$$NO_2^- \rightarrow NH_3, HCO_3^- \rightarrow CH_4, SO_4^{2-} \rightarrow H_2S. \tag{1.9}$$

1.3   The photosynthetic revolution

The Earth is energetically an open system and a substantial flux of solar radiation has reached it since the moment of its creation. Along with the rotational motion of the planet, the flux has powered the machinery that produces oceanic and atmospheric motions. The primary energy sources for the newly emerged life on Earth were nucleotide triphosphates and exhaustible supplies of small organic molecules such as monosaccharides. Life became energetically independent only when organisms learned how to harness practically inexhaustible solar energy or, more precisely, the fraction of it that reaches the surfaces of the oceans.

The possibility of utilizing solar energy by living cells is linked to the use of *chlorophyll* as a photoreceptor (Nitschke and Rutherford, 1991). The chlorophyll molecule contains an unsaturated carbon–nitrogen porphyrin ring (see Figure 1.4) with a built-in Mg^{2+} ion and phytol, a long saturated hydrophobic carbohydrate chain. The molecules of chlorophyll are easily excited in the optical range and easily transfer this excitation among each other, creating a light harvesting system in an appropriate protein matrix. The last chlorophyll molecule in such a chain can

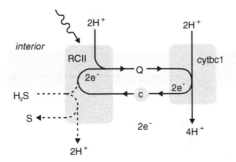

FIGURE 1.7
A proton pump in purple bacteria utilizing solar radiation energy. In the first
protein complex (type II reaction center or RC II), an electron from an excited
chlorophyll molecule (the primary donor) is transferred to a molecule of quinone
derivative (Q) along with a proton from the cell interior. The quinone derivative
molecule carries two hydrogen atoms formed this way to another protein complex
containing cytochrome bc1. Hydrogen atoms in the complex are again separated.
A proton is moved outside the cell while an electron reduces a molecule of the
water-soluble cytochrome c, which carries it back to the primary donor. An
alternative source of electrons (broken line) for sulfur purple bacteria can be a
molecule of sulfurated hydrogen (H_2S).

become an electron donor and replace the NADH + H^+ fuel in a proton pump (see
Figure 1.5b).

The first organisms to avail themselves of this possibility were probably purple
bacteria. Their proton pumps are two protein complexes built into the cell membrane
(see Figure 1.7). In the protein complex called the type-II reaction center (RC),
two electrons from the excited chlorophyll are transferred with two protons from the
cell interior to a *quinone* derivative Q with a long carbohydrate tail. Q is soluble
in the membrane. When reduced to quinol QH_2, it carries the two hydrogen atoms
inside the membrane to the next complex that contains a protein macromolecule
called cytochrome bc1. The macromolecule catalyzes the electron transfer from each
hydrogen atom onto another macromolecule called cytochrome c while the remaining
proton moves to the extracellular medium.

Cytochrome proteins contain a *heme* in the form of a porphyrin ring with a built-in
Fe^{2+} ion that may also exist in a form oxidized to Fe^{3+}. Cytochrome c is a water-
soluble protein that removes electrons outside the cell membrane and returns them to
the reaction center. This completes the cyclical process during which two protons are
carried from inside the cell to the outside. Alternative sources of electrons needed
to restore the initial state of the reaction center used, for example, in sulfur purple
bacteria may be the molecules of sulfurated hydrogen H_2S. Contrary to the oxidation
of NADH, oxidation of H_2S to pure sulfur is an endoergic reaction (consuming and
not providing free energy) and it cannot be used in proton pumps.

The proton concentration difference on each side of the cell membrane is further
used by purple bacteria to produce ATP the same way it is produced by nonphotosynthetic bacteria. Green bacteria found an alternative way of using solar energy

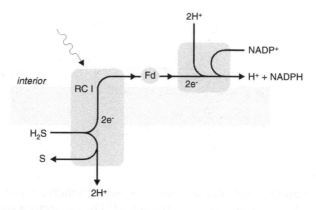

FIGURE 1.8
**The utilization of solar energy by green sulfur bacteria. In the first protein
complex (type I reaction center or RC I), an electron from an excited chlorophyll
molecule is transferred to a water-soluble protein molecule of ferredoxin (Fd)
which carries it to the complex of $NADP^+$ (nicotinamide adenine dinucleotide
phosphate) reductase. The deficit electron in the initial chlorophyll is compensated in the process of oxidation of sulfurated hydrogen (H_2S). The reduced
hydrogen carrier ($NADPH + H^+$) serves as fuel in the Calvin cycle synthesizing
sugar from water and carbon dioxide.**

(see Figure 1.8). In the protein complex called the type I reaction center, an electron
from photoexcited chlorophyll is transferred to a water-soluble protein called *ferredoxin*. The lack of electrons in the chlorophyll molecule is compensated uncyclically
from sulfurated hydrogen decomposition.

The electron carrier in ferredoxin is the iron–sulfur center composed of four Fe
atoms directly and covalently bound to four S atoms. After the reduction of iron,
ferredoxin carries electrons to the next protein complex where the electrons bind
to protons moving from the cell interior and reducing the molecules of $NADP^+$
(nicotinamide adenine dinucleotide phosphate) to $NADPH + H^+$. The entire system is
not really a proton pump since no net proton transport occurs across the cell membrane.
The system transforms light energy into fuel energy in the molecules of NADPH
together with hydrated protons H^+ that carry the original charge of $NADPH^+$. This
fuel is used in the synthesis of glucose from CO_2 and H_2O in the *Calvin cycle* whose
overall balance equation takes the form:

$$[6\ CO_2 + 12\ NADPH + 12\ H^+] + [18\ ATP + 18\ H_2O] \rightarrow [C_6H_{12}O_6$$
$$+ 6\ H_2O + 12\ NADP^+] + [18\ ADP + 18\ P_i + 18\ H^+]. \qquad (1.10)$$

This cycle is in a sense a reverse of the Krebs cycle. Analogously to the Krebs
cycle, we used square brackets to denote summary component reactions in order to
show more clearly the net reaction. ATP is also used in the Calvin cycle. After the
oxidation of glucose in the same way as for nonphotosynthetic bacteria, an excess of
ATP is produced.

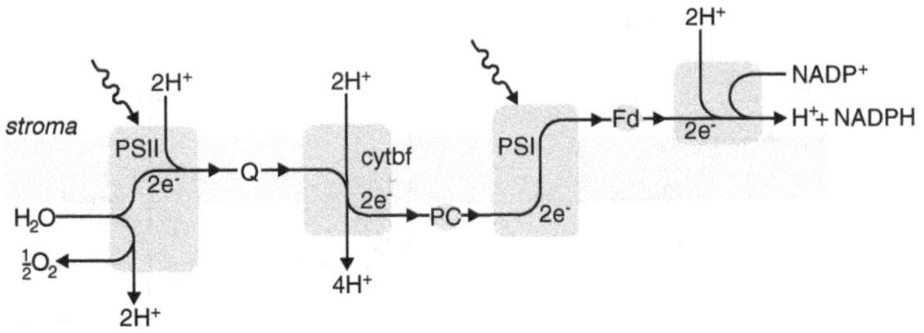

FIGURE 1.9

A proton pump using solar energy in cyanobacteria can be thought of as a combination of the proton pump in purple bacteria (type II reaction center, now called photosystem II or PS II) and the photosynthetic system of green bacteria (type I reaction center, now called photosystem I or PS I). The coupling of the two systems is done by a water-soluble molecule of plastocyanin (PC) with a copper ion serving as an electron carrier. The final electron donor is water which, after donating electrons and protons, becomes molecular O_2. The proton concentration difference between the two sides of the cell membrane is used to produce ATP via H^+ATPase (see Figure 1.5a) working in the reverse direction. In principle, the photosynthetic systems in the tylakoid membranes of chloroplasts that are organelles of eukaryotic plant cells are identical structures.

Combining the two methods of using solar energy offers the optimal solution. In cyanobacteria, cytochrome c1 was replaced by cytochrome f, whereas cytochrome c was replaced by *plastocyanin* (PC) and used as an electron carrier between type II and type I reaction centers. The centers are now known as *photosystem II* (PS II) and *photosystem I* (PS I), respectively (Figure 1.9). The electron carrier in plastocyanin is the Cu^{2+} copper ion which is reducible to Cu^+ and directly bound via four covalent bonds to four amino acids: cysteine, methionine, and two histidines.

The greatest breakthrough resulted not from the combination of the two photosystems, but from the utilization of water as the final electron donor (and a proton donor, hence a hydrogen donor). The dissociation of hydrogen atoms from a water molecule turned it into a highly reactive molecular O_2 gas that was toxic to the early biological environment. Initially, it oxidized only Fe^{2+} ions that were soluble in great quantities in contemporary ocean water. As a result of this oxidation, poorly soluble Fe^{3+} ions were formed. They sedimented, giving rise to modern iron ore deposits.

The increased production of sugars from CO_2 and H_2O reduced ocean acidity and caused a transformation of acidic anions of HCO_3^- into neutral CO_3^{2-} ions. The CO_3^{2-} reacted with the Ca^{2+} ions initially present in high concentrations, leading to sedimentation of insoluble calcium carbonate $CaCO_3$. The membranes of cyanobacteria captured the calcium carbonate and produced a paleobiological record of these processes in the form of fossils called stromatolites.

The formation of calcified stromatolites depleted the atmosphere from CO_2. When a deficit of compounds capable of further oxidation occurred, molecular O_2 started to be released into the atmosphere. Along with molecular N_2 formed by the reduction of nitrates, the O_2 brought about the contemporary oxygen–nitrogen based atmosphere

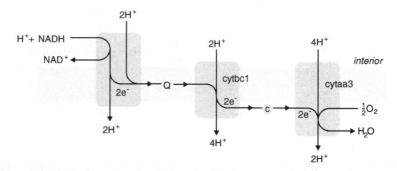

FIGURE 1.10
**Protein pump of heterotrophic aerobic bacteria. Electrons from the fuel in
the form NADH + H⁺ (produced in glycolysis and in the Krebs cycle) are
transferred via quinone (Q) to the protein complex with cytochrome bc1 and
then via cytochrome c to the complex with cytochrome aa3. The final electron
acceptor is molecular O_2. During the transfer of two electrons along the membrane, eight protons are pumped across it. The proton concentration difference
between the two sides of the membrane is used to produce ATP by H⁺ATPase (see
Figure 1.5a) working in reverse. This is in principle identical to the mechanism of
oxidative phosphorylation in the mitochondrial membrane, which is an organelle
present in all eukaryotic cells.**

containing only trace quantities of carbon dioxide. Life had to develop in a toxic
oxygen environment from that point onward. The problem was solved by the mechanism of oxidative phosphorylation used by modern aerobic bacteria and all higher
organisms. A proton pump that used inorganic anions as final electron acceptors (see
Figure 1.5b) was replaced by a pump in which the final electron acceptor is molecular
oxygen (see Figure 1.10).

The cytochrome bc1 transfers electrons from quinone Q to water-soluble
cytochrome c, a mechanism utilized earlier by purple bacteria (see Figure 1.7).
The source of electrons transferred to the quinone can be the NADH + H⁺ generated by glycolysis and in the Krebs cycle or, directly, FADH₂ (reduced flavin adenine dinucleotide) produced in one stage of the Krebs cycle (oxidation of succinate
to fumarate). Electrons can also come from an inorganic source (chemotrophy).
For example, nitrifying bacteria can oxidize ammonia to nitrate using molecular
oxygen:

$$NH_3 \rightarrow NO_2^- \rightarrow NO_3^-. \tag{1.11}$$

Nature demonstrates here, as it has many times, its ability to use environmental
pollution to its advantage. It will be interesting to see, for example, what use it finds
for the countless tons of plastic bottles deposited in modern garbage dumps.

1.4   Origins of diploidal eukaryotic cells

In its 19th century interpretation, Darwin's theory of natural selection favoring survival of the fittest could be readily associated with the contemporary struggle for survival in the early capitalist economy of that time. Both endeavors were ruled by the law of the jungle that became anathema to the ideological doctrines of many totalitarian regimes on the 20th century political landscape. The great biologist, Lynn Margulis (1998), emphasized the fact that survival can be accomplished through struggle or through peaceful coexistence (symbiosis is the biological term). Many clues support the significance of symbiosis in the formation of modern eukaryotic cells.

Figure 1.11 illustrates a simplified phylogenetic tree of living organisms. It shows a clear division between archaic bacteria (*Archaebacteria*) and true bacteria (*Eubacteria*) that may have emerged in the earliest periods of life on Earth. The history of subsequent differentiation of prokaryotic organisms within these two groups, however, is not all that clear. The modern phylogenetic tree is based on differences in DNA sequences coding the same functional enzymes or ribozymes. The more differences found in the DNA sequences, the earlier the two branches of the compared species must have divided. The results obtained by comparing, for example, ribosomal RNA with the genes of the proteins in the photosynthetic chain differed greatly and led to very dissimilar reconstructions of the history of the evolution of photosynthesis (Doolittle, 1999; Xiong et al., 2000). The reason for this ambiguity is the lateral gene transfer process by which genes are borrowed by one organism from another. Branches can split away and merge over time. Therefore, the phylogenetic tree of *Eubacteria* shown in Figure 1.11 must be viewed with caution, especially since only the kingdoms essential to our discussion are depicted.

Gram-positive bacteria have only single external membranes and are potentially sensitive to antibiotics. Fortunately, the group includes most of the pathogenic bacteria. Spirochetes have developed mechanisms of internal motion for entire cells. Photosynthetic purple bacteria with type II reaction centers can be sulfuric or non-sulfuric. They must be evolutionarily close to aerobic bacteria because they utilize the same mechanism of reduction of cytochrome c through the protein complex with cytochrome bc1 (see Figures 1.6 through 1.10). Biologists do not distinguish kingdoms of bacteria by this characteristiic. Most aerobic bacteria, including the common *Escherichia coli*, can survive in oxygen-deprived conditions. Green bacteria and cyanobacteria have in common the mechanism of sugar photosynthesis via the use of type I reaction centers.

Lateral gene transfer can be accomplished when several simple eukaryotic cells merge into one supercell. According to Margulis, this is how eukaryotic cells first formed. Most probably, a thermophilic bacterium with a stable genomic organization whose DNA was protected by proteinaceous histones that combined to form a chromatin prototype entered into a symbiotic arrangement with a spirochete containing a motile apparatus formed from microtubules (see Figure 1.11). This combination gave rise to a *mitotic* mechanism of cell division (Solomon et al., 1993). Chromatin with a doubled amount of genetic material organized itself after replication into

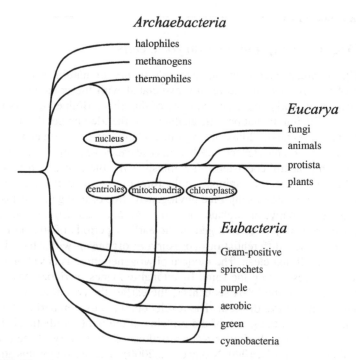

FIGURE 1.11
Simplified and somewhat hypothetical phylogenetic tree of living organisms.
The earliest are the two groups of prokaryotic organisms, *Archaebacteria* and
Eubacteria. As a result of the merger of prokaryotic cells with different properties, eukaryotic cells (*Eucarya*) were formed. *Eucarya* further evolved in
undifferentiated forms as single cell organisms (protista kingdom) or differentiated into multicellular organisms (the kingdoms of heterotrophic fungi with
multinuclear cells, heterotrophic animals, and phototrophic plants).

chromosomes pulled in opposite directions by a karyokinetic spindle formed from
centrioles by self-assembling microtubules that consumed GTP as fuel.
    In the next stage, cells with nuclei that contained chromatin assimilated
several oxygen bacteria (see Figure 1.11). The oxygen bacteria were transformed into
mitochondria, the power plants of cells that synthesized ATP via the phosphoryla-
tive oxidation mechanism shown in Figure 1.10. The formed *Eucarya* cells continued
evolving (Figure 1.11) in undifferentiated forms as single-cell organisms (the *protista*
kingdom) or in differentiated forms as multicellular organisms (*fungi* and *animal* kingdoms). All the original organisms were heterotrophs. The assimilation of prokaryotic
cells of cyanobacteria as *chloroplasts* led to the formation of phototrophic single-cell
organisms and multicellular *plants*.
    So far, we have only discussed the symbiosis of prokaryotic cells. An encounter
of two organisms belonging to the same species will lead to cannibalism or

symbiosis. According to Margulis, symbiotic encounters led to the emergence of *sex*. A symbiotic cell becomes diploidal, i.e., contains two slightly different copies of the same genome. Obviously, reproductive cells nurtured by a parent organism before entering into new symbiotic arrangements are haploidal and contain only one copy of genetic material. A reduction of the genetic information took place in the process of generating reproductive cells when *meiotic* division replaced mitotic division. The evolutionary advantage of sexual reproduction is due to the crossing-over of maternal and paternal genes in meiotic division. As a result, the genetic material undergoes a much faster variability compared to random point *mutations* and such recombinations are seldom lethal.

Figure 1.12 illustrates a eukaryotic cell composed of a system of lipid membranes that confine its various organelles (Solomon et al., 1993): the *nucleus, mitochondria, smooth* and *rough endoplasmic reticulum, Golgi apparatus*, and *lysosomes*. The *centrioles* organize the motile system. The illustration corresponds to an animal cell. Plant cells (see Figure 1.13) contain additional cell *walls, vacuoles* that store water, and various additional substances in the form of grains and *chloroplasts* — large organelles consisting of three layers of membranes that facilitate photosynthesis. The internal flattened chloroplast bubbles are called *thylakoids*; they can be viewed as removed mitochondrial *crista* (combs).

Three types of organelles of eukaryotic organisms contain their own genetic material, which is different from that of the nucleus. As mentioned earlier, eukaryotic systems emerged in the process of evolution due to the assimilation of previously developed prokaryotic cells. These are mitochondria that originated from oxygen bacteria, the centrioles originating from spirochetes and chloroplasts formed from cyanobacteria.

Figure 1.4 depicts the broad scheme of metabolism; it shows only the most important biochemical reactions. We now know that almost 100 times as many reactions take place in a cell (Stryer, 1995). Substrates are denoted by dots. A unique enzyme catalyzes each reaction (represented by a single or bidirectional arrow). Worth mentioning are vertical chains of sugar transformations (left side), fatty acids (right side), and a characteristic closed Krebs cycle. The diagram also shows how connections are made with more complex reaction systems transforming other important bio-organic compounds such as nucleotides, amino acids, sterols, and porphyrins.

Specific reactions occur in each cell compartment. Biopolymer hydrolysis takes place in lysosomes. Krebs reaction cycles and fatty acid and amino acid degradation can be found in the mitochondrial matrix. Oxidative phosphorylation reactions proceed on internal membranes of mitochondrial. Protein synthesis takes place on membranes of the rough endoplasmic reticulum and lipid synthesis can be observed on membranes of the smooth endoplasmic reticulum. Compound sugar synthesis occurs in the interior of the Golgi apparatus.

After merging with membrane proteins, sugars synthesized in the Golgi apparatus can be transported outside cells in a process called exocytosis. The external cytoplasmic membrane armed with such glycoproteins recognizes and selectively transports substances from the external environments of the cells. Transformations of simple sugars, amino acids, and mononucleotides take place between various organelles in

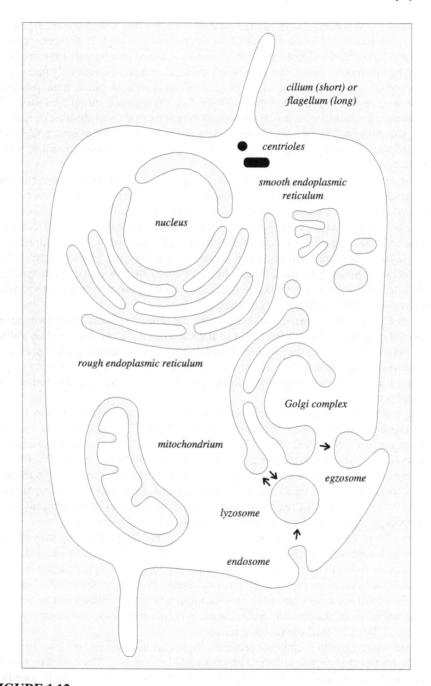

FIGURE 1.12
Compartments of a eukaryotic cell. Solid lines represent membranes formed
from two phospholipid bilayers and the neighboring compartments differ in
degrees of shading.

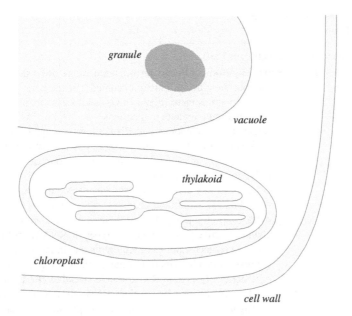

FIGURE 1.13
Additional organelles of a plant cell.

the cellular interior. The region is filled simultaneously with supramolecular structures of the protein cytoskeleton that provide the cell with motile machinery.

1.5   Summary: further stages of evolution

We will summarize here the most important stages in the evolution of life on Earth
from the Big Bombardment to the achievement of supracellular levels of organization
(Solomon et al., 1993). Some of the elements of the puzzle are well known; they left
visible traces that were precisely dated. Some elements are somewhat hypothetical
and have not yet been backed by solid discoveries and analyses. They are indicated
by question marks.

Time Frame	Event
4.6 billion years ago	Creation of Earth
3.9 billion years ago	End of Big Bombardment; surface temperature of Earth is lowered; early gaseous envelope of volcanic, cometic, or meteoric origin differentiates into atmosphere (mainly CO_2) and ocean (mainly H_2O); appearance of simple organic compounds
?	Emergence of nucleotides and RNA allows storage of information and self-replication; Darwinian selection based on survival of the fittest; triphosphates of nucleotides become key sources of free energy for polymerization reactions
?	Emergence of key elements of cellular machinery for prokaryotic cells (membranes, membrane channels, ribosomes, DNA and RNA polymerases, proton pumps); phospholipid bilayers with selective protein channels actively isolate various types of RNA from their surroundings and protect supplies of required nucleotides; primitive ribosomes express protein information contained in RNA; proteins become more efficient catalysts than RNA; double-stranded DNA becomes a more stable information carrier than RNA; protein enzymes allow use of sugars as free energy sources in phosphorylation reactions in which diphosphate nucleotides are recycled into triphosphate forms first through fermentation (substrate phosphorylation) and then more efficiently by membrane phosphorylation with inorganic anions acting as oxidants
3.5 billion years ago	Phosphorylation is linked to dissociation of water; resultant molecular oxygen oxidates Fe^{2+} to a poorly soluble Fe^{3+}; sugar synthesis from CO_2 and H_2O lowers ocean acidity, leading to elimination of insoluble calcium carbonate
2 billion years ago	Molecular oxygen starts to accumulate in large quantities in atmosphere; oxidative phosphorylation is discovered
?	Thermophilic bacteria possessing DNA protected by histone proteins enter into a symbiotic relationship with motile bacteria possessing micro tubular cytoskeleton; cell nuclei are formed; first mitotic cell division
1.5 billion years ago	Symbiotic coexistence of cells containing nuclei and oxygen bacteria is established; bacteria play the role of mitochondria; cyanobacteria may act as chloroplasts; the modern eukaryotic cell is born
?	Sex emerges; the amount of genetic information doubles as a diploidal cell forms; this is linked with meiosis and possibility of recombination
1 billion years ago	Cells arising from cell division stop dividing, thus leading to the emergence of an embryo that differentiates into a multicellular animal or plant organism
550 million years ago	Animals start developing skeletons from calcium carbonate or silicate, thus enabling the formation of fossils that produced lasting chronological records; beginning of Paleolithic Era (Cambrian explosion)
450 million years ago	Symbiosis of plants and fungi allows their emergence on land
100 million years ago	Perfection of the most modern and effective ways of protecting an embryo (angiospermous plants and mammals with placenta)
5 million years ago	First hominid forms appear
100,000 years ago	First *homo sapiens* walk upright

References

Biebricher, Ch.K. and Gardiner, W.C., *Biophys. Chem.,* 66, 179, 1997.

Cech, C.R., *Sci. Amer.,* 255, 76, 1986.

Dawkins, R., *The Blind Watchmaker*, Norton, New York, 1986.

Doolittle, W.F., *Science*, 284, 2124, 1999.

Eigen, M. and Schuster, P., *The Hypercycle*, Springer, Berlin, 1979.

Gesteland, R.E., Cech, T.R., and Atkins, J.F., Eds., *The RNA World*, Cold Spring Harbor Laboratory Press, Cold Spring Harbor, New York, 1999.

Margulis, L., *Symbiotic Planet: A New Look at Evolution*, Sciencewriters, Amherst, MA, 1998.

Nieselt-Struwe, K., *Biophys. Chem.*, 66, 111, 1997.

Nitschke, W. and Rutherford, A.W., *TIBS*, 23, 208, 1998.

Orgel, L.E., *TIBS*, 23, 491, 1998.

Ramakrishnan, V. and White, S.W., *TIBS*, 23, 208, 1998.

Schopf, J.W., *Cradle of Life. The Discovery of Earth's Earliest Fossils*, Priceton University Press, Princeton, NJ, 1999.

Solomon, E.P., Berg, L.R., Martin, D.W., and Villee, C.A., *Biology*, 3rd ed., Saunders College Publishing, Philadelphia, 1993.

Stryer, L., *Biochemistry*, W.H. Freeman and Co., San Francisco, 1995.

Woese, C., *Proc. Natl. Acad. Sci. USA,* 95, 6854, 1998.

Xiong, J. et al., *Science*, 289, 1727, 2000.

2

Structures of Biomolecules

There is still lack of knowledge about this advanced level of organization
of matter, we call 'life' and its novel nonmaterial consequences. However,
at this stage it is simply 'lack of knowledge' and not 'discrepancy' with
the present concepts of physics.

Manfred Eigen

2.1 Elementary building blocks

Animate matter is almost exclusively built from six elements: hydrogen (H, 60.5%),
oxygen (O, 25.7%), carbon (C, 10.7%), nitrogen (N, 2.4%), phosphorus (P, 0.17%),
and sulfur (S, 0.13%). The percentages of atomic abundance in the soft tissues of the
mature human body are shown (Bergethon, 1998). The remaining 0.4% are elements
that act as electrolytes: calcium (Ca^{2+}, 0.23%), sodium (Na^+, 0.07%), potassium
(K^+, 0.04%), magnesium (Mg^{2+}, 0.01%), and chloride (Cl^-, 0.03%). Two transition
metal ions that carry electrons are iron ($Fe^{3+} \rightarrow Fe^{2+}$) and copper ($Cu^{2+} \rightarrow Cu^+$).
Trace elements such as Mn, Zn, Co, Mo, and Se are much less abundant.

At the lowest level of chemical organization, animate matter is composed of a
limited number of standard building blocks. One can divide them into seven classes:

1. *Carboxylic acids* (general formula: R—COOH), dissociated to anions
   (R—COO$^-$) in neutral water environment
2. *Alcohols* (general formula: R—OH)
3. *Monosaccharides* (general formula: $(CH_2O)_n$), mainly with $n = 5$ (pentoses)
   or $n = 6$ (hexoses)
4. *Amines* (general formula: NH$_2$—R), protonated to cations (NH$_3{}^+$—R) in neutral water environment
5. *Nitrogenous heterocycles*
6. *Phosphates* (general formula: R—O—PO$_3^{2-}$), doubly dissociated in neutral
   water environment
7. *Hydrosulfides* (general formula: R—SH)

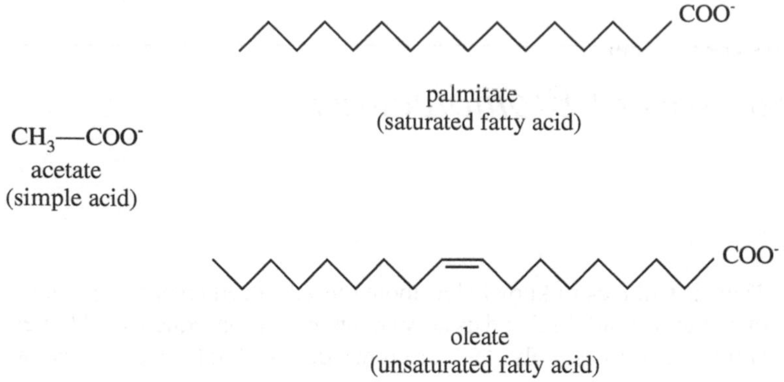

palmitate
(saturated fatty acid)

CH_3—COO^-
acetate
(simple acid)

oleate
(unsaturated fatty acid)

FIGURE 2.1
Examples of the three most important kinds of carboxylic acids with purely carbohydrate substituents.

C_2H_5—OH
ethanol
(simple alcohol)

glycerol
(triple alcohol)

cholesterol
(sterol)

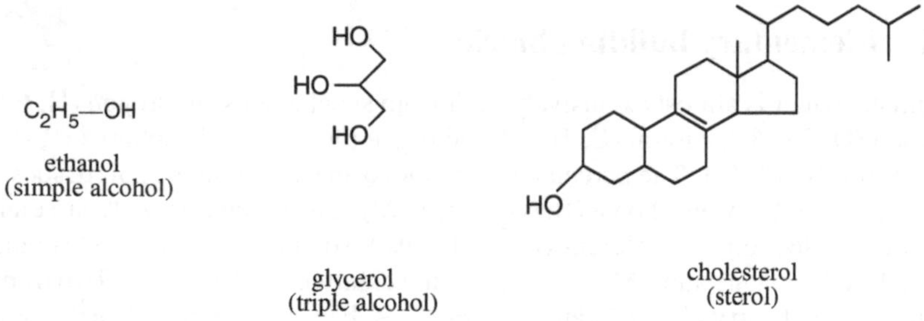

FIGURE 2.2
Examples of the three most important kinds of alcohols with purely carbohydrate substituents.

The Rs denote substituents additionally characterizing individual entities. They can be shorter or longer hydrocarbon chains or ring structures as shown for carboxylic acids in Figure 2.1 and for alcohols in Figure 2.2.

A chemical compound can be definitively identified only if its system of chemical bonds (*constitution*) is explicitly described, because two or more different compounds (*isomers*) can have the same atomic composition. To simplify notation of structural formulas we omit the carbon and hydrogen atom symbols (C and H, respectively) in larger molecules and leave only a lattice of covalent bonds between carbon atoms. It is assumed that at all vertices of such lattices, the carbon atoms are completed by an appropriate number of hydrogen atoms to preserve the four-fold carbon valency.

Figure 2.3 shows the three most important examples of monosaccharides. In a neutral water environment, a monosaccharide forms a ring structure closed by an oxygen atom from an *aldehyde* —CHO group or a *ketone* —CO— group. Nitrogenous heterocycles are modifications of aromatic hydrocarbons in which a hydrogenated carbon (CH) is replaced by N. Figure 2.4 shows the most important examples.

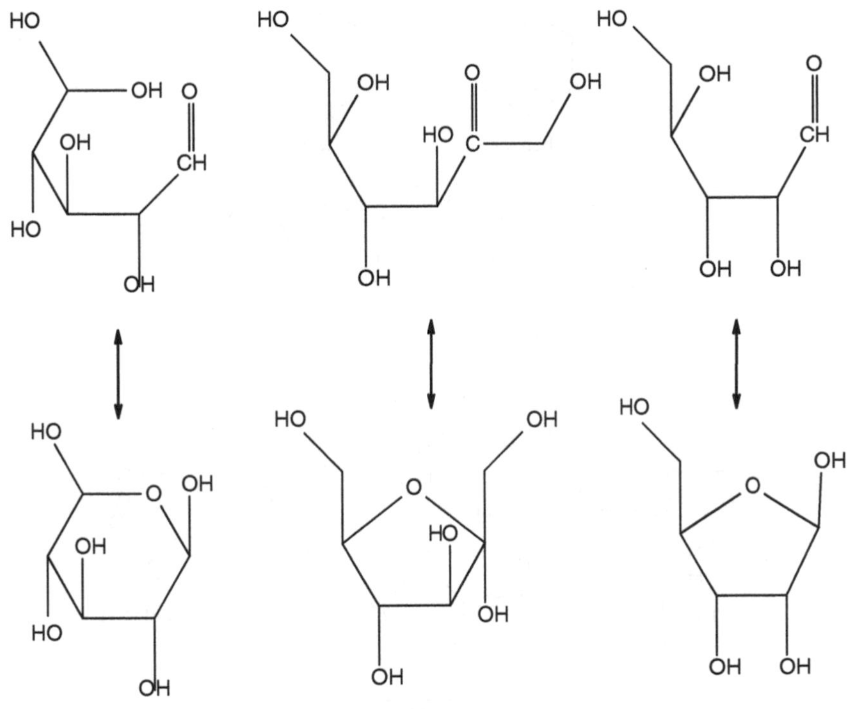

glucose   fructose   ribose

FIGURE 2.3
Three examples of monosaccharides: glucose, fructose (hexose), and ribose (pentose). The rings of glucose and ribose are closed by an oxygen coming from an aldehyde group. The ring of fructose is closed by an oxygen coming from a ketone group.

The simplest example of a phosphate is the *orthophosphate* (H—O—PO_3^-), a doubly dissociated anion of orthophosphoric acid known in biochemistry as an *inorganic phosphate* (P_i). Phosphates will be considered further in the context of phosphodiester bonds. Hydrosulfides (R—SH) are structurally similar to hydroxides (alcohols, R—OH), but differ by forming not only thioester (R—CO—S—R') but also (after reduction) disulfide (R—S—S—R') bonds.

Very often elementary organic building blocks belong simultaneously to two or even three classes listed above. This means that the substituent R, besides a hydrocarbon, consists also of an additional functional group or it serves as such a group. Thus, we can have simple carboxylic acids with ketonic (—CO—), hydroxylic (—OH), and protonated aminic (—NH_3^+) groups (*keto, hydroxy,* and *amino acids,* respectively), amines of heterocycles, or phosphates of saccharides and alcohols (Figure 2.5).

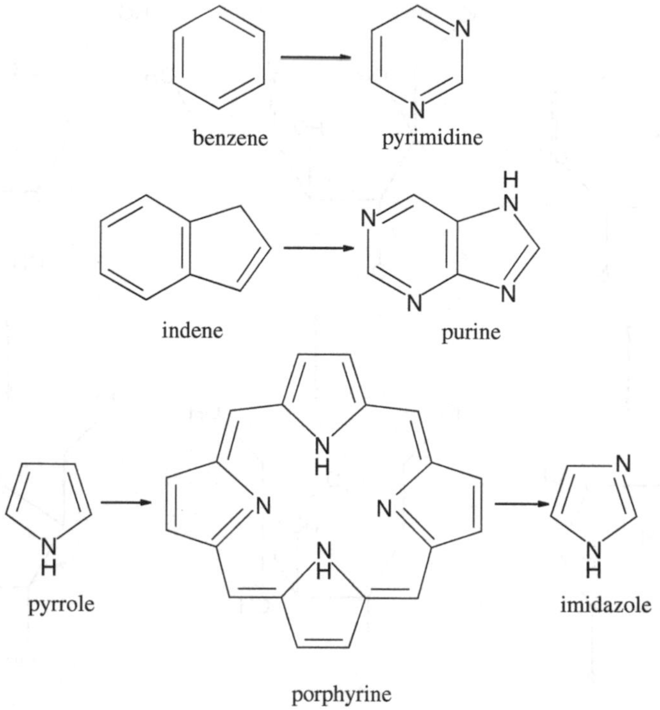

FIGURE 2.4
Most important examples of heterocycles. Pyrimidine is a direct modification of benzene, and purine is a direct modification of indene. No purely hydrocarbon homologues of imidazole and pyrrole exist. Porphyrine, the main component of heme and chlorophyll, is formed from four pyrrole rings.

2.2   Generalized ester bonds

Of special importance in organic chemistry is the reaction between a carboxylic acid and an alcohol, the product of which is referred to as an *ester*, with a characterististic *ester bond* (—COO—). See Figure 2.6 for an illustration of the ester bond. The alcohol in the esterification reaction can be replaced by amine to form *amide* with an *amide bond* (—CONH—). Carboxylic acid, on the other hand, combines with phosphate and forms a *phosphodiester bond*. *Lipids* are esters of fatty acids and glycerol (Figure 2.7). In *phospholipids*, one fatty acid chain is replaced by a phosphate with such a phosphodiester bond (Figure 2.7).

Monosaccharides behave as both alcohols and carboxylic acids. From their open structures (Figure 2.3), it follows that the carboxylic properties have —OH groups bound to the carbon atom neighboring the oxygen atom closing the saccharide ring. The generalized ester bond formed by such a group with another hydroxylic —OH

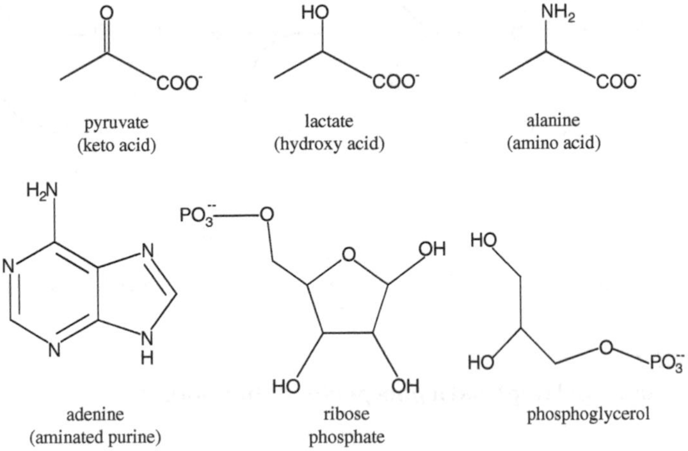

FIGURE 2.5
Examples of organic compounds with two functional groups.

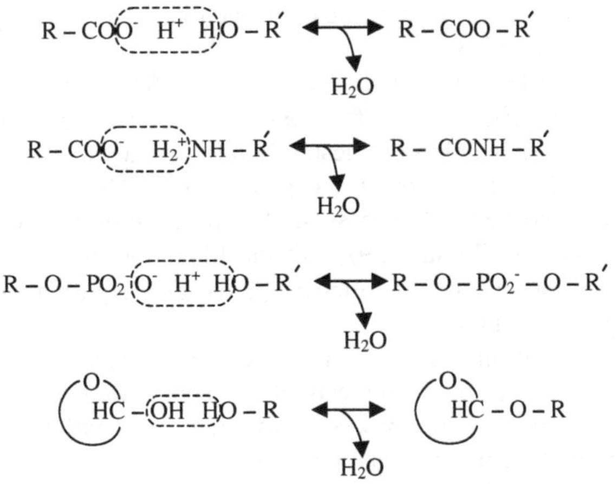

FIGURE 2.6
Formation of the ester bond (top) and its generalization to amide, phosphodiester, and glycosidic (bottom) bonds. The circular ring represents any monosaccharide ring closed by an oxygen atom neighboring a carbon atom taking part in the glycosidic bond. The carboxylic and phosphate groups are assumed to be dissociated.

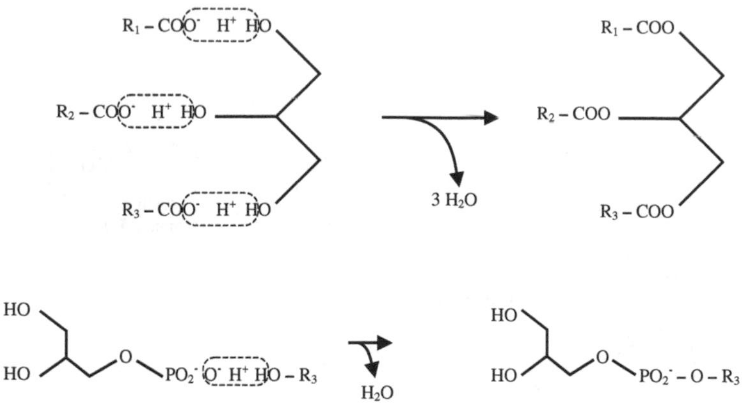

FIGURE 2.7
Formation of a lipid (top) and a phospholipid (bottom).

group of the alcoholic properties is referred to as a *glycosidic bond* (Figure 2.6). The alcohol hydroxylic group can be replaced by the *imino* groups (—NH—) of some nitrogen heterocycles that are derivatives of pyrymidine and purine (Figure 2.4). Compounds of ribose forming glycosidic bonds with such *nitrogenous bases* (imino eject groups are good proton acceptors) are of great biological importance and are called *nucleosides*. Compounds of ribose phosphate with nitrogenous bases are referred to as *nucleotides*. Figure 2.8 shows AMP (adenosine monophosphate), the most common nucleotide. All canonical nitrogenous bases entering nucleotides will be discussed in Section 2.8.

Binding of a nucleoside *monophosphate*, e.g., AMP, to *pyrophosphate* HO—$P_2O_5^{3-}$ (triply dissociated anion of pyrophosphoric acid, referred to as *inorganic diphosphate* or PP$_i$) by phosphodiester bonds produces, after liberation of the water molecule, ATP or adenosine triphosphate (Figure 2.9). ATP is a universal donor of free energy in biochemical processes. The reaction of its hydrolysis to ADP or adenosine *diphosphate* (Figure 2.9) with the liberation of *orthophosphate* (HO—PO_3^{2-} or inorganic phosphate P$_i$) is often coupled to endoergic reactions which without such coupling could not proceed.

The equilibrium of all the reactions presented in Figure 2.6 is strongly shifted toward hydrolysis rather than synthesis of the bond. Hence, the formation of the generalized ester bonds usually proceeds simultaneously with the hydrolysis of ATP along different reaction pathways as shown in Figure 2.6. The formation of any phosphate (R—O—PO_3^{2-}) takes part in the generation of ATP, but not an orthophosphate (Figure 2.9). The formation of phosphodiester bonds does not follow directly the scheme in Figure 2.6. In the presence of nucleotide triphosphates, it follows the two-step scheme presented in Figure 2.9, with the liberation of nucleotide monophosphate and pyrophosphate.

Three kinds of biological macromolecules — polysaccharides, proteins, and nucleic acids — are built from elementary entities: monosaccharides, amino acids, and nucleotides, respectively, linked by generalized ester bonds.

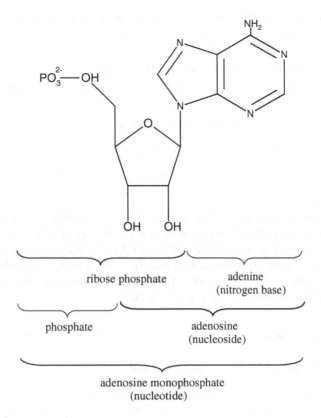

FIGURE 2.8
**An example of a nucleotide, AMP (adenosine monophosphate). A nucleotide is
a phosphate of nucleoside; a nucleoside is a monosaccharide ribose linked by a
glycosidic bond to a nitrogenous base.**

Polysaccharides are linear or branched polymers of monosaccharides linked by
glycosidic bonds (Figure 2.10a). Small polysaccharides are called *oligosaccharides*;
very small ones are called *disaccharides, trisaccharides*, etc.

Proteins are polymers of amino acids NH_3^+—R_iCH—COO$^-$ (see Figure 2.5)
linked into the linear chain by amide bonds (Figure 2.10b). Canonical side chains R_i
of amino acids will be discussed in Section 2.6. Small proteins are called *peptides*,
and amide bonds are often called peptide bonds.

Nucleic acids are linear polymers of nucleotides linked by a phosphodiester bond
(Figure 2.10c). Note that *dinucleotides* often appear as cofactors of protein enzymes
that carry hydrogen electrons jointly with protons, e.g., NAD$^+$ (nicotinamide adenine
dinucleotide) and FAD (flavin adenine dinucleotide) or small molecular groups (e.g.,
coenzyme A). They are two nucleotides linked by phosphotriester rather than diester
bonds (see Figure 2.10).

A more detailed introduction to the foundations of organic chemistry can be found
in many standard textbooks. We recommend, e.g., Applequist et al. (1982).

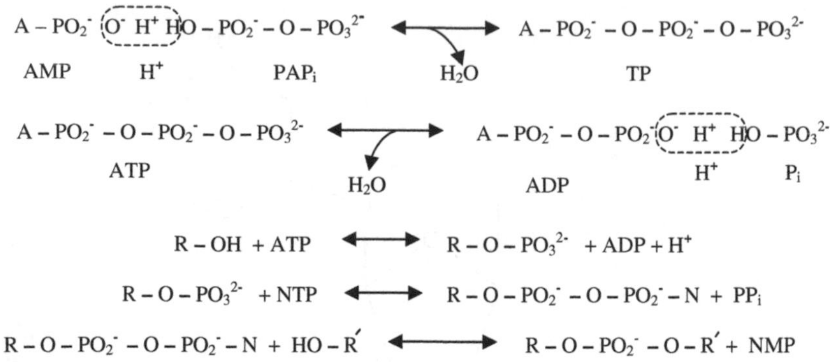

$$A - PO_2^- \quad \boxed{O^- \ H^+ \ H} O - PO_2^- - O - PO_3^{2-} \quad \longleftrightarrow \quad A - PO_2^- - O - PO_2^- - O - PO_3^{2-}$$

AMP H^+ PAP_i H_2O TP

$$A - PO_2^- - O - PO_2^- - O - PO_3^{2-} \quad \longleftrightarrow \quad A - PO_2^- - O - PO_2^- \boxed{O^- \ H^+ \ H} O - PO_3^{2-}$$

ATP H_2O ADP H^+ P_i

$$R - OH + ATP \quad \longleftrightarrow \quad R - O - PO_3^{2-} + ADP + H^+$$

$$R - O - PO_3^{2-} + NTP \quad \longleftrightarrow \quad R - O - PO_2^- - O - PO_2^- - N + PP_i$$

$$R - O - PO_2^- - O - PO_2^- - N + HO - R' \quad \longleftrightarrow \quad R - O - PO_2^- - O - R' + NMP$$

FIGURE 2.9
Formation of ATP from AMP (top) and of ADP from ATP. Phosphoester bonds and phosphodiester bonds are created in the presence of ATP (middle) or, more generally, NTP (bottom). AMP, ADP and ATP = adenosine monophosphate, diphosphate, and triphosphate, respectively. NMP and NTP = nucleoside mono- and triphosphates. P_i = orthophosphate anion (inorganic phosphate). PP_i = pyrophosphate anion (inorganic diphosphate).

2.3   Directionality of chemical bonds

The spatial structures of biomolecules, so important for their function, are related to directional properties of chemical bonds. These properties are, in turn, determined by the spatial distribution of electronic states of the constituent atoms. The hydrogen atom contributes only one s-orbital to chemical bonding while carbon, nitrogen and oxygen atoms contribute one s-orbital and three p-orbitals, and phosphorus and sulfur atoms can contribute another five d-orbitals (Atkins, 1998).

Figure 2.11a outlines the spatial distribution of the electron probability density in the s- and p-orbitals. One s-orbital and three p-orbitals can hybridize into four orbitals of tetrahedral symmetry (sp^3 hybridization); one s-orbital and two p-orbitals can hybridize into three orbitals of trigonal symmetry (sp^2 hybridization); and one s-orbital and one p-orbital can hybridize into two orbitals of linear symmetry (sp hybridization). Figure 2.11b shows the three types of hybridized orbitals.

The carbon atom has four electrons in the outer shell that are organized into four sp^3 hybridized orbitals. It needs four additional electrons for those orbitals to be completely filled. As a consequence, it can bind four hydrogen atoms, each giving one electron and admixing its own s-orbital to the common bonding σ-orbital, each of axial symmetry (see Figure 2.12).

The nitrogen atom has five electrons in four sp^3 hybridized orbitals. It needs only three additional electrons and thus binds three hydrogen atoms. The ammonia molecule formed, like the methane molecule, has tetrahedral angles between the bonds. However, one bond is now replaced by a lone electron pair (see Figure 2.12).

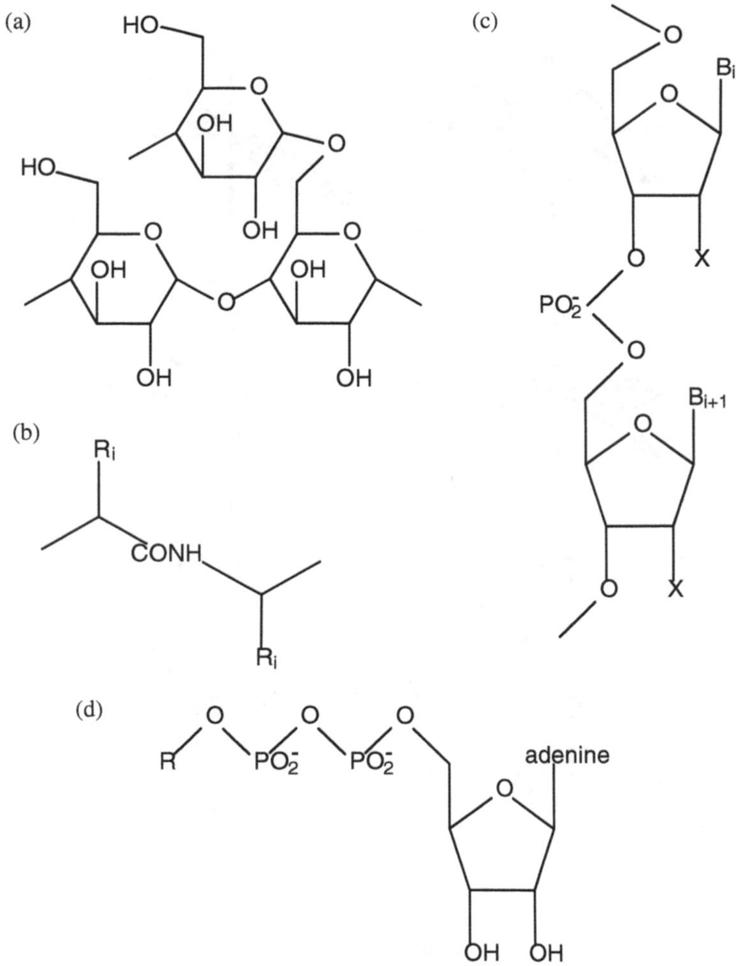

FIGURE 2.10
(a) Polysaccharides (hydrocarbons) are polymers of monosaccharides linked by glycosidic bonds, nonbranched as in the case of amylase, a type of starch and the main storage polysaccharide of plants, or branched as in the case of amylopectin, another type of starch or glycogen, the main storage polysaccharide of animals. (b) Proteins are polymers of amino acids linked by amide (peptide) bonds. (c) Nucleic acids are linear polymers of nucleotides linked by phosphodiester bonds. In RNA, X denotes a hydoxyl (OH) group; in DNA, it denotes H. (d) Phosphotriester bonds in dinucleotides usually form from adenosine and another, usually highly modified, nucleotide (R).

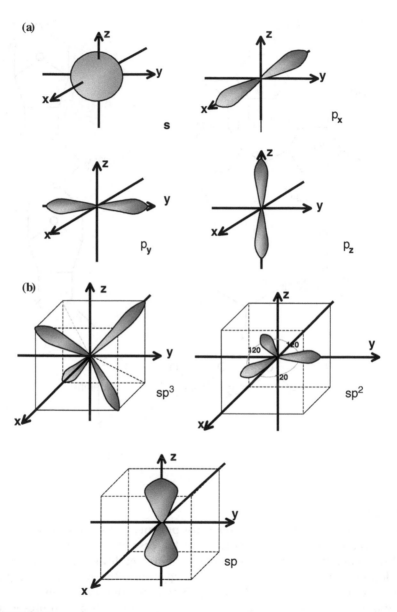

FIGURE 2.11
(a) Outline of the spatial distribution of electron probability density in an
s-orbital and three p-orbitals. (b) Four orbitals of tetrahedral sp^3 hybridization, three of trigonal sp^2 hybridization, and two of linear sp hybridization,
respectively.

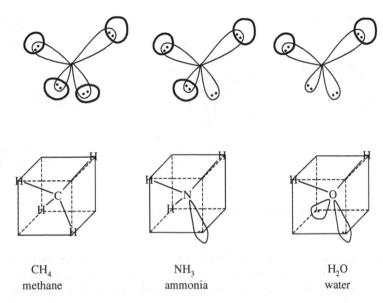

CH$_4$
methane

NH$_3$
ammonia

H$_2$O
water

FIGURE 2.12
Spatial structures of methane, ammonia, and water are determined by tetrahedral sp^3 hybridization of the central atom orbitals. A "cloud" with two dots denotes a lone electron pair.

The oxygen atom has six electrons in four sp^3 hybridized orbitals. It binds two hydrogen atoms and has two lone electron pairs (Figure 2.12). As a consequence of sp^3 hybridization, the nitrogen and oxygen atom orbitals have highly directed negative charge distributions leading to hydrogen bonding needed to achieve the spatial structures of all biomolecules.

By replacing one H atom in methane, ammonia, and water by the methyl group CH$_3$, one obtains ethane, methylamine, and methanol, respectively (Figure 2.13). The CH$_3$ can rotate freely around the C—C, C—N, or C—O bonds. The lowest value of the potential energy of the molecule is found for the electron densities of one triple of atoms or lone electron pairs located between the electron densities of the other triple of atoms when looking along the rotational axis.

In the cases of ethane, methylamine, and methanol, a 120-degree rotation does not lead to a structural change of the molecule. However, if the other triple has free electron pairs or atoms other than hydrogen, a rotation around the central covalent bond may lead to a new *conformational state* of the molecule. Such a state cannot be reconstructed from the original state by translation or rigid body rotation. Different conformational states become geometrically significant for long molecular chains, e.g., carbohydrates, which can exist in maximally stretched linear forms and as folded clusters. A rotation of each covalent bond in the chain allows one conformational *trans* state and two *gauche* states (Figure 2.14). If we ignore the steric constraints (excluded volume effects) that rapidly emerge in longer chains and constitute a separate significant problem, the differences between equivalent conformational states of

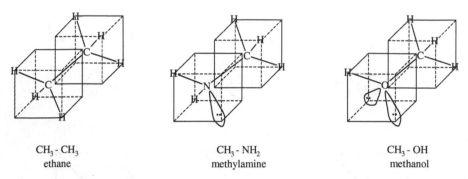

CH₃ - CH₃	CH₃ - NH₂	CH₃ - OH
CH_3 - CH_3	CH_3 - NH_2	CH_3 - OH
ethane	methylamine	methanol

FIGURE 2.13
Spatial structures of ethane, methylamine, and methanol.

a single bond amount to several kJ/mol while the potential energy barrier height is on the order of 10 to 20 kJ/mol (see Figure 2.14). This corresponds to four to eight times the value of the average thermal energy at physiological temperature $k_B T = 2.5$ kJ/mol. The probability of a random accumulation of such an amount of energy in one degree of freedom defined by the Boltzmann factor, i.e., the exponential of its ratio to $k_B T$, $\exp(-\Delta/k_B T)$ equals 10^{-2} to 10^{-4}. The latter value multiplied by the average frequency of thermal oscillations, 10^{13} s^{-1}, gives 10^{11} to 10^9 random local conformational transitions per second at physiological temperatures. A conformational state of a molecule is, therefore, not very stable.

In the case of closed chains (unsaturated cyclic carbohydrates or monosaccharides; see Figure 2.3), 120-degree rotations around individual bonds are not possible without breaking them. Hence, conformational transitions involve much smaller rotations that are simultaneously applied to many bonds. The process is called ring puckering since an entirely flat conformation becomes energetically unstable. In the case of a five-atom ring we distinguish the conformations of an *envelope* and a *half-chair*. In the case of a six-atom ring, we see the conformations of a *chair*, a *twist*, and a *boat* (Figure 2.15).

Not all transitions between different energetically stable geometrical structures can be achieved by simple rotations around covalent bonds involving small energy barrier crossings. A mutual exchange of hydrogens with hydroxyl groups of monosaccharide rings (see Figure 2.3) is not possible without covalent bond breaking and subsequent restoration and hence requires energy of about 300 kJ/mol (see Table 2.1). This is 1.5 orders of magnitude greater than the energies mentioned earlier. Various geometrical forms of chemical molecules that have the same summary formulas but require bond breaking and restoration are called *isomers*. It is easy to imagine a large diversity of isomers even for simple monosaccharides. Isomers differ in the locations of hydrogen atoms and hydroxyl groups and in the positions of the oxygen bridges. Hexoses can form both five- and six-atom rings (see the structures of glucose and fructose in Figure 2.3).

Half the monosaccharides are simple mirror images of their counterparts. Such isomers are called *enantiomers* and the phenomenon is called *chirality*. This is a

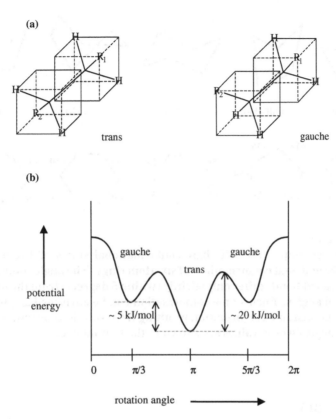

FIGURE 2.14
**Conformational states related to the rotation around one covalent bond (top).
R_1 and R_2 denote atoms or molecular groups (in the case of longer chains) other
than atomic H. The potential energy related to the rotation of a single covalent
bond (bottom). The three minima correspond to the three conformational states:
one *trans* and two *gauche*. They are separated by energy barriers with heights
of 10 to 20 kJ/mol.**

symmetry breaking of the handedness (left/right invariance) analogous to the human
hands. The origin of *chirality* comes from the Greek word *cheir* meaning *hand*.

Ignoring specific relations to double bonds, stereoisometry is usually linked to the
existence of at least one carbon atom in the molecule whose all four covalent bonds are
inequivalent in that they lead to different atoms or groups of atoms. Chiral molecules
are optically active and they twist the light polarization plane to the left or right.

Enantioners are divided into D- and L-types. Their definitions are based on
the simplest monosaccharide, i.e., glyceraldehyde (Figure 2.16). D-type monosaccharides are derivatives of D-glyceraldehyde. Similarly, L-type monosaccharides
are derivatives of L-glyceraldehyde. By definition the D-glyceraldehyde twists the
polarization plane to the right, but this is not the case for all D-type monosaccharides.
Also, all amino acids except glycine are chiral which is linked to the fact that the

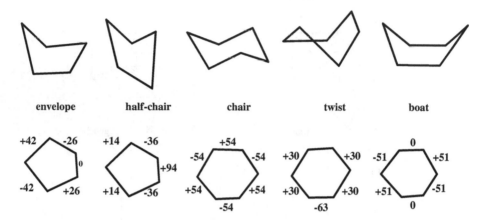

FIGURE 2.15
Top row: Envelope and half-chair conformational states of five-atom rings.
Chair, twist and boat conformations of six-atom rings. The angles between neighboring covalent bonds differ only slightly (within 5 degrees) from the undistorted
tetrahedral angles. Each stable conformation can be envisaged as puckering of
a planar but unstable conformation resulting from rotations around subsequent
bonds by angles whose values are shown in the bottom row.

TABLE 2.1
Properties of Biologically Important Bonds

Type of Bond	Bond Distance (nm)	Bond Energy (kJ mol^{-1})
C—C	0.154	350
C=C	0.133	610
N—N	0.145	161
N=N	0.124	161
C—N	0.147	290
C=N	0.126	610
C—O	0.143	350
C=O	0.114	720
C—S	0.182	260
O—H	0.096	463
N—H	0.101	390
S—H	0.135	340
C—H	0.109	410
H—H	0.074	430

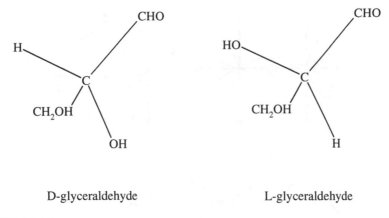

D-glyceraldehyde   L-glyceraldehyde

FIGURE 2.16
Two enantiomers of glyceraldehyde.

central carbon C^α has four inequivalent bonds. Therefore, we have both D-amino acids and L-amino acids. Chemists adopted an unambivalent way of transferring the determination of D and L-type enantioners from the defining glyceraldehydes onto other compounds different from monosaccharides. This is a little too complicated for our purposes, so we will not dwell on it. We only wish to note that all biologically active monosaccharides are D-types while all biologically active amino acids are L-types.

Let us now consider the bonds formed by electron orbitals in the process of trigonal hybridization sp.² Using these orbitals, two carbon atoms can form σ bonds (with axial symmetry) among themselves and with the other four atoms that lie in the same plane, for example, hydrogen. The remaining p-orbitals perpendicular to this plane, one for each carbon atom, then form a second, slightly weaker bond between these atoms. This is called a π bond (see Figure 2.17). A *planar* molecule of ethylene is formed in which two carbon atoms are bound together via a *double* bond. If one carbon atom is replaced by a nitrogen atom, with an extra electron, and simultaneously one hydrogen atom is replaced by a lone electron pair, we obtain an imine molecule. If, on the other hand, one carbon atom is replaced by an oxygen atom with two excess electrons and simultaneously two hydrogen atoms by two free electron pairs, we obtain a formaldehyde molecule (Figure 2.17). Table 2.1 lists some of the most biologically important single and double bonds together with their bond lengths and bond energies.

Double bonds can exist in longer chain molecules, for example, unsaturated carbohydrates. For each double bond, a structure can have two energetically stable spatial structures, each obtained from the other by a 180-degree rotation. A rotation around a double bond requires a temporary breakage of the π bond, and hence an energy of approximately 200 kJ/mol. Both structures are isomers rather than conformers. They are called *trans* and *cis configurational isomers* (see Figure 2.18).

The π bonds can be found generally in delocalized forms. In the case of the benzene molecule C_6H_6 (Figure 2.19a), one can imagine two symmetrical Kekule

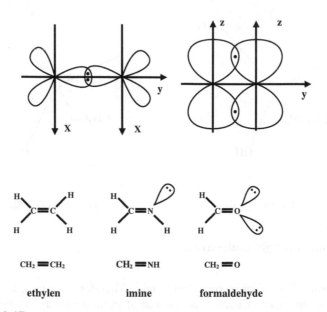

CH₂ $=$ CH₂   CH₂ $=$ NH   CH₂ $=$ O

ethylen   imine   formaldehyde

FIGURE 2.17
Top: A double bond consists of a σ bond formed by orbitals in the process of trigonal hybridization sp² and a π bond formed from orbitals perpendicular to previous ones. Bottom: Planar particles of ethylene, imine, and formaldehyde.

structures whose every other carbon–carbon bond is a double bond. Which of these two structures is actually adopted by the molecule? Are both represented statistically in the population of identical molecules with the same probability? Answers to these questions come from quantum mechanics. The state of each benzene molecule is a quantum mechanical linear combination of the two states defined by Kekule structures. The π bond is not localized on every other carbon pair; it is delocalized over the entire ring. An electric current circulates around the ring and can be induced by magnetic fields and other electric currents. Due to such planar interactions in benzene rings and also in other aromatic heterocycles (e.g., in purine and pyrimidine bases), their compounds have a tendency for parallel *stacking*. Effectively, each carbon pair is allocated one half a π bond, which justifies the notation using a broken line (Figure 2.19a).

The π bond between carbon and oxygen in a dissociated carboxyl group —COO⁻ is also delocalized (Figure 2.19b). This group can be envisaged as a planar, negatively charged plate that has a rotational degree of freedom around the axis that connects one carbon atom to the next atom. The phosphodiester bond (Figure 2.19c) exhibits similar behavior, except the phosphorus atom is connected to the rest of the molecule via two single bonds, each of which provides a rotational axis.

The delocalization of the π bond in the case of the orthophosphate group (—PO₃²⁻) is more complicated. A double bond can be established between carbon and each of the three oxygen atoms. Another possibility is a state involving transfer of an electron

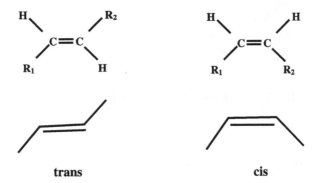

trans   cis

FIGURE 2.18
Configurational *trans* and *cis* isomers are related via rotation through one double bond. R_1 and R_2 denote atoms or molecular groups (in the case of longer chains) other than hydrogen atoms.

from the phosphorus atom onto a hitherto neutral oxygen atom. The actual state of the group is a linear combination of all four possibilities (see Figure 2.19d). Instead of drawing broken lines for partial bonds between phosphorus and the three equivalent, partially negatively charged oxygen atoms, biochemists represent the entire orthophosphate group with a *P* in a small circle.

The delocalization of the π bond in an orthophosphate group is related to a high energy which is stored in a phosphodiester bond in ATP. As a result of ATP hydrolysis into ADP and an inorganic phosphate P_i, two phosphodiester bonds and one orthophosphate group are replaced by one phosphodiester bond and two orthophosphate groups (Figure 2.19b). In this arrangement, the π electrons are more delocalized. The negative delocalization energy in connection with additional negative energy of hydration causes the products of ATP hydrolysis to have less energy than the products of the hydrolysis of other general ester bonds.

An electron transfer also takes place in the case of amide (peptide) bonds (see Figure 2.19e). Due to partial delocalization of the π bond, the amide bond structure becomes planar. The O, C, N, and H atoms lie in one plane which is an important element of the protein structure discussed later in this chapter.

2.4   Weaker intratomic interactions

Chemical bonds are the major forces holding atoms together within molecules. They may be classified according to their energy levels or how the bonds are made. Chemical bonds usually have specific interatomic distances (bond lengths) and maintain specific angles relative to each other. However, for bonds with smaller energies, these restrictions become less strict. This is especially true for macromolecules, the molecules commonly found in biological systems. In macromolecules, the collective contributions from a large number of bonds with small energies become important.

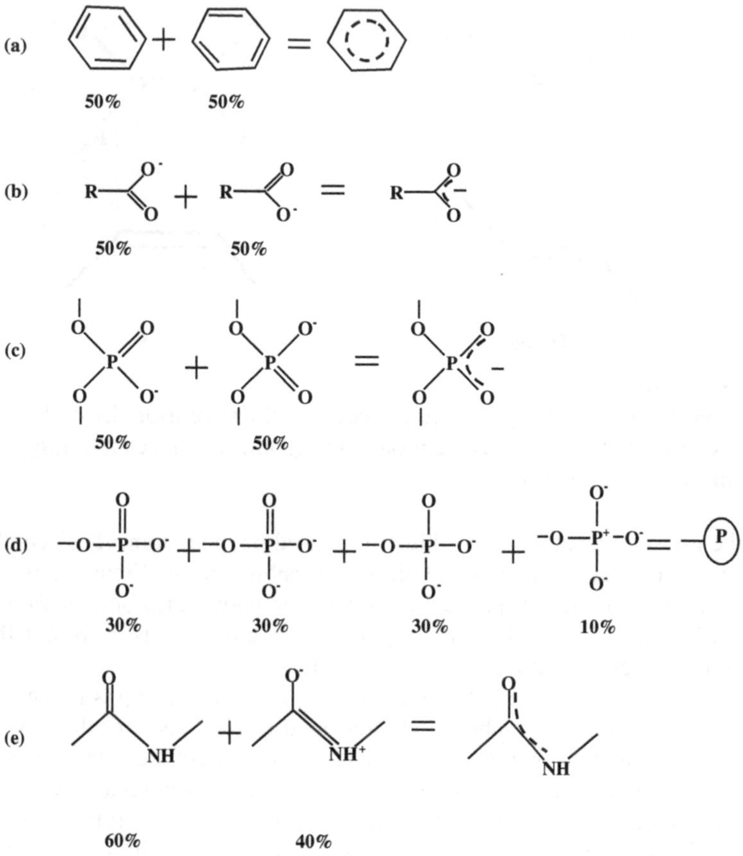

FIGURE 2.19
Delocalization of double bonds in the case of benzene (a), carboxylic anion (b), phosphodiester bond (c), orthophosphate anion (d), and amide (peptide) bond (e).

2.4.1 Ionic interactions

Forming ionic interactions is probably one of the most basic activities of matter. Ionic bonding involves ions of opposite charge that form a bound state as a result of Coulomb attraction forces. The respective energies range up to 10 eV. This is strong enough to resist dissociation at room temperature and only temperatures of thousands of kelvins (certainly above physiological range) can break ionic bonds. If two charges of q_1 and q_2 are separated by distance r in an area of space with dielectric constant ϵ, the force F between them is:

$$F = k\frac{q_1 q_2}{\epsilon\, r^2}.$$ (2.1)

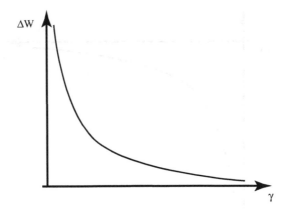

FIGURE 2.20
Amount of work necessary to bring two charges of the same sign together from infinity to a finite distance.

If the charges are opposite, the force will be attractive and vice versa. The potential energy between the two charges may be considered as the amount of work necessary to bring the charges together from infinite distance to finite distance r (see Figure 2.20 for two charges of the same sign). For example, the binding energy of two charged particles such as Na^+ and Cl^- is calculated from:

$$\Delta W = -k\frac{q_1 q_2}{\epsilon r} \tag{2.2}$$

where r is 2.4 Å giving $\Delta W = 5.5$ eV. This is an example of an ionic bond. The degree of polarity of a bond depends on the differences in the electronegativities of the two elements forming a molecule. When the two charges have opposite signs, the potential energy will be negative at a finite distance and zero at infinity (see Figure 2.21). When two ions become very close, repulsive forces due to the overlapping electron clouds and opposing nuclear charges will be generated. The potential energy giving these repulsive forces can be expressed by:

$$PE = be^{-\frac{r}{a}} \tag{2.3}$$

where a and b are constants. Figure 2.22 is a graphic example of this function. The net potential energy U is:

$$U = \left(k\frac{q_1 q_2}{\epsilon r}\right) + be^{-\frac{a}{r}} \tag{2.4}$$

and this is shown graphically for NaCl in Figure 2.23.

The equilibrium distance and corresponding potential energy vary according to the state of NaCl. The potential energy U in the crystalline form is greater than that in vapor, as are the bond distances between Na and Cl ions, because of the increased repulsive energy between ions locked into the crystal lattice (see Table 2.2).

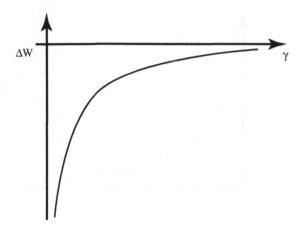

FIGURE 2.21
Potential energy where charges have opposite signs.

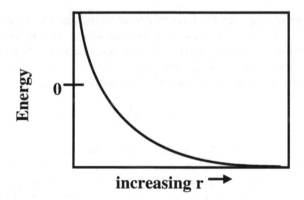

FIGURE 2.22
Potential energy curve for two ions close to each other.

2.4.2   Covalent bonds

Covalent bonds arise in pairs of neutral atoms such as two H or two C atoms. This interaction is electromagnetic and results from a charge separation between the nucleus and its electronic cloud. The two atoms share some electrons such that the energy and average separation between their nuclei are reduced. All the elements linked by covalent bonds are either electropositive (Li, Na, K, Rb, Be, Mg, and Ca) or electronegative (O, S, Se, F, Cl, Br, and I). When a covalent bond is formed between two atoms of the same element, the electron pair is shared equally by both atoms.

When a covalent bond is formed between two atoms of unequal electronegativity, the electron pair is attracted more toward the atom possessing higher electronegativity, e.g., in the C—Cl bond, the pair of electrons in the covalent bond is closer to Cl than to C. This leads to a partial positive charge on C and a partial negative charge on

TABLE 2.2
Potential Energies and Bond Distance
of Na Cl

	In Vapor	In Crystal
r	2.51 Å	2.81 Å
U	−118 Kcal/mole	−181 Kcal/mole

Cl, giving a partial ionic character to such covalent bonds. The energy of covalent bonds is still quite high. For example, the H—H bond has an energy of 4.75 eV while the C—C binding energy is 3.6 eV, which is of the same order of magnitude as the NaCl binding energy. Other covalent bonds are in the same range of several eV. Thus, we see that the covalent bond has a binding energy slightly lower but comparable to that of the ionic bond in vacuum. Both types of bonds have strengths which at room temperature are large compared to thermal energy $k_B T$. The covalent bond is responsible for maintaining the structural integrity of DNA, proteins, and polysaccharides (chain-like sugar molecules). In a covalent bond, electrons shuttle back and forth constantly between positively charged nuclei and hence the covalent bond is electrostatic in origin with a characteristic distance d = 1 Å.

In 1916, W. Kossel proposed the theory of the electrovalent bond, i.e., formation of bonds by loss or gain of electrons. G.N. Lewis (1916) proposed the idea of nonpolar bond formation by sharing of electrons between atoms. Subsequent development of quantum mechanics explained the quantitization of electron energy levels in atoms and the structures of atomic orbitals and consequently provided a basis for the theory of molecular orbitals by W. Heitler and F. London in 1927. The distribution of electrons in an atom was determined through the Pauli exclusion principle formulated in 1924. The theory states that no electron can have the same quantum state in an atom if the energies of the electronic orbitals are the same. Since the spin of the electron is quantized, and it can have only either $+\frac{1}{2}$ or $-\frac{1}{2}$ projection values, each electron orbital can be occupied by no more than two electrons. For an atom in its ground state, the electronic energy levels will be filled starting with the lowest energy first. For example, a carbon atom has six electrons and they may be distributed as:

$$C^{12}: \quad (1s)^2(2s)^2(2p)^2 --- (1s)^2(2s)^2(2p_x^1, 2p_y^1) \tag{2.5}$$

Alternatively, for an excited electron, the distribution of electron density can overlap that of other energy levels, giving rise to a new electron density distribution:

$$C^{12}: \quad (1s)^2(2s)^2(2p)^2 --- (1s)^2(2s)^1(2p_x^1, 2p_y^1, 2p_z^1) \tag{2.6}$$

demonstrating that the outermost electrons may form different types of electronic orbitals. When two atoms are bound together in covalent bonds by sharing electrons, the outermost electrons of each atom may be found in bonding orbitals, antibonding orbitals and nonbonding orbitals.

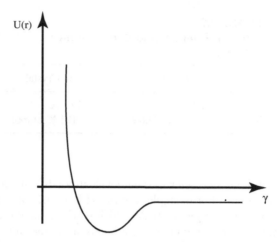

FIGURE 2.23
Potential energy of NaCl as a function of ionic separation r.

For example, when two hydrogen atoms form a hydrogen molecule, fusion of two s-orbitals forms a σ-orbital. The newly created wave functions for the bonding and antibonding may be expressed by the method of *linear combination of atomic orbitals* (LCAO), as $\Psi(A) + \Psi(B)$ for bonding. For antibonding, $\Psi(A) - \Psi(B)$. When the covalent bond is formed with two $2p_x$ orbitals, they will form $s2p$ and $s*2p$ orbitals. By pairing $2p_z$ or $2p_y$ orbitals, they may form p_y2p and $p*_y2p$ orbitals.

Consider bond formation between C and H. The outermost electrons of the C atom may have different configurations. When C interacts with four H atoms, it may be assumed that all the molecular bonding orbitals must be alike. Thus, we need four electrons of similar energies.

$$C^{12}: \quad (1s)^2(2s)^1(2p_x)^1(2p_y)^1(2p_z)^1 \tag{2.7}$$

Since it is assumed that $2s$ and $2p$ electrons have about the same energy levels, the wave functions of all four electrons may mix and four new hybrid electronic orbitals are formed by sp hybridization. The wave functions of these orbitals are of the form:

$$\Psi(t_1) = a\Psi(2s) + b\Psi(2p_x). \tag{2.8}$$

where a and b are the electron density distribution constants for each wave function. Four of these wave functions represent four electrons and they are directed as far as possible from each other in a tetrahedral arrangement.

When a C atom interacts with nonequivalent atoms, it may experience another form known as trigonal hybridization, in which $2s$, $2p_x$, and $2p_y$ electrons are involved:

$$C^{12}: \quad (1s)^2(2s)^1(2p_x)^1(2p_y)^1(2p_z)^1 \tag{2.9}$$

TABLE 2.3
Relative Ionization

Molecule	HF	HCl	HBr	HI
% Ionic character	60	17	11	0.5

TABLE 2.4
Electronic Distribution for Halogen Atoms

Atom	Distribution
F	$(1s)^2 (2s)^2 (2p)^5$
Cl	$(1s)^2 (2s)^2 (2p)^6 (3s)^2 (3p)^5$
Br	$(1s)^2 (2s)^2 (2p)^6 (3s)^2 (3p)^6$ $(\text{three-dimensional})^{10} (4s)^2 (4p)^5$
I	$K - L - M (4s)^2 (4p)^6 (4d)^{10} (5s)^2 (5p)^5$

The molecular orbital functions for these electrons may be represented by:

$$\Psi(h_1) = a_i \Psi(2s) + b_i \Psi(2p_x). \tag{2.10}$$

The electrons in the $2p_z$ orbital are free to form resonating π orbitals. The angle between planar orbitals is 120 degrees.

In covalent bonds, the electron density may be uniformly distributed about the bonding orbitals. If electronegative or electropositive atoms are present, electron density will be unevenly distributed. The result will be a bond possessing covalent and ionic bond characteristics. Consider the bonding of H and Cl atoms. Expressing electrons with the usual Lewis diagram dots, resonance will exist between the two structures:

$$H : Cl < - > H + Cl- \tag{2.11}$$

The bonding wave function may be expressed as:

$$\Psi = \Psi(\text{covalent}) + a\Psi(\text{ionic}) \tag{2.12}$$

The relative amount of ionization depends on the type of atom to which hydrogen is bound. Examples are shown in Table 2.3. Obviously, the degree of ionic character depends on the electronegativity of the atoms and the trend may be understood by examining the distribution of electrons in these atoms. Table 2.4 shows electronic distribution for some atoms in order of decreasing electronegativity.

The tendency of these atoms is to attract additional electrons to satisfy the sixth electron position in the p orbitals. The strength of this attraction (electronegativity) will be greater in atoms with fewer electrons which will shield the effects of

TABLE 2.5
Electron Distribution for Biologically
Relevant Atoms

Atom	Distribution
H	$(1s)^1$
C	$(1s)^2 (2s)^2 (2p)^2$
N	$(1s)^2 (2s)^2 (2p)^3$
O	$(1s)^2 (2s)^2 (2p)^4$
P	$K-(3s)^2(3p)^3$

the nuclear charge. Table 2.5 shows similar electron distributions for atoms common in biological systems.

Both covalent and ionic bonds have bonding energies as high as hundreds of kcal/mol in vacuum. While the ionic bonds may be affected by the condition of their environment, in particular the dielectric constant, the electron sharing covalent bonds experience few effects from environmental conditions. For covalent bonds, distances and angles are well defined. Depending on the number of electrons shared by the atoms, the structure will include single, double, and triple bonds. The rotation of atoms around double and triple bonds is restricted.

2.4.3   Free radicals

During chemical reactions, a covalent bond may be cleaved in various ways. When the bond breaks, the shared electron pair may remain with either atom or may be divided between them. The former situation is referred to as heterolytic cleavage; the latter is called homolytic cleavage. Atoms possessing unpaired electrons are called free radicals and are highly reactive in homolytic cleavage. The unpaired electron in a free radical possesses a magnetic moment and the radiation of microwave frequency induces transitions between magnetic energy levels of these particles. Electron spin resonance (ESR) study of these magnetic energy levels and transitions can provide valuable information about a system.

When a bond becomes similar to a covalent bond, the influences of other atoms of the environment diminish. The replacement of an electron acceptor by other molecules forces the donor molecule to form hydrogen bonds with the new acceptor molecules. While ionic and covalent bonds play major roles in constructing an architectural framework of macromolecules, other types of bonds with significantly smaller energies also contribute to the structures of macromolecules simply because there are so many of them.

2.4.4   Van der Waals bonds

Van der Waals forces are interactions among neutral molecules. Energies involved in Van der Waals interactions are in the range of 0.5 to 4.5 kJ/mol, i.e., some 100 times weaker than a covalent bond and 5 to 10 times weaker than a hydrogen bond,

TABLE 2.6
Physical Data for Different Types of Bonds

Donor-Acceptor	Molecules	Distance (Å)	Energy (kcal/mole)
–OH...O <	H_2O	2.76	4.5
–OH...O $=$ C <	$CH_3 COOH$	2.8	8.2
>NH...O $=$ C <	Peptides	2.9	
>NH...N<	Nucleotides	3.1	1.3
>NH F$^-$	NH_4F	2.63	5.0
>NH...S <		3.7	

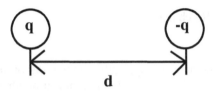

FIGURE 2.24
Dipole due to two opposite charges.

but they are important in producing cohesion of molecules in liquids and some solids. They can be classified as (1) Keesom forces when they are due to two permanent dipoles that are attractive and favor parallel arrangement along the axis joining the two molecules, (2) Debye forces between dipolar molecules that induce dipole moments of nonpolar molecules leading to an attraction between them, and (3) London or dispersive forces arising from induced dipole moment formation in nonpolar molecules, e.g., group IV halides such as CF_4 and CCl_4. The London interactions may have bond energies comparable to dipole–dipole energy. They are the main sources of attraction between nonpolar or weakly polar chemical species. London dispersion interactions appear to arise from nonstatistical time-dependent distribution of charges on molecules that creates temporarily induced dipoles. The exact nature of this interaction is not well understood, but the magnitude of its potential energy follows $1/r^6$. See Table 2.6.

We must introduce some background in order to describe dipolar interactions physically. As seen with ionic interactions, a dipole interacts with ionic charges and other dipoles in its vicinity. Consider a dipole due to two opposite charges q and $-q$ separated by a distance d (Figure 2.24). The dipole moment (p) is:

$$p = qd \qquad (2.13)$$

If the magnitude of the elementary charges is taken as $e = 1.6 \times 10^{-19}$ C and the distance separating them is $d = 1$ Å $= 10^{-10}$ m, the dipole moment is: $p = 1.6 \times 10^{-29}$ cm. Since 3.33×10^{-30} cm $= 1$ D, $p = 4.8$ D where 1D is called a debye, a common unit of dipole strength.

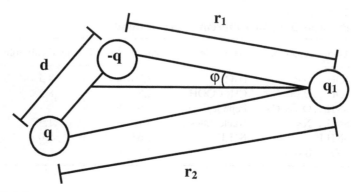

FIGURE 2.25
Dipole interacting with charge q$_1$.

Molecular dipole moments are typically on the order of one elementary charge e times 1 Å, i.e., about 10^{-29} cm. The dipole moment of ethanol is 5.6×10^{-30} cm. Water is polar and has a more complex charge distribution than ethanol, namely a negatively charged oxygen with two positively charged hydrogen atoms. The dipole moment of a water molecule amounts to 1.85 D which arises from 1.6 D for O—H bonds situated 104 degrees from each other. In addition to permanent dipole moments present in numerous bonds, induced dipole moments can arise after exposure to external electric fields. Larger symmetrical molecules such as methane may contain bonds with dipole moments that cancel out due to their symmetrical arrangement. A dipole may interact with ions and other dipoles with some definable potential energy as shown below. Consider first a dipole interacting with a charge q_1 (Figure 2.25).

The potential energies between the dipole and charge are contributions of interactions between charges $-q$ and q_1, and q and q_1:

$$V = k\frac{q_1\,(-q)}{\epsilon r_1} + k\frac{q_1 q}{\epsilon r_2} \tag{2.14}$$

Assuming that d is significantly smaller than r, we find that:

$$r_1 = r - \left(\frac{d}{2}\right)\cos\varphi \tag{2.15}$$

and

$$r_2 = r + \left(\frac{d}{2}\right)\cos\varphi \tag{2.16}$$

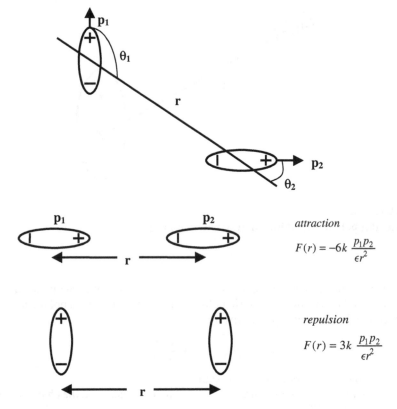

FIGURE 2.26
Interaction of two permanent dipoles.

Substituting these relations to the previous equation and making further approximations, we obtain:

$$V = -k\left(\frac{q_1 q}{\epsilon r^2}\right) d \cos\varphi \quad \text{or} \quad V = -k\left(\frac{pq_1}{\epsilon r^2}\right) \cos\varphi \tag{2.17}$$

The potential energy between two dipole moments, p_1 and p_2, may be calculated in a similar manner as:

$$V = k\frac{\vec{p}_1 \vec{p}_2}{\epsilon r^3} - \frac{3(\vec{p}_1 \vec{r})(\vec{p}_2 \vec{r})}{\epsilon r^5} \tag{2.18}$$

and the corresponding force as:

$$F = -\frac{3kp_1 p_2(2\cos\theta, \cos\theta_2 - \sin\theta, \sin\theta_2)}{\epsilon r^4} \tag{2.19}$$

when $d \ll r$. It is noteworthy that the potential energy is inversely proportional to r^3. For two permanent dipoles at an arbitrary angle to each other, their interaction is illustrated in Figure 2.26. The symbol r represents the distance between two permanent electric dipoles, p_1 and p_2 and they are θ_2 to the axis joining them, respectively.

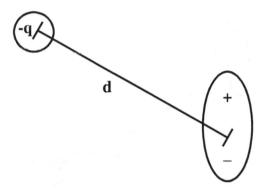

FIGURE 2.27
Induced dipole moment interacting with a change.

Molecules and bulk matter that have diffusive charges may create induced dipoles when exposed to an external electric field, such as that caused by a point charge. The induced dipole moment, p_{ind}, is given by:

$$p_{ind} = a \left(\frac{q}{\epsilon r^2} \right) \tag{2.20}$$

where a is the polarizability of the medium.

The potential energy between the induced dipole moment and the charge acting on it is (see Figure 2.27):

$$V = - \left(\frac{p_{ind}q}{\epsilon r^2} \right) = - \left(\frac{aq^2}{\epsilon^2 r^4} \right) \tag{2.21}$$

The influence of distance is now $1/r^4$. The magnitude of potential energy diminishes more rapidly as distance between the dipoles increases.

Important biomolecules such as carbohydrates are neither charged nor dipolar. They interact through a net-induced dipole-induced dipole attraction that results from distortion of electron distributions. This is called a Van der Waals interaction and it is always attractive. Two molecules separated by a distance R attract each other by an attractive potential energy:

$$U_{VW}(R) = - \frac{A}{R^6} \tag{2.22}$$

The parameter A depends on the chemical structures of the two molecules but its value is generally on the order of 10^{-77} Jm^6. Table 2.7 shows values of A for water and other molecules.

The table shows molecules in increasing sizes. The bigger the molecule, the larger the A value. When two atoms or molecules are so close their electron clouds overlap, the electrostatic interaction changes character and becomes highly repulsive, a force

TABLE 2.7
Coefficient A Values for Different Molecules

Molecule	A in 10^{-77} Jm^6
Water	2.1
Benzol	4.3
Phenol	6.5
Diphenylaniline	14.4

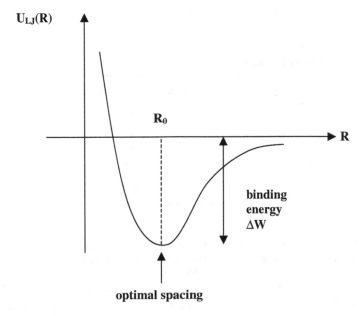

FIGURE 2.28
A sketch of the Lennard–Jones potential.

known as steric repulsion. The short range repulsive force can be included by adding a repulsive potential energy to the Van der Waals potential giving:

$$U_{VW}(R) = -\frac{A}{R^6} + \frac{B}{R^{12}} \tag{2.23}$$

We call this combination formula the Lennard–Jones potential (see Figure 2.28). The empirical values of the coefficients A and B for some very important atomic pairs are listed in Table 2.8.

The depth of the potential energy curve $U_{LJ}(R)$ at the minimum point, about 1 kcal/mol, is the strength of the Van der Waals bonds. They are somewhat weaker than polar and hydrogen bonds but play similar roles in molecular biology, namely as sources of weak, temporary links. Van der Waals interactions are nonspecific and

TABLE 2.8
Empirical Values of the Van der Waals Terms with
Coefficients A and B

Bond	R_{min} (Å)	A (kcal/mol)	B (kcal/mol) 10^{-4}
C—C	3.40	370	28.6
C—N	3.25	366	21.6
C—O	3.22	667	20.0
N—N	3.10	363	16.0
C—H	2.90	128	3.3

TABLE 2.9
Van der Waals Radii

Atom	Van der Waals Radius (Å)
H	1.2
C	2.0
N	1.5
O	1.4
S	1.85
P	1.9

involve any two atoms when they are approximately 3 to 4 Å apart. To find a contact distance between any two atoms, it is sufficient to add their two respective Van der Waals radii, some of which are listed in Table 2.9.

The Van der Waals bond energy for a pair of atoms is on the order of 1 kcal/mol which is only slightly more than the average thermal energy at room temperature (0.6 kcal/mol). Consequently, they are important only when a sufficient number of bonds are created simultaneously.

2.5   Hydrogen bonds and hydrophobic interactions

The one- or two-electron pairs possessed by nitrogen and oxygen atoms, respectively, bear enormous consequences for spatial organization of the four most important classes of biomolecules: lipids, hydrocarbons, proteins, and nucleic acids. The negative lone electron pairs distributed on tetrahedrally oriented σ-orbitals attract positively polarized hydrogen atoms of other molecules and form *hydrogen bonds* with them.

The energy of hydrogen bonds is comparable to the potential barrier heights for rotations around single covalent bonds, i.e., it ranges between 10 and 20 kJ/mol. The processes of reorganization of the system of hydrogen bonds, their breakage, and

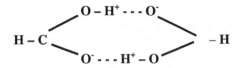

FIGURE 2.29
Dimerization of formic acid.

restoration possibly in new locations, take place at rates comparable to conformational transitions and for all intents and purposes are indistinguishable from them. Among different forms of noncovalent bonds, the hydrogen bond is one of the most familiar. It can be intermolecular or intramolecular. Dimerization of formic acid is a good example of an intermolecular hydrogen bond (Figure 2.29).

An H atom is normally placed between relatively weakly electronegative atoms (N, C, F). Despite its large electronegativity, Cl does not form a hydrogen bond. The hydrogen bond originates from the interaction between a dipole and a charge. However, their closeness results in the quantum mechanical interaction in which electrons are shared on both electronegative atoms. The bond distances and angles among the atoms are well defined. In small molecules, the atoms are oriented linearly, due primarily to the repulsive forces of the electronegative atoms. In larger molecules, the strain may be sufficient to bend them.

Hydrogen bonds are formed between a hydrogen nucleus of one molecule and the unbonded electrons of the electronegative atom of another molecule. Generally, the H bond is characterized as a proton shared by two electron pairs. The formation of H bonds is an exothermic process while the opposite is true of the dissociation of H bonds. The strengths of these bonds are in the fraction of an eV range, i.e., hydrogen bonds are some 10–20 times weaker than ionic or covalent bonds. For example, the energy of the hydrogen bond in the C–H...N complex is about 0.2 eV. For the CH...O complex, it is 0.25 eV. For H–F...H, it is 10 kcal/mol. For H–O...H, it is 7.2 kcal/mol and for H–N...H, it is 2 kcal/mol. Typically the energy of an H bond is in the range of 2 to 8 kcal/mol, i.e., it is intermediate between van der Waals and covalent bond strengths. The bond lengths vary from 2.3 to 3.4 Å as shown in Table 2.10.

All molecules can be classified into four groups:

1. Molecules with one or more donor groups and no acceptor groups (e.g., acetylenes)
2. Molecules with one or more acceptor groups and no donor groups (e.g., ketones, ethers, esters, aromatics)
3. Molecules with both donor and acceptor groups (alcohols, water, phenols)
4. Molecules with neither donor nor acceptor groups (e.g., saturated hydrocarbons)

Only saturated hydrocarbons are unable to make H bonds. Molecules with no donor or acceptor groups can self-associate by H bonding with themselves. The two types

TABLE 2.10
Hydrogen Bond Lengths for Various Bond Types

Bond Type	Bond Length (Å)	Location
O—H...N	2.78	
O—H...O	2.72	Often found in proteins as hydroxyl–hydroxyl or hydroxyl–carbonyl bonds
N—H...N	3.07	Found in proteins as amide–imidazole bonds
N—H...O	2.93	Found in proteins as amide–hydroxyl and amide–carbonyl bonds
N—H...S	3.40	Found in proteins as amide–sulfur bonds
C—H...O	3.23	

of H bonds are intramolecular and intermolecular. Intermolecular bonding results in association of molecules and is affected by steric factors. It leads to changes in the numbers, masses, shapes, and electronic structures of the molecular systems involved. See Figure 2.30.

Interactions of biomolecules take place often via H bonds that are very important in biological systems due to the abundance of water in cells. One key role of H bonds is in the protein folding phenomenon that will be discussed later in this chapter.

Water constitutes a large percentage of the surface of the Earth and the bodies of living organisms. In fact, the presence of water in liquid and ice forms makes the Earth unique. Water has a high heat of fusion (80 cal/g) and a large dielectric constant (80 at 10°C). Having a large number of hydrogen bonds also makes water possess a high surface tension. The density of water reaches maximum, about 1.0 g/ml, at 4°C. All these properties make water an ideal solvent for organic and inorganic substances and an excellent supporting base for living organisms.

Water molecules carry dipole moments and interact through the dipolar electrical fields they produce. As a consequence, the electric field of a charged protein polarizes the water molecules in the neighborhood of the protein which affects the nature of the electric field of the protein. In addition, both blood and the fluid inside living cells (cytoplasm) are electrolytes. The free ions of electrolytes further modify the electrical field. Since the dielectric constant of water is near 80, the electrostatic interactions between ions in a water environment are reduced accordingly. This has important consequences, one of which is that ionic binding energy is reduced by a factor of 80. For instance, instead of $8.8 \times 10^{-19} J$ for the binding energy of NaCl in a vacuum, in water we find only about $10^{-20} J$ or 5 kcal/mol, using $d = 2.4$ Å and $q_1 = -q_2 = -e$. The ionic binding energy in water is only about 10 times the thermal energy $k_B T$. In many cases, biochemical reactions involve ATP to ADP conversion as a source of energy which releases about 7.3 kcal/mole — enough to break an ionic bond in water but not sufficient to break an ionic bond in vacuum.

Water is unique among the family of R—H_n compounds. Within this family, water has a high melting point, boiling point, heat of vaporization, specific heat capacity, heat of vaporization, surface tension, etc. All these effects may be attributed to the high percentage of hydrogen bonds in water (Table 2.11).

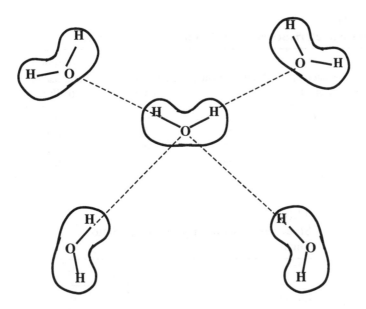

FIGURE 2.30
Hydration of H_2O by H_2O: hydrogen bond.

An important distance relating the magnitude of the electrostatic energy in water to thermal energy k_BT is the so-called Bjerrum length (l_B) defined as the distance at which the electrostatic energies of two monovalent ions are equal to k_BT:

$$\frac{k}{\epsilon}\frac{e^2}{l_B} = k_BT \tag{2.24}$$

For water at room temperature, the Bjerrum length is about 7.2 Å. The electrostatic interaction energy between two monovalent ions is negligible compared to the thermal energy if their separation is large compared to the Bjerrum length.

While the ionic (or polar) bond is greatly reduced by surrounding water molecules, this is not so true for the covalent bond. The strength of the covalent bond in water is about the same as in vacuum because the covalent bond involves the motion of an electron between two closely adjacent nuclei such that surrounding water molecules do not affect it much. As a result, two very different energy scales apply to biomolecules: (1) covalent bonds with energies in the 50 to 100 kcal/mole range and (2) ionic bonds in the 5 kcal/mole range. Covalent bonds cannot be broken by thermal fluctuations and the covalent bonding energy is large compared to the chemical energy stored in one ATP-to-ADP conversion. Covalent bonds are indeed used as permanent links to keep macromolecules together. The much weaker polar bonds can be broken by an ATP-to-ADP conversion by thermal fluctuations and light adsorption. They are useful as temporary links, e.g., to control the folding or mutual binding of proteins or the binding of proteins to DNA.

A study of the heat of sublimation revealed that the energy required to remove water molecules from ice is 12.2 kcal/mol. For methane, which has no hydrogen bonding, 2 kcal/mol is required. This energy may be from the contribution of London dispersion

TABLE 2.11
Physical Characteristics of Several Common Compounds

Compound	Molecular Weight	Melting point (°C)	Boiling point (°C)	Heat of Vaporization (cal/g)	Heat Capacity (cal/g)
CH_4	16.04	−182	−162	149.0	
NH_3	17.03	−78	−33	326.4	1.23
H_2O	18.02	−0	100	661.9	1.00
H_2S	34.08	−86	−61	160.4	

TABLE 2.12
Structure of Water in Liquid and Solid Form

	a_1	a_2	b
Liquid	0.96 Å	1.41 Å	104°
Ice	0.99	1.61	109°

forces or translational enthalpy. The energy contributed by hydrogen bonds should equal 10 kcal/mol for water or 5 kcal/mol for each hydrogen bond. The accepted value now is closer to 4.5 kcal/mol.

Water is the simplest system in which the structure and dynamics of hydrogen bonds play fundamental roles. Each oxygen atom, in addition to covalent bonds within the same water molecule, can form two hydrogen bonds with other water molecules (Figure 2.31). The structure of crystallized ice with completely saturated hydrogen bonds is highly ordered but has low density. To determine which structure is more stable at a given temperature T, one uses the condition of free energy F minimum (see Chapter 5):

$$F = E - TS \qquad (2.25)$$

where the internal energy E favoring order competes with the entropy S that favors disorder. The higher the density of the system, the lower the internal energy E, and the more hydrogen bonds are formed, the lower the entropy S, hence the greater the value of the term $-TS$ (see Figure 2.31). Table 2.12 summarizes characteristics of water in liquid and ice form. See Figure 2.32 for the explanation of symbols used.

When each water molecule has the most extended and straight hydrogen bond arrangement around tetrahedrally configured oxygen, the structure of ice can be seen with crystallography, and a bulk density of 0.9 g/cm^3 is predicted. As the temperature and pressure change, this conformation collapses and the density increases. While the temperature is increasing, the collapse of hydrogen bonds is compensated by random thermal motion of water molecules and the bulk volume tends to increase.

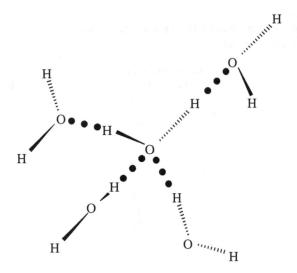

FIGURE 2.31
Hydrogen bonds in water.

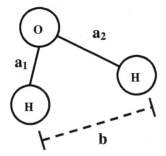

FIGURE 2.32
Structures of water in liquid and ice forms.

Thus, the highest density of water is at 4°C. It is also possible that this density reflects the formation of a clathrate with one water molecule in its center. Some structural orderliness is maintained even at temperatures as high as at 83°C. The source of this orderliness, after long-standing disagreement about it, is now assumed to come from quickly exchanging formation of water between the cluster and surrounding free forms. The size of the cluster at different temperatures has also been estimated (see Table 2.13).

Formation of ice distorts both the bond distances and angle *b*. The molecular orbital configuration is tetrahedral around each *O* atom with

$$O : (1s)^2(2s)^2(2p)^4$$
$$H : (1s)^1 \tag{2.26}$$

TABLE 2.13
Estimates of Cluster Sizes of Water and Heavy Water
at Various Temperatures

Temperature (°C)	Number of Molecules per Cluster		Mole Fraction	
	H_2O	D_2O	H_2O	D_2O
0	91		0.24	
4	117		0.23	
10	72	97	0.27	0.25
20	75	72	0.29	0.27
30	47	56	0.32	0.30
40	38	44	0.34	0.32
50	32	35	0.36	0.34
60	28	29	0.38	0.36

TABLE 2.14
Changes of Thermodynamic Parameters
of Organic Molecules Added to Water

	Methane into Water	Benzene into Water
ΔG	3.0 kcal/mol	4.3 kcal/mol
ΔS	−18 cal/mol deg	−14 cal/mol deg
ΔH	−2.6 kcal/mol	0 to 0.6 kcal/mol

So that eight electrons (six in $2s$ and $2p$ orbitals of oxygen plus two in the $1s$ orbital of each of the two hydrogen atoms) circulate around the oxygen atom. Two pairs of electrons near the oxygen give the region a $2 \times (-0.17\ e)$ charge. Two pairs of bonding hydrogens lose electrons to the electronegative oxygen, and consequently a charge of $+0.17e$ will be near each hydrogen. Thus, water molecules can become hydrogen donors and acceptors for hydrogen bond formation purposes.

Thermodynamically, the interactions between nonpolar groups and polar groups are unfavored. Nonpolar groups in water tend to stick together to reduce the energy of mixing. This unfavorable mixing, however, introduces interesting thermodynamic changes in water, e.g., a decrease in entropy in response to the more ordered form and change in its molar heat capacity arising from the increased hydrogen bonds. Table 2.14 illustrates the change of thermodynamic parameters when an organic molecule is introduced into water. The entropic nature, rather than the enthalpic nature, makes the mixing of a nonpolar molecule in water nonspontaneous.

As early as 1945, formation of icebergs around hydrophobes was postulated leading to the suggestion that the iceberg around a molecule may contribute to its stabilization. Nemethy and Scheraga (1962) proposed that a water molecule can have a coordination of five, i.e., four hydrogen bonds and one hydrophobic pocket (Figure 2.33). The potential to form four hydrogen bonds may be interrupted when the water molecule

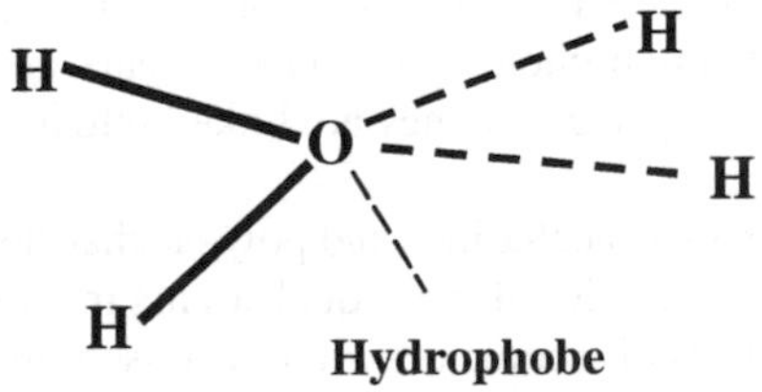

FIGURE 2.33
Water molecule with four hydrogen bonds and one hydrophobic pocket.

TABLE 2.15
Hydrogen Bonds in Water Molecules

Number of Bonds	Percent in Pure Water	Percent in Nonpolar Solution
4	23	43
3	20	6
2	4	18
1	23	12
0	29	21

is mixed with other water molecules or a nonpolar solvent. The number of hydrogen bonds for a water molecule may vary as shown in Table 2.15.

The system of hydrogen bonds in water determines its key biologically significant properties: high specific heat capacity, high latent heat of melting, high electric permittivity, and specific dynamic properties that will be discussed in the next chapter. This system also determines the water solubility properties of various molecules. Molecules capable of forming hydrogen bonds with water, e.g., sugars or alcohols, increase the disorder in the system of hydrogen bonds and hence increase entropy leading to free energy reduction. The process of solvation is thermodynamically favorable in this case. Molecules that do not form hydrogen bonds but are electrically charged or at least have high dipole moments reduce the electrostatic energy of the system and their solvation is also thermodynamically favorable in spite of introducing order into the hydrogen bond distribution.

Molecules that do not form hydrogen bonds and are uncharged and nonpolar, e.g., long carbohydrate chains or aromatic rings, only order the water environment (Figure 2.33) but do not contribute to the system's energy and hence are not water soluble. We refer to them as *hydrophobic* (fearing water), in contrast to *hydrophilic* soluble particles.

2.5.1 Polysaccharides

The ring oxygen atom of a carbohydrate is an H bond acceptor and each hydroxyl group is associated with two H bonds: a donor and an acceptor. A single sugar molecule or monosaccharide has the chemical formula $(CH_2O)_n$. The n varies; 5 or 6 are the most common choices for cells. Sugars are the primary sources of energy. If not used as energy sources, they are linked into long branched chains called polysaccharides.

H bonds are common in disaccharides and polysaccharides. H bond distances vary between 2.7 and 3.0 Å. The order of individual sugars in a long chain is not precisely determined or encoded. Their sequences are not passed on genetically. Specialized enzymes repeatedly add sugars to the ends of growing chains. Each type of sugar is linked to a different type of enzyme. Polysaccharides are neutral and unreactive so they are ideal energy storage molecules. Sugars also act as structural elements, for example, as cellulose fibers in wood and, of course, as components of DNA molecules. Sucrose and lactose are examples of disaccharides. Glucose is a common food molecule in cells and can be stored as a polysaccharide glycogen in animal cells or as starch in plant cells. Cellulose fibrils aggregate into microfibrils held to one another by extensive hydrogen bonds. The core of a microfibril possesses a perfect three-dimensional crystalline lattice.

2.6   Amphiphilic molecules in water environments

Interesting physical phenomena take place when *amphiphilic* molecules, containing both a hydrophilic and a hydrophobic moiety, are placed in an aqueous environment (Tanford, 1980). Among biological systems such amphiphilic molecules are phospholipids (see Figure 2.7) and glycolipids. Their polar *head* is hydrophilic and two hydrocarbon *tails* are hydrophobic. In a water environment, to minimize free energy, amphiphilic molecules spontaneously organize into spherical micelles (at low concentrations) or *bilayers* (at higher concentrations). These structures allow the molecules to have their hydrophilic head groups facing outside and hydrophobic tails inside (Figure 2.34). Bilayers close up to form three-dimensional *vesicles*, which, when sufficiently large and containing a hierarchy of internal vesicles, are referred to as *liposomes*. When the amount of solvent becomes too small, liposomes unfold to form *lamellae*, in which subsequent bilayers are placed parallel to each other (Figure 2.34).

Vesicles and liposomes are *lyophilic* (they like solvents) *colloidal* 100–500 nm diameter particles and when dispersed in water form a spatially inhomogeneous *dispersive structure* called *sol*. A decrease of water content results in an unfolding process of liposomes into lamellae and a transition of sol into a spatially homogeneous *lamellar phase* (Figure 2.35). In the lamellar structure, molecular orientation is ordered but their spatial arrangement is not. The lamellar phase is a special example of a liquid crystal. Removing more water molecules from the systems causes excessive proximity of the polar or similarly charged hydrophilic groups, which destabilizes the lamellar phase. Smaller surfaces of the head groups in contact form a *cubic phase* or *hegagonal phase*, in which bilayers are again replaced by spherical (in the first case) or cylindrical (in the second case) micelles (see Figures 2.35 and 2.36).

Both in the sol phase and the three liquid crystalline phases individual molecules retain their translational degrees of freedom within micellar or bilayer structures. In

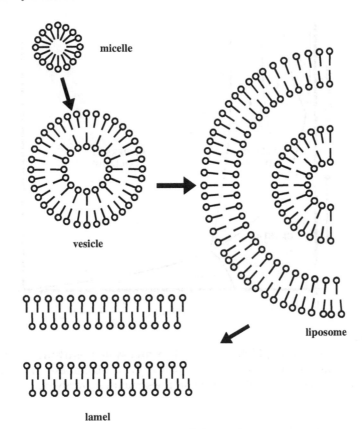

FIGURE 2.34
Structures created by amphiphilic molecules in water environment from micelles through vesicles, to liposomes and lamellae. Circles denote hydrophilic head groups; line segments denote one or two hydrophobic tails.

the sol phase, additional motion of entire micelles, vesicles, or liposomes is allowed. A lowering of the temperature causes a freezing of translational degrees of freedom. Sol undergoes a phase transition into *gel* and a liquid crystal into a normal crystal (see Figure 2.35).

For biological systems, the sol phase is optimal since it contains vesicles or liposomes of appropriate sizes. The stability and mechanical properties of a bilayer at physiological temperatures are controlled by an appropriate chemical composition of phospholipids and glycolipids and, additionally, by cholesterol (see Figure 2.2), which stiffens the bilayer. Such a lipid bilayer integrated with the built-in protein molecules forms a *biological membrane*. Proteins of the membrane can perform various functions such as those of immobilized enzymes, channels, pumps, receptors, signal generators, and constituents of a membrane skeleton. An important

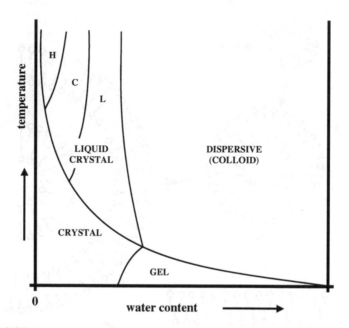

FIGURE 2.35
Phase diagram of an amphiphilic molecule–water system as a function of temper-
ature and water content. L = lamellar phase; C = cubic phase; H = hexagonal
phase. See Figure 2.36 for illustration of the structures involved.

immunological role is played by carbohydrates of glycolipids and glycoproteins of
the external face of the cytoplasmic membrane.

2.7   Structures of proteins

There are 20 "canonical" amino acids; others occur very rarely. Figure 2.37 shows
the side chains of these 20 canonical compounds. Three amino acids play special
roles in the spatial structures of proteins. We refer to them as *structural* amino acids.
   The first is *glycine* (Gly or G); it has a side chain reduced to a single hydrogen
atom. The small side chain produces no serious steric hindrance and enables almost
free rotation of the main chain about the neighboring C^α—N and C^α—C bonds. The
site where glycine is situated behaves like a ball joint in the polypeptide chain.
   Proline (Pro or P) is an imino acid. It attaches the main chain through the two bonds
and makes it locally rigid and looped in a defined manner. The third structural amino
acid, *cysteine* (Cys or C), forms relatively strong covalent bonds (disulfide bonds)
after oxidation and pieces together distant sites of a single polypeptide chain and
separate chains composing the protein macromolecule. See Figure 2.37 for graphical
illustration.

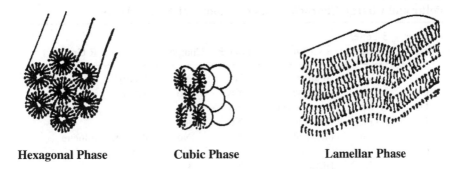

Hexagonal Phase   **Cubic Phase**   **Lamellar Phase**

FIGURE 2.36
Three possible liquid crystalline phases of the amphiphilic molecule–water system.

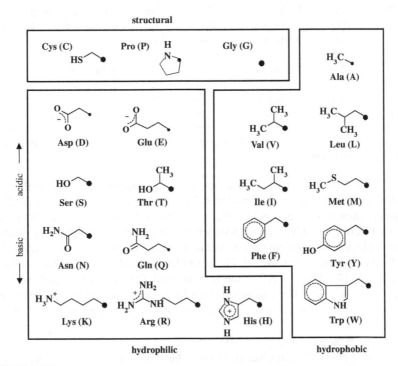

FIGURE 2.37
Side chains of 20 canonical amino acids and their partition into three main groups. Dots denote central C^{α} atoms.

TABLE 2.16
Polar and Charge Characteristics of Canonical Amino Acids

1-Letter Code	3-Letter Code	Name	Polar?	Charge?	R Group
A	Ala	Alanine	No	No	Methyl
C	Cys	Cysteine	Yes	No	Thiol
D	Asp	Aspartic acid	No	−e	
E	Glu	Glutamic acid	No	−e	
F	Phe	Phenylalanine	No	No	Aromatic, toluene
G	Gly	Glycine	No	No	Proton
H	His	Histidine	No	+e[a]	
I	Ile	Isoleucine	No	No	Sec-butyl
K	Lys	Lysine	No	+e	
L	Leu	Leucine	No	No	Isobutyl
M	Met	Methionine	No	No	
N	Asn	Asparagine	Yes	No	Amide connected at $C\alpha$
P	Pro	Proline	No	No	Cyclopentyl amine
Q	Gln	Glutamine	Yes	No	Amide connected at $C\beta$
R	Arg	Arginine	No	+e	
S	Ser	Serine	Yes	No	1 Alcohol
T	Thr	Threonine	Yes	No	2 Alcohol
V	Val	Valine	No	No	Isopropyl
W	Trp	Tryptophan	No	No	
Y	Tyr	Tyrosine	Yes	No	Aromatic, para-methylphenol

[a]Histidine has a pKa of 6.5 and will be protonated and positively charged if the pH of the cytoplasm dips below 6.5.

As Table 2.16 illustrates, some amino acids are charged and hydrophilic. Others are rich in carbon and strongly hydrophobic. Some are rigid and some are flexible. Some are chemically reactive and others are neutral. Each amino acid has the general chemical formula: $RCH(NH_2)COOH$ where R represents a residue. By giving up a water molecule at each connection, amino acids can form a valence-bonded chain in the form of a peptide or a protein. For example, myoglobin has the chemical formula $C_{738}H_{1166}FeN_{203}O_{208}S_2$. All amino acids are built from a central α carbon bonded to four different groups. The α indicates the priority position from which the numbering follows for all subordinate groups. The four substituents connected to C^α are:

1. An α proton or hydrogen (−H)
2. A side chain −R that gives rise to the chemical variety of amino acids
3. A carboxylic acid functional group (−COOH)
4. An amino functional group (−NH)

The α carbon is the asymmetric center of the molecule for all the canonical amino acids except glycine, which has only a proton as its side chain. The configuration about the α carbon center must be the L-isomer for proteins synthesized on ribosomes (see Figure 2.38).

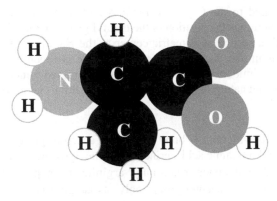

FIGURE 2.38
Four substituents connected to C^{α}.

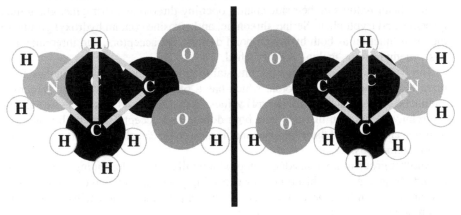

FIGURE 2.39
Naturally occurring and synthetic amino acids.

All but one of the amino acids can be written in the form NH_2—RCH—COOH. The variation in the 20 amino acids designed by nature is in the side chain. Some side chains are hydrophilic and others are hydrophobic. Since these side chains extend from the backbone of the molecule, they help determine the properties of the protein they make. Most naturally occurring amino acids are the L-forms. Synthetically produced amino acids are 50:50 mixtures. Since these molecules are mirror images, it is impossible to rotate one molecule to make it look like the other (see Figure 2.39)

The side chains exhibit wide chemical variety and can be grouped into three categories: nonpolar, uncharged polar, and charged polar. The simplest amino acid is glycine. The side chains of alanine, valine, leucine, isoleucine, and proline are entirely aliphatic. Alanine has a methyl group as its side chain. Valine has two methyl groups connected to the β-carbon, and this residue is said to be β-branched. Leucine has one more carbon atom in the side chain than valine, so that two methyl groups are attached to the C^γ. Leucine and isoleucine are isomers whose only difference in structure is the position of the methyl group. Isoleucine is a β-branched amino acid with a second asymmetric center at the β-carbon. Proline contains an aliphatic side chain that is covalently bonded to the nitrogen atom of the α-amino group, forming an imide bond and leading to a constrained 5-member ring.

Depending on the nature of the side chain groups, three classes of amino acids can be distinguished:

1. Amino acids with both H bond donor and acceptor groups: histidine, lysine, aspartic acid, glutamic acid, serine, threonine, tyrosine, and hydroxyproline
2. Amino acids with donor groups only: arginine, tryptophan and cysteine
3. Amino acids with acceptor groups only: asparagine and glutamine

Nonpolar side chains generally have low solubility in water because they can engage only in van der Waals interactions with water molecules. The remaining amino acids contain hetero-atoms in their side chains, opening threonine, asparagine, glutamine, tyrosine, and tryptophan. Serine, threonine, and tyrosine contain hydroxyl groups so they can function as both hydrogen bond donors and acceptors, and threonine also has a methyl group, making it β-branched.

The benzene ring in tyrosine permits stabilization of the anionic phenolate form upon loss of the hydroxyl proton, which has a pKa near 10. Serine and threonine cannot be deprotonated at ordinary pH values. Asparagine and glutamine side chains are relatively polar in that they can both donate and accept hydrogen bonds. The nitrogen and proton of the tryptophan indole side chain can also take part in hydrogen bond interactions.

Another group of polar residues can produce full charges, depending on pH. These include lysine, arginine, histidine, aspartic acid, glutamic acid, and cysteine. Lysine and arginine can have positive charges at the ends of their side chains. The lysine α-amino group has a pKa value near 10.

Histidine is another basic residue with its side chain organized into a closed ring structure containing two nitrogen atoms. One of these nitrogens already has a proton. The other has an available position that can take up an extra proton and form a positively charged histidine group with a pKa of about 6. Aspartic acid and glutamic acid differ only in the number of methylene ($—CH_2^-$) groups in the side chain, with one and two methylene groups, respectively. Their carboxylate groups are extremely polar, can donate and accept hydrogen bonds, and have pKa values near 4.5.

A group of three amino acids that all have aromatic side chains are phenylalanine, tryptophan, and tyrosine. The aromatic ring of phenylalanine is like that of benzene. It is very hydrophobic and chemically reactive only under extreme conditions, although its ring electrons are readily polarized. The side chain of histidine is arguably aromatic.

Finally, the α-amino and α-carboxylate groups of amino acids can also ionize, with pKa values of 6.8 to 7.9 and 3.5 to 4.3, respectively, for the aliphatic amino acids. Nearby charged side chains can alter the pKa levels of these groups. Each amino acid incorporated into a polypeptide chain is referred to as a residue. Only the amino- and carboxyl-terminal residues possess available α-amino and α-carboxylate groups, respectively.

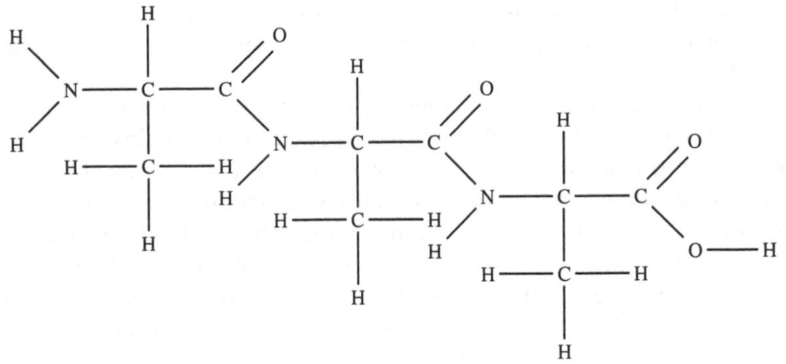

FIGURE 2.40
Polypeptide chain.

The α-carboxyl group of one amino acid is joined to the α-carboxyl group of another amino acid by a peptide bond (also called an amide bond). To form a protein, its amino acids engage peptide bonds involving the group O = C—N—H through dehydration synthesis. Some proteins contain only one polypeptide chain while others, such as hemoglobin, contain several polypeptide chains twisted together. The sequence of amino acids in each polypeptide or protein is unique, producing its own unique three-dimensional shape or native conformation.

2.7.1 Polypeptide chains

In order to form a polymeric chain, amino acids are condensed with one another through dehydration synthesis. This reaction occurs when water is lost between the carboxylic functional group of one amino acid and the amino functional group of the next to form a C—N bond. These polymerization reactions are not spontaneous. They can be set to occur through the energy-driven action of the ribosomes. Ribosomes are complexes of proteins and RNA that translate a gene sequence in the form of mRNA into a protein sequence. The 20 canonical amino acids are encoded by genes and incorporated by the ribosomal machinery during protein synthesis.

A polypeptide chain consists of a regularly repeating section called the main chain and a variable section consisting of distinctive side chains (Figure 2.40). The main chain is called the backbone. In some cases, a change in only one amino acid in the sequence can alter a protein's ability to function. For example, sickle cell anemia is caused by a change in only one nucleotide in the DNA sequence that causes one amino acid in one of the hemoglobin polypeptide molecules to be different. Due to this alteration, whole red blood cell ends are deformed and unable to carry oxygen properly.

The peptide unit is rigid and planar. The hydrogen of the amino group is almost always *trans* to the oxygen of the carbonyl group with no freedom of rotation about the bond between the carbonyl carbon atom and the nitrogen atom of the peptide unit due to the double bond whose length is 1.32 Å. However, a lot of rotational freedom exists around the single bonds involving the carbonyl carbon group, an α-carbon, and a nitrogen. As a polypeptide chain forms, it twists and bends into its native conformation.

The peptide bond between two amino acids is a special case of an amide bond flanked on both sides by α-carbon atoms. Peptide bond angles and lengths are well known from many direct observations of protein and peptide structures. The bond lengths and angles reflect the distributions of electrons between atoms due to differences in polarity of the atoms and the hybridization of their bonding orbitals.

The more electronegative O and N atoms can bear partial negative charges, and the less electronegative C and H atoms can bear partial positive charges. The peptide group consisting of these four atoms can be thought of as a resonance structure. The peptide bond has partial double bond character, accounting for its intermediate length. The formation of a peptide bond is an endoergic reaction and needs free energy that is usually released via GTP (guanine triphosphate) hydrolysis.

Like any double bond, rotation about the peptide bond angle ω is restricted, with an energy barrier of $\sim$3 kcal/mole between *cis* and *trans* forms. These two isomers are defined by the path of the polypeptide chain across the bond. While rotation about the peptide bond is restricted, the four bonds to the α-carbon of each residue have free rotation. Two of these rotations are of particular relevance for the structure of the polypeptide backbone. Successive α-carbons in the chain ($i, i + 1$) are on the same side of the bond in the *cis* isomer as opposed to the staggered conformation of the *trans* isomer.

For all amino acids but proline, the *cis* configuration is greatly disfavored because of steric hindrance between adjacent side chains. Ring closure in the proline side chain draws the β-carbon away from the preceding residue, leading to lower steric hindrance across the X-pro peptide bond. In most residues, the *trans* to *cis* distribution about this bond is about 90:10. With proline, the *trans* to *cis* distribution is about 70:30. As with other double bonds, certain atoms are confined to a single plane about the peptide bond. The bond from the α-carbon to the carbonyl carbon of that residue is given the name ψ. Similarly, the bond from the α-carbon to the amino group of that residue is given the name ϕ. Because C^α is one of the six planar atoms of the peptide group, rotation about ϕ or ψ flanking C^α rotates the entire plane of the peptide group. Since the entire plane rotates on either side of Cα, certain values of the angles ϕ and ψ cannot be achieved due to steric occlusion. The allowed regions of ϕ, ψ space differ for each amino acid because of the restriction due to C^α and its substituents.

2.7.2 Proteins

Proteins are the most abundant macromolecules present in membranes, cytosol, organelles, and chromosomes of living cells. They constitute about 50% of the dry mass of a living cell. Like nucleic acids, proteins are long, unbranched molecular chains. They are polymers consisting of combinations of any of the 20 amino acids, each of which is a monomer. Proteins are the most structurally complex macromolecules known.

Each type of protein has a unique structure and function. Human hair, skin, and fingernails consist largely of a protein called keratin. Fibroin is the protein contained in the silk of cocoons and spider webs. Sclerotin forms the external skeletons of insects. Structural proteins such as collagen and elastin and adhesive proteins such as fibronectin and laminin are abundant in cells. The collagen protein is really a family of 15 or more different types, most of which bundle together to form fibrils whose diameters range from 10 to 300 nm. Collagen is the most abundant protein in mammals. Elastin forms sheets and cross-linked networks to give connective tissue its elasticity. Adhesive proteins are found in matrices. Their purpose is to bind matrices to cells (fibronectin) and bind connective tissues to the laminae (laminin). These are only a few examples of structural proteins.

Functional proteins play crucial roles in dynamic processes within living cells. Insulin regulates the metabolism of glucose; rhodopsin converts incoming light in the retina of the eye to ionic signals in the optic nerve. Actin and myosin generate forces in muscle cells. Dynein is an energy-producing component of the cytoskeleton. Kinesin performs intracellular transport. Functional proteins catalyze nearly all cell metabolic processes. Each functional protein has a specific conformation suited for its function.

The three-dimensional shape adopted by a protein in water is crucial to its function. For example, enzymes are proteins which, by positioning a target molecule next to the proper reactive group, regulate chemical reactions usually increasing their rates by large factors. Structural proteins, on the other hand, have binding properties that promote the formation of large aggregates for strengthening and support roles in cells. Proteins also act as carriers of cargo for directed transport inside cells.

Many hormones are proteins that send molecular messages to specific locations in the body. Proteins, due to their diversity and functionality, perform most of the typical tasks of cells. While a typical bacterium may contain a few thousand types of proteins, the human body has tens of thousands of protein types. All their structural and functional properties derive from the chemical properties of the polypeptide chains.

Proteins are linear polymers of amino acids (to be more exact, they are L-enantiomers of α-amino acids, with characteristic side chains radiating from a central C^α atom) linked by amide (peptide) bonds —CONH— (see Figure 2.41). In each polypeptide chain, we distinguish the N (amine) and the C (carboxyl) termini. Successive amino acids are numbered starting from the N terminus. Most proteins are heteropolymeric (i.e., they contain most or all the different amino acids). Only rarely do regions of proteins consist of sequences composed of only a few amino acids. Any region of a typical protein will therefore have a chemically heterogenous environment. The levels of structural organization of globular proteins are:

Primary (amino acid) sequence $\Rightarrow$ secondary sequence

$\Rightarrow$ supersecondary sequence $\Rightarrow$ domain $\Rightarrow$ globular protein $\Rightarrow$ aggregate

The *primary structure* is defined as the linear sequence of amino acids in a polypeptide chain. It represents the covalent backbone and linear sequence of amino acid residues in the peptide chain. The sequences in which individual amino acids of definite side chains appear along the main chain are strictly fixed and genetically determined. The concept of a primary structure is identical to that of a chemical structure (constitution). It determines the system of covalent bonds in the protein macromolecule and includes information about disulfide bridges that form spontaneously during protein synthesis on ribosomes.

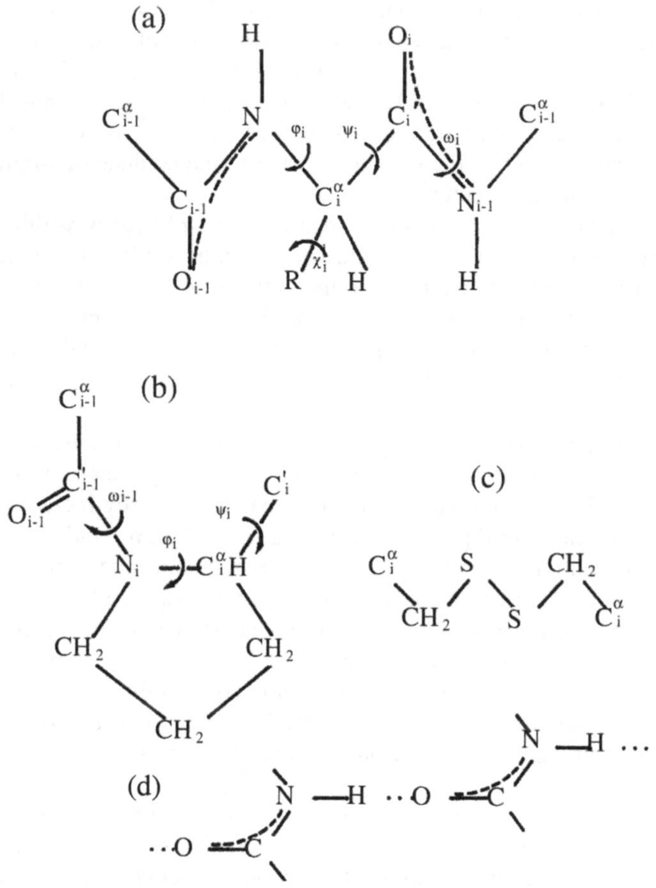

FIGURE 2.41
Angles of notation in the chemical structure of a protein.

The *secondary structure* is characterized by spatial organization in terms of α helices, β sheets, random coils, triple helices, and other motifs usually in localized regions of macromolecules. The secondary structure is a certain regular geometric figure of a chain created by H bonding between the C—O and N—H groups of peptide bonds. The β-pleated sheets are more common in structural proteins. In globular proteins, they are interconnected by α-helix segments. Such combinations of secondary structures are called supersecondary structures.

Further folding of polypeptide chains results in the formation of *tertiary structures*. They are produced by long-range contacts within the chains and stabilized by a combination of van der Waals, hydrogen, electrostatic, and disulfide bonds. A *quaternary structure* is an organization of protein subunits or two or more independent polypeptide chains. Quaternary structures represent aggregates in which a number of globular proteins are bound by noncovalent interactions and spontaneously form oligomers of sizes ranging from dimers to dozens of monomers.

After completion of the protein synthesis process, most proteins are subjected to additional enzymatic modifications. The modifications may involve cutting off fragments of the main chain, methylation of charged side chains (leading to charge neutralization), or phosphorylation of side chains ending with hydroxyl groups that endow the originally neutral chains with negative charges. Enzymatic proteins often form permanent bonds with different prosthetic groups (coenzymes), while the external proteins of cell membranes undergo glycolization to a high degree. The process of protein biosynthesis ends with spontaneous formation of supermolecular structures, e.g., multienzymatic complexes.

The spatial structure of a protein macromolecule with a given primary structure is determined by local conformations of the main and side chains and a system of noncovalent (secondary) bonds. The commonly accepted notation for dihedral angles in the polypeptide chain is shown in Figure 2.41. In the absence of steric hindrances, rotation about the single bonds C—C, C—N, C—O, and C—S allows three stable local conformations (one *trans* and two *gauche*; cf. Figure 2.14) unless the rotation involves planar carboxylic, amidic, or aromatic rings. This would allow two stable conformations. Individual conformations can differ in energy by a few kJ/mol and are separated by barriers whose heights range from 10 to 20 kJ/mol originating from van der Waals and electrostatic multipolar interactions.

Only rotations of covalent bonds distant from C^α atoms (branching sites of polypeptide chains) are usually not hindered if one does not take into account the long-range excluded volume effects. This applies to angles $\chi^1, \chi^2, \chi^3, \ldots$ describing internal conformations of longer side chains if not branched and to the ω angle describing the local conformations of peptide bonds.

The peptide bond C—N, as discussed in Section 2.3, is in part a double bond and may exist as two configurational *trans* and *cis* isomers (Figure 2.18). The potential energy barrier height is determined by delocalization energy and approaches 80 kJ/mol. This is why peptide bonds occur as single local configurational isomers, almost exclusively in *trans* form. The only exception is a peptide bond neighboring a proline residue (Figure 2.19) in which the part of the polar structure is smaller and results in lowering the *trans*-to-*cis* transformation barrier to 50 kJ/mol.

An energy barrier exists between the α-helical region of the ϕ, ψ space and the β-strand region. Direct conversion between α and β structures is restricted even though most residues are allowed in both regions. Since the restrictions on ϕ, ψ space arise in part from steric hindrance between side chains and the backbone, this same steric hindrance is the origin of α and β secondary structures. The steric restrictions of the α and β spaces have no sequence dependence because ϕ, ψ restrictions arise within each residue rather than between residues. However, a sequence of residues that all have similar ϕ, ψ spaces can give rise to a chain segment that forms α or β structures.

These secondary structures owe their formations to backbone and side chain steric restrictions. The allowed regions of ϕ, ψ space for each amino acid are displayed on so-called Ramachandran plots (Schulz and Schirmer, 1979). The allowed regions can be defined in terms of the energetic cost required to enter a disallowed region or limitation of the so-called hard-sphere boundary when atoms clash. For β-branched residues the restrictions are severe, and only a small fraction of ϕ, ψ space is allowed. Valine and isoleucine have access to only about 5% of all ϕ, ψ spaces. However, all

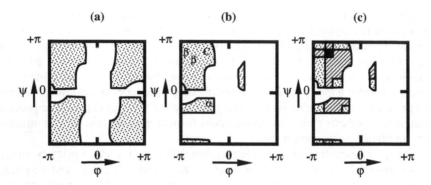

FIGURE 2.42
Left: Sterically allowed regions for angles φ and ψ (Ramachandran map) for glycine. Middle: Arbitrary side chain different from glycine. Configurations of α helix, parallel and antiparallel β sheets, and triple helix of collagen (C) are indicated. Right: Number of sterically allowed conformations for the angle χ^1 according to the value of φ and ψ. For residues with one side carbon atom third from C^α, blackened, heavily, and lightly shaded regions correspond to three, two and one conformations, respectively. For residues with two side carbon atoms third from C^α (threonine, valine and isoleucine), shadings correspond to three, two, and no conformations, respectively. For alanine three rotational conformations are allowed everywhere.

residues have access to at least part of the most favorable regions of ϕ, ψ spaces in the upper and lower left parts of the plot. These two regions correspond to combinations of ϕ and ψ angles that characterize the two common regular secondary structures that can be adopted by the polypeptide backbone: the α helix and the β strand. The presence of bulky side chains in all amino acids always affects steric hindrance for angles φ and ψ, making the corresponding local conformations mutually dependent.

Figure 2.42 shows the region sterically allowed for angles φ and ψ (the Ramachandran map) for glycine and an arbitrary side chain different from glycine. Glycine behaves like a ball joint. For the remaining side chains, the three distinct, much smaller, sterically allowed regions correspond to three distinct, already cooperative conformations of φ, ψ. Because rotation about the angle χ^1 is hindered for some values of angles φ and ψ, the whole triple (φ, ψ, χ^1) should be treated as a single unit. A number of sterically allowed local conformations for angle χ^1 is shown on a background of the Ramachandran map in Figure 2.42.

The secondary bonds in proteins are mainly hydrogen bonds. Their energy, 10 to 20 kJ/mol, is comparable to potential barriers for transitions between local conformations of protein chains. It is reasonable to distinguish hydrogen bonds within polypeptide backbones from those formed by side chains. The hydrogen bonds within the main chains link nitrogen and oxygen atoms of distinct peptide groups (see Figure 2.19e).

A regular pattern in which hydrogen bonds are organized in the polypeptide backbone without reference to side chain types is traditionally known as a *secondary structure* of protein. Two main secondary structures are distinguished. The α *helix*

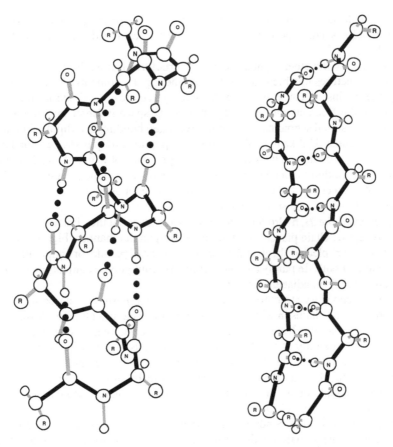

FIGURE 2.43
Structure of α helix (left) and antiparallel β-pleated sheet (right).

is a helical arrangement of a single polypeptide chain with hydrogen bonds between each carbonyl group at position i and a peptide amine at position $i + 4$ (Figure 2.43). The helix structure looks like a spring. The most common shape is a right handed α helix defined by a repeat length of 3.6 amino acid residues and a rise of 5.4 Å per turn. Residues $i + 3$ and $i + 4$ are closest to residue i in the helix. The pitch and dimensions of the helix also bring the peptide dipole moments of successive residues into proximity so that their opposite charges neutralize each other substantially in the middle of a helix. The pitch and dimensions also bring the amide proton of residue $i + 3$ or $i + 4$ into proximity to the carbonyl oxygen of residue i so that a hydrogen bond can form.

All peptide group hydrogen bond donors and acceptors are satisfied in the central part of the helical segment, but not at the ends. While structural evidence indicates that these hydrogen bonds are highly populated in helical segments of proteins, their contribution to helix stability is less clear since donors and acceptors would be satisfied by hydrogen bonding to water in nonhelical structures. The structure of an α helix was first deduced by Pauling and Corey before it was observed in x-ray data. The collagen helix is a third periodic structure and it is specialized to provide high tensile strength.

β strands are the other regular secondary structures formed by proteins. They are extended structures in which successive peptide dipole moments alternate directions along the chains. Because they are extended structures, ϕ, ψ steric hindrance is reduced in the $β$ strand and the $β$ region of ϕ, ψ space is larger than the α region. Two or more strand segments can pair by hydrogen bonding and dipolar interactions to form a $β$ sheet. Unlike helical segments, all peptide group hydrogen bond donors and acceptors are satisfied between $β$ strand segments, not within them. Individual $β$ strands do not exhibit independent existence. Also unlike helical segments, adjacent strands of a sheet can come from sequentially distant segments of the chain; this occurs rarely within one strand of a sheet.

In *β-pleated sheets*, hydrogen bonds are also evident but they link different strands of a polypeptide chain placed in a parallel or antiparallel manner. Figure 2.43 shows an antiparallel $β$-pleated sheet. A $β$ sheet can consist of parallel or antiparallel strands or a mixture of both. In purely antiparallel sheets, contiguous segments in the primary structure often form adjacent strands.

Another secondary structure, the collagen *triple superhelix*, can be formed only from polypeptide chains in which every third amino acid is a glycine. However, even when forming hairpins from contiguous chain segments, linearly distant residues are brought into proximity at the N and C terminal ends. Thus, while a $β$ strand is a secondary structure element because of its geometrically regular features, a $β$ sheet can be thought of as a tertiary structural feature because it is intrinsically non-local. This example shows that distinctions between secondary and tertiary structural features are not entirely clear.

Assuming all peptide bonds occur in *trans* conformations ($\omega_i = 180$ degrees), the values of the remaining dihedral backbone angles are $\varphi_i = -57$ degrees, $\psi_i = -47$ degrees for the α helix, $\varphi_i = -119$ degrees, $\psi_i = 113$ degrees for the parallel $β$-pleated sheet, $\varphi_i = -139$ degrees, and $\psi_i = 135$ degrees for the antiparallel $β$-pleated sheet. These positions lie within two distinct sterically allowed regions of the angles φ and ψ and coincide almost exactly with two main minima of the potential energy as a function of these angles although it is energetically profitable for planar $β$ sheet structures to be slightly twisted. The systems of hydrogen bonds in the α helix and $β$ sheet structures considerably stabilize local backbone conformations. This is why fragments of both secondary structures are so abundant in protein macromolecules.

The so-called *turn structures* are also classified as secondary structural elements. Unlike helices and strands, they do not form repeating, regular patterns. They can have well defined spatial dispositions defined by certain values of ϕ and ψ angles that often require specific residue types and/or sequences and fixed hydrogen bonding patterns. Most turns are local in the primary structure. The Ω loops can have large numbers of intervening residues lacking defined geometries, with turns defined by the conformations of residues that form the constrictions that give loops their name. Turns are essential for allowing a polypeptide chain to fold back upon itself to form tertiary interactions. Such interactions are generally long range and result in compaction of the protein into a globular, often approximately spherical, form. The turn regions are generally located outside the globular structure, with helices and/or sheets forming its core. Turns on protein surfaces exhibit a wide range of dynamics, from quite mobile in cases where they form few interactions with the underlying protein surface to fixed due to extensive tertiary contacts.

The side chains of all the amino acids involved help determine the native conformation of a protein. Since some amino acid side chains are hydrophobic and some are hydrophilic, all the hydrophobic side chains try to concentrate in the centers of molecules, away from the aqueous environment. Hydrophilic side chains are attracted to the outsides of molecules in order to lower energy by making contact with the liquid environment.

Some hydrophilic side chains have attached groups of atoms that make them acidic, while others have attached groups that make them basic. Side chains with acidic ends are attracted to side chains with basic ends and can form ionic bonds. The side chains interacting with each other help hold a protein in its native conformation. The side chains project outward from both α helical and β strand structures and are available for interactions with other surfaces through hydrophobic contacts and various bonding interactions to form tertiary structures. In a helix, the side chains project radially outward. In a strand, successive side chains project alternately up and down. Rotations about bonds in the side chains project alternately up and down and they are also restricted. The same steric hindrance that limits backbone conformation also limits side chain conformation about the C_α—C_β bond to preferred rotamers defined by rotation angle χ^1 and is restricted by side chain packing in the tertiary structure.

If secondary structural elements result from steric restrictions in ϕ,ψ space, it is less obvious why *tertiary structures* form. Proteins with highly organized tertiary structures generally have well developed cores of hydrophobic residues contributed by most or all secondary structure elements in the chain. Secondary and tertiary structures are in general intimately interconnected. The buried residues do not form liquid-like oily interiors. They are usually well packed and have extensive rotamer restrictions.

In aqueous solvents, the hydrophobic effects drive the chain toward compaction to relieve unfavorable solvation of these exposed side chains, but compaction and internal organization are entropically costly due to loss of chain flexibility, and these competing effects nearly cancel each other energetically. On the other hand, upon compaction, bonding interactions with solvent molecules are replaced by intramolecular partners, with a likely net gain in favorable energetic contributions due to several effects. Hydrogen bonding is favored within secondary structures because these are partially preorganized by ϕ, ψ restrictions into configurations that permit bonding at little additional entropic cost. In the case of β sheet formation, an additional favorable effect may result when two β strands are brought into contact.

The view presented above suggests that secondary and tertiary structures are interdependent and it is possible that this interdependence is the molecular origin of structural stability whereby protein secondary and tertiary structures are lost in an all-or-none manner upon changes in the environment that disfavor the folded state, such as higher temperatures or solvent additives.

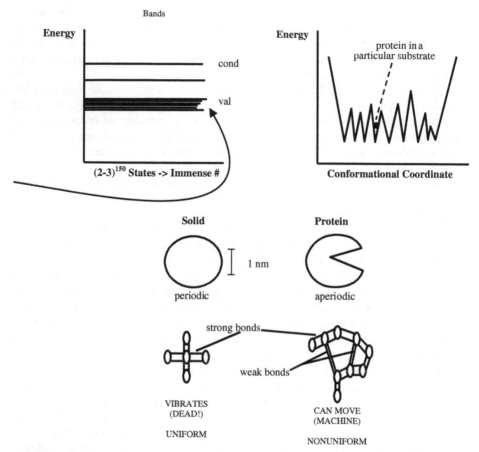

FIGURE 2.44
A schematic comparison between a solid and a protein.

The highest level of protein structural organization is the *quaternary structure*. The subunits that associate may or may not be identical and their organizations may or may not be symmetric. In general, a quaternary structure results from association of independent tertiary structural units through surface interactions, such as the formation of a hemoglobin tetramer from myoglobin-like monomers. Subunit assembly is a necessary step in tertiary structure formation. The codependence of tertiary and quaternary structures parallels the codependence of secondary and tertiary structures, and suggests that the distinction among these levels of protein structure organizational hierarchy may be misplaced. In the past, there have been comparisons made between crystal solids and proteins. However, as Figure 2.44 shows, such analogies are not well-founded on the physical properties of proteins.

TABLE 2.17
Energy Contributions to Native State Stabilization

Folding	Unfolding
Hydrophobic collapse	Conformational entropy
Intramolecular H bonding	H bonding to solvent (water)
van der Waals interactions	

Contributions to Free Energy of U $\Rightarrow$ N Reaction

-200 kcal/mole	$+190$ kcal/mole

2.7.3 Protein folding

The spontaneous act of protein folding is remarkable in that the complex motion of a protein's structural elements, i.e., amino acids, transfers the one-dimensional sequence of data into a three-dimensional object. The process results from thermally activated Brownian motion and resembles a phase transition (see Appendix B) such as crumpling, but is much more complex and harder to model. The net stabilization of the native state conformation of a protein results from a balance of large forces that favor both folding and unfolding. For a hypothetical protein, the energetic contributions to native state stabilization may be distributed as shown in Table 2.17. The net free energy of folding is 10 kcal/mol.

The overall tendency of side chains of nonpolar amino acid residues is to locate in the interior of a protein, a dynamic process known as the hydrophobic effect. This is a combined effect of the hydrophobicity of some side chains of amino acids and the structure of water. Water molecules form hydrogen bonds among themselves and form hexagonal structures on the surfaces of proteins. The order of hydrophobicity and hydrophilicity for the amino acids is:

Hydrophobic: Phe > Ala > Val > Gly > Leu > Cys

Hydrophilic: Tyr > Ser > Asp > Glu > Asn > Gln > Arg

Consequently, hydrophobic side chains coalesce in the interiors of protein structures. Much of the free energy of protein folding is entropic. The molecular explanation is that solvent-exposed surface area is reduced and fewer water molecules must be ordered. The folding behavior of proteins is well approximated by a hetero-polymer model composed of two residue types (hydrophobic and hydrophilic) in a "poor" solvent (Figure 2.45). Conformational entropy is meant to define the entropy associated with the multiplicity of conformational states of a disordered polypeptide chain (Figure 2.46).

To determine how long it would take a 100-residue polypeptide to complete a random search for the native state, we assume each residue has three possible conformational states and that it takes 10^{-13} s to interconvert between states. A 100-residue polypeptide has 3^{100} or 5×10^{47} possible conformational states. Assuming a single unique native state conformation, it would take $(5 \times 10^{47})(10^{-13})$ s $= 5 \times 10^{34}$ sec

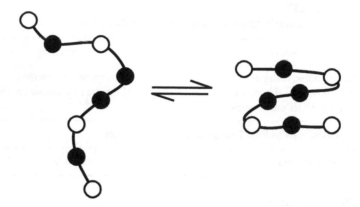

FIGURE 2.45
Heteropolymer model composed of two residue types, hydrophobic and hydrophilic, in a "poor" solvent.

or 1.6×10^{27} yr. This absurd result often called the *Levinthal paradox* clearly shows that protein folding does not occur by random search.

A protein can assume a large number of related but distinct conformational states whose distribution can be described by a so-called energy landscape. Each substate is a valley in a $3N - 6$ dimensional hyperspace where N is the number of atoms forming the protein. The energy barriers between different substates range from 0.2 to about 70 kJ/mol. It is believed that this energy landscape has a hierarchical structure with different tiers having widely separated barrier heights. Recent *in vitro* studies demonstrate that protein folding follows paths characterized by retention of partially correct intermediates.

The efficiency of protein folding can be compromised by aggregation of folding intermediates that have exposed hydrophobic surfaces. Such aggregate formation is essentially irreversible. Molecular chaperone proteins bind reversibly to folding intermediates, prevent aggregation, and promote their passage down the primary folding path. Molecular chaperones also are known as heat shock proteins because they are synthesized in much greater amounts by cells subjected to a wide variety of stresses, including elevated temperature and oxidative stress.

Specialized enzymes catalyze key steps in protein folding, for example peptidyl-prolyl *cis/trans* isomerase (PPIase) catalyzes the *cis/trans* isomerization of X-pro peptide bonds, where X is any amino acid. Steric interference between neighboring residues in an amino acid sequence is an example of a local interaction. The conformational behavior of an unfolded polypeptide is dominated by local interactions; long range interactions stabilize secondary and tertiary structures in native proteins. In an unfolded polypeptide, a given peptide bond is much more likely to be *trans* (96%) than *cis* (4%); except in the case of an X-pro peptide bond where the likelihood of the *cis* conformer is significantly increased (20%). This relatively high probability of *cis* presents a significant barrier to the folding of a protein for which the native secondary and tertiary structure demands the *trans* conformation.

Denaturation represents loss of native conformation. Several factors can denature proteins including changes of temperature, pH, and salt concentration of the solution

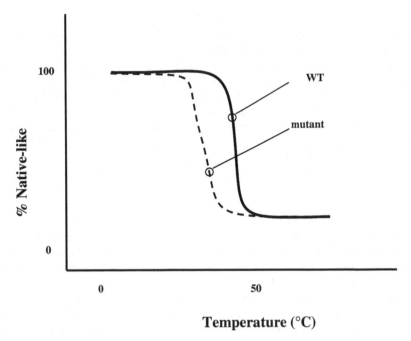

FIGURE 2.46
Change in percentage of native-like character as temperature rises.

or putting the protein into a hydrophobic solvent where amino acids with hydrophobic side chains try to move to the outside of the molecule. Those amino acids with hydrophilic side chains cluster in the center of the molecule. If a protein remains water-soluble when denatured, it can return to its native conformation if placed back in a "normal" environment.

Proteins are chemically and biologically stable unless they are deliberately depolymerized. The decomposition of a polypeptide chain into individual amino acids can also be facilitated by hydrolytic enzymes. The thermal melting behavior of protein structure indicates a phase transition to the unfolded state, a characteristic of highly cooperative systems (see Appendix B). A typical point mutant causes little or no change in the structure of the protein, but reduces stability as indicated by lower melting temperature. Many amino acid substitutions in protein sequences have no detectable effects on protein function. Mutations causing functional defects usually occur in an enzyme active site or interfere with intermolecular assembly.

2.7.4 Electrophoresis of proteins

Electrophoresis is the transport of charged macromolecules, e.g., proteins or DNA, through an electrolyte under an applied voltage difference and is a basic biochemistry laboratory tool. The idea behind the technique is to segregate macromolecules according to size. Electrophoresis is the phenomenon of migration of small electrically-charged particles, suspended in aqueous electrolytic solution, under the influence of an externally applied electric field.

In biological applications, electrophoresis is used to characterize various objects ranging from bacterial cells to viruses to globular proteins to DNA. This is possible due to the net electric charges of these objects. The important clinical applications include separation of various components, e.g., proteins in blood plasma. As a result, abnormal patterns of blood composition can be identified through electrophoresis. Protein amino acid sequence variations lead to different folding patterns and, consequently, functional properties in the body. The net charge on a protein may vary from -100 to $+100$ elementary charges and is largely determined by the pH of the solution in which the protein is suspended. A charged protein with a total electric charge Q is subjected to the electric force $F_e = QE$ where E is the applied electric field and the drag force $F_D = -6\pi\eta av$ where η is the viscosity of the medium, v is the velocity of the migrating protein, and a is its radius. Counterions in the solution present an additional complication. They shield the protein by surrounding it like a "cloud." The shielding cloud moves in the opposite direction under the influence of field E. We summarize the results as follows.

The mobility of a protein (velocity v divided by electric field intensity) is proportional to the net charge on the protein, inversely proportional to the viscosity, the radius a squared, and the ionic strength of the solution. For example, a protein of charge 10 e placed in an aqueous solution of ionic strength of 0.1 mol/L subjected to an electric field of 1 V/cm will attain a drift velocity of approximately 10^{-4} cm/s requiring 3 hr for displacement over 1 cm. The distance traveled over a given time will be proportional to the charge on objects of the same size. Electrophoresis is useful for separating biological materials in a solution according to charge and size. If some components possess dipole moments and no net charge, they can be separated by an analogous method called dielectrophoresis in which an electric field gradient provides force on dipole moments.

2.7.5 Protein interactions with environment

A structured water medium surrounds all the protein within a cell. This is the solvation problem and is a result of the formation of hydrogen bonds with water by a protein and its charged surface. Ions are formed also within the cytoplasm which may influence conduction properties by their interactions with the side chains of proteins or even their localizations within charged pockets of the protein. Consequently, the study of dry proteins does not give an accurate description of their properties.

The interactions of metal ions and proteins with peptide groups are of great importance in the biophysical chemistry of proteins. Such interactions are involved in conformational changes of macromolecular structure and are the basis of one class of gated ion channels. They may also affect physical properties and the chemical reactivities of biomolecules *in vivo*. Interactions with metal ions may also play important roles in genetic expression, metalloenzyme activities, and metal–nucleic acid processes.

The most common of all such reactions is simply the association of a proton that drastically changes the electrostatics. At low pH when carbonyl groups are protonated, the structure of the protein differs somewhat from its high pH form. Rather than changing the band gap, interactions with metal ions may allow the protein to donate electrons to the metal ions, thereby creating holes in the conduction band. Protons (H^+) seem to accept electrons much more readily than lithium, sodium and larger metallic ions. This property led to studies of electronic conduction where protons act like ferries that carry electrons in only one direction.

The three significant types of charge carriers are electrons, heavy ions, and protons. The electrons have high mobility. Many familiar materials such as metals are capable of supporting their conduction. Heavy ions can easily carry charges but are immobile within solids. Protons fill the intermediate place. While they present many experimental difficulties, they can exist within both liquid and solid media.

Although three orders of magnitude heavier than an electron, a bare proton or H^+ ion is much smaller than any other possible charge carrier precisely because it does not carry electrons. As a result, it has better mobility than other ion species. Its involvement may be its ability to "hop along" the outside of a protein and become localized at negatively charged pockets of the protein. Potential barriers are required at electron injection and ejection points to prevent short circuiting. Protons may also act as carriers of negatively charged ions or electrons. The injected protons in this mechanism of conduction would come from chemical reactions, redox reactions, or a proton reservoir with high chemical potential such as the interior of a mitochondrion.

2.7.6   Electron transfers in proteins

A body of experimental evidence explains the mechanism of electron transfer. The transfer of electrons along a protein is accomplished by a series of redox centers incorporated into the protein structure. This allows for the directionality of electron transfer over distances of 0.3 to 3.0 nm. The transfer is modeled by electron or nuclear tunneling but we should not exclude the possibility of electron transfer along a chain of conjugated orbitals.

The protein–electron carrier system consists of a protein, possibly localized within a lipid bilayer, which has a single redox center. Redox centers are usually prosthetic groups containing molecules of nonprotein origin that have conjugated orbital systems and often incorporate metal ions. The prosthetic groups have multiple functions. They ensure the fixation of charges and dipoles in their microsurroundings and catalyze electron transfers. The orientation of the prosthetic groups relative to other proteins may enhance recognition of the redox center. Prosthetic groups may act to influence the electronic states of associated proteins or vice versa, and produce a degree of isolation from the polar solvent. Finally, the groups may control electron transport through oxidation or reduction that alters carrier concentration.

It has been argued that electron transfer in biological systems occurs via a redox scheme, but reports of electron tunneling over distances of 3.0 to 7.0 nm have been disputed. Instead, it has been proposed that protein is an insulator at physiological temperature and that electron transport is mediated by protons. In this model, an electron is ferried along the protein backbone by protons. There is no net movement of protons as they hop back and forth, but they carry electrons in one direction only, producing conduction. One benefit of the redox scheme is that the reduction and oxidation of the protein other than those that result in electron transfer can act as dopants as installed in traditional semiconductors. This structural flexibility may allow cells to tune their electrical properties to the desired biological function.

Proteins and glasses are similar in at least one aspect, i.e., both possess large numbers of isoenergetic minima leading to huge numbers of thermally assisted transitions between these degenerate minimum energy states. The frequency of these transitions is lowered with the reduction of the temperature but some motions occur even at 100 mK. No clear consensus has been reached on the usefulness of this energy landscape in physiology. It should also be mentioned that the conformational dynamics of a protein is slaved to the solvent. The protein–solvent structure should be treated as a single system whose behavior can be controlled by the environment.

Fluctuations at equilibrium and relaxations from nonequilibrium states are essential characteristics of protein dynamics. Fluctuations correspond to equilibrium transitions between various conformational substates in the ground state manifold while relaxations are transitions toward the ground state manifold from higher energy states. The striking features of these processes in proteins are very broad distributions of relaxation times and the hierarchical natures of these processes. While agreement seems to exist regarding the functional importance of these processes, it is still unclear how a protein structure is related to its energy landscape and relaxation dynamics.

A more detailed discussion of the relaxation processes in biomolecules is given in Chapter 3.

2.8   Structures of nucleic acids

Nucleic acids exist in cells in two varieties: DNA and RNA. RNA differs from DNA by having one additional oxygen atom on each sugar and one missing carbon atom on each thymine base. DNA and RNA perform distinct functions in cells. DNA is more stable and better suited for information storage. RNA is less stable and more useful in information transfer as a messenger, a translator, and a synthetic machine (Saenger, 1984).

Nucleic acids are linear polymers of nucleotides linked by phosphodiester bonds —O—PO_2^-—O—. Ribose sugar is found only in RNA. In DNA, it is replaced by deoxyribose. Five carbon atoms in the pentose ring are numbered from $1'$ to $5'$. In deoxyribose, the OH group bound to carbon $2'$ is replaced by H. In each nucleic acid chain, we distinguish the $5'$ (phosphate) and $3'$ (hydroxyl) termini. Successive nucleic acids are numbered starting from the $5'$ terminus.

The formation of a phosphodiester bond is an endoergic reaction and needs free energy release in nucleotide triphosphate hydrolysis (Figure 2.47). The sequence in which the individual nucleotides of definite side chains appear along the nucleic acid chain is strictly fixed and genetically determined. We refer to it as a *primary structure* of the nucleic acid.

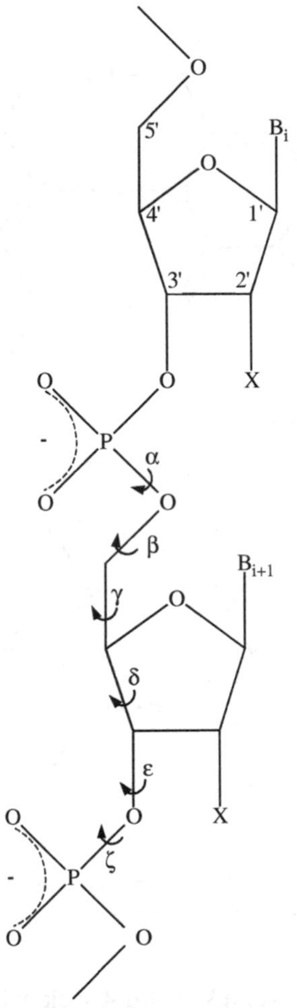

FIGURE 2.47
Notations for angles in the structure of nucleic acids.

DNA has four canonical nitrogenous bases. Guanine (G) and adenine (A) are derivatives of pyrimidine. Cytosine (C) and thymine (T) are derivatives of purine. In RNA, T is replaced by uracil (U). A watershed event in molecular biology was the discovery by Watson and Crick in 1953 that specific pairs of nitrogenous bases can make unique hydrogen bonds (Figure 2.48).

Nucleic acids store and transmit genetic information in cells. They are composed of long chains of nucleotides, each of which is composed of a sugar–phosphate group and a disk-shaped base group. The sugar–phosphate groups are connected to form a hydrophilic backbone (phosphates have negative electric charges). The mainly hydrophobic bases are located off the sides of the chains. In a water environment, one side of the chain is protected by the hydrophilic backbone while the other side is exposed. The edges of the four bases are chemically complementary. A forms

cytosine (C)   guanine (G)

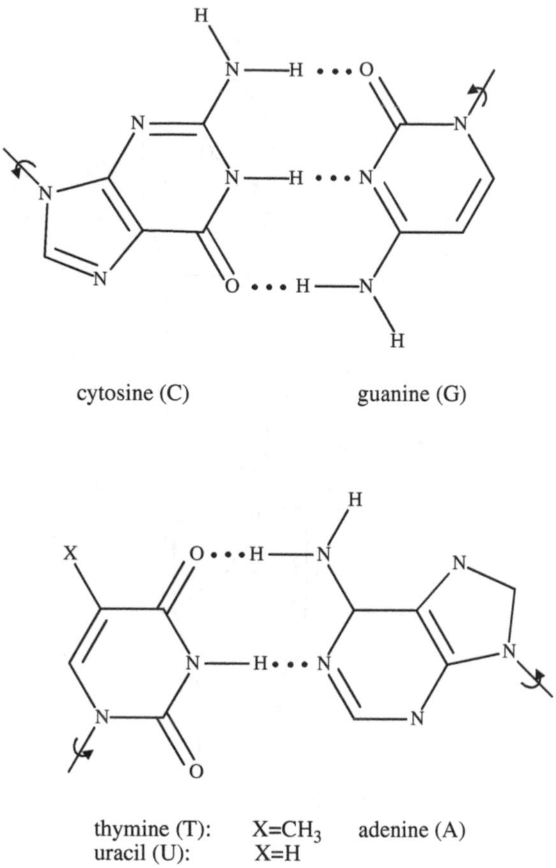

thymine (T):   X=CH₃   adenine (A)
uracil (U):   X=H

FIGURE 2.48
Canonical nitrogenous bases and Watson–Crick pairing.

two hydrogen bonds with T and C forms three hydrogen bonds with G. No other combinations of base pairs lead to bond formation. The matching patterns of hydrogen bonds allow a second strand of DNA, provided it has the proper base sequence, to nestle up to the first and form a stable complex, the famous double helix whose bases are protected from the water environment.

RNA directs protein synthesis in cells. Its bases are adenine, guanine, cytosine, and uracil. RNA and DNA differ in their associated sugars. The bases of double-stranded DNA lie in the interior of the helix and are held by hydrogen bonds. The origin of these bonds lies in the attraction of the H^+ atom of adenine to the O^- atom of thymine, for example. The reaction of a sugar with a base releases water (–OH from the sugar and H^+ from the base) and produces a sugar–base combination called a nucleoside. Adding a phosphate to a nucleoside releases a water molecule and produces a nucleotide.

All living things pass on genetic information from generation to generation via chromosomes composed of genes. The information needed to produce a particular type of protein molecule is contained within each gene. Genetic information

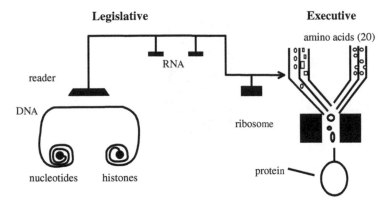

FIGURE 2.49
Each gene contains the information necessary to produce a specific protein.

in a gene is built into the principal molecule of a chromosome, namely its DNA (see Figure 2.49). DNA in a chromosome consists of two long DNA chains wrapped around each other in the form of a double helix. Figure 2.50 illustrates a section of a DNA double helix. The particular helical arrangement occurs when a chromosome replicates itself just before cell division. The arrangement of A paired to T and G opposite C ensures that genetic information is passed accurately to the next generation. Figure 2.51 depicts this process. The two strands or helices of DNA separate with the help of enzymes, leaving the charged parts of the bases fully exposed. The enzymes too operate via electrostatic forces.

To see how the correct order of bases occurs, we focus on the G molecule indicated by an arrow on the lowest strand in Figure 2.51. Unattached nucleotide bases of all four kinds move around in the cellular fluid. Only one of the four bases (C) will experience attraction to G when it approaches. The charges on the other three bases are arranged so they do not get into close proximity of the Gs and therefore no significant attractive force will be exerted on them. The forces decrease rapidly with distance between molecules. Since A, T, and C are hardly attracted, they tend to be knocked away by collisions with other molecules before enzymes can attach them to the growing chain.

The electrostatic force will often ensure a C opposite G long enough so for an enzyme to attach the C to the growing end of the new chain. The electrostatic forces hold the two helical chains together and also select the bases in the right order during replication. In Figure 2.51, the new number four strand has the same order of bases as the old number 1 strand. The two new helices, 1—3 and 2—4, are identical to the original 1—2 helix. If a T molecule were incorporated in a new chain opposite a G by accident, an error would occur. However, this occurs infrequently. The error rate of incorporating a T into a new chain opposite a G is 1 in 10^4 and may be deemed a spontaneous mutation resulting in a possible change in some characteristic of the organism. If the organism is to survive, the error must have a low rate but it cannot be zero if evolution is to take place.

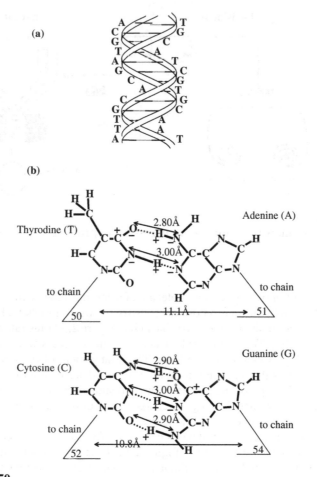

FIGURE 2.50
**(a) Section of DNA double helix. (b) Close-up showing how A and T and C and
G are always paired.**

DNA replication is often portrayed as occurring in a clockwork fashion, i.e., as if
each molecule knew its role and went to its designated place. This is not the case.
The forces of attraction between electric charges are rather weak and only become
significant when the molecules approach one another. If the shapes of the molecules
are not correct, little electrostatic attraction is generated. This is why errors are rare.

2.8.1   Electrostatic potential of DNA

DNA is a highly charged molecule and carries two fully ionized monovalent PO_4^-
groups per base pair on its outer surface. To model DNA, we assume the charge per
unit length is λ, which is about one negative charge per 1.7 Å of length. For simplicity,
we assume that DNA is a long, charged, cylindrical molecule. The cylinder diameter

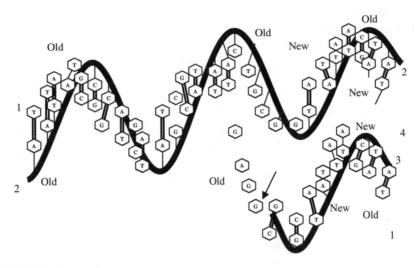

FIGURE 2.51
Replication of DNA.

is about 20 Å. The electrical field of a long strand of DNA in water depends on the perpendicular distance r from the strand and on λ.

The electric field lines must point inward because DNA is negatively charged. Denoting the electric field at a distance r from the cylinder by $E(r)$, we observe that it is inversely proportional to the dielectric constant ϵ and inversely proportional to some power, say r^α, of r, hence we write:

$$E(r) = const\, k\frac{\lambda}{\epsilon r^\alpha} \tag{2.27}$$

In vacuum, the electric field of a monopole of charge Q a distance R away is kQ/R^2. Since λ is a charge per unit length, $\alpha = 1$, hence:

$$E(r) = 2k\frac{\lambda}{\epsilon r} \tag{2.28}$$

Since $E(r) = -dV/dr$, we have an equivalent form of Equation (2.28).

$$\frac{dV(r)}{dr} = -2k\frac{\lambda}{\epsilon r} \tag{2.29}$$

Integrating Equation (2.29), we see that:

$$V(r) = -2k\frac{\lambda}{\epsilon}\ln r + k_0 \tag{2.30}$$

In Equation (2.30), k_0 is a constant conveniently chosen to be zero.

Suppose a positive monovalent ion moves in from a distance of 1μm to the surface of the DNA. The work done or ΔW will be:

$$\Delta W = |e|\,V(r = 1\mu m) - |e|V(r = 10\text{Å}) \tag{2.31}$$

TABLE 2.18
The Genetic Code

3′ Terminal Base (First)	Middle or Second Base				5′ Terminal Base (Third)
	U	C	A	G	
U	Phe	Ser	Tyr	Cys	U
	Phe	Ser	Tyr	Cys	C
	Leu	Ser	Stop	Stop	A
	Leu	Ser	Stop	Trp	G
C	Leu	Pro	His	Arg	U
	Leu	Pro	His	Arg	C
	Leu	Pro	Gln	Arg	A
	Leu	Pro	Gln	Arg	G
A	Ile	Thr	Asn	Ser	U
	Ile	Thr	Asn	Ser	C
	Ile	Thr	Lys	Arg	A
	Met	Thr	Lys	Arg	G
G	Val	Ala	Asp	Gly	U
	Val	Ala	Asp	Gly	C
	Val	Ala	Glu	Gly	A
	Val	Ala	Glu	Gly	G

Hence:

$$\Delta W = -2k\frac{\lambda}{\epsilon}\,|e|\,\ln\left(\frac{10^{-6}}{10^{-9}}\right) = 10 \cdot 2 \cdot 10^{-20}J \tag{2.32}$$

Using $k = 8.99 \times 10^9$ m^2/C^2, $\epsilon = 80.4$, $\lambda = -|e|$ per 1.7×10^{-10}m, and $e = 1.6 \times 10^{-19}$C from Equation (2.32), we find that ΔW exceeds by about 20 times the thermal energy $k_B T$ where k_B is Boltzmann's constant and T is the absolute temperature. This estimate suggests that a monovalent ion stays near the surface of DNA and we expect DNA to be surrounded by a cloud of Na$^+$ ions.

2.8.2 DNA: information and damage

Genetic coding can be discussed using information theory. DNA is the primary genetic material in living cells and specifies the information to be passed from one generation of cells to the next. It is also intimately involved with the synthesis of new material. The so-called genetic code is a detailed prescription regarding the synthesis of proteins according to the algorithm in Table 2.18.

For a long time, we did not know how information from DNA was transferred to other parts of cells so that the production of protein could occur. When the two strands of DNA are exposed, the genetic code contained in it can be expressed since it is believed that specific sequences of nucleotides code for specific sequences of amino acids that build up to form proteins.

We saw earlier that DNA contains only four distinct nucleotides. Finding any one of these in a specific location along a strand increases the information by $\log_2 4 = 2$ bits of information. Based on 20 amino acids, identification of a particular amino acid requires $\log_2 20 = 4.22$ bits. We therefore conclude that coding must be done by at least three nucleotides arranged in order.

DNA in cells is too valuable to do the work of synthesizing proteins directly. Transcription of coded DNA is accomplished by producing from it what are called messenger RNA molecules. The production of proteins is directed by the RNA molecules when they travel to the ribosomes. They specify the sequence in which amino acids will be linked. Four nucleotides can be coded onto the messenger RNA by the DNA strand, namely A, C, G, and U. To identify the groups of nucleotides coded for specific amino acids, scientists produced synthetic messenger RNA and examined the protein synthesized as a result. They found that if the RNA contained only uracil, for example, the protein was composed only of a phenylalanine chain containing only lysine. The code is based on combinations of three nucleotides and any amino acid can be called up in a number of ways.

Three of the 64 possible combinations do not code for any amino acid and are believed to act as terminators. When a ribosome reaches this portion of the messenger RNA chain, protein growth is halted. It is also believed that the code is universal and applies to all organisms. If a mutation or error occurs in the DNA, all copies will carry it and incorrect synthesis will occur at the ribosomes.

As an example, GAA and GAG code for glutamic acid, GUA, and GUG for valine. If an error of inserting U for A occurs in the portion of a strand of DNA that provides information for producing hemoglobin, it is believed that valine is built instead of glutamic acid. The resulting hemoglobin is called sickle cell hemoglobin. The name arises from the distorted shapes of the red blood cells. People who have sickle cell disease suffer severely from anemia. Many other hereditary diseases are likely to be caused by minor coding errors of this sort.

Electromagnetic radiation, especially in the ultraviolet (UV) range of 200 to 350 nm, can be harmful to living systems. Cellular injury, mutation, and lethality are known to result from high intensity UV exposure. One effect seen in samples of DNA extracted from UV-irradiated cells is breaking of bonds known as local denaturation and entanglement with proteins called protein cross-linking (Figure 2.52).

2.8.3   Fluorescence in biomolecules

In many plant materials the phenomenon of fluorescence has been known for more than a century, e.g., in quinone. For many years fluorescence has been used to check the presence of trace elements, drugs, and vitamins. It is surprising that fluorescent assays have only recently been used extensively for identification and location studies, particularly in view of their sensitivity and the ability of most molecules after suitable chemical treatment to fluoresce.

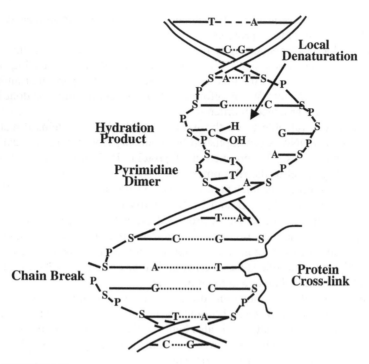

FIGURE 2.52
Defects found in DNA irradiated with ultraviolet light.

The method affords many of the same advantages of radioactive techniques without the associated hazards. Many *in vivo* radioactive assays require laboriously synthesized substances. The same investigations can be undertaken more simply and with equal accuracy using fluorescence. When it is combined with separation techniques like electrophoresis, it can detect impurities at concentrations below 1 in 10^{10}.

When a substance to be studied is in solution, the container, the solvent, and other components may also fluoresce. Clearly, no solvent that fluoresces in the same general region should be used. The light used as the stimulant should be strictly monochromatic even if filters are required. Elaborate studies may require a spectrometer to analyze the fluorescent light and the wavelengths of the various components measured before identification becomes possible. In biochemistry, the sensitivity of fluorescence means it is used regularly to investigate the locations of enzymes and coenzymes in organisms and to follow the fates of drugs and their metabolic products. Fluorescence can also be used to determine the amino acid sequences in proteins.

To understand fluorescence we must augment the elementary theory of quantum energy levels for electrons in atoms by vibrational degrees of freedom present in the energy structures of molecules. Vibrational energy levels arise because a molecular bond acts like a system of two masses connected by an elastic spring that has a natural

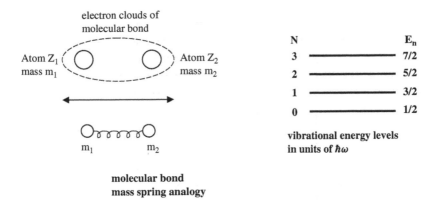

FIGURE 2.53
Quantization of vibration energy levels.

frequency of vibration ω. The energies are quantized (Figure 2.53) according to the formula:

$$E_n = \hbar\omega\left(n + \frac{1}{2}\right) \tag{2.33}$$

Transitions between vibrational energy levels can be induced by photons provided $\Delta n = \mp 1$. Each electronic state may have a number of vibrational states. At room temperature, most molecules will occupy the ground electronic state and lowest vibrational energy level.

If a molecule enters a high vibrational level, it will, after several collisions, drop to a lower level and give up its excess energy to other molecules in its vicinity. This is called vibrational relaxation. A molecule in an excited electronic and vibrational state will undergo vibrational relaxation followed by emission of a photon of energy $h\nu'$ and return to the ground state. When the emitted photon is in the visible range, the process is called fluorescence. The cycle of excitation by a photon of energy $h\nu$, vibrational relaxation, and fluorescence takes typically on the order of 10^{-7} s. Figure 2.54 illustrates the quantum mechanical origin of fluorescence.

When we reincorporate spin in electronic states, the ground state is usually a so-called singlet state, i.e., it has a net spin of zero for the two electrons occupying it. Some excited states are also singlet states. Some involve a net spin of unity, which is called a triplet state. We show the distinction in Figure 2.55.

This gives rise to the phenomenon of phosphorescence in which the return of a photoexcited electron to the ground state involves an intermediate transition to an excited triplet state. A distinguishing feature of phosphorescence is that it can appear even several seconds after photon absorption. A phosphorescent photon has an energy $h\nu^*$ which is always lower than the absorbed photon $h\nu$ or fluorescent photon $h\nu'$ (see Figure 2.56).

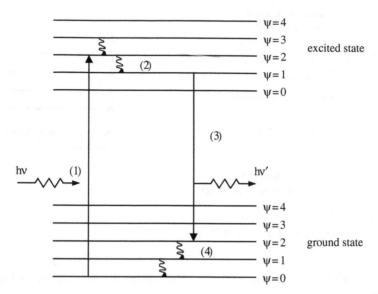

FIGURE 2.54
Quantum mechanical origin of fluorescence.

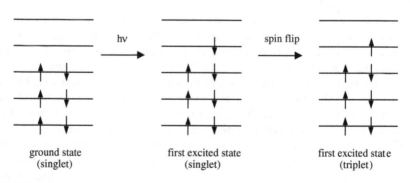

FIGURE 2.55
Singlet and triplet states of molecules.

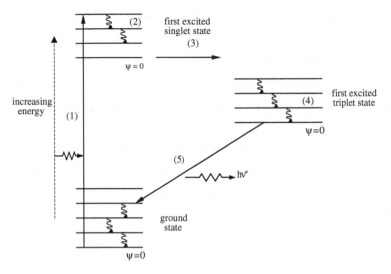

FIGURE 2.56
Quantum mechanical origin of phosphorescence.

References

Applequist, D.E., Depuy, C.H., and Rinehart, K.L., *Introduction to Organic Chemistry*, John Wiley & Sons, New York, 1982.

Atkins, P.W., *Physical Chemistry*, 6th ed., Oxford University Press, Oxford, U.K., 1998.

Benedek, G.B. and Villars, F.M.H., *Physics with Illustrative Examples from Medicine and Biology*, Springer, Berlin, 2000.

Bergethon, P.R. and Simons, E.R., *Biophysical Chemistry: Molecules to Membranes*, Springer, Berlin, 1990.

Bruinsma, R., *Physics*, International Thomson Publishing, New York, 1998.

Gennis, R.B., *Biomembranes: Molecular Structure and Function*, Springer, Berlin, 1989.

Heitler, W. and London, F., *Zeit. f. Physik*, 44, 455, 1927.

Klotz, I.M., *Protein Sci.* 2, 1992, 1993.

Kossel, W., *Ann. der Physik*, 49, 229, 1916.

Lewis, G.N., *J. Am. Chem. Soc.*, 38, 762, 1918.

Némethy, G. and Scheraga, H.A., *J. Phys. Chem.*, 66, 1773, 1962.

Nichols, A., Sharp, K., and Honig, B., *Struct. Function Genet.*, 11, 281, 1991.

Pauli, W., *Naturwiss.*, 12, 741, 1924.

Pauling, L., and Corey, R.B., *Nature*, 171, 346, 1953.

Saenger, W., *Principles of Nucleic Acid Structure*, Springer, New York, 1984.

Schulz, G.E. and Schirmer, R.H., *Principles of Protein Structure*, Springer, Berlin, 1979.

Sinden, R., *DNA Structure and Function*, Academic Press, San Diego, 1990.

Stryer, L., *Biochemistry*, 4th ed. W.H. Freeman & Co., San Francisco, 1981.

Tanford, C., *The Hydrophobic Effect: Formation of Micelles and Biological Membranes*, 2nd ed., John Wiley & Sons, 1980.

3

Dynamics of Biomolecules

> The kind of stability that is displayed by the living organism is of a nature somewhat different from the stability of atoms or crystals. It is a stability of process or function rather than the stability of form.
>
> Werner Heisenberg

3.1 Diffusion

Diffusion is macroscopically associated with a gradient of concentration as shown in Figure 3.1. In contrast to the mass flow of liquids, diffusion involves random spontaneous movements of individual molecules. This process can be quantified by a constant known as the *diffusion coefficient*, D, of the material, given in general by the *Stokes-Einstein equation*:

$$D = \frac{k_B T}{f},$$
(3.1)

where k_B is the Boltzmann constant, T is the absolute temperature in K, and f is a frictional coefficient.

If the diffusing molecule is spherical, is in low concentration, is larger than the solvent molecules and does not attract a layer of solvent molecules to itself, the frictional coefficient

$$f = 6\pi \eta r,$$
(3.2)

where r is the radius of the diffusing molecule in m, and η is the coefficient of viscosity of the solvent expressed in Ns/m^2. As a consequence:

$$D = \frac{k_B T}{6\pi \eta r} \quad \text{or} \quad r = \frac{k_B T}{6\pi \eta D}.$$
(3.3)

The diffusion constant of a particular molecular species depends on the nature of the molecule and the solvent. Large molecules have smaller diffusion constants. The diffusion constant D depends also on temperature. Table 3.1 lists diffusion constants for several molecules in water at room temperature.

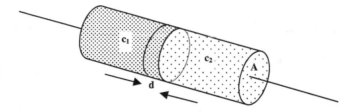

FIGURE 3.1
Diffusion process due to a concentration gradient in a cylindrical container.

TABLE 3.1
Diffusion Constants in Water at Room Temperature

Molecule	Diffusion Constant D (m²/s)
Water	2×10^{-9}
Oxygen	8×10^{-10}
Glucose	6×10^{-10}
Tobacco mosaic virus	3×10^{-12}
DNA (molar mass: 5 million g)	1×10^{-12}
Protein (order of magnitude)	1×10^{-10}
Hemoglobin	6.9×10^{-12}

The volume of a sphere $V = \frac{4}{3}\pi r^3$, the mass of a molecule $m = \rho V$, and the molar mass $M = mN_A$, where N_A is Avogadro's number (6.02×10^{23} molecules/mol). Combining the above equations, we obtain:

$$M = \rho \frac{4}{3}\pi \left(\frac{kT}{6\pi \eta D} \right)^3 N_A. \tag{3.4}$$

If all quantities in this equation are shown in SI units, the molar mass will be in kg/mol.

As indicated in Table 3.1 for biologically important molecules diffusing through water at room temperature, values of D range from 1 to 100×10^{-11} m²/s; the corresponding range of molecular weights is about 10^4. The diffusion constant may be related to the temperature and viscosity of the liquid by Equation (3.3). In Equation (3.4), the radius of a sphere is proportional to the cube root of its mass. We therefore conclude that D is inversely proportional to the cube root of the mass (see Equation 3.3). This explains why the values of D for a wide range of molecular weights fall within a comparatively small range. This result does not hold for gases. D becomes inversely proportional to the square root of the mass of the solute particles. An important characteristic of diffusion processes is the proportionality of the average value of the squared displacement of a diffusing particle to the time elapsed during the process (see Appendix A):

$$\bar{R}^2 = 6Dt. \tag{3.5}$$

This equation helps us to determine whether cellular processes such as transcription and translation are physically feasible. For example, one might ask whether the rates of diffusion of $D = 2 \times 10^{-9}$ m²/s are sufficient to allow 50 amino acids per second to be made into proteins at the ribosomes. Taking the distance to be the length of a bacterial cell (i.e. 3 μm), gives:

$$t = \frac{\bar{R}^2}{6D} = \frac{(3 \times 10^{-6})^2}{6 \times 2 \times 10^{-9}}\text{s} = 7.5 \times 10^{-4}\text{s}. \tag{3.6}$$

The process in a real bacterial cell would not be quite as rapid because the cytoplasm is about five times as viscous as water. This decreases D to one fifth the value used, with the result that time will be increased by a factor of five to 3.8×10^{-3} s which is still very fast. Thus diffusion, while a slow process in a bulk liquid, is rapid within the confines of a cell.

Fick's law states that the rate of diffusion per unit area in a direction perpendicular to the area is proportional to the gradient of concentration of solute in that direction. The concentration is the mass of solute per unit volume, and the gradient of concentration is the change in concentration per unit distance. If the concentration changes from c_1 to a lower value of c_2 over a short length (d) of the pipe (Figure 3.1), then the mass (m) of the solute diffusing down the pipe in time (t) is

$$\frac{m}{t} = DA\frac{c_1 - c_2}{d}. \tag{3.7}$$

This is a simplified version of Fick's law whose differential form is

$$J = -D\frac{dc}{dx} \tag{3.8}$$

This equation states that J, the flux of particles (number of particles passing through an imaginary normal surface of unit area per unit of time) is related to the force that pushes them ($-dc/dx$).

Osmosis is usually defined as the transport of molecules in a fluid through a semi-permeable membrane due to an imbalance in its concentration on either side of the membrane. Osmosis may be by diffusion, but it may also be a bulk flow through pores in a membrane. In either case, water moves from a region of high concentration to a region of low concentration. In Figure 3.2, the pressure on the right is then greater than the pressure on the left by an amount $h\rho g$, where ρ is the density of the liquid on the right and is called the relative osmotic pressure. The general formula for the osmotic pressure P of a solution containing n moles of solute per volume V of solvent is

$$P = \frac{n}{V}RT = cRT. \tag{3.9}$$

The net osmotic pressure exerted on a semipermeable membrane separating the two compartments is thus the difference between the osmotic pressures of both compartments. This equation is known as 'tHoff's law and it looks exactly like the ideal gas law. A detailed study of osmosis in plant cells began in the middle of the 19th century. If a plant cell is put into a concentrated solution of sugar, for example,

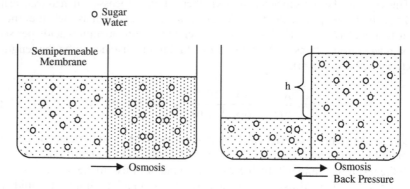

FIGURE 3.2
Osmosis.

the living portions of the cells (the protoplasts) contract away from the walls. If the cells thus treated are removed and placed in pure water, the protoplasts expand again. This phenomenon is easily observed under a microscope and is known as plasmolysis.

The osmotic pressure, P, is proportional to the concentration of solute, i.e., inversely proportional to the volume of the solution and also proportional to the absolute temperature. Hence the law of osmosis may be written

$$PV = R'T, \tag{3.10}$$

where R' is a constant depending only on the mass of the solute present:

$$R' = nR. \tag{3.11}$$

The solute behaves like a perfect gas so osmotic pressure arises from the bombardment of the walls of the container by the molecules of sugar in the solution. At higher concentrations, Equation (3.10) is not valid and the reasons are analogous to those that cause the breakdown of the simple gas laws that must be reformulated as a Van der Waals equation. Reverse osmosis may take place in situations where external pressure is applied to one compartment such that it exceeds the osmotic pressure (see Figure 3.3).

Osmosis between roots and groundwater is thought to be responsible for the transfer of water into many plants. Groundwater is purer and hence has a higher water concentration than sap, so osmosis moves water into the roots. Water in sap is then transferred by osmosis into cells, causing them to swell with increased pressure. This pressure is called turgor and is partly responsible for the ability of many plants to stand up. Relative osmotic pressure is not large enough to cause sap to rise to the top of a tall tree, however. To do so, the sap would have to contain more dissolved materials than found to be the case.

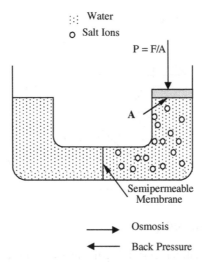

FIGURE 3.3
Reverse osmosis occurs when a pressure greater than the osmotic pressure is applied.

3.1.1   Diffusional flow across membranes

Biological membranes are highly complex, dynamic structures that regulate the flow of compounds into cells. Structurally, the conventional model (the so-called Singer–Nicolson model discussed in Chapter 4) describes the membrane as a bipolar, lipid layer, interspersed with regions of ion pumps, globular proteins, and pores.

A complete analysis of particle flow across the membrane must consider the influence of pressure gradients, Coulomb attraction, and concentration differences. Each exerts a varying degree of influence. For example, the flow of water across a membrane (osmosis) is explained as the result of a pressure gradient. This flow is influenced by the other two mechanisms, but they are not as dominant. The phenomenon of active transport is often described as the result of a Coulomb attraction between the solute and the membrane. It occurs frequently, but the primary mechanism for transport of solute is a difference of concentration at the membrane. The flow resulting from this difference (diffusion) dominates cellular processes such as metabolism (energy-yielding processes essential for life) and respiration (the flow of oxygen into cells). Because of the predominance of diffusion in solute flow, we will discuss diffusion across membranes next.

Our discussion is divided into two subsections that cover cells without sources and cells with sources. The first subsection emphasizes essential features of transport across the membrane. The second focuses on limiting effects of cells due to diffusion, size, and concentration. Specifically, a region about the center of the cell that consumes no oxygen has been mathematically predicted and experimentally verified.

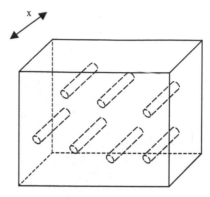

FIGURE 3.4
An idealized membrane with pores.

   The fundamental equations describing the process of diffusion are the continuity equation, Fick's law, and the diffusion equation. In cells without sources, these are written as:

$$\frac{\partial C}{\partial t} = -\nabla j \qquad j = -D\nabla C \qquad \frac{\partial C}{\partial t} = D\nabla^2 C \qquad (3.12)$$

In cells with sources, the diffusion equation is altered to accommodate a (presumed) constant source Q:

$$\frac{\partial C}{\partial t} = D\nabla^2 C + Q \qquad (3.13)$$

3.1.2   Cells without sources

Figure 3.4 illustrates an idealized porous membrane assumed to be shaped like a thin rectangular slab, interspersed with cylindrical pores. To calculate the flow rate through this membrane, two further assumptions are required. The first assumption is that concentrations on both sides of the membrane are uniform. This is accurate if the fluids are stirred or circulated, as is the case for blood plasma in a capillary. The second assumption is that the profile of the concentration in the pore is linear. This is the quasistationary situation, for if $\frac{\partial C}{\partial t} \cong 0$, then $\frac{\partial^2 C}{\partial x^2} \cong 0$ (from the diffusion equation), and so, $\frac{\partial C}{\partial x} \cong const$. Figure 3.5 shows the concentration profile with these assumptions. Inside the pore:

$$\frac{\partial C}{\partial x} = -\frac{C_1 - C_2}{\Delta x} = -\frac{\Delta C}{\Delta x} \qquad (3.14)$$

Using Fick's law, the flow of solute through a single pore is:

$$\frac{j(x,t)}{pore} = -D\frac{\partial C}{\partial x} = D\frac{\Delta C}{\Delta x} \qquad (3.15)$$

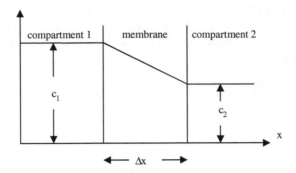

FIGURE 3.5
Concentration profile across a membrane.

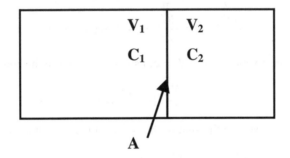

FIGURE 3.6
Two compartments separated by a membrane.

Commonly, the flow rate or number of particles flowing per second is described by J. The relation between the current density $j(x, t)$ and J is simply:

$$J = j(x, t) \cdot (\text{cross} - \text{sectional area}) = j(x, t) \cdot \pi a^2 \qquad (3.16)$$

(for cylindrical pores). Therefore, for n pores:

$$J = n\pi a^2 \frac{D}{\Delta x} \Delta C \quad \text{or} \quad J = \wp \Delta C \quad \text{where} \quad \wp = n\pi a^2 \frac{D}{\Delta x} \qquad (3.17)$$

Thus, the flow rate is proportional to the difference of concentration, where the constant of proportionality, $\wp$, is called permeability. Essentially, this constant dictates whether a solute will pass through a membrane. As may be expected, permeability is dependent on pore size, thickness of the membrane, and diffusion constant.

Now that we have a rough idea of the flow rate, the time required for the concentrations to equilibrate can be determined. Consider the flow across a membrane separating two compartments of varied volumes and concentrations (as shown in Figure 3.6). Let us assume that $C_1 > C_2$, and let N_1 be the number of particles in

(1) and N_2 be the number of particles in (2). The total solute flow per second for the entire membrane is $(J \times A)$. J is defined above and A is the membrane area. From the conservation of particles, the particle decrease in (1) must correspond to the particle increase in (2):

$$\frac{dN_1}{dt} = -\frac{dN_2}{dt} = -J \cdot A = -A\wp(C_1 - C_2) \tag{3.18}$$

However:

$$N_1(t) = C_1(t)V_1$$
$$N_2(t) = C_2(t)V_2 \tag{3.19}$$

so:

$$\frac{dN_1}{dt} = V_1\frac{dC_1}{dt} = -A\wp(C_1 - C_2)$$
$$\frac{dN_2}{dt} = V_2\frac{dC_2}{dt} = +A\wp(C_1 - C_2) \tag{3.20}$$

With particle conservation, $N_1 + N_2 = $ constant, we may use the condition

$$\frac{dN_1}{dt} + \frac{dN_2}{dt} = 0 \tag{3.21}$$

Substituting the above into the sum of the two formulas in Equation (3.20), we get:

$$V_1\frac{dC_1}{dt} + V_2\frac{dC_2}{dt} = \frac{d}{dt}(V_1C_1 + V_2C_2) = 0 \tag{3.22}$$

That is, $(V_1C_1 + V_2C_2)$ is a constant in time. Let the initial concentrations be $C_1(0)$ and $C_2(0)$, and the final common concentration be C_∞:

$$C_\infty = \frac{V_1}{V_1 + V_2}C_1(0) + \frac{V_2}{V_1 + V_2}C_2(0) \tag{3.23}$$

Using Equation (3.20) in which we subtract the two formulas from each other gives:

$$\frac{dC_1}{dt} - \frac{dC_2}{dt} = \frac{d}{dt}(C_1 - C_2) = -A\wp\frac{V_1 + V_2}{V_1 V_2}(C_1 - C_2) \tag{3.24}$$

Defining the constant, τ_0, as the characteristic time for concentration equilibrium:

$$\tau_0^{-1} = A\wp\frac{V_1 + V_2}{V_1 V_2}, \quad \Delta C(t) = \Delta C(0)e^{(-\frac{t}{\tau_0})} \tag{3.25}$$

we obtain an exponential equilibration process for the concentration function of time. This means that for large values of τ_0, a great deal of time is required to equilibrate the concentrations. As a check, as the area increases, one expects an increase of total solute flow, and hence, a lower equilibrium time. The schematic plot of $C(t)$ is shown in Figure 3.7.

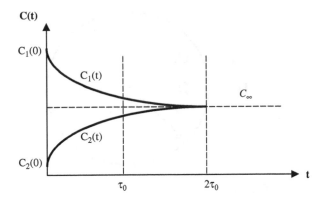

FIGURE 3.7
Plot of $C(t)$ based on Equation (3.26).

Solving for $C_1(t)$ and $C_2(t)$, we finally obtain:

$$C_1(t) = C_\infty + [C_1(0) - C_\infty]e^{\left(-\frac{t}{\tau_0}\right)}$$
$$C_2(t) = C_\infty + [C_2(0) - C_\infty]e^{\left(-\frac{t}{\tau_0}\right)} \qquad (3.26)$$

Thus, the two concentrations exponentially approach a common value C_∞ with a characteristic equilibration time with the magnitude of τ_0 given by Equation (3.25).

3.1.3 Cells with sources

Assuming a cell has a constant source, it can be described by the modified diffusion equation:

$$\frac{\partial C}{\partial t} = \nabla^2 C + Q \qquad (3.27)$$

where, if $Q > 0$, a solute is produced. If, on the other hand, $Q < 0$, the solute is consumed. To simplify the analysis, the cell will be assumed to be spherical. This is admittedly a crude approximation but one that provides good insight into the nature of the diffusion processes in cells. In addition, a quasistationary situation will again be assumed so that $\frac{\partial C}{\partial t} \cong 0$.

Let the concentration and the diffusion constant inside the cell be C_i and D_i, respectively, and those outside be C_e and D_e, respectively (as depicted in Figure 3.8). In addition, let the solute concentration far from the cell be designated as C_o. Assume the cell is placed in a medium of concentration C_o, so that the concentration varies

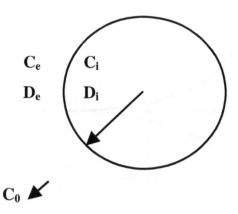

FIGURE 3.8
Concentration distribution at the beginning of diffusion.

near the membrane and nowhere else. C_o may be any nonnegative value including zero. The diffusion equations are then:

$$D_i \nabla^2 C_i + Q = 0$$

$$D_e \nabla^2 C_e = 0 \tag{3.28}$$

with the condition that whatever enters or leaves the cell, leaves or enters the external environment, at $r = r_0$:

$$D_i \frac{dC_i}{dr} = D_e \frac{dC_e}{dr} \quad \text{and} \quad -D_i \frac{dC_i}{dr} = k\Delta C = k(C_i - C_e) \tag{3.29}$$

where $k = \wp A$. By making the substitution:

$$C_i^* = C_i r \quad \text{and} \quad C_e^* = C_e r \tag{3.30}$$

in the diffusion equations (for spherical cells) we obtain:

$$\frac{d^2 C_i^*}{dr^2} = -\frac{Q}{D_i} r$$

$$\frac{d^2 C_e^*}{dr^2} = 0 \tag{3.31}$$

Integrating the above equations and solving for C_i and C_e subject to the boundary conditions, we find:

$$C_i = C_0 + \frac{Qr_0}{3k} + \frac{Q}{6D_i}(r_0^2 - r^2) + \frac{Qr_0^2}{3D_e}$$

$$C_e = C_0 + \frac{Qr_0^3}{3D_e}\frac{1}{r} \tag{3.32}$$

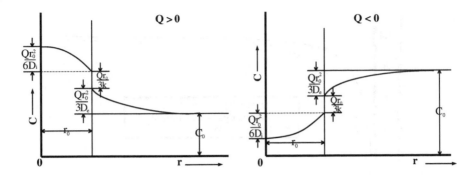

FIGURE 3.9
Concentration profiles as functions of the spherical radius for $Q > 0$ (left) and $Q < 0$ (right).

Figure 3.9 shows variations of the concentrations C_i and C_e for $Q > 0$ and for $Q < 0$. Notice that if $Q < 0$ and $r = 0$, $C_i = 0$ when:

$$C_0 = C^* = |Q| \left(\frac{r_0}{3k} + \frac{r_0^2}{6D_i} + \frac{r_0^2}{3D_e} \right) \tag{3.33}$$

However, if $C_o < C^*$, $C_i < 0$ and $r = 0$ which is physically impossible! To alleviate this problem, assume the cell has a region at a radius r_1 where no consumption occurs. That is, for $r = r_1$:

$$\frac{dC_i}{dr} = 0$$

$$C_i = 0 \tag{3.34}$$

For $r > r_1$, the same conditions apply. Resolving the diffusion equations with these new conditions, we now find:

$$C_i = C_0 + \frac{Qr_0}{3k} + \frac{Q}{6D_i}(r_0^2 - r^2) + \frac{Qr_0^2}{3D_e}$$

$$- \frac{1}{3} \left(\frac{1}{D_e r_0} - \frac{1}{D_i r_0} + \frac{1}{kr_0^2} \right) Qr_1^3 - \frac{Qr_1^3}{3D_i} \frac{1}{r}$$

$$C_e = C_0 + \frac{Q}{3D_e}(r_0^3 - r^3) \frac{1}{r} \tag{3.35}$$

Using the boundary conditions at $r = r_1$, a cubic equation in r_1 is obtained.

$$C_0 - C^* - \frac{Qr_1^2}{2D_i} - \frac{1}{3} \left(\frac{1}{D_e r_0} - \frac{1}{D_i r_0} + \frac{1}{kr_0^2} \right) Qr_1^3 = 0 \tag{3.36}$$

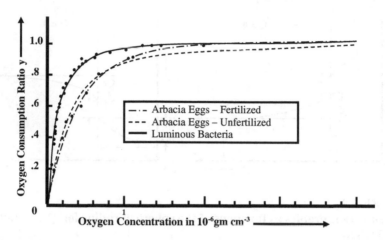

FIGURE 3.10
Graphic comparison of typical data obtained in a study of oxygen consumption in *Arbacia* eggs and bacteria and the theoretical prediction in Equation (3.41).

Solving this equation for r_1, and introducing it into Equation (3.35), the diffusion equation will thus be solved. This leads to a lengthy formula of no real practical use. Instead, a comparison with experimental data is made. For $C_0 < C^*$, the total rate of consumption is given by:

$$Q_{tot} = \frac{4}{3}\pi(r_0^3 - r_1^3)Q \tag{3.37}$$

since all consumption is assumed to take place within a shell of volume:

$$\frac{4}{3}\pi(r_0^3 - r_1^3) \tag{3.38}$$

For $C_0 \geq C^*$, Q_{tot} is equal to:

$$Q_{tot} = \frac{4}{3}\pi r_0^3 Q \tag{3.39}$$

With C_0 decreasing from C^* to zero, Q_{tot} also decreases to zero. A theoretical comparison of solute consumption to C_0 is possible with the following definition:

$$\gamma = \frac{r_i}{r_0} = \left[1 - \frac{Q_{tot}}{Q_0}\right]^{1/3} \tag{3.40}$$

For, now, the cubic equation in r_1 becomes:

$$C_o = C^* + \frac{Qr_0^2}{2D_i}\gamma^2 - \left(C^* + \frac{Qr_0^2}{2D_i}\right)\gamma^3 \tag{3.41}$$

Figure 3.10 is a graphic comparison of data obtained by studying oxygen consumption in *Arbacia* eggs and luminous bacteria to the theoretical prediction. The accuracy is remarkable, considering the assumptions made.

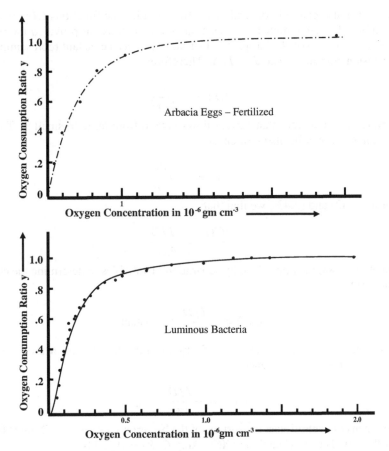

FIGURE 3.11
**Oxygen consumption versus oxygen concentration for cells of *Arbacia* eggs (top)
and luminous bacteria (bottom).**

Cell respiration is a detailed process that shows similar results as indicated by
Figure 3.11. As a particular example, consider oxygen consumption by aerobic
bacteria. Oxygen is supplied by diffusion for the metabolism of aerobic bacteria.
Consider a single bacterium in an aqueous solution that contains a certain amount
of dissolved oxygen. Suppose the concentration of oxygen far from the bacterium
has its equilibrium value, C_0. For oxygen dissolved in water, it is about 0.2 mol/m^3.
Assume the bacterium is spherical, has a radius R, and constantly consumes oxygen.
Oxygen molecules are adsorbed very efficiently at its surface. As a result, the oxygen
concentration just outside the bacterium must be close to zero.

The concentration difference between the surface of the bacterium and infinity will
set up a diffusion current I of oxygen from infinity to the bacterium. The oxygen
concentration $C(r)$, where $r = 0$ is the origin of the coordinate system, must depend
only on the distance r from the origin. At the surface of the bacterium where $r = R$,
the concentration must vanish so $C(R) = 0$. The oxygen molecule diffusion current

density $J(r)$ also can only depend on r. To determine the functions $C(r)$ and $J(r)$, the number of oxygen molecules was assumed to be constant provided the product of $J(r)$ and the area of A of a spherical surface at r was a constant representing only the diffusion current I since $J = I/A$. Therefore:

$$J(r) = -\frac{I}{4\pi r^2} \tag{3.42}$$

The minus sign indicates that current flows inward from large to small r. The concentration $C(r)$ may be also written as:

$$J(r) = -D\frac{dC}{dt} \tag{3.43}$$

Equating (3.42) and (3.43), we find that:

$$\frac{dC(r)}{dr} = \frac{I/D}{4\pi r^2} \tag{3.44}$$

Equation (3.44) may be solved by separation of variables to determine the concentration $C(r)$:

$$C(r) = -\frac{I/D}{4\pi r} + \text{constant} \tag{3.45}$$

We find the constant in Equation (3.45) by recalling that far from the bacterium, i.e., $r \to \infty$, $C(r) = C_0$. Hence:

$$C(r) = -\frac{I/D}{4\pi r} + C_0 \tag{3.46}$$

The oxygen concentration at the surface of the bacterium, i.e., at $r = R$, must be zero so $C(R) = 0$. Thus in Equation (3.46), using $C(R) = 0$ yields:

$$I = 4\pi RDC_0 \tag{3.47}$$

Equation (3.47) determines the maximum number of oxygen molecules per second that can be taken up by the bacterium. The number of oxygen molecules required per unit volume by a bacterium is called the metabolic rate M and is about 20 mol/m^{3}s. The incoming diffusion current I must equal or exceed M times the volume, $V = 4\pi R^3/3$, of the bacterium for it to function normally. Hence:

$$I > \frac{4}{3}\pi R^3 M \tag{3.48}$$

Combining Equation (3.47) with Equation (3.48) yields an inequality for the size of R. That is:

$$R < \sqrt{\frac{3DC_0}{M}} \tag{3.49}$$

which provides an upper bound for the bacterium size. Inserting values for the parameters, we find that $R < 10 \ \mu m$. Bacteria larger than this will simply not get enough oxygen by pure diffusion. This limit is, in fact, about the size of a typical spherical bacterium. Larger cells can be produced by making them very long while keeping the diameter small, i.e., cylindrical. This greatly increases diffusion current and diffusion imposes no size limitation on the lengths of cylindrical bacteria. Large bacteria are cylindrical rather than spherical (Figure 3.11).

In conclusion, several interesting results were obtained by limiting analysis to only the concentration difference mechanisms. For cells without sources, the permeability characteristic of time governs the diffusion process while pore size, membrane thickness, diffusion constant, and membranous area were shown also to be factors. For cells with sources, the existence of a cell region that did not consume oxygen was theoretically predicted and experimentally verified. Not all cells have this property, but for those that do, a study of diffusion reveals a possible explanation.

A more in-depth analysis of the transport of particles across a membrane should include the pressure gradient and the Coulomb attraction. In addition, the nonstationary case should be investigated with and without the other two mechanisms. Since chaos theory is becoming more prominent (see Appendix C), a closer study of the interplay of these mechanisms would most likely show signs of chaotic dynamics. Nonetheless, limiting the analysis only to diffusion produced rather interesting conclusions.

3.2 Vibrations versus conformational transitions

Since atomic nuclei are at least 2000 times more massive than electrons, we can separate the nuclear dynamics from the electronic dynamics and describe them with the help of so-called *Born–Oppenheimer adiabatic potentials* for individual quantum mechanical electronic states of a molecule (Atkins, 1998). We will assume the ground electronic state to be well separated in energy from the excited states.

A molecule consisting of N atoms has $3N$ nuclear degrees of freedom. Six of them are external degrees of freedom characterizing the translational and rotational motion of a whole molecule as a rigid body. The remaining $3N - 6$ internal degrees of freedom — a huge number in the case of biological macromolecules — can be identified by covalent bond lengths and angles and dihedral angles of rotation about the bonds (see Figure 3.12a). The ability to perform such rotations (limited only to a degree by steric hindrances) combined with the possibility of hydrogen bond break-up and reformation makes the landscape of the ground electronic state potential energy of internal degrees of freedom extremely complex.

A general feature of this landscape is the presence of an astronomical number of local minima separated by higher or lower energy barriers of noncovalent nature. As in the stereochemistry of low molecular weight organic compounds, regions of the configurational space surrounding the local minima can be referred to as a biomolecule's *conformational states* (*substates* in certain contexts). In a reasonable approximation (Kurzynski, 1998) the dynamics of a biomolecule can be decomposed into *vibrations* within particular conformational states and *conformational transitions*

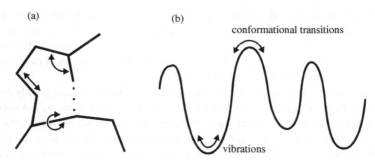

FIGURE 3.12
(a) The internal dynamics of biomolecules consists of changing the values
of covalent bond lengths, angles, and dihedral angles of rotations about the
bonds.   Break-up and reformation of weak hydrogen bonds is important.
(b) In a many-dimensional landscape of the configurational potential energy
of a biomolecule, we can distinguish conformational substates characterized by
the local energy minima separated by higher or lower energy barriers. Assuming interconformational barriers are high enough, the intramolecular dynamics
can be reasonably decomposed into vibrations within particular conformational
substates and conformational transitions.

(Figure 3.12b). The vibrations are damped harmonic oscillations subjected randomly
to stochastic perturbations and the conformational transitions are purely stochastic-
activated processes.

A typical structural subunit of biological macromolecules has approximately
10^4 internal degrees of freedom. This corresponds to a small protein or a protein
domain consisting of some 200 amino acids, a polysaccharide crystallite composed
of 150 monosaccharides, or a transfer RNA molecule of less than 100 nucleotides. Its
spectrum of vibrational periods, whose number is equal to the number of degrees of
freedom, ranges from 10^{-14} s for weakly damped localized C—H or N—H stretching modes directly observed in the infrared spectroscopy to 10^{-11} s for overdamped
collective modes involving a whole subunit that is a counterpart of acoustic phonons
in regular crystals.

This spectrum can be conveniently divided into two ranges (see Figure 3.13). A
reasonable dividing point is the period 2×10^{-13} s since equilibrium vibrations with
this period have an energy that corresponds to room temperature (300 K). Modes
in the high frequency range involve mainly the stretching and bending of bonds.
Those in the low frequency range involve collective torsional motions in dihedral
angles about the bonds. Because of the effect of energy quantization, only low frequency vibrations are, in principle, thermally excited. Thus, only they contribute to
the thermal properties of biomolecules (determining the values of entropy and specific
heat).

The conformational transition dynamics, on the other hand, is characterized by a
spectrum of relaxation times whose number is equal to the number of conformational
states. Each *global* conformational state of a biomolecule is represented by a sequence

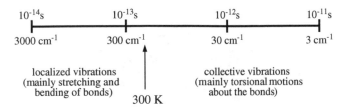

FIGURE 3.13
**Frequency spectrum of biomolecule vibrations and its conventional division into
two ranges. Particular vibrational periods are related to the corresponding wave
numbers of infrared spectra.**

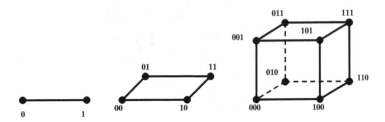

FIGURE 3.14
**For local states labeled by a single bit $s_i = 0, 1$, the global states of a system
composed of two units represent vertices of a square. Global states of a system composed of three units represent vertices of a 3-dimensional cube and,
in general, those of a system composed of M units represent 2^M vertices of an
M-dimensional cube.**

$(s_1, s_2, s_3, \ldots s_M)$ of *local* conformational states, each described by a generalized
spin variable (s_i) that can take two or more discrete values $s_i = 0, 1, 2, \ldots m_i - 1$ (all
m_i values are not necessarily equal). For $s_i = 0, 1$ (the local states labeled by a single
bit), the global states represent vertices of an M-dimensional cube (see Figure 3.14);
for larger values of m_i, they represent sites of a more complex M-dimensional lattice.
The index i can label particular monomers comprising the macromolecule or its larger
structural subunit. Then m_i values are of the order of 10 (e.g., two conformational
states of the peptide bond times five conformational states of the side chain in the
case of proteins).

For $M = 100$ we obtain via this method the astronomical number of 10^{100} global
conformational states. However, only a minute fraction of them are occupied
under physiological conditions. The first reason for this reduction is very simple.
Namely, because of sterical constraints, energies of many global conformations are
exceedingly high. A second reason is subtler. A characteristic feature of biologically
active macromolecules under physiological conditions is their well defined spatially
organized structures. Spatial organization (known as folding in the case of proteins)
is a discontinuous phase transition process (see Chapter 2 and Appendix B). Thus,
all the particular conformational states under physiological conditions can, in

FIGURE 3.15
Cross-section of a universal unit of biochemical processes, a supramolecular multienzyme protein complex. Heavily shaded areas are solid-like fragments of secondary structures; medium shaded areas are nonpolar liquid-like regions; weakly shaded areas are polar liquid-like regions. Black areas are individual catalytic centers usually localized at two neighboring solid-like elements. (After Kurzynski, 1998).

principle, be divided unambiguously between native (folded) and nonnative (unorganized, unfolded) states.

Until the end of the 1970s, the native state of a biological macromolecule was commonly considered a single conformational state, identified by its tertiary structure. In the 1980s and the 1990s, increasing experimental evidence demonstrated rich stochastic dynamics of conformational transitions within the native states of most biological macromolecules (McCammon and Harvey, 1987; Brooks et al., 1988; Frauenfelder et al., 1991, 1999; Kurzynski, 1998). Conformational transitions do not take place in the entire body of the macromolecule (Kurzynski, 1997a). They are limited to liquid-like amorphous regions surrounding solid-like fragments of secondary structures or larger structural subunits in the case of proteins (Figure 3.15) and nucleic acids or microcrystallites in the case of polysaccharides.

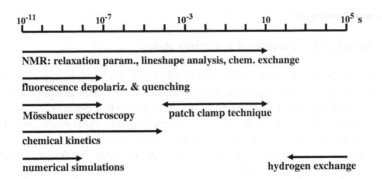

FIGURE 3.16
Time scales of conformational transitions in the protein native state observed via various experimental techniques. Time period 10^{-11} s at the left edge characterizes localized conformational transitions on the protein surface. Time period 10^5 s at the right edge is an underestimated value of the waiting time for spontaneous unfolding of the protein under physiological conditions. A typical reciprocal turnover of enzymatic reactions (10^{-3} s) is in the middle of the scale. Time scale of conformational transitions noted in chemical kinetics experiments at low temperatures is extrapolated to physiological temperatures. (After Kurzynski, 1998.)

DNA consists of a single secondary structure, the double helix and it is relatively rigid in terms of conformational transitions. Polysaccharides of medium size are usually networked, include a single microcrystallite, and hence are conformationally rigid. The two main classes of biomolecules with nontrivial conformational transition dynamics are proteins and RNAs. RNAs differ from proteins in that they contain phosphodiester bonds that need counterions to be electrically neutralized and a lot of sugar rings experiencing repuckering conformational transitions. The general character of their conformational transition dynamics is very similar to that of proteins (McCammon and Harvey, 1987). Because much more research has been done on protein dynamics than on RNA dynamics, we confine our attention mainly to protein dynamics.

Figure 3.16 shows various experimental techniques presented on a background of time scales of protein conformational transitions (Kurzynski, 1998). The time scales are very broad, ranging from 10^{-11} s to 10^5 s or more, making the task of integrating time dependence a genuine challenge.

The experimentally established picture of conformational transition dynamics is, however, still far from complete. We know the time scales but do not know the number of conformational substates comprising the native state of a given protein. In typical experiments only a few, often only two, conformational states are discernible. Observations of nonexponential time courses of biochemical processes involving proteins appear to point to the existence of a quasicontinuum of conformational states. The interpretation of the data is based on the stochastic theory of reaction rates and it will be the topic of the next section.

3.3   Stochastic theory of reaction rates

The basic assumption of the stochastic theory of reaction rates is that a molecule can exist in a number of substates, the transitions between which are purely stochastic. The origins of this theory extend back to Smoluchowski's description of diffusion-controlled coagulation and Kramers' one-dimensional theory of reactions in the over-damped limit. Montroll and Shuler (1958) devised a general formulation of the theory. Clear discussions of the key concepts can be found in papers by Widom (1965, 1971) and Northrup and Hynes (1980).

Let us begin with a simple picture exemplifying a realization of the microscopic (or rather mesoscopic) stochastic dynamics underlying a unimolecular reaction

$$R \rightleftharpoons P$$

between two chemical species R and P of a given molecule. We assume that the molecule fluctuates among several substates that can be divided into two subsets corresponding to the chemical species R and P (Figure 3.17). The chemical reaction is realized through transitions between selected substates in R jointly forming what is called a *transition state* $R^{\ddagger}$ and selected substates in P, jointly forming a transition state $P^{\ddagger}$.

Two limiting cases can be distinguished: (1) where both transition states comprise all the substates in R and P, known as a reaction with *fluctuating barriers* (each substate is related to a generally different free energy barrier for the reactive transition), and (2) where the transition states reduce to single conformational substates, jointly forming a gate; this is known as a *gated reaction*.

In Smoluchowski's theory of coagulation, the subset R is a three-dimensional region accessible to the translational diffusion of a molecule. $R^{\dagger}$ is formed from uniformly distributed traps and P is reduced to the totally absorbing and irreversible "limbo state" (van Kampen, 2001). The reaction can be considered gated when we restrict R to what is distributed on the average to one trap. In Kramers' theory of reaction rates, all substates lie along a one-dimensional *reaction coordinate* and the transition state is a single substate with the highest free energy.

The model dynamics may be described generally by a system of master equations (see Appendix D):

$$\dot{p}_l(t) = \sum_{l'} [w_{ll'} p_{l'}(t) - w_{l'l} p_l(t)].\tag{3.50}$$

The quantity $p_l(t)$ denotes the probability that the molecule will be in the l substate at time t. The dot is a derivative with respect to time and the transition probabilities per unit time $w_{ll'}$ are assumed to satisfy the *detailed balance condition*:

$$w_{l'l} p_l^{eq} = w_{ll'} p_{l'}^{eq},\tag{3.51}$$

where p_l^{eq} denotes the equilibrium solution to Equation (3.50). Using appropriate linear combinations of probabilities

$$Y_k(t) = \sum_l y_{kl} p_l(t) = \langle \eta_k(t) \rangle\tag{3.52}$$

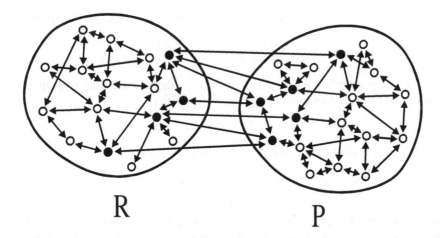

R

P

FIGURE 3.17

Model of intramolecular dynamics underlying the unimolecular reaction R ↔ P. The chemical states R and P of a molecule are composed of many substates (circles). The intramolecular dynamics involves purely stochastic transitions between the states (arrows). Chemical reaction is achieved through transitions between selected substates in R, jointly forming what is called transition state R‡ (black circles) and selected substates in P that jointly form transition state P‡. If the transition states comprise all the substates in R and P, the reaction has fluctuating barriers. If the transition states are reduced to single conformational substates, the reaction is gated.

the system of linear Equation (3.50) is decoupled into the system of independent linear equations

$$\dot{Y}_k(t) = \tau_k^{-1} Y_k(t). \tag{3.53}$$

If Equation (3.51) is satisfied, the coefficients τ_k^{-1} are real and positive and represent reciprocal *relaxation times*. The normal modes of relaxation (Equation 3.52) are written in such a way that they can be interpreted as the mean values of certain physical quantities (real functions on the set of substates) η_k.

The molar fractions of individual species, proportional to the molar concentrations, are the sums of probabilities

$$P_R(t) = \sum_{l \in R} p_l(t), \quad P_P(t) = \sum_{l \in P} p_l(t). \tag{3.54}$$

These can be rewritten as the mean values.

$$P_R(t) = \langle \chi_R(t) \rangle, \quad P_P(t) = \langle P_P(t) \rangle \tag{3.55}$$

(see Equation 3.52) of the characteristic functions of subsets R and P, respectively:

$$\chi_{Rl} = \begin{cases} 1 & \text{if } l \in \text{R} \\ 0 & \text{if } l \in \text{P} \end{cases} \qquad \chi_{Pl} = \begin{cases} 1 & \text{if } l \in \text{P} \\ 0 & \text{if } l \in \text{R} \end{cases}. \tag{3.56}$$

Following the normalization of probability to unity, the mole fractions (Equation 3.55) are related through the equation:

$$P_R + P_P = 1. \tag{3.57}$$

The molar fractions (Equation 3.54) satisfy the equation

$$\dot{P}_R(t) = -\dot{P}_P(t) = - \sum_{l \in R^{\ddagger}, l' \in P^{\ddagger}} [w_{l'l} p_l(t) - w_{ll'} p_{l'}(t)], \tag{3.58}$$

where $R^{\ddagger}$ and $P^{\ddagger}$ are appropriate transition states. In general, the solution to Equation (3.58) is nonexponential and depends on the initial values of all the probabilities p_l. The situation simplifies when the reaction is an *activated process*, i.e., as a result of a bottleneck of energetic or entropic origin, transitions between both subsets are not very probable. For such reactions, equilibration of substates *within* individual chemical species proceeds much faster than equilibration *between* species. A consequence is a time-scale separation in the system and a gap in the spectrum of reciprocal relaxation times between the longest and the next shortest relaxation times τ_1 and τ_2, respectively (see Figure 3.18). After the elapsed time τ_2 (an *initial stage* of reaction) Equation 3.58 takes the form of the usual kinetic equation

$$\dot{P}_R(t) = -\dot{P}_P(t) = -k_+ P_R(t) + k_- P_P(t)$$
$$= -\tau_1^{-1}(P_R - P_R^{eq}) = \tau_1^{-1}(P_P - P_P^{eq}) \tag{3.59}$$

with an exponential solution. For given equilibrium values of the molar fractions P_R^{eq} and P_P^{eq}, the longest *chemical relaxation* time τ_1 determines in a unique way the *forward* and *reverse reaction rate constants* k_+ and k_-, respectively, through the equations

$$\tau_1^{-1} = k_+ + k_- \tag{3.60}$$

and

$$\frac{k_+}{k_-} = \frac{P_P^{eq}}{P_R^{eq}} = K, \tag{3.61}$$

where K is called the *equilibrium constant*. P_R and P_P do not have to coincide exactly (only up to some multiplicative and additive constants) with the slowest variable of the system Y_1 (Equation 3.52). If it holds, the kinetic equation (3.59) is valid at any time scale including the initial stage of the reaction.

Because of the special properties of the characteristic functions:

$$P_R^2 = P_R, \qquad P_P^2 = P_P, \qquad P_R P_P = 0, \tag{3.62}$$

the thermodynamic perturbation theory for the problem can be applied exactly, up to an arbitrary order (Kurzynski, 1990). This results in the following exact expression

FIGURE 3.18
A schematic spectrum of reciprocal relaxation times characterizing the con-
formational transition dynamics of a molecule and a chemical transformation
involving it. A gap in this spectrum between the reciprocals of the longest and
the next shorter relaxation times (τ_1^{-1} **and** τ_2^{-1}), **respectively, testifies to the**
existence of time-scale separation. The ground value of the spectrum equal
to zero (infinite relaxation time) is related to the sum of all probabilities that
remains constant.

(valid arbitrarily far from equilibrium) for the reaction rate constant k_+ in terms of
the equilibrium time correlation function of fluxes (Northrup and Hynes, 1980):

$$k_+ = \int_0^\infty \frac{dt' \langle \dot{\chi}_R(t') \dot{\chi}_R(0) \rangle^{\mathrm{eq}}}{\langle \chi_R \rangle^{\mathrm{eq}}}. \tag{3.63}$$

A similar formula determines the reverse reaction rate constant k_-. After integration
over time, Equation (3.63) can be rewritten as the limit

$$k_+ = \lim_{t \to \infty} J_+(t) \tag{3.64}$$

of the reactive flux

$$J_+(t) = \frac{\langle \chi_R(t) \dot{\chi}_R(0) \rangle^{\mathrm{eq}}}{P_R^{\mathrm{eq}}}. \tag{3.65}$$

Similarly, k_- can be rewritten as a limit of $J_-(t)$ given by a formula analogous to Equation (3.65).

To determine the reactive fluxes explicitly in terms of the dynamics described by Equation (3.50) we write solutions to those equations in terms of conditional probabilities:

$$p_l(t) = \sum_{l_0} p_{l|l_0}(t) p_{l_0}(0). \tag{3.66}$$

The $p_{l|l_0}(t)$ denotes the conditional probability that the molecule will be in substate l at time t if it was in the substate l_0 at time 0. This way we obtain

$$J_+(t) = \sum_{l''} \sum_{l' \in P^\ddagger} \sum_{l \in R^\ddagger} \frac{p_{l''|l}(t) w_{l'l} p_l^{eq}}{P_R^{eq}}. \tag{3.67}$$

The initial value of the expression in Equation (3.67):

$$J_+(0) = \sum_{l' \in P^\ddagger} \sum_{l \in R^\ddagger} \frac{w_{l'l} p_l^{eq}}{P_R^{eq}}. \tag{3.68}$$

coincides with the value of the reaction rate constant provided by the *transition state theory* (Atkins, 1998) that can be seen easily after rewriting it in the form

$$k_+^{eq} = \nu \frac{P_{R^\ddagger}^{eq}}{P_R^{eq}} = \nu e^{-\frac{\Delta G_R^\ddagger}{k_B T}}. \tag{3.69}$$

The ν denotes a *mean frequency of transitions*, P_R^{eq} is the equilibrium occupation of the transition state, and $\Delta G_R^\ddagger$ denotes the *free energy of activation*.

The assumption regarding time-scale separation corresponds to the plateau value behavior of $J_+(t)$ and $J_-(t)$ (see Figure 3.19). Note the possibility of a faster pace of the reaction in the initial stage and the necessity of cutting long-time exponential decay by the appropriate regularization factor in the integral (Equation 3.63). The reactive flux vanishes for $t < 0$ which underlies the necessity for a careful treatment of the lower limit in the integral (Equation 3.63) such that the moment $t = 0$ should be an internal point in the interval of integration. The jump at $t = 0$ is related to a Dirac *delta* component of the time correlation function of fluxes that appears to have the form

$$\frac{\langle \chi_R(t) \chi_R(0) \rangle^{eq}}{\langle \chi_R \rangle^{eq}} = k_+^{eq} \delta(t) + S_+(t). \tag{3.70}$$

Equations (3.63) and (3.70) state clearly that the core of transition state theory is the assumption that the flux $\chi_R(t)$ is δ-correlated white noise. To determine the transition state theory rate constant (Equation 3.69), no knowledge of intramolecular dynamics is needed. The finite correlation-time component $S_+(t)$ in the sum (Equation 3.70) arises from intramolecular dynamical processes.

If the transition states $R^\ddagger$ and $P^\ddagger$ are short-lived intermediates, the reciprocal rate constants can be decomposed into three time components (see Chapter 5):

$$k_+^{-1} = (k_+^{eq})^{-1} + \tau_R + K^{-1}\tau_P \tag{3.71}$$

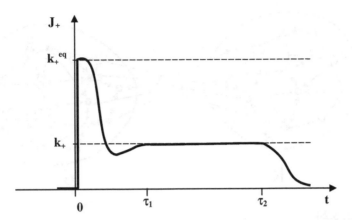

FIGURE 3.19
Variation of the reactive flux over time. The plateau value behavior is characteristic for the activated process of the reaction. The transition state theory approximates the reactive flux time course by the Heaviside step function. The long-term behavior of the reactive flux is drawn on a more compressed time scale.

and similarly for k_-^{-1} related to k_+^{-1} by Equation (3.61). The first component in Equation (3.71) determines the time needed to cross the boundary under the assumption in the transition state theory that $R^{\ddagger}$ is in local equilibrium with the rest of microstates composing R. However, as a result of the transition, this equilibrium is disturbed. The second component in Equation (3.71) determines the time needed for restoring this equilibrium from the side of the R species. The third component determines the time needed for the same process from the side of the P species (recrossing the border).

From Equation (3.71), it follows that k_+^{eq} is always larger than the exact rate constant k_+ (see Figure 3.19). If all the three components in Equation (3.71) are comparable (as for reactions of small molecules in a gas phase), the reaction rate constant is well described by the transition state theory, possibly with a certain transmission coefficient smaller than unity. The initial stage of the reaction is then practically absent. If, on the contrary, the second and the third components prevail, the reaction is considered *controlled* by processes of intramolecular dynamics and the transition state theory fails. In that case, the initial stage of the reaction can even appear to dominate.

The rate constants k_+ and k_- are not the probabilities per unit time of an R molecule making the R→P transition and a P molecule making the P→R transition. The values $k_+ P_R$ and $k_- P_P$ are not the separate P→R and R→P fluxes (Widom, 1965). This holds only for *imagined* irreversible reactions R→P or P→R with *absorbing* boundaries between the R and P subsets of microstates that can be achieved by adding a fictitious totally absorbing *limbo state* (see Figure 3.20).

The stochastic theory of such fictitious or real irreversible reactions is identical to the first passage time problem for corresponding stochastic processes (Montroll and Shuler, 1958; van Kampen, 2001; Gardiner, 1996). For irreversible reactions, the

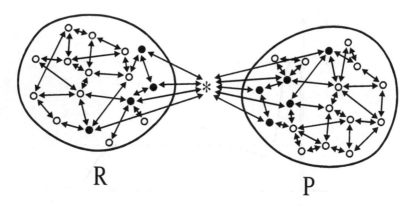

R   P

FIGURE 3.20
**A reversible reaction can be divided into two irreversible reactions after intro-
ducing a fictitious *limbo state**.**

addition of the limbo state introduces the smallest finite value τ_1^{-1} into the spectrum
of reciprocal relaxation times (the dynamics of the set R or P alone is characterized by
the next larger value τ_2^{-1}; see Figure 3.20). We will later consider only the irreversible
reaction R $\rightarrow$ P since the reaction P $\rightarrow$ R is analogous.

By definition, transition probabilities per unit time from the limbo state * to any
microstate l in the transition state $R^{\ddagger}$ vanish:

$$w_{l,*} = 0. \tag{3.72}$$

Consequently, the occupation probability of the limbo state tends asymptotically over
time to unity:

$$\lim_{t\to\infty} p_*(t) = 1. \tag{3.73}$$

In the presence of the limbo state, the quantity

$$P(t|l_0) = \sum_{l\in R} p_{l|l_0}(t) = 1 - p_{*|l_0}(t) \tag{3.74}$$

determines the *survival probability* in R through time t, i.e., the probability that at
time t the system started from state l_0 is still in R. In various contexts time t in
Equation 3.74 is known as *dwell time* in R, *waiting time* for transition to P, *first exit
time* from R, or *first passage time* to the limbo state. The $1 - P(t|l_0)$ is the cumulative
probability that the first passage time is shorter than t, thus its derivative:

$$-\dot{P}(t|l_0) = f(t|l_0) \tag{3.75}$$

represents the *first passage time distribution density*. After the density is defined, we
can calculate the *mean first passage time*:

$$\tau(l_0) = \int_0^\infty dt\, t\, f(t|l_0) = -\int_0^\infty t\, dt \frac{dP(t|l_0)}{dt} = \int_0^\infty dt\, P(t|l_0) \tag{3.76}$$

provided it is finite. In the last equality in Equation (3.76), we applied integration by parts.

The mole fraction $P(t)$ of the molecules R that survives through time t is the survival probability $P(t|l_0)$ averaged over the initial distribution of states $p_{l_0}(0)$:

$$P(t) = \sum_{l \in R} p_{l_0}(0) P(t|l_0). \tag{3.77}$$

Following Equation (3.75), the most general equation determining the time variation of $C(t)$ has the form

$$\dot{P}(t) = -f(f), \tag{3.78}$$

where $f(t)$ denotes the first passage time distribution density to the limbo state averaged over the initial distribution of conformational substates $p_{l_0}(0)$. In general, the average survival probability P does not obey, at least in the beginning, a kinetic equation. However, one can always formally determine a *time-dependent rate parameter* $k(t)$ through the equation:

$$\dot{P}(t) = -f(t) = -k(t)P(t). \tag{3.79}$$

Equivalently

$$k(t) = \frac{f(t)}{P(t)}. \tag{3.80}$$

If the reaction considered is the activated process, $k(t)$ in Equation (3.80) reaches the long-lasting stationary value:

$$k = \frac{f(t)^{st}}{P(t)^{st}}. \tag{3.81}$$

The flux overpopulation formula (Equation 3.81) is usually simpler in application than the time correlation function formula (Equation 3.63) that requires calculation of the full reactive flux (Equation 3.67). This method was used in the pioneering work of Smoluchowski and Kramers (Hänggi et al., 1990).

Equation (3.81) includes crossing the boundary on assuming the local equilibrium conditions and restoring this equilibrium from the R side. It neglects the process of recrossing the boundary (compare Equation 3.71). We can take into account the effects of the latter process on considering the reverse irreversible reaction P $\rightarrow$ R. Because the forward and reverse reaction rates of the transition state theory are related by Equation (3.61) the reciprocal reaction rates for both irreversible reactions are of the forms

$$k_R^{-1} = (k_R^{eq})^{-1} + \tau_R, \qquad k_P^{-1} = K(k_R^{eq})^{-1} + \tau_P. \tag{3.82}$$

Knowing k_R^{eq}, we can express τ_R and τ_P in terms of k_R and k_P. After substitution to Equations (3.61) and (3.71), we obtain complete reciprocal reaction rate constants:

$$k_+^{-1} = k_R^{-1} + K^{-1}k_P^{-1} - (k_R^{eq})^{-1} \tag{3.83}$$

and

$$k_-^{-1} = k_P^{-1} + K k_R^{-1} - K (k_R^{eq})^{-1}. \tag{3.84}$$

3.4   Conformational transitions of proteins

The control of biochemical reactions by the intramolecular dynamics of transitions among a number of conformational substates of proteins follows from observations of nonexponential initial reaction stages. The first historically important experiment performed a quarter century ago by Frauenfelder and coworkers (Austin et al., 1975; Frauenfelder et al., 1991) concerned the kinetics of ligand binding to myoglobin.

Myoglobin, like a more complex hemoglobin, is a protein that stores molecular oxygen. The replacement of oxygen with carbon monoxide (CO) poisons the organism. This is related to the fact that the CO binding reaction, as opposed to the O_2 binding one, is irreversible (see Figure 3.21a). The researchers broke up the heme—CO bond in a nonthermal way using a laser flash and observed ligand rebinding to heme under various conditions after photolysis (see Figure 3.21b). At 300 K, only the bimolecular reaction of binding from the solution was observed with a normal exponential time course. The essential novelty of the experiment was the study of the process at low nonphysiological temperatures. Under such conditions, the curve of the time course of the bimolecular reaction revealed the evidently nonexponential time dependence of the unimolecular reaction of ligand binding from the protein matrix.

A second type of experiment involved a powerful technique called patch clamp (Sackmann and Naher, 1985) and revealed fluctuations of ionic current flowing through single protein channels (Figure 3.22). It appeared that most channels occurred in two discrete states: open and closed. Tracking opening and closing times often showed nonexponential distribution density (Sansom et al., 1989).

In standard kinetic experiments with ensembles of molecules, the initial distribution of microstates is not specifically prepared and usually not much different from the local equilibrium distribution. This leads to the absence of a preexponential stage of the reaction, even if the reaction rate is controlled by intramolecular dynamics. However, in experiments with the patch clamp technique, a single protein channel molecule can be seen stochastically changing its state from open to closed. As a result, experiments with single molecules yielded first passage time distribution densities $f(t)$ separately for the forward and backward reactions, each treated formally as irreversible. After each reactive transition, the molecule starts its microscopic evolution from a conformational substate within the transition state of the return reaction.

The initial distribution of conformational substates confined only to the transition state is attained also in the first experiment in which an ensemble of molecules initially in a thermodynamically stable state is nonthermally excited to the unstable state P. The presence of nonexponential initial stages in the time courses discussed implies that some, if not all, biochemical processes are controlled by the intramolecular dynamics of the proteins involved.

(a)   (b)

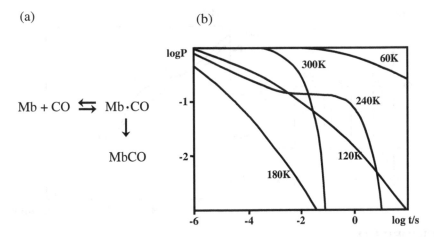

$$Mb + CO \rightleftharpoons Mb \cdot CO$$
$$\downarrow$$
$$MbCO$$

FIGURE 3.21
(a) Irreversible carbon monoxide binding to myoglobin consists of two steps: a reversible bimolecular reaction of ligand adsorption from the solution and an irreversible unimolecular reaction of the ligand covalent binding to heme from the protein interior. (b) Draft of the time dependence of rebinding of CO molecules after photodissociation of CO-bound sperm whale myoglobin at various temperatures, after Austin et al. (1975), $P(t)$ represents the fractions of the myoglobin molecules that have not rebound CO at time t after the laser flash. At low temperatures, only the unimolecular reaction of CO rebinding from the protein interior is observed. Its time course is evidently nonexponential. The exponential stage observed at 240 K and higher temperatures is attributed to the bimolecular reaction of CO rebinding from the solution. The biomolecular reaction masks the exponential stage of the unimolecular reaction of CO rebinding from the protein interior (incomplete masking of horse myoglobin was observed by Post et al., 1993).

Because experiments to date have not elucidated the nature of the conformational transition dynamics within the protein native state in detail, the problem of modeling the dynamics is to an extent open to speculation. In two classes of models cited in the literature, the speculative element appears within reasonable limits. We refer to them as the *protein–glass* and *protein–machine* models (Kurzynski, 1998).

In essence, the question concerns the form of the reciprocal relaxation time spectrum above the gap (Figure 3.23). The nonexponential time courses of the processes discussed indicate the spectrum is quasicontinuous, at least in the range from 10^{-11} to 10^{-7}s. The simplest way to approach problems without a well defined time scale separation is to assume that the dynamics of a system looks alike on all time scales, i.e., the spectrum of relaxation times has a self-similarity symmetry. This assumption is the core of any protein–glass model (Figure 3.23a).

An alternative is the protein–machine class of models in which the variety of conformations composing the native state is supposed to be labeled with only a few *mechanical* variables. Then, the reciprocal relaxation time spectrum is a sum of several more or less equidistant subspectra (Figure 3.23b).

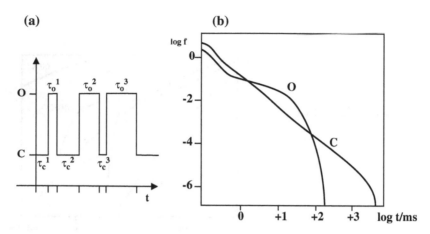

FIGURE 3.22
(a) Telegraphic noise recorded via the patch clamp technique. The ionic
current flowing through a single protein channel fluctuates between the two
values corresponding to chemical states of the protein: open (O) and closed (C).
(b) Draft of the time dependence of closed time and open time distribution density $f(t)$, observed with the patch clamp technique for the protein K^+ channel of
NG 108-15 cells. Both curves show short-term nonexponential behavior. (After
Sansom et al., 1989.)

3.4.1 Protein–glass model

Time scaling, considered a generic property of glassy materials, can originate from
a hierarchy of barrier heights in a potential energy landscape or from a hierarchy of
bottlenecks in the network-joining conformations between which direct transitions
take place. A hierarchy of interconformational barrier heights (Figure 3.24) was
proposed over 10 years ago by Frauenfelder et al. (1991), to combine the results of
various experiments involving ligand binding to myoglobin. A reasonable mathematical realization of such a hierarchy in the context of applications to proteins are spin
glasses (Stein, 1992).

   The mathematical realizations of hierarchical networks are lattices with effective dimensions between 1 and 2, e.g., geometrical fractals or percolation lattices
(Kurzynski et al., 1998). The process of diffusion on fractal lattices can (but does
not have to) be interpreted as simulating structural defect motions in the liquid-like
regions between solid-like fragments of the secondary structure.

3.4.2 Protein–machine model

This model was proposed 30 years ago by Chernavsky, Khurgin and Shnol (1987).
While speculative when proposed, the model has been justified experimentally
(Kurzynski, 1997). Simply put, mechanical variables can be identified by the

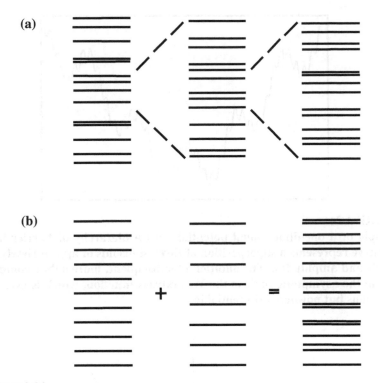

FIGURE 3.23
Spectra of reciprocal relaxation times of conformational transition dynamics within the native state of protein. (a) Protein–glass model. The spectra look somewhat alike on several successive time scales. (b) Protein–machine model. The spectra consist of a few more or less equidistant subspectra.

angles describing the mutual orientations of rigid fragments of the secondary structure or larger structural elements (see Figure 3.25).

A mechanical coordinate may be also identified with a reaction coordinate, if one can be determined. Strong evidence indicates that the mechanical coordinates are related to the "essential modes" of motion studied extensively in recent years with the help of molecular dynamics simulations (Kitao and Go, 1999). In the continuum limit, successive conformational transitions along a given mechanical coordinate are to be approximated by diffusion in an effective potential; the simplest is parabolic (Agmon and Hopfield, 1983; Kurzynski, 1997). The reciprocal relaxation time spectrum for such a model is exactly equidistant (see Figure 3.23b).

Each model of conformational transition dynamics discussed may be true to an extent, but one class of model of the protein–glass type, namely the random walk on fractal lattices, seems to offer the greatest theoretical power.

FIGURE 3.24
Example of a one-dimensional potential with a hierarchy of barrier heights.
The curve represents a superposition of three sinusoids of appropriately scaled
periods and amplitudes. The addition of subsequent, more subtle components
leads in the asymptotic limit to the Weierstrass function, which is everywhere
continuous but nowhere differentiable.

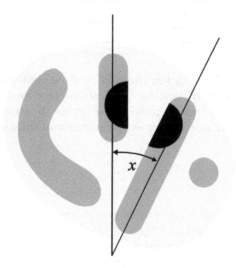

FIGURE 3.25
Cross-section of the fundamental structural unit of protein domain. The meaning of shading as on Figure 3.15. In models of the protein–machine type, the
dynamics of conformational transitions is treated as a quasi-continuous relative
diffusion of solid-like elements along a mechanical coordinate, identified here by
the angle x. (After Kurzynski, 1997.)

3.5 Models of random walks on fractal lattices

A set of points is called a *lattice* if one can define for this set the notion of the nearest neighborhood. A stochastic process of values in a certain lattice is designated a *random walk* or *diffusion* (see Appendix A) if the only transitions possible are between nearest neighbors; otherwise the designation is a *random fly*.

Figures 3.26 and 3.27 show fractal lattices (see Appendix B) with hierarchies of bottlenecks: the planar Sierpinski gasket and the planar percolation cluster. A *fractal* is defined as an object with a fractional value of the fractal dimension. The hierarchical properties of lattices relate to the spectral rather than fractal dimension. Both concepts are clearly worth an explanation.

The notion of the *fractal (Hausdorff-Besicovitch) dimension d* of a given lattice is simple (Mandelbrot, 1988). It is the exponent in the power law determining how the number of sites n changes with the scale (size) s:

$$n = s^{\bar{d}}. \tag{3.85}$$

Consequently

$$\bar{d} = \frac{\log n}{\log s}. \tag{3.86}$$

For the planar Sierpinski gasket shown in Figure 3.26, the two-fold change of the scale entails a three-fold increase of the number of sites thus $\bar{d} = \log 3 / \log 2 \approx 1.585$.

In addition to *geometrical* fractals, one can define *statistical* fractals whose structures are obtained through stochastic processes instead of strict algorithms. An example of a statistical fractal is a percolation cluster (Figure 3.27). Such clusters offer better insights into the conformational dynamics of proteins (Kurzynski, 1997a).

The concept of *spectral* or *fracton dimension* is more complex (Nakayama et al., 1994). It resorts to the functional dependence of the density of vibrational normal modes versus frequency when a given lattice is considered to consist of massive points with an elastic coupling between the nearest neighbors. The Hamiltonian dynamics of a system of coupled harmonic oscillators is described generally as

$$\dot{a} = -i\Omega a, \tag{3.87}$$

where a is a vector of complex numbers with real and imaginary parts corresponding to positions and momenta, respectively, of particular harmonic oscillators:

$$a_l = \frac{1}{\sqrt{2}}(q_l + ip_l) \tag{3.88}$$

FIGURE 3.26
Planar Sierpinski gasket. Three small equilateral triangles are combined into a larger triangle, three larger triangles into an even larger one etc. The Sierpinski gasket of a finite order is shown with imposed periodic boundary conditions (identification of outgoing bonds of one external vertex with those incoming to two other external vertices).

and Ω is the eigenfrequency matrix. In the coordinates of the normal modes of vibrations, the frequency matrix becomes diagonal and the set represented by Equation (3.87) is decoupled into a set of independent equations:

$$\dot{a}_k = -i\omega_k a_k. \tag{3.89}$$

If the density of vibrational modes in the spectrum of frequencies ω behaves regularly, according to a certain power law

$$\rho(\omega) \propto \omega^{\tilde{d}-1}, \tag{3.90}$$

the number $\tilde{d}$ is referred to as the spectral dimension of the lattice. The relation in Equation (3.90) can be considered a generalization of the Debye relation for acoustic phonons in crystal lattices of the integer Euclidean dimension $\tilde{d}$ (Kittel, 1996). The normal modes of vibration in lattices of a fractional dimension $\tilde{d}$ are referred to as *fractons* (Nakayama et al., 1994).

The set of master equations (3.50) describing a random walk on a given lattice can be rewritten in a form analogous to Equation (3.87):

$$\dot{P} = -\Gamma(p - p^{eq}). \tag{3.91}$$

Γ is the matrix composed of transition probabilities w. The corresponding set of decoupled equations for the relaxational normal modes reads (cf. Equation 3.52):

$$\dot{p}_k = -\gamma_k(p_k - p_k^{eq}). \tag{3.92}$$

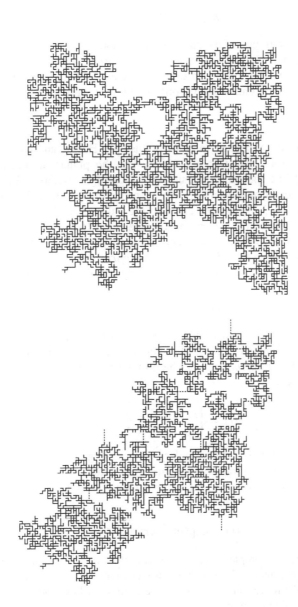

FIGURE 3.27
Percolation cluster. Bonds on a square lattice are achieved stochastically with the probability of one half. Clusters not connected to the largest one are removed. Note the hierarchical structures of bottlenecks resulting in time scaling: the equilibration is completed first within subclusters of a lower order and only then, on a longer time scale within the subclusters of a higher order. A finite number of conformational substates in real proteins needs the hierarchy to be bounded both from above and below (with the imposed periodic or reflecting boundary conditions).

There are, in fact, twice as many Equations (3.89) as Equations (3.92) because a_ks are complex variables and p_k s are real. Thus, following Equation (3.90), the density of relaxational modes in the spectrum of reciprocal relaxation times γ should behave as:

$$\rho(\gamma) \propto \gamma^{\frac{\tilde{d}}{2}-1}. \tag{3.93}$$

This is an alternative definition of the spectral dimension.

Time scaling takes place only if the density of relaxational normal modes increases with a decreasing reciprocal relaxation time γ (Figure 3.23a). Consequently, the hierarchy of bottlenecks is characteristic only of lattices with spectral dimensions smaller than 2:

$$\tilde{d} < 2. \tag{3.94}$$

For a Sierpinski gasket embedded in a d-dimensional Euclidean space, $\tilde{d} = \log(d + 1)/\log(d + 3)$ (Nakayama et al., 1994). For the planar Sierpinski gasket ($d = 2$, Figure 3.26), the spectral dimension $\tilde{d} = \log3/\log5 \approx 1.365$. The spectral dimension of any percolation cluster (in particular, one embedded in 2-dimensional Euclidean space, Figure 3.27) is very close to the value $\tilde{d} = 4/3$ supporting the so-called Alexander–Orbach conjecture (Nakayama et al., 1994).

The spectral dimension influences two very important physical quantities. The first is the *probability to return to the original point of departure*, which in the case of free diffusion (without boundary conditions), behaves asymptotically in time as

$$p_{0|0}(t) \propto t^{-\frac{\tilde{d}}{2}}. \tag{3.95}$$

This equation is a generalization of the well known result for free diffusion in Euclidean spaces (van Kampen, 2001). The second quantity is the *mean number of distinct sites visited by a random walker*, which in the case of free diffusion behaves asymptotically in time as

$$S(t) \propto \begin{cases} t^{\frac{\tilde{d}}{2}} & \text{if } \tilde{d} < 2 \\ t & \text{if } \tilde{d} > 2 \end{cases}. \tag{3.96}$$

The preexponential initial stage of a large class of gated reactions with the intramolecular conformational transition dynamics of a random walk type on fractal lattices and the initial substate reduced to its very gate is well described by a simple expression (Kurzynski et al., 1998)

$$P_{\text{ini}}(t) = \exp(\eta t)^{2\alpha} \text{erfc}(\eta t)^{\alpha} \tag{3.97}$$

in which erfc denotes the complementary error function, η^{-1} is a characteristic unit of time, and α is an exponent whose value is lower than unity. In the limit of short times, Equation (3.97) represents the stretched exponential law, and in the limit of long times, the algebraic power law:

$$C_{\text{ini}}(t) \approx \begin{cases} \exp[\frac{-2(\eta)^{\alpha}}{\sqrt{\pi}}] & \text{for } t \ll \eta^{-1} \\ \frac{(\eta t)^{-\alpha}}{\sqrt{\pi}} & \text{for } t \gg \eta^{-1} \end{cases}. \tag{3.98}$$

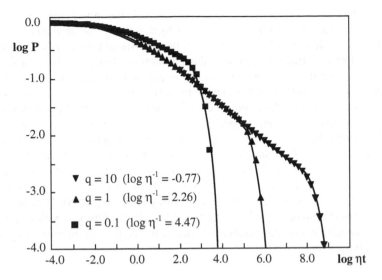

FIGURE 3.28
Fit of the results of computer simulations of a random walk on the planar Sierpizoki gasket with periodic boundary conditions (Figure 3.27) to the combined analytical formulae [Equations (3.97) and (3.99)]. The fitted curves are plotted as continuous lines. The simulation data are represented by points. Not involved in the fitting procedure, the fixed value of $\alpha = 1 - \log 3 / \log 5 = 0.317$ was assumed. Different values of the ratio q of the probability of leaving the lattice to the probability of transition between the neighboring sites determine different values of the time unit η^{-1}. (After Kurzynski et al., 1998.)

The crossover from nonexponential decay to the exponential decay with the chemical relaxation time κ^{-1} equal to the reciprocal of the irreversible reaction rate constant is related to the finite number of conformational substates and can be described with the help of the formula

$$C(t) = [(1 - a)C_{\text{ini}}(t) + a]e^{-\kappa t} \qquad (3.99)$$

with a denoting the level (concentration) where exponential decay begins.

Equation (3.97), with the exponent $\alpha = 1/2$, is the exact solution of the continuous one-dimensional problem (Kurzynski, 1998). For $\alpha > 1/2$, it was carefully verified by computer simulations of random walks on fractal lattices (Kurzynski et al., 1998). The α exponent was found to be related to the spectral dimension of the lattice $\tilde{d}$:

$$\alpha = 1 - \frac{\tilde{d}}{2} \qquad (3.100)$$

The approximation of the results of simulations with the help of the combined analytical formulae [Equations (3.97) and (3.99)] is very good (Figure 3.28).

Depending on the value of the time constant η^{-1}, the initial stage of the reaction described by Equation (3.97) can proceed by following the stretched exponential law, the algebraic power law, or both. All three types of behavior were observed in CO-rebinding experiments after laser flash photolysis (Figure 3.21) and patch clamp experiments (Figure 3.22). Apart from the α exponent, Equations (3.97) and (3.99) comprise two dimensionless parameters, the level (concentration) a related somehow to the conformational relaxation time γ^{-1}, determined by the fractal (not spectral) dimension of the lattice and responsible for a possible plateau preceding exponential decay, and the ratio $b \equiv \kappa/\eta$. Both parameters depend on temperature via the Arrhenius relation. One should have no problem describing in such terms a time course of any experimentally observed reaction including its variation with temperature, especially when taking into account certain time variations of α and β allowed by the model of a somewhat extended gate (Kurzynski et al., 1998). However, the success of the fit should be treated very cautiously, as similar fitting capacities are yielded by the alternative one-dimensional model with fluctuating barriers (Agmon and Sastry 1996).

3.6 Introduction to polymer biophysics

As discussed earlier in Chapter 2, the molecular components of most structural elements in cells contain polymers in the form of linear chains composed of amino acids (peptides), nucleic acids (DNA and RNA), or phospholipids in membranes. These chains may form more complicated geometries by folding into three-dimensional globules or forming sheets. To understand the processes leading to biomolecular structure formation, it is important to discuss the principles of polymer biophysics.

A flexible linear chain may exhibit various three-dimensional configurations that can be characterized by several physical parameters. Describing each of the N segments of a chain of length a_i and a given direction, we use the vector $\vec{a}$ and construct an end-to-end displacement:

$$\vec{r}_{ee} = \sum_{i=1}^{N} \vec{a}_i \tag{3.101}$$

Another quantity that characterizes the position of a chain is the center of a mass vector given by:

$$\vec{r}_{cm} = \sum_{i} \frac{\vec{r}_i}{N+1} \tag{3.102}$$

where $\vec{r}_i$ is the position vector of each of the $N+1$ vertices in the chain. A measure of the spatial extent of the chain is the so-called radius of gyration defined with respect to the center-of-mass position vector $\vec{r}_{cm}$ as:

$$R_g^2 = \frac{\sum_{i=1}^{N+1}(\vec{r}_i - \vec{r}_{cm})^2}{N+1} \tag{3.103}$$

For an absolutely straight chain of N segments, we have $r_{ee} = Na$ where a is the length of each segment. On the other hand, for a random chain, we calculate its end-to-end displacement according to:

$$\langle r_{ee}^2 \rangle = \sum_i \sum_j \langle \vec{a}_i \vec{a}_j \rangle \tag{3.104}$$

For a random orientation, the scalar product average in the expression above vanishes for i different from j. The only terms in the sum that survive the averaging process are the diagonal ones. As a result, we obtain:

$$\langle r_{ee}^2 \rangle = Na^2 \tag{3.105}$$

a relationship that results in the so-called ideal scaling law where $\langle r_{ee}^2 \rangle^{1/2}$ proceeds as $\sqrt{N}$.

Polymer chains have an additional property called self-avoidance that prevents their displacement vectors from crossing one another and producing an excluded volume effect. In 1953, Flory proposed a model to determine the scaling exponent between the end-to-end distance and the number of monomers in the chain for self-avoiding chains. First, the power law dependence of the free energy of the effective chain size r and the number of segments N is calculated. The value of r that minimizes free energy is found as a function of N. For short distances, steric repulsion between the segments causes the chain to swell. This repulsive energy is proportional to the product of the concentration of segments and the volume over which they interact. Denoting the interaction volume by v_{ex}, it can be shown that the free energy at short distances scales as:

$$F_{short} = kT v_{ex} \frac{N^2}{r^d} \tag{3.106}$$

where k is the Boltzmann constant and d the dimension of the physical space. As the chain is stretched, the probability of finding a given end-to-end distance r decreases exponentially as:

$$e^{-\frac{dr^2}{2Na^2}} \tag{3.107}$$

Since entropy S is proportional to the logarithm of probability, we find that, up to a multiplicative constant:

$$\frac{S}{k} = -\frac{dr^2}{2Na^2} \tag{3.108}$$

TABLE 3.2
Scaling Exponents for the Relationship of $\langle R_g^2 \rangle^{1/2}$ and N^ν for Various Geometries and Dimensionalities

Configuration	$d = 2$	$d = 3$	$d = 4$
Ideal chains	0.5	0.5	0.5
Self-avoiding chains	0.75	0.59	0.5
Branched polymers	0.64	0.5	
Collapsed chains	0.5	0.33	0.25

Consequently, we find the free energy in the long distance regime is:

$$F_{long} = \frac{kT dr^2}{2Na^2} \tag{3.109}$$

Combining the two expressions for the free energy in the two regimes, we find that the free energy behaves according to:

$$F = kT \left(\frac{v_{ex} N^2}{r^d} + \frac{dr^2}{2Na^2} \right) \tag{3.110}$$

We can then minimize it with respect to the chain radius and obtain the scaling behavior for a self-avoiding chain governed by:

$$R \sim N^{\left(\frac{3}{2+d}\right)} \tag{3.111}$$

The $3/(2 + d)$ operation is called the Flory exponent. Table 3.2 summarizes the ν exponents in the scaling law $\langle R_g^2 \rangle^{1/2}$ proportional to N^ν. Since the branched polymers have more than two ends, their measure of spatial extent is the radius of gyration. Collapsed objects occupy the smallest volume possible and they are also included in the Table 3.2.

To discuss elastic properties of polymer chains, we must introduce several new physical properties. The effective bond length B_{eff} is defined as:

$$\langle r_{ee}^2 \rangle^{1/2} = B_{eff} N^{1/2} \tag{3.112}$$

This quantity describes the level of alignment between the various segments. If θ is the bond angle between two subsequent segments that can freely rotate about their axis, then it can be demonstrated that:

$$\langle \vec{a}_i \vec{a}_{i+k} \rangle = a^2 (-\cos \theta)^k \tag{3.113}$$

and consequently it follows for the end-to-end distance that:

$$\langle r_{ee}^2 \rangle = Na^2 \frac{(1 - \cos \theta)}{(1 + \cos \theta)} \tag{3.114}$$

TABLE 3.3
Three Parametrizations for Characteristic Polymer Lengths and Relationships

Parameterization	$\langle r_{ee}^2 \rangle$	L_c
Effective bond length B_{eff}	NB_{eff}^2	N_a
Kuhn Length L_K	$N_K L_K^2$	$N_K L_K$
Persistence length ξ	$2N_p \xi^2$	$N_p \xi$

Note: N_K is the number of Kuhn lengths in the contour.

It is easy to see that the effective bond length for freely rotating chains is given by:

$$B_{eff} = a \left[\frac{1 - \cos \theta}{1 + \cos \theta} \right]^{1/2} \tag{3.115}$$

Another measure of the effective segment size is the distance along the chain over which the orientation of the bond vector becomes uncorrelated. This is called the persistence length ξ. Defining the unit vector along the segment as:

$$u_i = \frac{\vec{a}}{|\vec{a}_i|} \tag{3.116}$$

the persistence length ξ can be defined as:

$$\langle u(s)u(0) \rangle = e^{-\frac{s}{\xi}} \tag{3.117}$$

where s is the length coordinate along the chain. The persistence length can be related to the contour length L_c and the end-to-end distance via:

$$\langle r_{ee}^2 \rangle = 2\xi L_c - 2\xi^2 [1 - e^{-\frac{L_c}{\xi}}] \tag{3.118}$$

which, in the limit of a short persistence length, is compared to the contour length via the relationship:

$$\langle r_{ee}^2 \rangle = 2\xi L_c \tag{3.119}$$

This leads to a third characterization of polymer chains, the so-called Kuhn length L_K defined by:

$$L_K = \frac{\langle r_{ee}^2 \rangle}{L_c} \tag{3.120}$$

Table 3.3 relates all three parameterizations one to another:

3.6.1   Elastic properties of polymers

We treat polymers as elastic rods whose bending energy is given by a quadratic functional in terms of $\frac{\partial \vec{u}}{\partial s}$ such that:

$$E_{bend} = \frac{YI}{2} \int_0^{L_c} ds \left[\frac{\partial \vec{u}}{\partial s} \right]^2 \tag{3.121}$$

where bending resistance or flexural rigidity is the product of Young's modulus Y and the geometrical moment of inertia I. The moment of inertia is given by:

$$I = \int y^2 \, dA \tag{3.122}$$

where y is the distance in the direction of the curvature and dA is an infinitesimal area element. For a uniform rod of radius R, $I = \frac{1}{4} \pi R^4$. Assuming thermal fluctuations with a probability governed by the Boltzmann factor leads to the conclusion that the persistence length is long at low temperatures and short at high temperatures with a functional dependence on YI given by:

$$\xi = \frac{YI}{kT} \tag{3.123}$$

Chain size is determined by a combination of elastic energy and entropic effects because $F = E - TS$. Far more chain configurations have end-to-end displacements close their to r_{ee} than close to their contour length L_c. The entropy is proportional to the logarithm of this number and it is reduced when the chain is stretched. Using a Gaussian probability distribution, the effective spring constant of a random chain in three dimensions is:

$$K_{sp} = \frac{3kT}{Na^2} \tag{3.124}$$

and it increases linearly with temperature T providing more resistance to stretching as the polymer is heated. The force f required to produce an extension x in the end-to-end displacement, according to the Gaussian approximation, is given by:

$$f = \left(\frac{3kT}{Na^2} \right) x \tag{3.125}$$

For flexible filaments, a more appropriate representation is the so-called worm-like chain picture for which the force extension formula is more complicated, namely:

$$f = \frac{kT}{4} \frac{[(1 - \frac{x}{L_c})^{-2} - \frac{1}{4} + \frac{x}{L_c}]}{\xi_p} \tag{3.126}$$

where ξ_p is the persistence length.

3.7 Cellular automata

Cellular automata are computer simulations that try to emulate the way laws of nature are supposed to work algorithmically. More specifically, a cellular automaton (CA) is an array of identically programmed cells that interact via a set of rules. The arrays usually form a 1-dimensional string of cells, a 2-dimensional grid, or a 3-dimensional solid. Most often, cells are arranged as a simple rectangular grid, but other arrangements, such as a honeycomb, are sometimes used. The essential features of a cellular automaton are:

1. A *state* is a variable that takes a different value for each cell. This can represent a number or a property. If each cell represents part of a landscape, the state might represent the number of objects at each location.
2. Its *neighborhood* is a set of cells with which a cell interacts. In a grid, they are normally the cells physically closest to the cell in question. The following represent simple neighborhoods (cells marked n) of a cell (C) in a 2-dimensional grid:

```
n       nnn     nnnn
nCn     nCn     nCnnnn
n       nnn     nnnn
```

3. Its *program* is the set of rules that defines how its state changes in response to its current value and that of its neighbors.

Cellular automata have the following properties:

1. *Self-organization.* When we plot successive states, a geometric pattern emerges. Even if the line of cells starts with a random arrangement of states, the rules force patterns to emerge.
2. *Life-like behavior.* Empirical studies by Wolfram and others show that even very simple linear automata behave in ways reminiscent of complex biological systems. For example, the fate of any initial configuration of a cellular automaton is to die out, become stable or cycle with fixed periods, grow indefinitely at fixed speed, or grow and contract irregularly.
3. *Thermal behavior.* Models that force a change of state for few configurations tend to freeze into fixed patterns, whereas models that change the cell state in most configurations tend to behave in a more active "gaseous" way; fixed patterns do not emerge.

3.7.1 Conway's game of life

The Game of Life invented by Cambridge mathematician John Conway originally began as an experiment to determine whether a simple system of rules could create a universal computer. Alan Turing invented the universal computer concept to denote a machine capable of emulating any kind of information processing by implementing a small set of simple operations. The game is a simple 2-dimensional analog of basic processes in living systems. It involves tracing changes through time in the patterns formed by sets of living cells arranged in a 2-dimensional grid.

Any cell in the grid may be in one of two states: alive or dead. The state of each cell changes from one generation to the next, depending on the state of its immediate neighbors. The rules governing these changes are designed to mimic population change:

1. A living cell with only one or no living neighbors dies from isolation.
2. A living cell with four or more living neighbors dies from overcrowding.
3. A dead cell with three living neighbors becomes alive.
4. All other cells remain unchanged.

Astonishingly complex patterns are created by these simple rules. Conway's prediction that the system was capable of computation was confirmed by the use of a "glider" or pattern that moved diagonally over the board to represent bit streams. The behavior was typical of the way many cellular automata reproduce features of living systems. That is, regularities in the model tend to produce order. Starting from an arbitrary initial configuration, order usually emerges quickly. Most configurations ultimately disappear entirely or break up into isolated patterns that are static or cycle among several different forms with fixed periods.

Even more fascinating are the similarities of cell patterns and organic life. Another of Conway's goals was to find self-reproducing organisms within a life system — agglomerations of cells that split and form several new organisms identical to the original over time. No such cell patterns have been found. We can imagine such organisms developing through evolutionary mechanisms and evolving into progressively fitter and more complex life forms.

The "game" of life is in the border area between chaos and order. Any change of the rules results in a static universe where cells die out or fill the entire game board or in a chaotic universe where no structures or patterns can be distinguished. The initial configurations tend to evolve into several periodic or static organisms, but the system is chaotic in that it is very sensitive to changes in the environment. Organisms are very unstable when subjected to small changes.

What is interesting about the "game" of life is that it tends to spontaneously develop organisms consisting of several cells in the same way natural laws seem to lead to more complex organisms. Higher life forms would never have developed in a universe with natural laws like the rules leading to chaos. The parallels to the development of life on Earth are intriguing.

In 1982, Stephen Wolfram set out to create a simpler, one-dimensional system. The main advantages of a one-dimensional automaton are that: (1) changes over time can be illustrated in a single, two-dimensional image and (2) each cell only has two neighbors. For a cellular automaton based on a three-cell local rule, if the state of a cell is dependent on its own and its two neighbors' states in the previous generation, it has eight configurations for which a result must be specified, for a total of 256 rules that make it possible to examine all the results of all rule sets. However, the sets are usually reduced to a total of 32 "legal" rules arising from the fact that cellular automata were used initially for modeling biological processes such as reproduction. The question was whether such a simplified system could display complex behavior. The answer found is: yes, it can.

All legal rules can be divided into four types according to their behavior when seeded with a single cell:

Type 1 rules disappear over time.
Type 2 rules merely copy the nonzero cell forever.
Type 3 rules yield completely uniform rows or pairs of rows.
Type 4 rules develop nontrivial patterns.

Eight rules are in each type. As in the Game of Life, some rules yielded chaos and some yielded static states. The most interesting rules were, of course, those that produced results between a state vaguely deemed "complex" and a state resembling biological life. Even Wolfram's simple systems proved capable of displaying complex behavior, although less spectacularly than life forms.

Gliders abound in most one-dimensional rule systems. They appear as diagonal lines in an image. Most rule systems create diagonal patterns. Those that produce gliders from everything are obviously those whose next generation only depends on the states of the left and right neighbors in the previous generation. The main reason one-dimensional systems seem so much more trivial and boring than their two-dimensional equivalents is probably that our vision is adapted to two-dimensional pictures. We find it easier to recognize patterns, organisms, and agglomerations in two dimensions. In summary, Wolfram's hypothesis that complexity can arise from almost arbitrarily simple rules seems to be confirmed.

3.8   Bioenergetics: the Davydov model

In general, the following mechanisms of intermolecular energy transfer must be considered in the context of cellular activities:

1. Energy transfer by radiation (discussed later in this chapter)
2. Energy transfer by charged waves (calcium waves or action potential propagation, discussed in Chapter 7)
3. Energy transfer by hopping of individual charges (discussed later in this chapter)
4. Energy transfer by elementary excitations such as phonons or excitons (discussed in this section)

It is generally accepted that energy transfer processes in living matter involve proteins (see Chapter 2). While protein structures are well understood, the same cannot be said about their functions, particularly energy transduction. We know the currency of biochemical energy is the ATP molecule and its analogs such as GTP.

ATP binds to a specific site on a protein, reacts with a water molecule, and releases 0.48 eV of energy in a process called hydrolysis.

While single energetic events pose no serious challenges to the theoretician, energy transport on length scale of protein filaments or DNA strands is still very much an open problem. A.S. Davydov (1982) developed a nonlinear theory of biological energy transfer that focuses on the so-called amide I bond of C$=$O in a peptide chain containing the H—N—C=O group and its vibrational and dipolar coupling. Lomdahl (1984) wrote a review of potential applications of solitons to biology.

Assume amide I vibration is excited on one of the three spines of an α helix (for example by the energy of ATP hydrolysis). The oscillating C$=$O dipole with dipole moment $\vec{d}$ interacts with the dipoles of neighboring peptide groups of the spine. $\frac{2|\vec{d}|^2}{R^3}$ is the interaction energy between two parallel dipoles positioned at a distance R from each other. The energy of the initial excitation does not remain localized; it propagates through the system. The excitation of the amide I vibration results in the deformation of the H—O bond since it is much weaker than the covalent bonds of the α helix.

The deformation of the hydrogen bond can be described as the deformation of a spring of length K. The initial deformation (excitation of the amide I vibration) of a spring will cause deformation of neighboring springs of the same strength and the disturbance will propagate along the chain of coupled springs (H—O bonds) as a longitudinal sound wave. According to Davydov, the longitudinal sound wave couples nonlinearly to the amide I vibration and thus acts as a potential well trapping the energy of the vibration. The energy remains localized and travels along the α helix chain as a soliton. The nonlinear relation between the amide I vibration and the sound wave can be expressed by:

$$\chi = \frac{dE}{dR} \tag{3.127}$$

where E is the amide I excitation energy.

Davydov used a semiclassical approach to model solitons traveling along the single chain of hydrogen-bonded peptide groups. The amide I vibrations were treated quantum mechanically and the longitudinal sound wave was treated classically. In the continuum approximation, this leads to the nonlinear Schrödinger equation:

$$i\hbar\frac{\partial a}{\partial t} + J\frac{\partial^2 a}{\partial^2 x} - E_0 a + \kappa|a|^2 a = 0 \tag{3.128}$$

where $\kappa = \frac{4\chi^2}{K(1-s)}$, $s = \frac{v}{v_s}$, and $v_s = R\sqrt{\frac{K}{M}}$ represents the velocity of sound, $|a(x,t)|^2$ is the probability amplitude that excitation will occur at position x in the chain of hydrogen-bonded peptide groups at time t, $J = \frac{2|\vec{d}|^2}{R^3}$, $E_0 = E - 2J + \frac{1}{2}\int_{-\infty}^{+\infty}[M(\frac{\partial u}{\partial t})^2 + K(\frac{\partial u}{\partial x})^2]dx$ is the excitation energy, M is the mass of the peptide group, E is the amide I excitation energy, and u is the displacement of the peptide group at position x.

For a stationary case, the solution of Equation (3.128) is:

$$a(x,t) = \frac{\chi}{\sqrt{(2KJ)}} \sec h\left[\frac{\chi^2}{KJ}(x - x_0)\right]\exp\left[-i/\hbar\left(E_0 - \frac{\chi^2}{K^2 J}\right)t\right] \tag{3.129}$$

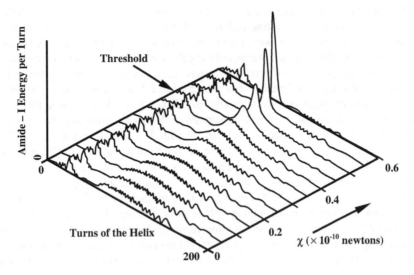

FIGURE 3.29
Total amide I energy (summed over all three spines of the α helix) as a function of χ.

where x_0 is the position of the maximum probability of amide I excitation along the chain. The energy of the stationary soliton is:

$$E_{sol} = E - 2J - \frac{\chi^4}{3K^2 J} \tag{3.130}$$

In Equation (3.129), the amplitude $a(x, t)$ represents a "bump-like" soliton (see Appendix C). In case of an absolutely rigid chain, i.e., for $K \to \infty$, the nonlinear Schrödinger equation reduces to the linear Schrödinger equation whose solutions have the forms of plane waves (Scott et al., 1973). That means that the energy of the amide I vibration is not localized and is uniformly distributed along the whole chain. According to Equation (3.130), the energy in the system is then larger than the soliton energy. Since the system tends to occupy the state with lower energy, the situation with the energy localized in the soliton is more probable. However, in reality the three spines of an α helix interact through the transverse dipole–dipole interaction. This problem can be solved by means of three coupled nonlinear Schrödinger equations.

On the basis of Davydov's ideas it has been suggested that the energy transport in proteins is carried out by localized pulse-like excitations. Some investigations of proteins support these considerations. Numerical computations of Davydov solitons on α helices performed in 1979 by Hyman, McLaughlin, and Scott (1981) revealed that solitons can arise only above some threshold value of the χ parameter. In other words, the nonlinear interaction between the dipole vibration of a peptide group of a spine of an α helix and its deformation caused by the vibration must be sufficiently strong. The results of the calculations are shown in Figure 3.29. We can clearly see how a soliton-like object arises at a χ value of about $0.45 \times 10^{-10} N$.

Another fact that supports the Davydov soliton theory resulted from studies of ACN (acetanilide), a crystalline polymer. Interest in this chemical arose because of the great similarity of its structure to the structure of the α helix. ACN consists of hydrogen-bonded peptide chains held together by Van der Waals forces. In 1984, Careri et al., found in the infrared absorption spectrum of ACN an anomalous line at 1650 cm^{-1} shifted only 15 cm^{-1} from the amide I line at 1665 cm^{-1}. Attempts to explain this phenomenon conventionally failed. Careri et al., suggested that solitons could have been responsible for the observation. It was assumed that the spectral line was caused by nonlinear coupling between the amide I vibration and an out-of-plane displacement of the hydrogen-bonded proton. Such coupling allows soliton excitation by direct electromagnetic radiation. The α helix soliton cannot be excited in this way because the displacement of the whole and thus much heavier peptide group occurs there. A.C. Scott developed his model (1992) in analogy to Davydov's model for the α helix and obtained acceptable values of χ. However, numerical simulations of Lomdahl et al. (1984) showed that the value of the coupling constant may be too small to support a soliton. Other arguments support the explanation of the anomalous line in the ACN absorption spectrum proposed by Careri and Scott. Experiments with polarized light appear to provide evidence for the existence of solitons in ACN crystals. If the line at 1650 cm^{-1} is due to a soliton, it should behave in the same manner as the amide I 1665 cm^{-1} line, e.g., it should have the same polarization. Observations confirmed these expectations.

Attempts have also been made to apply the soliton model of biochemical energy transfer to globular proteins. While globular proteins lack translational symmetry, α-helix protein chains may still allow solitons to travel. It has been argued that the Davydov model can be generalized to this case, taking into account the full geometry of the protein. Promising results of model calculations performed for globular proteins have been obtained, but experimental evidence remains elusive.

3.9 Biological coherence: the Froehlich model

While Davydov tried to find spatial localization of vibrational energy in biological systems such as DNA and peptides, Herbert Froehlich, another famous physicist, sought evidence for frequency selection in biological systems. A conceptual link exists between these two approaches as shown by Tuszynski et al. (1983). It involves self-focusing in reciprocal (Froehlich) or real space (Davydov), respectively.

Starting in the late 1960s and continuing until his death in 1991, Froehlich developed a theory of biological coherence based on quantum interactions between dipolar constituents of biomolecules. He advocated momentum-space correlations within living systems such as enzymes, membranes, cells, and organisms. The dynamic order emerging from such a system would be a characteristic feature distinguishing

DISORDERED STATE (Unpolarized)

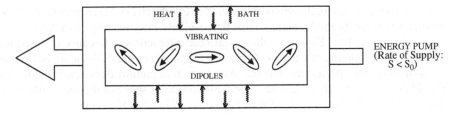

ORDERED STATE (Polarized)

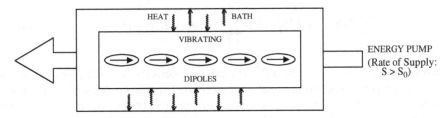

FIGURE 3.30
The Froehlich model in a "nutshell."

living systems from inanimate matter. The key assumptions of Froehlich's theory can be listed as follows (1980):

1. A continuous supply of metabolic energy (energy pumping) above a threshold level
2. Thermal noise due to physiological temperature
3. Internal structural organization that promotes functional features
4. Large transmembrane potential difference
5. Nonlinear interactions between two or more types of degrees of freedom

As a result of these nonlinear interactions, in addition to the global minimum characterizing a biological system in a nonliving state, a metastable energy minimum was predicted to emerge in the living state.

Froehlich's model of biological coherence is based on a condensate of quanta of collective polar vibrations (1968). It is a nonequilibrium property due to the interactions of the system with the surrounding heat bath and an energy supply (see Figure 3.30). The energy is channeled into a single collective mode that becomes strongly excited. Most importantly, the model relies on the nonlinearity of internal vibrational mode interactions and in this respect is somewhat reminiscent of the laser action principle.

Associated with this dynamically ordered, macroscopic quantum state is the emergence of polarization due to the ordering of dipoles in biomolecules such as membrane head groups. Froehlich predicted the generation of coherent modes of excitation such as dipole oscillations in the microwave frequency range. Nonlinear interactions between dynamic degrees of freedom were predicted to result in the local stability of the polarized state and long-range frequency-selective interactions between two identical systems.

In terms of concrete realizations of this model, Froehlich (1972) emphasized dipole moments in many biomolecular systems that oscillate in synchrony in the frequency range of 10^{11} to 10^{12} Hz due to their nonlinear interactions. Because of resonant dipole–dipole coupling in a narrow frequency range, the entire biological system can be seen as a giant oscillating dipole. An alternative picture developed within the Froehlich theory was Bose–Einstein condensation in the space of dipole oscillations. The Hamiltonian postulated by Wu and Austin (1977) takes the form:

$$H = \sum_i \omega_i a_i^\dagger a_i + \sum_i \Omega_i b_i^\dagger b_i + \sum_i \theta_i P_i^\dagger P_i + \frac{1}{2} \sum_{i,j,k} (\chi a_i^\dagger a_j b_k + \chi^* a_j a_i^\dagger b_k)$$

$$+ \sum_{i,j} (\lambda b_i a_j^\dagger + \lambda^* b_i^\dagger a_j) + \sum_{i,j} (\xi P_i a_j^\dagger + \xi^* P_i^\dagger a_j) \qquad (3.131)$$

where $(a_i^\dagger, a_i)$, $(b_i^\dagger, b_i)$, and $(P_i^\dagger, P_i)$ are, respectively, the cell, heat bath, and energy pump creation and annihilation (boson-type) operators. A kinetic rate equation derived for this model indicates Bose-type condensation in the frequency domain with a stationary occupation number dependence of the dipole modes given by:

$$N_i = [e^{\beta(\omega_i - \mu)} - 1]^{-1} \qquad (3.132)$$

The nonlinear coupling comes from the dipole–phonon interaction proportional to χ. Provided that the oscillating dipoles are within a narrow band of resonance frequencies ($\omega_{min} \leq \omega_i \leq \omega_{max}$) and that the coupling constants χ, λ, and ξ are large enough, strong, long-range ($\sim 1\ \mu$m) attractive forces are expected to act between the dipoles. The effective Froehlich potential between any two interacting dipoles that initially vibrate with frequencies ω_1 and ω_2 is:

$$U(r) = -\frac{E}{r^3} + \frac{F}{r^6} \qquad (3.133)$$

where F is the London–van der Walls coefficient and E is the long range constant given by:

$$E = \frac{\hbar e^2 Z |\bar{A}| \gamma}{4 \bar{\omega} M} \left(\frac{1}{\varepsilon'(\omega_+)} - \frac{1}{\varepsilon'(\omega_-)} \right) \qquad (3.134)$$

where M is the mass of a dipole, r is the dipole–dipole spacing, e is the electron charge, Z is the number of elementary charges on each dipole, $\bar{A}$ is an angle constant, and $\varepsilon'(\omega)$ is the real part of the frequency-dependent dielectric constant. The quantities $\omega_\pm$ are the new dipole vibration frequencies:

$$\omega_\pm = \left(\frac{1}{2}(\omega_1^2 + \omega_2^2) \pm \left(\frac{1}{4}(\omega_1^2 - \omega_2^2) + \frac{\beta_0^4}{(\varepsilon'_\pm)^2} \right)^{1/2} \right)^{1/2} \qquad (3.135)$$

where $\beta_0 = \gamma^2 e^2 \frac{\sqrt{Z_1 Z_2}}{Mr^3}$. In the resonant frequency case, the effective interaction energy between two oscillating dipoles was found to be of long-range type depending on the distance as r^{-3}.

Most of the expected condensation of dipolar vibrations was foreseen by Froehlich to occur in cell membranes due to their strong potential on the order of 10 to 100 mV across a thickness of 5 to 10 nm, giving an electric field intensity of 1 to 20×10^6 V/m. The resultant dipole–dipole interactions were calculated to show a resonant long-range order at a high frequency range of 10^{11} to 10^{12} Hz with a propagation velocity of about 10^3 m/s. In addition to membrane dipoles, several other candidates for Froehlich coherence were considered, namely, double ionic layers, dipoles of DNA and RNA molecules, plasmon oscillations of free ions in cytoplasm, etc.

Applications of the Froehlich theory were subsequently made to cancer proliferation where a shift in resonant frequency was seen to affect cell–cell signaling, brain waves, and enzymatic chemical reactions to name but a few areas of investigation.

Over the past two decades, several experiments appeared to demonstrate the sensitivity of metabolic processes to certain frequencies of electromagnetic radiation above the expected Boltzmann probability level. Raman scattering experiments of Webb (1980) revealed nonthermal effects in *E. coli* but could not be reproduced by other laboratories. Irradiation by millimeter waves of yeast cells showed increased growth at specific frequencies (Grundler, 1983). Rouleaux formation of human erythrocytes (Rowlands et al., 1982) was explained in terms of Froehlich's resonant dipole–dipole attraction (Paul et al., 1983) but did not rule out standard coagulation processes. While some experiments illustrate nonthermal effects in living matter that would require nonlinear and nonequilibrium interactions for explanation, to date no unambiguous experimental proof has been furnished to support Froehlich's hypothesis.

3.10 Ionic currents through electrolytes

Electrical current in cells and organisms is not carried by electrons. Instead, it is transported by the mobile ions of electrolytic solutions. The relationship between electromotive force E and electrolytic current I is:

$$E = IR \tag{3.136}$$

i.e., Ohm's Law remains valid for electrolytic conduction. As for metallic conductors, resistance depends on the dimensions of the electrolytic cell. If two poles of an electrolytic cell are in the form of two parallel plates placed a distance L apart, the resistance R is:

$$R = \rho L / A \tag{3.137}$$

where A is the cross-section of the plates, ρ is the resistivity of the electrolytic cell, and conductance $\kappa = 1/\rho$. The typical order of magnitude of the resistivity for body fluids is about 1 ohm-meter. This is nine orders of magnitude larger than the resistivity of copper, so electrical conduction by ions is less effective than conduction by electrons.

TABLE 3.4
Molar Conductance at Infinite Dilution
of Various Salt Solutions

Salt	Λ_0 (Ohm·meter^{-1} per molar)
NaCl	12.8
KCl	14.9
NaNO$_3$	12.3
KNO$_3$	14.5
NaOH	24.6
KOH	27.1

TABLE 3.5
Values of Λ_0 for Various Ions

Ion	Λ_0
H$^+$	34.9
OH$^-$	19.8
Na$^+$	5.0
Cl$^-$	7.6
K$^+$	7.4

The conductivity σ of the solution as a function of salt concentration c, obeys the relationship (Benedek and Villars, 2000):

$$\sigma(c) = \Lambda_0 c - k_c c^{\frac{3}{2}} + \ldots \tag{3.138}$$

with both Λ_0 and k_c positive and independent of concentration. The conductivity of an electrolytic solution is proportional to ionic concentration for low salt concentrations. For a given temperature, the constant Λ_0 called molar conductance at infinite dilution only depends on the kind of salt used. Table 3.4 shows molar conductance at infinite dilution where the concentration is expressed in molars. According to the so-called Kohlrausch Law (Bruinsma, 1998), the molar conductance of a salt is the sum of the conductivities of the ions comprising the salt. Table 3.5 lists the values of Λ_0 for several key ions.

For electrolytic conduction the electric force, $\vec{F} = q\vec{E}$, applied to the ion is balanced by Stokes friction $\vec{F}_H = 6\pi\eta r\vec{v}$. Consequently, velocity can be computed as:

$$\vec{v} = \mu q\vec{E} \tag{3.139}$$

where μ is given by:

$$\mu = \frac{1}{6\pi\eta r} \tag{3.140}$$

and is called electrophoretic mobility. Since current passing through a cross-sectional area A is proportional to q, A, c, and v where c is ionic concentration, combining positive and negative ionic mobilities μ^+ and μ^-, we obtain:

$$I = Ac(\mu^+ + \mu^-)eE = \frac{A}{L}c(\mu^+ + \mu^-)e\Delta V \qquad (3.141)$$

which is in the same form as Ohm's law. Molar conductance at low dilution is an additive quantity:

$$\Lambda_0 = e(\mu^+ + \mu^-) \qquad (3.142)$$

3.11 Electron conduction and tunneling

Energy transfer by charger carrier is the most common reaction in metabolic processes. It can follow very different courses. The redox process is a classic example. It consists basically of a transfer of one or two electrons from a donor to an acceptor system. The donor becomes oxidized and the acceptor reduced. This apparently simple scheme conceals a number of complicated subroutines that have not yet been completely resolved.

To achieve electron transfer, donor and acceptor molecules must be in exactly defined positions and at a minimum distance so that overlapping of respective electron orbitals can occur. In the right positions, donor and acceptor form a complex of highly specific steric configuration called a *charge transfer complex*. Complex formation, which occasionally requires steric transformations of both molecules, causes the transfer. It occurs at lower rates than energy transfer by induction. Hence, the charge transfer complex is an activated transition state that enables redox processes to occur between highly specific reaction partners in the enzyme systems of cellular metabolism. Because of the oscillating nature of electron transfer, this coupling of two molecules is strengthened by additional electrostatic forces sometimes called *charge transfer forces* (Figure 3.31).

In the process of energy transfer, differences between energetic potentials of donor and acceptors play an important role. An uphill transfer of electrons is only possibly through an input of external radiation energy. The differences are slight (in the region of about 1.5 eV) when compared with the absolute values of the ionization energy. What is the scale of these energy gradients? Szent-Gyorgyi (1960) proposed a scale of so-called *bipotentials* based on the ionization energy of water (12.56 eV) — the basic molecule of life — in an opposite direction to the scale of ionization energy. The scale has never been used widely, but it is illustrative. In electrochemistry, a more common scale is used, namely the scale of redox potentials based on measurements with hydrogen electrodes, i.e., platinum electrodes surrounded by hydrogen gas. A potential of 0.82 V exists between a hydrogen electrode and an oxygen electrode generating water. This corresponds to the reference point of Szent-Gyorgyi's biopotenials.

Many biomolecular complexes such as peptide chains, α helices, and protein filaments represent complex polymer structures with periodically located structural

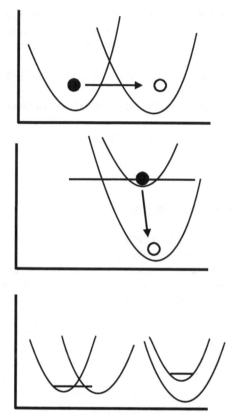

FIGURE 3.31
Plots of nuclear configuration versus potential energy of electron transfer reactions.

units. It is therefore natural to expect the periodic structures to exhibit semi-conducting properties at the very least. Table 3.6 summarizes conducting properties of common biopolymers (See Chapter 4 for their characterization from the cell biology and biochemistry view). As shown in Figure 3.32, the natures of the conducting processes in biopolymers such as protein filaments differ significantly from those of solid state systems such as semiconductors or metals.

A common motif in biomolecular complexes is a hydrogen-bonded unit. See Figures 3.33 and 3.34. Hydrogen-bonded chains provide an opportunity for another type of charge transfer mechanism, namely an electron–proton couple that proceeds along the following steps:

1. Transfer of proton impurity H^+
2. Acceptance of electron ejected by donor molecule D
3. Formation and transport of charged radical H^+
4. Ejection of electron from protein molecule with charged radical H^+

TABLE 3.6
Dark Conductivity Dry Polymers

Biopolymer	$\sigma[Scm^{-1}]$	$\Delta E[eV]$
K Polyadenylate (KPolyA)	7.9×10^1	2.20
Na desoxyribonucleate (NaDNA)	2.2×10^2	2.36
Ribonucleic acid (RNA)	2.3×10^2	2.36
Cytochrome C	6.4×10^4	2.60
Lysozyme	8.0×10^4	2.62
Hemoglobin (natural)	1.0×10^5	2.66
Hemoglobin (denaturated)	1.3×10^6	2.89
Globin	1.0×10^5	2.97
Bovine plasma albumin	2.5×10^6	2.90
Hemoglobin (denaturated)	4.0×10^6	2.99
Poly L-tyrosine (random coil)	1.3×10^6	2.98
Polyglycine	2.0×10^6	3.12

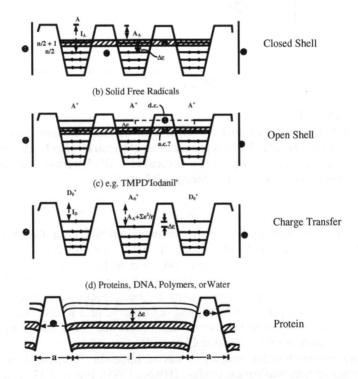

FIGURE 3.32
Comparison of the energy landscapes for the electronic conductivity in closed shell, open shell, charge transfer and protein–polymer complexes.

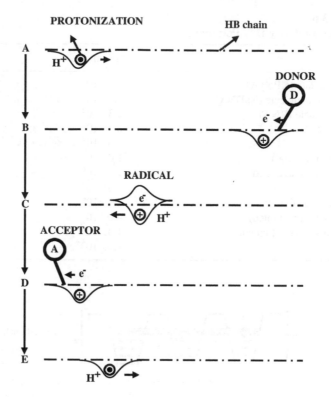

FIGURE 3.33
Electron–proton transport in hydrogen-bonded chains: (A) Transfer of proton impurity H^+. (B) Acceptance of an electron ejected by donor molecule D. (C) Formation and transport of a radical H^+. (D) Ejection of an electron from the protein molecule with the charged H^+ radical.

Electric current effects in cells related to various biological phenomena have been studied for decades (Jaffe and Nuccitelli, 1977). Most focused on growth and differentiation processes and cell division. Current fluxes of 0.03/20 mÅ/cm^2 were measured in cell growth regions and found to flow across cytoplasmic bridges during cell division. The onset of conductivity has been speculated to function as a trigger for mitosis, a trigger for chromosome segregation, or a special control mechanism for orientation of centriolar microtubule triplets that act as blades by coordinating positional navigation of cells (see Chapter 4).

The electron transfer mechanism has an important function in the operation of enzymes such as cytochrome oxidase (Harris, 1995). Figure 3.35 illustrates electron transfer between cyt c_1, cyt c, and cytochrome oxidase via the following stages: (1) orientation, (2) docking, (3) relaxation to the functioning configuration, and (4) release and rotation. These mechanisms are certainly at the heart of biological sensing in naturally occurring subcellular structures such as enzymes. We expect to see

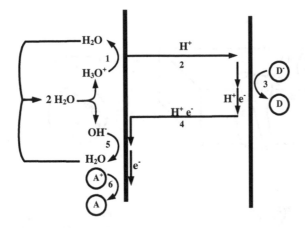

(1) **PROTONATION**
   $H_3O \longrightarrow H_2O + H^+$ (200 kcal/mol)

(2) **PROTON TRANSFER**

(3) **ELECTRON DONATION**
   $\text{\textcircled{$D^{\cdot}$}} \longrightarrow \text{\textcircled{D}} + e^-$

(4) **ELECTRON-PROTON TRANSFER**

(5) **DEPROTONATION**
   $OH^- + H^+ \longrightarrow H_2O$

(6) **ELECTRON ACCEPTING**
   $\text{\textcircled{$A^+$}} + e^- \longrightarrow \text{\textcircled{A}}$

FIGURE 3.34
Basic reaction processes of electron–proton transport in proteins.

development of devices that can mimic this behavior and operate in a controllable manner.

Several approaches are likely to include enzyme immobilization in conducting polymers on ultramicroelectrodes, surface design using plasmon resonance sensing, microelectronic protein patterning, surface preparation based on biocompatibility treatments, and design of integrated electrochemical and optical immunosensors. Nanotechnology is well suited for this function because the dimensions of fabricated structures and biological molecules are similar. The challenges, of course, are control of biochemical processes and operation in a wet environment.

3.12 Proton transport

Protonic conduction differs from electronic conduction in several respects. First, a proton has a positive charge and is three orders of magnitude more massive than an

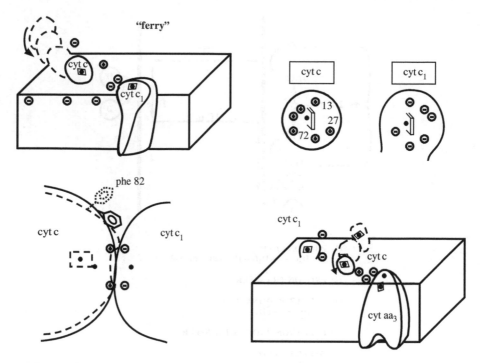

FIGURE 3.35
Electron transfer in cytochrome oxidase.

electron. Protons are abundant in water complexes surrounding all biopolymers. They constitute mobile units in common hydrogen-bonded structures in peptides, proteins, and DNA. Finally, they are freed in hydrolysis reactions of the energy-releasing ATP and GTP molecules (see Chapter 6).

We can therefore expect to find protonic conduction in biochemical processes at subcellular levels. Protonic conduction usually manifests itself through mechanisms of fault and defect migration in linear and closed (ring-like) organic polymers (see Figures 3.36 and 3.37). The two types of faults in hydrogen-bonding systems are *D* faults (presence of an extra proton) and L faults (proton deficiency). See Figures 3.38 and 3.39. Table 3.7 summarizes potential applications of biomolecular complexes as future bioelectronic devices.

3.13 Interactions with electromagnetic radiation

In 1922, A.G. Gurwitsch found evidence of a weak photon emission of a few counts/(s· cm^2) in the optical range from a biological system and indicated that it stimulated cell division. After the development of photomultiplier techniques during the second world war, many researchers rediscovered ultraweak light

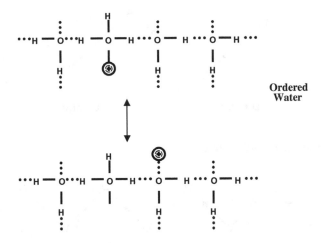

FIGURE 3.36
Protonic conduction in chain-like and ring-like organic polymers.

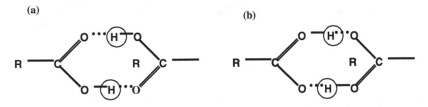

FIGURE 3.37
Dimer rings.

emissions from living tissues (Popp et al., 1988). Most groups involved in biophoton research believed that spontaneous photon emission originates from radical reactions within cells.

Popp et al., maintain that biophoton emission is due to a coherent photon field responsible for intracellular and intercellular communication and regulation of biological functions such as biochemical activities, cell growth, and differentiation in living systems (1994). Popp et al. (1981) and Li and Popp (1983) showed that biophoton emission can be traced back to DNA as the most likely source, and delayed luminescence (DL) — the long-term afterglow of living systems after exposure to external light — corresponds to excited states of the biophoton field.

When living systems relax in darkness into the quasistationary states of biophoton emission, DL follows a hyperbolic-like relaxation function rather than an exponential one, indicating under ergodic conditions a fully coherent field (Popp and Li, 1993; Bajpai, 1999). DL and biophoton emissions display identical spectral distribution. They have in common the Poissonian photocount statistics (PCS), at least down to preset intervals as low as 10^{-5} s. All correlations between DL and biophoton emission and biological functions such as cell growth, cell differentiation, biological rhythms, and cancer development appear consistent with the coherence hypothesis.

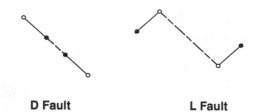

D Fault   **L Fault**

FIGURE 3.38
Faults in hydrogen bond system.

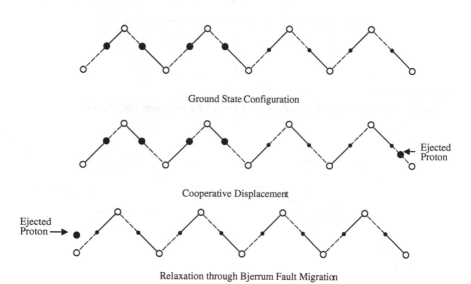

Ground State Configuration

Cooperative Displacement

Relaxation through Bjerrum Fault Migration

FIGURE 3.39
Proton migration in hydrogen-bonded chains.

Numerous biochemical reactions are regulated by light energy. Every biochemical reaction of this type takes place if and only if at least one photon excites the initial states of molecules reacting to a transition state that decays finally into a stable final product by releasing the overshoot energy of at least one photon. The availability of suitable photons determines reaction rate. Figure 3.40 shows a simple photosensitive biochemical reaction.

Of particular importance are photoactivated enzymatic reactions in which enzymatic components link parts of the enzyme via covalent bonds. They can be activated by photon absorption and deactivated by photon emission that changes the functional characteristics of the enzyme in a dramatic way (see Figure 3.41)

Electromagnetic effects on cells and cell division include (1) sensitivity of growing tissues and organs to magnetic (5 to 15 mT) and electric fields; (2) modulation by electromagnetic fields arising from Ca^{2+} transport, hormone receptor activation, protein kinase activity, melatonin synthesis, and microtubule (MT) assembly; (3) alignment

TABLE 3.7
Potential Functions of Molecular Complexes as Bioelectronic Materials

Function	Molecular Complexes
Wiring	Polyene anitibiotics, conductive polymers: actin, MTs
Storage	Bacteriorhodopsin, reaction centers (PSII), cytochromes, blue proteins, ferritin, collagen, DNA
Gates and switches	Bacteriorhodopsin, photosynthetic systems, cell receptors, ATPase
Input/output devices	Photosensitive proteins, enzymes, receptors, metal–protein complexes.

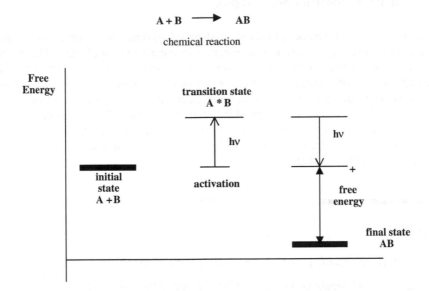

FIGURE 3.40
Simple biochemical reaction activated by light energy quanta.

of MTs in electric fields at 20 and 50 V/m; (4) alignment in magnetic fields at 20 T; and (5) influence on mitosis and meiosis at 60 Hz and 400 V/m.

Energy transfer by radiation occurs when an excited molecule emits fluorescent radiation which matches exactly the absorption spectrum of the neighboring molecule and consequently excites it. Such mechanisms can transfer energy over comparatively large distances in contrast to other processes described in this context. However, the efficiency of the process is low and declines sharply with increasing distance. This mechanism does not play a significant role in most biological processes.

However, an important factor in photosynthesis is the transfer of energy by an inductive processes, namely the so-called *resonance transfer*. This form of molecular energy transfer is *non-radiant* because no fluorescent light is involved. The mechanism can be envisioned as a type of coupling between oscillating dipoles. The

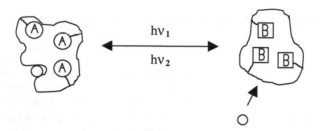

FIGURE 3.41
Photoactivation and deactivation of an enzyme by covalently linked photosensitive components. The binding site is accessible in the undistorted protein (left) but not in the distorted protein (right).

excited electron of the donor molecule undergoes oscillations and returns to its basic state, thus inducing excitation of an electron in the acceptor molecule. The process requires overlapping of fluorescent bands of the donor and absorption bands of the acceptor, i.e., on the resonances of both oscillators. The smaller the difference between the characteristic frequencies of donor and acceptor, the faster the transfer will be. Singlet and triplet states can be involved. So-called strong dipole–dipole couplings are possible to distances up to 5 nm. A $S_1 \rightarrow S_0$ transition in the donor molecule induces $S_0 \rightarrow S_1$ excitation in the acceptor. Inductions through triplet states are only effective over short distances.

References

Agmon, N. and Hopfield, J.J., *J. Chem. Phys.*, 79, 2042, 1983.

Agmon, N. and Madhavi-Sastry, G., *Chem. Phys.*, 212, 207, 1996.

Atkins, P.W. *Physical Chemistry*, 6th ed., Oxford University Press, U.K., 1998.

Austin, R.H. et al., *Biochemistry* 14, 5355, 1975.

Bajpai, R.P., *J. Theor. Biol.*, 19, 287, 1999.

Benedek, G.B. and Villars, F.M.H., *Physics with Illustrative Examples from Medicine and Biology*, Springer, Berlin, 2000.

Brooks, C.L., III, Karplus, M., and Pettitt, B.M., *Proteins: A Theoretical Perspective of Dynamics, Structure and Thermodynamics, Advances in Chemical Physics series*, Vol. 71, John Wiley & Sons, New York, 1988.

Bruinsma, R., *Physics,* International Thomson Publishing, New York, 1998.

Careri, G. et al., *Phys. Rev.,* B 30, 4689, 1984.

Chernavsky, D.S., Khurgin, Y.I., and Shnol, S.E., *Mol. Biol.*, 20, 1356–1368, 1987.

Conway, J.H., *Sci. Am.,* October, 120–123, 1970.

Davydov, A.S. *Biology and Quantum Mechanics,* Pergamon Press, Oxford, U.K., 1982.

Flory, P.J., *Principles of Polymer Chemistry,* Cornell University Press, Ithaca, NY, 1953.

Frauenfelder, H., Sligar, S.G., and Wolynes, P.G., *Science 254,* 1998, 1991.

Frauenfelder, H., Wolynes, P.G., and Austin, R.H. *Rev. Mod. Phys.* 71, S419, 1999.

Froehlich, H., *Int. J. Quant. Chem.,* 2, 641, 1968.

Froehlich, H., *Phys. Lett.,* 39A: 153, 1972.

Froehlich, H., *Adv.Electr. Electr. Phys.,* 53, 85, 1980.

Gardiner, C.W. *Handbook of Stochastic Methods for Physics, Chemistry and the Natural Sciences,* Series in Synergetics, Vol. 13, Springer, Berlin, 1996.

Grundler, W., Eisenberg, C.P., and Sewchand, L.S., *J. Biol. Phys.,* 11, 1, 1983.

Gurwitsch, A.G., *Arch. Entw. Mech. Org.,* 51, 383, 1922.

Hänggi, P., Talkner, P., and Borkovec, M., *Rev. Mod. Phys.,* 62, 251, 1990.

Harris, D.A., *Bioenergetics at a Glance,* Blackwell Science, Cambridge, 1995.

Hillson, N., Onuchi, J.N., and García, A.E., *Proc. Natl. Acad. Sci. U.S.A.,* 96, 14848, 1999.

Hummer, G., García, A.E., and Garde, S., *Phys. Rev. Lett.,* 85, 2637, 2000.

Hyman, J.M., McLaughlin, D.W., and Scott, A.C., *Physica, D*3, 23, 1981.

Kitao, A. and Go, N., *Curr. Opinion Struct. Biol.,* 9, 164, 1999.

Kittel, V.C., *Introduction to Solid State Physics,* John Wiley & Sons, New York, 1996.

Kurzynski, M., *Biophys. Chem.,* 65, 1, 1997.

Kurzynski, M., *Progr. Biophys. Molec. Biol.,* 69, 23, 1998.

Kurzynski, M., Palacz, K., and Chelminiak, P., *Proc. Natl. Acad. Sci. USA,* 95, 11685, 1998.

Li, K.H. and Popp, F.A., *Phys. Lett.,* 93A, 626, 1983.

Lomdahl, P.S. *Los Alamos Sci.,* 10, 27, 1984.

Lomdahl, P.S., Layne, S.P., and Bigio, I.J., *Los Alamos Sci.,* 10, 2, 1984.

Mandelbrot, B. *Fractal Geometry of Nature,* W.H. Freeman, San Francisco, 1988.

McCammon, J.A. and Harvey, S.C., *Dynamics of Proteins and Nucleic Acids,* Cambridge University Press, Cambridge, U.K., 1987.

Montroll, E.W. and Shuler, K.E., *Adv. Chem. Phys.,* 1, 361, 1958.

Nakayama, T., Yakubo K., and Orbach, R., *Rev. Mod. Phys.*, 66, 381, 1994.

Northrup, S.H. and Hynes, J.T., *J. Chem. Phys.*, 73, 2700, 1980.

Jaffe, L.F. and Nuccitelli, R., *Ann. Rev. Biophys. Bioelec.*, 6, 445, 1977.

Nymeyer, H., García, A.E., and Onuchic, J.N., *Proc. Natl. Acad. Sci. U.S.A.*, 95, 5921, 1998.

Paul, R. et al., *J. Theor. Biol.*, 104, 169, 1983.

Popp, F.A. and Li, K.H., *Intl. J. Theor. Phys.*, 32, 1573, 1993.

Popp, F.A., Gu, Q., and Li, K.H., *Mod. Phys. Lett. B*8, 1269, 1994.

Popp, F.A. et al., *Collective Phen.*, 3, 187, 1981.

Popp, F.A. et al., *Experientia*, 44, 543, 1988.

Post et al., *Biophys.*, 64, 1833, 1993.

Rowlands, S., Sewchand, L.S., and Enns, E.G., *Can. J. Physiol. Pharmacol.*, 60, 52, 1982.

Sackmann B. and Naher, E., *Single-Channel Recording*, 2nd ed., Plenum, New York, 1995.

Sansom, M.P.S. et al., *Biophys. J.*, 56, 1229, 1989.

Scott, A.C., *Phys. Rep.*, 217, 1, 1992.

Scott, A.C., Chu, F.Y.F., and McLaughlin, D.W., *Proc. IEEE*, 61, 1443, 1973.

Stein, D.L., *Spin Glasses and Biology*, World Scientific, Singapore, 1992.

Szent-Györgi, A., *Introduction to a Submolecular Biology*, Academic Press, New York, 1960.

Tuszynski, J.A. and Dixon, J.M., *Applications of Physics to Biology and Medicine*, John Wiley & Sons, New York, 2002.

Tuszynski, J.A. et al., *Phys. Rev.*, A30, 2666, 1983.

van Kampen, N.G. *Stochastic Processes in Physics and Chemistry*, Rev. Ed., North-Holland, Amsterdam, 2001.

Webb, S.J., *Phys. Rep.*, 60: 201, 1980.

Widom, B., *J. Chem. Phys.*, 43, 3898, 1965.

Widom, B., *J. Chem. Phys.*, 55, 44, 1971.

Wolfram, S., *Rev. Mod. Phys.*, 55, 601, 1983.

Wu, T.M., and Austin, S., *Phys.Lett.*, A64, 151, 1977.

4

Structure of a Biological Cell

The central problem of cell biology now
is not so much the gathering of information
but the comprehension of it.

If cell biology becomes a part of physics,
it will have only itself to blame.

J. Maddox

What Remains To Be Discovered

4.1   General characteristics of a cell

Life is the ultimate example of a complex dynamic system. A living organism develops through a sequence of interlocking transformations involving an immense number of components composed of molecular subsystems. When combined into a larger functioning unit (e.g., a cell), emergent properties arise.

For the past several decades, biologists have greatly advanced the understanding of how living systems work by focusing on the structures and functions of constituent molecules such as DNA. Understanding the compositions of the parts of a complex machine, however, does not explain how it works. Scientific analysis of living systems poses an enormous challenge and today we are more prepared than ever to tackle this enormous task. Conceptual advances in physics (through the development of non-linear paradigms; see Appendix C), vast improvements in the experimental techniques of molecular and cell biology (electron microscopy, atomic force microscopy, etc.), and exponential progress in computational techniques brought us to a unique point in the history of science when the expertise of researchers representing many areas can be brought to bear on the main unsolved puzzle of life, namely how cells live, divide, and eventually die.

Cells are the key building blocks of living systems. Some are self-sufficient while others function as parts of multicellular organisms. The human body is composed of approximately 10^{13} cells of 200 different types. A typical cell size is on the order of 10 μm and its dry weight equals about 7×10^{-16} kg. In a natural state, water molecules constitute 70% of cell content. The fluid content, known as the cytoplasm,

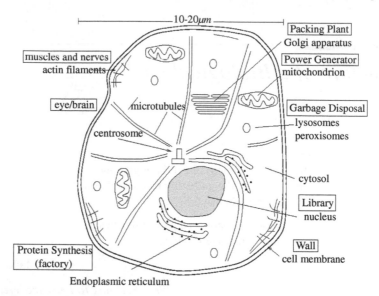

FIGURE 4.1
Generalized animal cell during interphase. Note major cellular organelles including nucleus, centrosome, and radiating microtubules. The endoplasmic reticulum is dotted with ribosomes.

is the liquid medium bound within a cell. The cytoskeleton is a lattice of filaments formed throughout the cytoplasm (Amos and Amos, 1991). Figure 4.1 is a cross-section of a simplified animal cell. The two major cell types are:

Prokaryotic — simple cells with no nuclei or compartments. Bacteria, e.g., *Escherichia coli* and blue-green algae belong to this group.

Eukaryotic — cells with nuclei and differentiated structures including compartmentalized organelles and filamentous cytoskeletons. Examples include higher order animal and plant cells, green algae, and fungi. Eukaryotic cells emerged about 1.5 billion years ago.

Bacteria have linear dimensions in the 1 to 10 μm range; the sizes of eukaryotic cells range from 10 to 100 μm. The interior of a bacterium may be under considerable pressure as high as several atmospheres. The pressure is maintained by a membrane which, except in Archaebacteria, is composed of layers of peptidoglycan sandwiched between two lipid bilayers, the innermost of which is a plasma membrane.

Plant cells have linear dimensions that vary from 10 to 100 μm. Each cell is bounded by a cell wall composed of cellulose whose thickness ranges from 0.1 to 10 μm. A plant cell has organelles, a nucleus, endoplasmic reticulum (ER), Golgi apparatus, and mitochondria. Unique to plant cells are chloroplasts and vacuoles. Unlike bacteria, plant cells possess cytoskeletal networks that add to their mechanical strength.

Animal cells tend to be smaller than plant cells since they do not have liquid-filled vacuoles. The organizing center for an animal cytoskeleton appears to be a cylindrical organelle called a centriole that is about 0.4 μm long. Instead of chloroplast photosynthesis sites, animal cells have mitochondria that produce the required energy supplies in the form of ATP molecules obtained from reactions involving oxygen and food molecules (e.g., glucose). Mitochondria are shaped like cylinders with rounded ends.

Eukaryotic cells have membrane-bound internal structures called organelles (De Robertis and De Robertis, 1980). Mitochondria produce energy. A *Golgi* apparatus that modifies, sorts, and packages macromolecules for secretion by the cell for distribution to other organelles) is shaped like a stack of disks. The ER surrounds the nucleus and is the principal site of protein synthesis. It is small compared to the cell surface area.

The nucleus is the residence of chromosomes and the site of DNA replication and transcription. All material in a cell except the nucleus is defined as the cytoplasm. Its liquid components are called the cytosol. The solid protein-based structures that float in the cytosol constitute the cytoskeleton (Amos and Amos, 1991). The main component of the cytosol is water. Most organelles are bound within their own membranes. Most DNA is housed within the nucleus and the nucleus is protected by the nuclear envelope. The nucleolus within the nucleus functions as the site of ribosomal RNA synthesis. The diameter of a nucleus ranges from 3 to 10 μm. Despite many differences, animal and plant cells have striking similarities (see Figure 4.2).

4.2 Membrane and membrane proteins

Amphipathic molecules adsorb themselves to air–water or oil–water interfaces such that their head groups face the water environment. They aggregate to form spherical micelles or liquid crystalline structures. In general, amphipathic molecules can be anionic, cationic, nonionic, or zwitterionic.

The relative concentrations of these surfactants in an aqueous solution affect their physical and chemical properties. At a specific value, called the critical micelle concentration (CMC), micelles containing 20 to 100 molecules are formed spontaneously in the solution with the hydrophilic head groups exposed and the hydrophobic tails hidden inside the micelles. The principal driving force for micelle formation is entropic due to a negative free energy change accompanying the liberation of water molecules from clathrates. When phospholipids are mixed in water, they form double-layered structures since their hydrophilic ends are in contact with water; the hydrophobic ends face inward and touch each other.

Biomembranes compartmentalize areas of different metabolic activities within cells. They also regulate the flow into and out of cells and cell compartments. Finally, membranes are sites of key biochemical reactions. They have unique amphipathic properties in that they possess hydrophobic and hydrophilic parts. A cell membrane is the thin nearly invisible structure that surrounds the cytoplasm. It is a continuous boundary region. It also connects the ER and the nuclear membrane.

Membranes are composed of phospholipids, glycolipids, sterols, fatty acid salts, and proteins. Figure 4.3 shows a phospholipid with a globular head group structure representing the hydrophilic (water loving) section. The tails extending from the

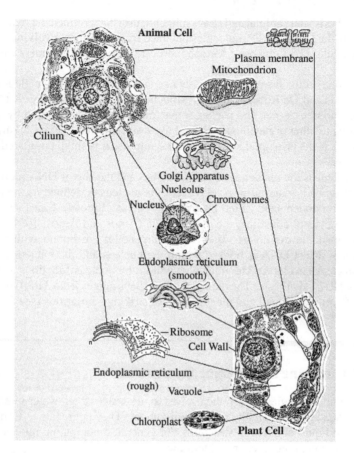

FIGURE 4.2
Comparison of structures of typical animal and plant cells.

sphere represent the hydrophobic (water fearing) end of the phospholipid. The two long chains coming off of the bottom of the molecule are made of carbon and hydrogen. Because both of these elements share their electrons evenly, the chains have no charge. Molecules with no charges are not attracted to water; as a result water molecules tend to push them out of the way as they are attracted to each other. This causes uncharged molecules not to dissolve in water.

At the other end of the phospholipid are a phosphate group and several double-bonded oxygens. The atoms at this end of the molecule are not shared equally. This end has a charge and is attracted to water. Figure 4.4 is a simplified representation of a cellular membrane. In a nutshell, a cell membrane:

1. Serves as a selectively permeable barrier between two predominantly aqueous compartments
2. Allows compartmentalization of various structures in the cell

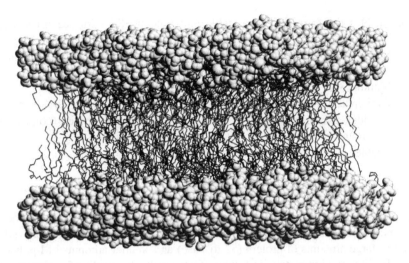

FIGURE 4.3
Phospholipid with a globular head group structure representing the hydrophilic (water loving) segment.

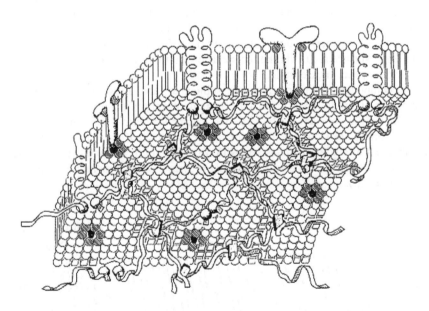

FIGURE 4.4
Typical cellular membrane shown schematically.

3. Enables the formation of a stable and fluid medium for reactions that are catalyzed
4. Provides a flexible boundary between the cell or organelle and its surrounding medium
5. Maintains an electric potential difference
6. Participates in signal transmission to the actin cycloskeleton via integrins
7. Provides adhesion forces to attach cells to their substrates (controlled by membrane elasticity).
8. Enables mass transport via ion channels

The fluid mosaic model of Singer and Nicolson (1972) views the membrane as a fluid bilayer of amphipathic complex lipids embedded with proteins as shown in Figure 4.4. The relative abundance of proteins in a membrane varies from species to species and correlates with metabolic activity. For example, the mitochondrial wall contains large amounts of protein (52 to 76%) and smaller amounts of lipids (24 to 48%) facilitating high metabolic activity. Conversely, the inactive membranes of the myelin sheaths in neurons contain only 18% proteins and 79% lipids.

As presented above, a double layer of phospholipid molecules with embedded proteins constitutes the plasma membrane or outer surface of a cell. However, the plasma membrane does not resemble the surface of a fluid or even the interface of two fluids. The reason is that the plasma membrane has an essentially fixed surface area, i.e., it has only a fixed number of phospholipid molecules and proteins which, when packed together, comprise the membrane. Each lipid molecule or protein has a preferred surface area. Unlike the surface of a fluid, the plasma membrane is for practical purposes inextensible.

The components of membranes are subject to rapid movements. Rapid lateral movement of lipids has a diffusion constant of approximately 10^{-8} cm^2/s while for proteins the constant is 10^{-10} to 10^{-12} cm^2/s. "Flip-flop" movements across the membrane are slow, on the order of 10^5 s. Indeed, the phospholipids of the membrane may undergo a phase transition from a gel to a liquid crystal phase as a result of a change in temperature, external pressure, or even membrane composition (see Figure 4.5).

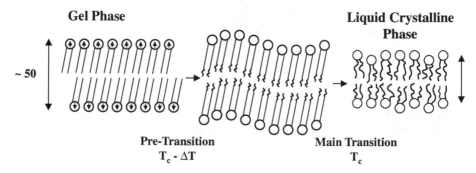

FIGURE 4.5
Membrane phase transitions.

Depending on prevailing conditions, membrane dynamics has been described via continuous Landau-Ginzburg models (see Appendix B) developed by Owicki et al. (1978) and de Gennes (1990). Alternatively, discrete variable models using a quasispin representing the state of each individual hydrocarbon chain have been proposed. An Ising model approach was advocated by Zuckerman et al. (1985) and Pink (1984) in which the Hamiltonian for the membrane is postulated in the form:

$$H_T = -1/2 \sum_n \sum_m J_{nm}(D_{nm}) L_{ni} L_{nj} + \sum_n (\pi A_n + E_n) L_{ni} \quad (4.1)$$

where i labels chain sites, n enumerates chains, L_n is chain length (depends on the number of twists), π is lateral pressure, and E_n is internal energy of the n-th chain. Typical of two-dimensional Ising models is the emergence of an order–disorder phase transition at a finite temperature that depends on the number of nearest neighbors, the strength of their interactions, and the applied external pressure (see Appendix B). Note that the chain-chain interaction constant J_{nm} is distance dependent, i.e., $J_{nm}(D_{nm})$.

4.2.1 Elastic pressure of membrane

If we increase the osmotic pressure of a cell, the cell will try to swell but this is prevented because the surface area of the plasma membrane is nearly fixed, i.e., elastic stress will build up inside the membrane. If it becomes too great, the cell will burst — a condition called lysis. To find the stress at which the membrane will burst, we will cut the membrane along some line of length ℓ.

To prevent the two sides of the cut from separating, force F which is proportional to ℓ must be present. In the expression for F, $F = \gamma \ell$, the proportionality constant γ is the elastic tension in the wall. This tension is not the surface tension T of a fluid but it plays a similar role. The excess pressure inside a bubble of radius R over and above atmospheric pressure is 2T/R. A similar relation may be used to compute the excess pressure ΔP inside a spherical pressure vessel if we replace T with γ:

$$\Delta P = \frac{2\gamma}{R} \quad (4.2)$$

The elastic stress σ inside the vessel wall is related to the elastic tension by:

$$\gamma = D\sigma \quad (4.3)$$

where D is the wall thickness. The reason is that the surface area of the cut is ℓD and hence the force per unit area on the surface of the cut is $F/\ell D = \gamma/D$ and this is the stress of the vessel wall. Following Equations (4.2) and (4.3), the elastic stress σ is given by (Bruinsma, 1998):

$$\sigma = \frac{R\Delta P}{2D} \quad (4.4)$$

The vessel will burst when this stress exceeds the fracture stress of the material from which the vessel was made.

4.2.2 Mass diffusion across membranes

A large amount of diffusion in biological organisms takes place through membranes. These membranes measure from 65×10^{-10} to 100×10^{-10} m in diameter. Most

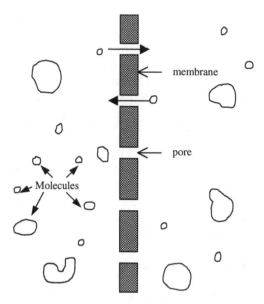

FIGURE 4.6
Semipermeable membrane.

membranes are selectively permeable; that is, they allow only certain substances to cross them and have pores through which substances diffuse. These pores are so small (7 $\times 10^{-10}$ to 10 $\times 10^{-10}$ m) that only small molecules pass through them.

Other factors contributing to the semipermeable nature of membranes relate to the chemistry of the membrane, cohesive and adhesive forces, charges on the ions involved, and carrier molecules. Diffusion in liquids and through membranes is a slow process. See Figure 4.6.

We can apply Fick's law to the transport of molecules across a biomembrane, e.g., the phospholipid layer surrounding a cell. Consider a container of sugar water divided by a membrane of thickness Δx. Assume the concentration of glucose on the left side is c_L and on the right side is c_R. The glucose diffusion current across the membrane, according to Fick's law, is (Benedek and Villars, 2000):

$$I = k \frac{D}{\Delta x} A \Delta c \qquad (4.5)$$

where $\Delta c = c_R - c_L$, k is the diffusion constant and A is the cross-sectional area of the membrane.

4.2.3 Membrane proteins

The function of a cell membrane generally revolves around its proteins. Floating around in the cell membrane are different kinds of proteins. They are generally globular. They are not held in a fixed pattern; they move about the phospholipid layer. Generally these proteins structurally fall into three categories: (1) carrier proteins that regulate transport and diffusion, (2) marker proteins that identify the cell to other cells, and (3) receptor proteins that allow the cell to receive instructions, communicate, transport proteins, and regulate what enters or leaves the cell.

Membrane proteins are (1) peripherical or (2) integral. Peripherical proteins are bound electrostatically to the exterior parts of head groups and hence can be easily extracted. Integral proteins are tightly bound to lipid tails and are insoluble in water. Steroids are sometimes components of cell membranes in the form of cholesterol. When it is present, it reduces the fluidity of the membrane. However, not all membranes contain cholesterol. Transport proteins come in two forms. Carrier proteins are peripherical proteins that do not extend all the way through the membrane. They bond and drag specific molecules through the lipid bilayer one at a time and release them on the opposite side.

Channel proteins extend through the lipid bilayer. They form a pore through the membrane that can move molecules in several ways. In some cases, channel proteins simply act as passive pores. Molecules randomly move through the openings via diffusion. This requires no energy and molecules move from an area of high concentration to low concentration. Symports also use the process of diffusion. In this case a molecule that moves naturally into a cell through diffusion is used to drag another molecule into the cell. For example, glucose "hitches a ride" with sodium.

Marker proteins extend across cell membranes and serve to identify the cells. The immune system uses these proteins to tell the organism's own cells from foreign invaders. The cell membrane can also engulf structures too large to fit through the pores in the membrane proteins. This process is known as endocytosis. The membrane wraps itself around the particle and pinches off a vesicle inside the cell. The opposite of endocytosis is exocytosis. Large molecules made in the cell are released through the membrane. A common example of this process is exocytosis of neurotransmitter molecules into the synapse region of a nerve cell.

4.2.4 Electrical potentials of cellular membranes

First measurements of electrical properties of cell membranes were made on red blood cells by H. Fricke and on sea urchin cells by K.S. Cole (1937). Membranes act as capacitors and maintain potential differences between oppositely charged surfaces composed mainly of phospholipids embedded with proteins. A typical value of capacitance per unit area C/A is about 1 $\mu F/cm^2$ for cell membranes. This relates to the dielectric constant ϵ via:

$$\frac{C}{A} = \frac{\epsilon \varepsilon_0}{d} \tag{4.6}$$

where $\varepsilon_0 = 8.85 \cdot 10^{-12} \ C/Nm^2$ giving a value of $\epsilon \cong 10$ which is greater than $\epsilon \cong 3$ for phospholipids resulting from the active presence of proteins. The cellular membrane is much more permeable (intercellular fluid contains primarily NaCl) in the normal resting state to potassium ions than sodium ions. This results in an outward flow of potassium ions; voltage inside the cell is –85mV. See Figure 4.7.

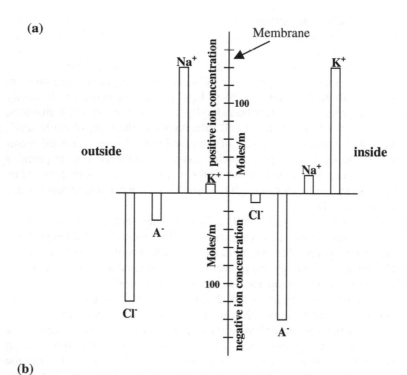

FIGURE 4.7
**(a) Ion concentrations outside and inside a cell. Positive concentrations are
graphed above the line and negative concentrations below. (b) The membrane
is permeable to K^+ and Cl^- ions, which diffuse in opposite directions. Small
arrows indicate that the Coulomb force resists the continued diffusion of both
K^+ and Cl^- ions.**

TABLE 4.1
Concentration Ratios for Cellular Membranes and
Corresponding Nernst Potentials

Ion	$c(r)/c(L)$	Valency	$\Delta V (mV)$
Na$^+$	10	1	62
K$^+$	0.03	1	−88
Cl$^-$	14	−1	−70

This voltage is called the resting potential. If the cell is stimulated by mechanical, chemical, or electrical means, sodium ions diffuse more readily into the cell since the stimulus changes membrane permeability. The inward diffusion of a small amount of sodium ions increases the interior voltage to +60 mV which is known as the action potential. The membrane again changes its permeability once the cell has achieved its action potential and potassium ions then readily diffuse outward so the cell returns to its resting potential.

Depending on the state of the cell, the interior voltage can vary from its resting potential of −85 mV to its action potential of +60 mV. This results in a net voltage change of 145 mV in the cell interior. The voltage difference between the two sides of the membrane is fixed by the concentration difference. Having a salt concentration difference across a membrane and allowing only one kind of ion to pass through the membrane produces a voltage difference given by the formula:

$$V_L - V_R = \frac{k_B T}{e} \ln \left(\frac{c_R}{c_L} \right) \tag{4.7}$$

and called the Nernst potential. This is the basic mechanism whereby electrical potential differences are generated inside organisms. The Nernst potential difference only depends on the concentration ratio. Table 4.1 shows concentration ratios for cellular membranes and the corresponding values of the Nernst potential computed from Equation (4.7). A more detailed discussion on ion diffusion across membranes of nerve cells appears in Chapter 7.

According to Froehlich (1980; see Chapter 3), the phospholipid head groups in membranes play important roles in cell–cell interactions due to their oscillating dipole moments as shown in Figure 4.8. He suggested that their mutual interactions would lead to cell-specific microwave frequency coherent oscillations that would resonantly couple to other oscillations of the same type cells in the vicinity. However, this hypothesis still awaits experimental confirmation. The key assumption is that the membrane has electric fields: E $\sim 10^{-5} V/cm$ that lead to coherent dipole oscillations: f $\sim 10^{11} Hz$ in the microwave range. As shown in Figure 4.9, dipole dynamics of membrane phospholipids is very complicated with many different modes of vibration assigned to different motions of their molecular groups. It is still unclear which of these groups may possibly lead to coherent dipole oscillations in the 10^{11}–$10^{12} Hz$ range as predicted by Froehlich.

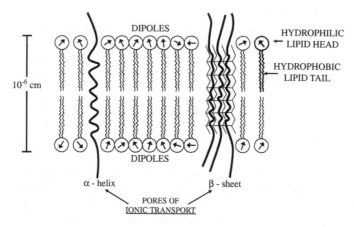

FIGURE 4.8
Membrane dipole dynamics as envisaged by Froehlich.

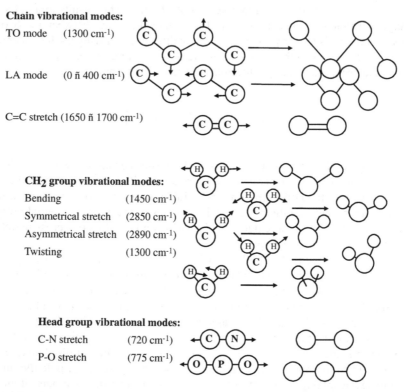

FIGURE 4.9
Some key vibrational modes of membrane molecular groups.

4.3 Ion channels and ion pumps

A cell membrane has pores or channels that allow selective passages of metabolites and ions in and out of the cell and can even drag molecules from areas of low concentration to high concentration working directly against diffusion. An example is the sodium–potassium pump. Most work of transporting ions across membranes is done via ion pumps such as sodium–potassium pumps.

The energy required for the functioning of the pump comes from the hydrolysis of ATP. A phosphorylated protein serves as an intermediate in the process. The hydrolysis of a phosphoprotein usually causes a conformational change that opens a pore that drives sodium and potassium transport. Some membrane proteins actively use energy from ATP in the cell to perform mechanical work. The energy of a phosphate is used to exchange sodium atoms for potassium atoms. The free energy change in the hydrolysis of a phosphoprotein with a value of 9.3 kJ/mol will drive a concentration gradient of 50:1 uphill. For each ATP molecule hydrolyzed, three sodium ions are pumped out and two potassium ions are pumped in.

As shown in Figure 4.10, ion channels fall into three general classes: (1) voltage-gated, (2) ligand-gated, and (3) gap junctions. They differ in their design geometry and in the physical and chemical mechanisms used to select ions for passage. Figure 4.11 shows how a conformational change in an ion channel leads to the opening and closing of the gate. Figure 4.12 illustrates an equivalent electrical circuit for an ion channel which is modulated by voltage.

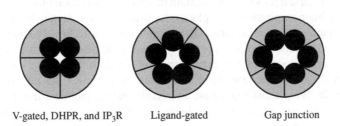

V-gated, DHPR, and IP$_3$R   Ligand-gated   Gap junction

FIGURE 4.10
Three key types of ion channels.

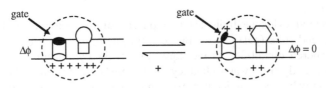

FIGURE 4.11
Functioning of a gate: modulation of a local electric field by an ion channel.

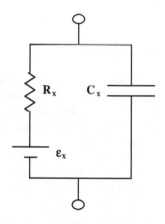

Equivalent Electrical Circuit

FIGURE 4.12
Equivalent electrical circuit of an ion channel.

4.4 Cytoplasm

The fluid contents of a cell are known as the cytoplasm. The cytoplasm is hugely important because it provides the medium in which fundamental biophysical processes such as cellular respiration take place. Its properties are somewhat different from those of dilute aqueous solutions. The contents must be accurately known for *in vitro* studies of enzymatic reactions, protein synthesis, and other cellular activities.

Typical constituents of the cytoplasm are listed in Tables 4.2 (ionic) and 4.3 (biomolecular). Most of the trace ions listed in Table 4.2 are positively charged. However, the cytoplasm cannot have an overall charge, and the difference is made up of the other constituents such as proteins, bicarbonate (HCO_3^-), phosphate (PO_4^{3-}), and other ions that for the most part are negatively charged, a few of which are significantly electronegative.

Most cells maintain neutral pH and their dry matter is composed of at least 50% protein (see Table 4.3). The remaining dry material consists of nucleic acids, trace ions, lipids, and carbohydrates. A few metallic ions are required for incorporation into metalloproteins but these ions such as Fe^{2+} are typically found in nanomolar concentrations.

Experimental evidence indicates two phases of the cytoplasm: the so-called liquid (sol) and solid (gel) phases. In the solid phase, the major constituents of the cell are rendered immobile. In the liquid phase, the cytoplasm viscosity does not differ significantly from water (Janmey et al., 1994). Diffusion in the cytoplasm is affected mainly by macromolecular crowding. In the solid phase, diffusion is slowed by a factor of three relative to diffusive movement in water. Such properties of the

TABLE 4.2
Major Components of Cytoplasm of Typical Mammalian Cell

Ions	Concentration	Nonionic Constituents	
K^+	140 mM	Protein	200–300 mg/mL
Na^+	12 mM	Actin	2–8 mg/mL
Cl^-	4 mM	Tubulin	4 mg/mL
Ca^{2+}	0.1 μM	pH	~7.2
Mg^{2+}	0.5 mM	Specific tissues may differ	

Source: Luby-Phelps, K., *Curr. Opin. Cell Biol.*, 6, 3, 1994. With permission.

TABLE 4.3
Percentage Content and Molecular Numbers of Key Cellular Components

Molecule	Content in Dry Weight (%)	No. of Molecules
DNA	5	2×10^4
RNA	10	4×10^4
Lipids	10	4.5×10^6
Polysaccharides	5	10^6
Proteins	70	4.7×10^6

cytoplasm seem to be regulated in some sense by the cytoskeleton, but the manner of regulation is unclear. It may involve the tangling and detangling of a mesh of various protein filaments. The important point is that once a cell has acted to organize itself, the transition to a solid phase can allow it to expend relatively minimal energy to maintain its organization (Boal, 2001).

Contrary to early perceptions, the cytoplasm is not a viscous, soup-like, amorphous substance; it is a highly organized, multicomponent, dynamic network of interconnected protein polymers suspended in a dielectrically polar liquid medium. This will become clear when we discuss the cytoskeleton in the next section.

4.4.1 Osmotic pressures of cells

As shown in Table 4.2 a variety of solute molecules are contained within cells (see Figure 4.13). The cellular fluid (cytosol) has a chemical composition of 140 mM K^+, 12 mM Na^+, 4 mM Cl^-, and 148 mM A^- where 1 mM represents a concentration of 10^{-3} mol/liter. The symbol A represents protein. Cell walls are semipermeable membranes and permit the transport of water but not of solute molecules. We can apply the osmotic pressure concept to cells, but because of the content of cellular fluid, we must find the osmotic pressure of a mixture of solute molecules.

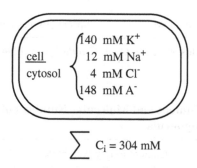

FIGURE 4.13
Solute molecules inside and outside a cell.

We use Dalton's law to determine osmotic pressure inside a cell. A mixture of chemicals with concentrations $c_1, c_2, c_3 \dots$ dissolved in water makes the total osmotic pressure equal to the sum of the partial osmotic pressures Π of each chemical (see Chapter 3). Thus:

$$\Pi = \Pi_1 + \Pi_2 + \Pi_3 + \dots = RT(c_1 + c_2 + c_3 + \dots) \qquad (4.8)$$

The total osmotic pressure inside a cell, Π_{in}, is therefore:

$$\Pi_{in} = RT \frac{(140 + 12 + 4 + 148) \times 10^{-3} \text{mol}}{1 \text{ liter}} = 7.8 \times 10^4 \text{Pa} \qquad (4.9)$$

where we used the concentrations above and a temperature of T = 310K since the gas constant is R = 8.31 J/molK. Cell walls would be expected to burst under such large pressures. However, they do not because the exterior fluid exerts osmotic pressure in the opposite direction. The cell exterior is composed of 4mM K^+, 150 mM Na^+, 120 mM Cl^-, and 34 mM A^-. As a consequence, the total osmotic pressure of the cell exterior Π_{out} is:

$$\Pi_{out} = RT \frac{(4 + 150 + 120 + 34) \times 10^{-3} \text{mol}}{1 \text{ liter}} = 7.9 \times 10^4 \text{Pa} \qquad (4.10)$$

Π_{out} is again a large osmotic pressure but because Π_{in} and Π_{out} are quite close in values, the osmotic pressure difference between the exterior and interior parts of the cell is very small; it is the net pressure exerted on the cell wall. For fragile animal cells, it therefore becomes vitally important to keep their interior and exterior osmotic pressures closely matched. Cells have sophisticated control mechanisms to do this.

4.4.2 Osmotic work

If two solutions have the same osmotic pressure, they are isoosmotic. If the pressures are different, the one at higher pressure is hypertonic and the one at lower pressure is hypotonic. When cells are placed in a solution and neither swell nor shrink, the solution is isotonic. In the tissues of most marine invertebrates, total osmotic concentration is near that of sea water.

The salt concentration of sea water is about 500 mM. As long as the salt concentration remains near this value, the blood of many crabs is isotonic with that of sea water. When it is outside this range, the system maintains the osmotic pressure difference across its membrane through the activity of ion pumps in a process known as osmoregulation.

Cell composition begins to drift away from its optimal mixture if the ion pumps are chemically destroyed. Across the cell wall, the osmotic pressure difference then rises, causing the cell to swell, become turgid and eventually explode. The cells of bacteria and plants are not osmotically regulated since their cell walls can withstand pressures in the range of 1 to 10 atm. The removal of interior salts by teleost fish and the importation of salt by freshwater fish requires work. The minimum work performed when n moles of solute are transferred from one solution with a concentration c_1 to a solution with concentration c_2 is given by:

$$W = nRT \ln \frac{c_1}{c_2} \tag{4.11}$$

where c_1 may be the salt concentration in the tissue of a fish and c_2 the salt concentration in sea water. In this case, $c_2 > c_1$ so the osmotic work done by the sea water should be negative. The physical reason is that energy is required to move salt molecules from a solution of low concentration to one of high concentration. The work done by the cells of a fish in excreting salt will be the reverse of that in Equation (4.11).

Osmotic pressure is also used by the cells of plants and, in particular, trees. Tree roots have high internal osmotic pressures which leads to absorption of water from the soil. A key role in plant growth may be played by osmotic pressure. The openings on the surfaces of cell leaves, called stomata, are bordered by guard cells that can regulate their internal pressure by controlling potassium concentration. Water absorption causes these cells to swell under osmotic pressure and the stomata are closed.

Contained within the cytoplasm are the components of the cytoskeleton and certain smaller compartments known as organelles that are specialized to perform their respective functions. We discuss the components of the cytoskeleton in the section that follows.

4.5 Cytoskeleton: the proteins participating in cytoskeletal organization

4.5.1 Cytoskeleton

One of the most important issues of molecular biophysics is the complex and multifunctional behavior of the cytoskeleton. Interiors of living cells are structurally organized by cytoskeleton networks of filamentous protein polymers: microtubules, actin (microfilaments), and intermediate filaments with motor proteins providing force and directionality needed for transport.

Unlike the hardened skeleton that supports mammals, the cytoskeleton is a dynamic structure that undergoes continuous reorganization. The cytoskeletal network of filaments has the responsibility of defining the cell shape (Frey et al., 1998), protecting the cell from changes in osmotic pressure, organizing its contents, providing cellular motility, and finally separating chromosomes during mitosis.

The cytoskeleton is unique to eukaryotic cells. It is a dynamic three-dimensional structure that fills the cytoplasm and acts as both muscle and skeleton to allow movement and provide stability. The long fibers of the cytoskeleton are polymers composed of protein subunits. The key proteins of the cytoskeleton are listed below (Alberts et al., 1994):

Tubulin and microtubules
Actin and actin filaments
Intermediate filaments
Families of motor proteins: kinesin, myosin, dynein
Specialized molecules in the two phases of the cytoplasm: gel promoters, sol
    promoters
Gelsolin
Cross-linkers: integrin, talin
Cross-linking, sequestering, and severing molecules:
    a-actin, filamin, cap-Z for actin; MAPs, e.g., tau, for microtubules

For example, in the epithelial (skin) cells of the intestine (see Figure 4.14), all three types of fibers (microfilaments, microtubules, and intermediate filaments) are present. Microfilaments project into the villi, giving shape to the cell surface. Microtubules grow out of the centrosome to the cell periphery. Intermediate filaments connect adjacent cells through desmosomes.

4.5.2 Biopolymers of cytoskeleton

Except for cellulose fibers of plant cell walls which are polysaccharides, the filaments of importance to cell are all composed of protein polymers (Flory, 1969). Some cells, such as auditory outer hair cells, contain strings of spectrin. The extracellular matrix

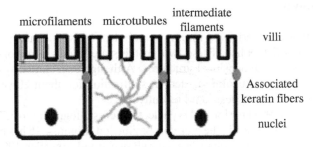

FIGURE 4.14
Cytoskeletal components of intestinal epithelial cells.

TABLE 4.4
Key Biomolecular Polymers in Cells

Polymer	Unit	Linear Density (Da/nm)	Persistence Length (nm)
Spectrin	Two-stranded filament	4,600	10–20
IF	Four-stranded protofilament	4,000	
Collagen (type I)	Three-stranded filament	1,000	
DNA	Double helix	1,900	51–55
F-actin	Filament	16,000	10,000–20,000
Microtubules	Thirteen protofilaments	160,000	2×10^6–6×10^6

of connective tissue is traversed by fibers of collagen; a family of proteins exhibiting a variety of forms.

Collagen present in connective tissues such as tendons is organized hierarchically into ropes and sheets. Type I collagen, one of the most common varieties, produces a single strand with a molecular mass of 100 kDa. Three such strands organize into tropocollagen macromolecules that are 300 nm long and 1.5 nm in diameter. A collagen fibril whose diameter is 10 to 300 nm contains numerous strands of tropocollagen. These fibrils are commonly further organized into a parallel formation referred to as a collagen fiber. In the cytoskeleton of the erythrocyte, two dimers of spectrin are connected end to end to form a tetramer about 200 nm long, whose monomer units have masses of 240 to 260 kDa. Table 4.4 summarizes physical characteristics of key cellular polymers (Flory, 1969).

We can readily conclude based on the above that persistence length increases with mass density. This can be deduced from the expression for persistence length of a uniform cylindrical rod as a function of its radius R:

$$\xi_p = \beta \, YI \frac{\pi R^4}{4} \qquad (4.12)$$

from which one finds that, for example, for microtubules and actin their Young's moduli range from 10^9 to 2×10^9 N/m^2, a value comparable to collagen but much smaller than that for steel.

4.5.3   Tubulin

The tubulin that polymerizes to form microtubules (MTs) is actually a heterodimer of α tubulin and β tubulin. These two proteins are highly homologous and have nearly identical three-dimensional structures. Although their similarity had long been suspected, the fact that tubulin resisted crystallization for about 20 years prevented confirmation of this hypothesis until recently.

Nogales et al. (1998) performed cryoelectron crystallography to 3.7 Å resolution on sheets of tubulin formed in the presence of zinc ion. Figure 4.15 has been produced from Nogales' data via the protein data bank (Bairoch and Apweiler, 1998; PDB entry: 1tub), using MOLSCRIPT (Kraulis et al., 1991) and it clearly shows the similarities of the two proteins.

Each monomer is composed of a peptide sequence more than 400 members long which is highly conserved between species. The amino acid sequences for these proteins may be compared. Table 4.5 lists the conventional one- and three-letter codes for the 20 naturally occurring amino acids. Codes should be read from left to right and spaces are inserted between each group of 10 residues for clarity.

The α and β tubulin monomer each consists of two β sheets flanked by α helices to each side. Each monomer can be divided into three domains: an N terminal nucleotide-binding domain, an intermediate domain that in β tubulin contains the taxol-binding site, and a C terminal domain thought to bind to motor proteins and other microtuble-associated proteins (MAPs). The monomer structures are strikingly similar to the crystal structure of the bacterial division protein FtsZ (Loew and Amos, 1998), which was determined to 2.8 Å resolution. FtsZ can form tubules *in vitro* and has been proposed as a prokaryotic predecessor of eukaryotic tubulins. The tubulins and FtsZ share a core structure of 10 β strands surrounded by 10 α helices.

Figure 4.16 shows electrostatic charge distribution in vacuum for a tubulin heterodimer that includes two C termini. Note that it is highly negative.

In humans, no less than six α isotypes and seven β isotypes, in addition to the well known γ tubulin have been found (Lu et al., 1998). Although the sequence of amino acids is highly conserved overall, certain regions of α_1 tubulin show divergence from α_2 tubulin and so on. Recent studies have shown that the differences in α tubulin are more subtle than those in β tubulin. Table 4.6 compares the main β tubulin isotypes in cows. The locations of cells expressing that particular variant of tubulin are listed along with the homology in percent with β_1 derived from a comparison of primary sequences (Lu et al., 1998). Tubulin isotypes and the relationship between their structures and functions present an intriguing scientific puzzle. For example, β_I has been largely conserved in evolution; in β_{IV}, divergences have been conserved for over 0.5 billion years. Correlations between distribution and function indicate that β_{II}, and β_{VI} are often found in blood cells. β_{III} is frequent in tumors. The $\alpha\,\beta_{III}$ dimer is more dynamic than $\alpha\beta$II or $\alpha\beta$IV, while β_{II} and β_{IV} cross-link easily. The mitotic spindle contains β_I, β_{II} and, β_{IV}; the nucleus contains β_{II}, and γ forms nucleation sites for microtubule assembly.

Several studies confirmed the different conformational states of tubulin; however we have little quantitative information on the structural changes. The first indication of more than a single conformational state of tubulin came simply from observing the

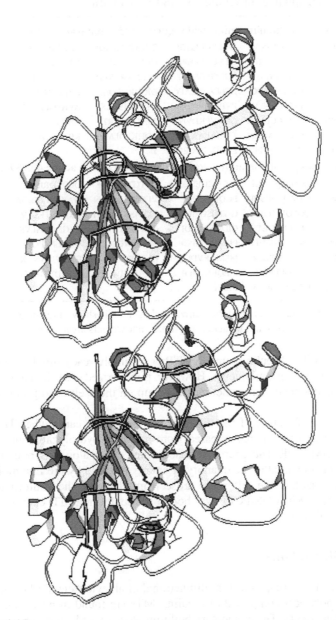

FIGURE 4.15
Tubulin molecule based on electron crystallography data of Nogales et al. (1998) shows similarities of the α (a) and β (b) subunits). The stick outlines near the bases of the subunits indicate locations of GTP when bound.

TABLE 4.5
Amino Acid Sequences of Human α_1 and β_1 Tabulin

Human α_1 tubulin amino acid sequence (451 amino acids; 50,157 Da)					
MRECISIHVG	QAGVQIGNAC	WELYCLEHGI	QPDGQMPSDK	TIGGGDDSFN	TFFSETGAGK
HVPRAVFVDL	EPTVIDEVRT	GTYRQLFHPE	QLITGKEDAA	NNYARGHYTI	GKEIIDLVLD
RIRKLADQCT	RLQGFLVFHS	FGGGTGSGFT	SLLMERLSVD	YGKKSKLEFS	IYPAPQVSTA
VVEPYNSILT	THTTLEHSDC	AFMVDNEAIY	DICRRNLDIE	RPTYTNLNRL	IGQIVSSITA
SLRFDGALNV	DLTEFQTNLV	PYPRIHFPLA	TYAPVISAEK	AYHEQLSVAE	ITNACFEPAN
QMVKCDPGHG	KYMACCLLYR	GDVVPKDVNA	AIATIKTKRT	IQFVDWCPTG	FKVGINYQPP
TVVPGGDLAK	VQRAVCMLSN	TTAIAEAWAR	LDHKFDLMYA	KRAFVHWYVG	GMEEGEFSE
AREDMAALEK	DYEEVGVHSV	EGEGEEEGEE	Y		

Human β_1 tubulin amino acid sequence (444 amino acids; 49,759 Da)					
MREIVHIQAG	QCGNQIGAKF	WEVISDEHGI	DPTGTYHGDS	DLQLDRISVY	YNEATGGKYV
PRAILVDLEP	GTMDSVRSGP	FGQIFRPDNF	VFGQSGAGNN	WAKGHYTEGA	ELVDSVLDVV
RKEAESCDCL	QGFQLTHSLG	GGTGSGMG	TL LISKIREEYP	DRIMNTFSVV	PSPKVSDTVV
EPYNATLSVH	QLVENTDETY	CIDNEALYDI	CFRTLRLTTP	TYGDLNHLVS	GTMECVTTCL
RFPGQLNADL	RKLAVNMVPF	PRLHFFMPGF	APLTSRGSQQ	YRALTVPDLT	QVFDAKNMM
AACDPRHGRY	LTVAAVFRGR	MSMKEVDEQM	LNVQNKNSSY	FVEWIPNNVK	TAVCDIPPRG
LKMAVTFIGN	STAIQELFKR	ISEQFTAMFR	RKAFLHWYTG	EGMDEMEFTE	AESNMNDLVS
EYQQYQDATA	EEEEDFGEEA	EEEA			

Note: The amino acid sequences of human α_1- and β_1- tubulin show a high degree of homology. The sequences, number of amino acids, and molecular weights are shown.

assembly of MTs. Tubulin bound to GTP or assembly-ready tubulin binds together and forms straight protofilaments. However, polymerized GDP-bound tubulin forms curved protofilaments that sometimes close up on themselves to produce oligomer rings.

Another manifestation of the different conformations came when Hyman et al. (1995) measured the energy released from a slowly hydrolyzable analog of GTP known as GMPCPP. The energy released when the analog was bound to tubulin was reduced compared to the quantity of energy released by the free molecule. The speculation was that the difference arose from changing the conformation of the tubulin dimer. When GMPCPP was bound to tubulin in a MT, energy release was further diminished.

4.5.4   Microtubules

The thickest and perhaps most multifunctional of all cytoskeletal filaments are MTs which are polymers comprised of tubulin. MTs are found in nearly all eukaryotic cells (Dustin, 1984). The elementary building block of a MT is an α–β heterodimer whose dimensions are 4 by 5 by 8 nm that assembles into a cylindrical structure that typically has 13 protofilaments (see Figure 4.17). The outer diameter of a MT is 25 nm and the inner diameter is 15 nm (Amos, 1995). The monomer mass is about 55 kDA. MTs are larger and more rigid than actin microfilaments (MFs) and intermediate filaments (IFs), and thus serve as major architectural struts of the cytoplasm.

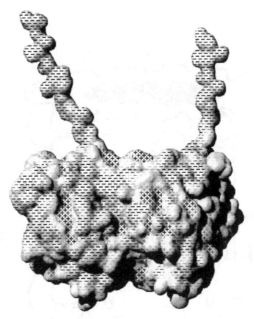

FIGURE 4.16
Electrostatic charge distribution on an α tubulin monomer microtubule.

TABLE 4.6
Localization and Homology of Bovine β Tubulin

Isotype	Localization	Homology	Abundance in Brain (%)
β_1	Everywhere, thymus	100.0	3
β_2	Brain	95.0	58
β_3	Brain, testis, tumors	91.4	25
β_4	Brain, retina, trachea	97.0	11

Indeed, MTs act as scaffolds to determine cell shape and provide "tracks" for cell organelles and vesicles to move on. MTs are involved in a number of specific cellular functions such as (1) organelle and particle transport inside cells (e.g., nerve axons) through the use of motor proteins, (2) signal transduction; (3) when arranged in geometric patterns inside flagella and cilia, they are used for locomotion or cell motility, (4) during cell division they form mitotic spindles required for chromosome segregation, and (5) organization of cell compartments (e.g., positioning of ER, Golgi apparatus, and mitochondria). MTs perform these tasks by careful control over assembly and disassembly (MT dynamics) and by interactions with microtubule-associated proteins (MAPs).

Ledbetter and Porter (1963) were the first to describe these tubules found within the cytoplasm whose structure was well established by light microscopy, immunofluoresence, and cryoelectron microscopy. The tube is composed of strongly bound

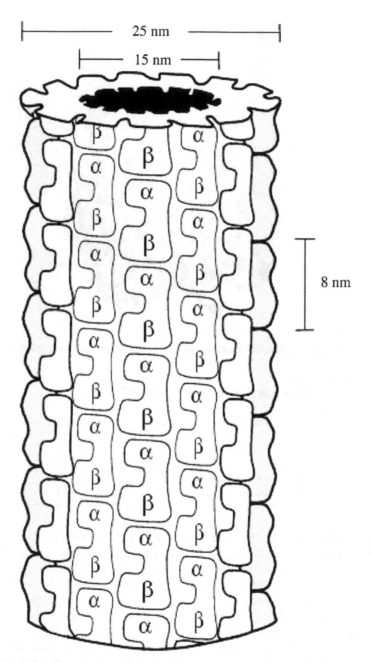

FIGURE 4.17
Section of a typical microtubule demonstrating the hollow interior filled with cytoplasm and the helical nature of its construction. Each vertical column is known as a protofilament and a typical microtubule has 13 protofilaments.

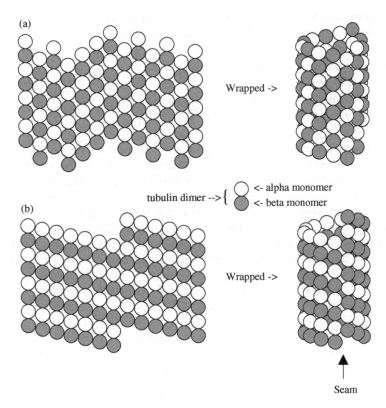

FIGURE 4.18
The 13A and 13B MT lattices: (a) in the A lattice, perfect helical spirals are formed; (b) the structural discontinuity in the B lattice is known as the seam.

linear polymers, known as protofilaments, that are connected via weaker lateral bonds to form a sheet that wraps to form a cylinder. While the electron crystallography of Nogales et al. (1998) has shown that the α and β monomers are nearly identical, this small difference on the monomer level allows the possibility of several lattice types.

In particular, the MT A and B lattices have been identified (see Figure 4.18). Moving around the MT in a left-handed sense, protofilaments of the A lattice exhibit a vertical shift of 4.92 nm upward relative to their neighbors. In the B lattice, this offset is only 0.92 nm because the α and β monomers have switched roles in alternating protofilaments. This change results in the development of a structural discontinuity in the B lattice known as the seam. In addition to lattice variation, it is also known from experimental observation that the number of protofilaments may differ from one MT to another. Although 13 protofilaments represent the most common situation *in vivo*, Chrétien et al., observed that the protofilament number need not be conserved along the length of an individual MT. This leads to the emergence of a structural defect.

MT formation takes place in relation to changes in cell shape, movement, and in preparation for mitosis under the following conditions:

Critical concentration of tubulin ~1 mg/ml
Optimal pH = 6.9
Strict ionic requirements (Mg, Ca, Li, etc.)
Assembly can be inhibited by cold (7°C)

The onset of assembly crucially depends on (1) temperature range (Fygenson et al., 1994), (2) concentration of tubulin in the cytoplasm, and (3) supply of biochemical energy in the form of GTP. Tubulin contains two bound GTP molecules, one of which is exchangeable and binds to the β tubulin molecules while the other that binds to α tubulin is fixed. Dimers polymerize head to tail into protofilaments (pfs), and each MT polymer is polarized. The plus end is a fast assembly end; the minus end is a slow assembly end (Tran et al., 1997). Organization of MTs *in vivo* is also polarized; minus ends are bound to a microtubule organizing center (MTOC) while plus ends extend outward to cell periphery.

The first stage of MT formation is nucleation, a process that requires tubulin, Mg^{++}, and GTP and also proceeds most rapidly at 37°C. This stage is relatively slow until the microtubule is initially formed. The second phase, elongation, proceeds much more rapidly. During nucleation, α and β tubulin molecules join to form heterodimers. They in turn attach to other dimers to form oligomers that elongate to form protofilaments. MT lengths may range from hundreds of nanometers to micrometers. When a tubulin molecule adds to the MT, the GTP is hydrolyzed to GDP. Eventually the oligomers join to form ringed MTs. The hydrolysis of GTP, of course, is facilitated at 37°C and stopped at 7°C and below.

MTs *in vitro* undergo irregular periods of growth, rapid shortening (catastrophe), and regrowth (rescue). They may vary in their rate of assembly and disassembly depending on prevailing conditions. See Figure 4.19. The assembly dynamics at each end of the MT differs also. The so-called plus end is three to five times as dynamic as the negative end. Both the growth and shortening of the positive end of the MT occur at rates at least three times those of the negative end, but both ends are dynamic in free MTs. Tubulin half-life is nearly a full day, but the half-life of a MT may be only 10 min. Thus, MTs are in a continued state of flux.

This process is believed to respond to the needs of the cell and is called dynamic instability (Mitchison and Kirschner, 1984; Horio and Hotani, 1986). This behavior is the consequence of GTP hydrolysis during MT polymerization. GTP caps on growing MTs are stable (Mandelkow et al., 1991), but GDP caps are unstable and lead to catastrophe events. MTs will form normally with nonhydrolyzable GTP analog molecules attached. However, they will not be able to depolymerize. Thus, the normal role of GTP hydrolysis may be to promote the constant growth of MTs as needed by cells.

Mathematical models of dynamic instability use the Monte Carlo approach (Bayley et al., 1990) or master equation formalism (see Appendix D for general exposition and Flyvbjerg et al., 1994, 1996 for details of application to microtubules). Polymerized and unpolymerized monomers bind GTP or GDP. Upon assembly, the energy

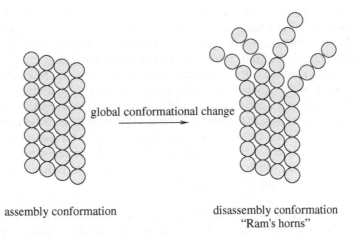

assembly conformation   disassembly conformation
                                                 "Ram's horns"

FIGURE 4.19
Global conformational change that occurs between assembly and disassembly phases. The once-straight protofilaments become curved after the individual tubulin subunits undergo structural changes.

of hydrolysis from GTP to GDP is imparted to the tubulin subunits, but its fate is unknown. We believe that at least part of this relatively large amount of energy is stored in MTs in the form of stacking fault energy which, when a critical amount is exceeded, can be released in the form of an earthquake-like collapse of the entire MT structure. This hypothesis is consistent with the observation by Hyman et al. (1992) of a structural change accompanying GTP hydrolysis. They found a length change in monomer spacing from 4.05 nm to 4.20 nm. Thus, elastic energy may be stored locally as lattice deformation. This 4% change in tubulin length resulted in a new moiré pattern when the MTs were imaged by electron cryomicroscopy and different positions of equivalent peaks between the x-ray crystallographic diffraction patterns of GDP-MTs and GMPCPP-MTs.

Tran et al. argued that three conformational states exist with a metastable intermediary state between growing and shrinking conformations. They seem to believe that these may be tubulin with GTP bound, with GDP·P_i bound, and finally with GDP bound at the exchangeable site. This is similar to the hypothesis of Semënov (1996) in his review of MT research. It appears that in addition to the multitude of α and β isotypes and numerous post-translational modifications, tubulin may also exist in other conformations.

Above a certain tubulin concentration threshold, MT ensembles show quasiperiodic, regular patterns of damped oscillations (Carlier et al., 1987). This indicates that interactions between individual microtubules must play a crucial role in affecting this drastic change in behavior. What is intriguing is how the stochastic individual behavior may change into smooth collective oscillations observed at high tubulin concentrations (Mandelkow and Mandelkow, 1992). The answer may simply be given by the application of statistical mechanics to ensemble averaging. A complete explanation of the above requires application of the master equation formalism or alternatively the use of chemical kinetics (Marx and Mandelkow, 1994; Houchmandzedeh and Vallade, 1996; Sept et al., 1999; see also Chapter 6). Moreover, MTs have been demonstrated to self-organize *in vitro*, leading to spatial pattern formation (Tabony and Job, 1990). A mathematical description of this process can be provided in terms of reaction–diffusion equations (Hess and Mikhailov, 1994).

Gittes et al. (1993) measured MT flexural rigidity by thermal fluctuations in MT shape and found that it increased when MTs were treated with MAP τ, such that the MTs were structurally stabilized by preventing GTP hydrolysis. In addition to the fact that hydrolysis of GTP can result in destabilizing the MT with respect to catastrophic disassembly, it also makes the polymer more flexible. It is interesting to note that the application of taxol, which reduces MT assembly dynamics, reduces the mechanical rigidity of MTs (Mickey and Howard, 1995).

The measured flexural rigidity of MTs corresponds to a Young's modulus of 1.4 GPa in normal MTs and can be raised by more than a factor of two to 3.4 GPa when hydrolysis is prevented. In an independent study, a Young's modulus of 4.6 GPa was derived from the buckling of microtubules that required the application of a 10-pN force.

The Young's modulus of F actin has also been measured and is of the same order of magnitude although conflicting measurements make actin both more and less rigid than MTs. The number is sufficiently large to imply that the cell must rely on depolymerization rather than deformation to effect a shape change. This measured value of Young's modulus indicates that bending of a protofilament into an arc with a radius of curvature of ~20 nm, as observed by Mandelkow et al., would require about 0.14 eV/dimer (3.2 kcal/mol) which is slightly less than the energy of GTP hydrolysis, 0.22 eV (5.1 kcal/mol). This is believed to explain the difference observed in the free energy release when free floating GTP is hydrolyzed compared to the hydrolysis of GTP bound to a MT.

Normal cell organization precludes the existence of free MTs since MTs become attached to centrosomes or organized within minutes. Microtubule organizing centers (MTOCs) contain MAPs that bind minus ends of MTs and nucleate MT assembly. Within the cell body, most MTs emanate from centrioles. The negative ends of MTs are anchored at these microtubule organizing centers. MTs *in situ* are interconnected and intraconnected by microtubule-associated proteins. MAPs exert stabilizing effects on the dynamics of MTs as mentioned above. At the base of cilia and flagella, MTOCs are called basal bodies or centrioles. In interphase cells, MTOCs are found in the centrosomes (cell centers) containing two centrioles plus pericentriolar material.

MTs are formed in a cell in an area near the nucleus called the aster. Usually the minus end is the anchor point. Regulatory processes appear to control MT assembly in cells. MT growth could be promoted in a dividing or moving cell, but would be better controlled in a stable, polarized cell that provides a GTP cap on the growing end to regulate further growth. This happens when tubulin molecules are added faster than GTP can be hydrolyzed. Thus, the MT becomes stable and does not depolymerize. Another way of capping it is to add a structure such as a cell membrane at its end.

Assembling microtubules exert pushing forces on chromosomes during mitosis. The force that a single MT can generate was measured by attaching MTs to a substrate at one end and causing them to push against a rigid barrier at the other end.

The subsequent buckling of the MTs was analyzed to determine the force on each microtubule end and the growth velocity.

MTs and their individual dimers possess net charges per dimer of approximately –40 e at neutral pH and dipole moments on the order of 1800 D. Thus, MTs are electrets, i.e., oriented assemblies of dipoles predicted to have piezoelectric properties due to the coupling of elastic and electric degrees of freedom of the protein (Brown and Tuszynski, 1997).

The predicted dipole orientation of tubulin units in a MT is almost perpendicular to the surface of the cylinder, resulting in a net near zero value of the dipole moment for the MT as a whole. It has been recently emphasized that microtubule networks play an essential role in cellular self-organization phenomena including reaction–diffusion instabilities in the mechanisms of cytoskeletal self-organization. We believe that electrostatic interactions are crucial to self-organization of the cytoskeleton that is so prominent in mitosis.

4.5.5 Microtubule-associated proteins

MAPs are tissue- and cell type-specific. Several classes with different functions have been identified. MAPs are high molecular weight proteins (200 to 300 kDa) or the so-called tau (20 to 60 kDa) proteins. General classes are minus end binding (MTOCs), plus end binding (kinetochores of mitotic chromosomes; Hays and Salmon, 1990), polymer severing, polymer stabilizing, cross-linking (MAP-2 and tau in neurons), and motor proteins (dyneins and kinesins; Amos and Cross, 1997). They are discussed later in this chapter.

One of their domains binds to tubulin polymers or to unpolymerized tubulin. This speeds up polymerization, facilitates assembly, and stabilizes the MTs. The other end binds to vesicles or granules. MAPs may vary with cell type. The best examples are found in neurons. Furthermore, it is believed that some MAPs may bind to special sites on the α tubulin that forms after it is assembled into a MT. These are sites where a specific molecule is acetylated or the tyrosine residue is removed from the carboxy terminal. They are important marker sites for stabilized microtubules, because they disappear when microtubules are depolymerized. In summary, MAPs accelerate polymerization, serve as "motors" for vesicles and granules, and essentially control cell compartmentalization.

4.5.6 Microfilaments

Microfilaments (MFs) have the smallest diameters of all the cytoskeletal filaments and are the most common. They are composed of a contractile protein known as actin and consequently are called actin filaments. Actin is the most abundant cellular protein. Microfilaments are fine, thread-like protein fibers 3 to 6 nm in diameter (see Figure 4.20). Their association with myosin is responsible for muscle contraction (see below). Microfilaments can also carry out cellular movements including gliding, contraction, and cytokinesis.

FIGURE 4.20
Microfilaments with fluorescent labels.

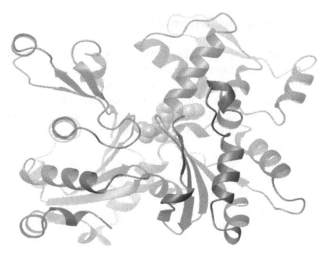

FIGURE 4.21
Structure of the G actin monomer.

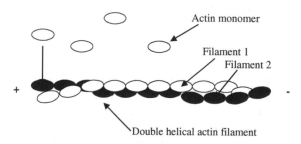

FIGURE 4.22
Formation of F actin polymer.

4.5.7 Actin

Actin is one of the most abundant proteins in eukaryotic cells. Globular (G) actin and filamentous (F) actin are reversible structural states of actin. F actin is a polymer of G actin. These strands are capable of forming stable and labile components within cells. F actin is the main component of thin filaments in muscle sarcomere. Actin can also be found in nonmuscle cells and often is associated with other proteins in the cytoskeleton. It provides a dynamic component of the cytoskeleton (Frieden et al., 1983).

G actin is a structural protein of 375 residues (see Figure 4.21) which has been highly conserved in evolution and whose globular monomer form has a weight of 42 kDa. It binds ATP and Mg^{2+}. G actin monomers assemble into F actin filaments of two-stranded geometry (see Figure 4.22) with a mass density three times as high as that of spectrin filaments. Polymerization of actin (Oosawa and Asakura, 1975) has two different rates at the poles of nucleating filament barbed and pointed ends. ATP actin binds faster to the barbed ends. ATP is then hydrolyzed to ADP after incorporation.

4.5.8 Actin filaments

The length distributions of actin filaments are exponential with a mean of about 7 μm. The polymerization of F actin from G actin is a largely monotonic process dependent on the concentration of ATP. The F actin assembles by and large according to a standard nucleation and elongation mechanism to a saturation value that depends on pH (Tobacman and Korn, 1983).

The length of an actin filament is independent of the initial concentration of actin monomer, an observation inconsistent with a simple nucleation–elongation mechanism. However, with the addition of physically reasonable rates of filament annealing and fragmenting, a nucleation–elongation mechanism can reproduce the observed average lengths of filaments in two types of experiments: (1) filaments formed from a wide range of highly purified actin monomer concentrations, and (2) filaments formed from 24 mM actin over a range of CapZ concentrations (see Chapter 6 for details of the mathematical formulation). Once assembled, microfilaments have diameters of about 8 nm (Figure 4.23).

In cells, F actin exhibits characteristic forms as follows:

Thin filaments in striated muscle
Individual filaments in cortical cytoplasm
Bundled filaments (stress fibrils) in interphase fibroblasts
Short actin bundles in microvilli
Filament bundles ending in focal contacts.
Filaments running parallel with membrane at the membrane–cytoplasm
    interface

Microfilaments are linked by actin-associated proteins and congregate into one of the three major forms (Figure 4.24). The polymerization dynamics and filament

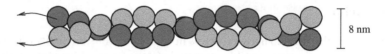

8 nm

FIGURE 4.23
Double-stranded helix of an actin filament. Each F actin strand has a polarity indicated by the arrows. Each sphere represents a monomer of G actin.

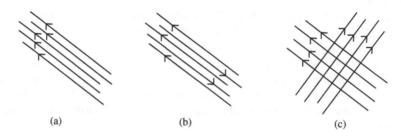

(a)   (b)   (c)

FIGURE 4.24
Three main arrangements of microfilament aggregation that form with the aid of actin cross-linking proteins: (a) parallel bundles, (b) contractile bundles, and (c) gel or lattice-like arrangements.

organization of actin into spatial structures has been modeled, for example, by Edelstein-Keshet (1998). In the configuration of parallel strands, microfilaments often form the cores of microvilli, while in an antiparallel arrangement, actin may act in conjunction with myosin to bring about muscle contraction in the presence of ATP (Geeves and Holmes, 1999). Microfilaments are often found with the lattice configuration near the leading edges of growing or motile cells where they provide greater stability to the newly formed regions. New actin filaments are nucleated at the leading edges of cell growth and trailing microfilaments are disassembled.

4.5.9   Actin-binding proteins

Actin filaments are known to interact with at least 48 other proteins. Cytosolic myosin motors drive actin filaments. Actinin interacts with actin and other proteins including integrins. Spectrin is another multifunctional actin-binding protein that binds actin filaments to transmembrane proteins.

Talin is a 230-kDa protein concentrated at regions where actin filaments attach to and transmit tension across the cytoplasmic membrane to the extracellular matrix. It is absent at cell–cell interaction sites. It is also thought to interact with integrins and vinculin, a 117-kDa protein found at all cell–cell and cell–matrix interaction sites. These are focal contacts (adherens junctions), dynamic structures related to cell adhesion and movement. Tensin is also found in close proximity to actin and integrins. Profilin is a small actin-binding protein with high expression in cytoplasm of nonmuscle cells. PIP and PIP2 cause dissociation of complexes and bind at the

TABLE 4.7
Summary of Actin-Binding Proteins

Function	Protein types
Bind monomeric actin; prevent polymerization	Profilins
Cap one end of filament	CapZ
Sever filaments; bind to barbed ends	Gelsolin
Bind laterally along filaments	Tropomyosin
Move along filaments	Myosins
Connect filaments and network bundles	α-Actinin, synapsin, villin

barbed ends of actin filaments. α Actinin is involved in nonmuscle cells where some actin bundles interact with the membrane.

Gelsolin caps fast-growing ends of filaments. It may also be involved in filament disassembly. It is a representative of a number of proteins forming a family believed to act in regulating actin polymerization. For example, cofilin participates in depolymerization of actin filaments further back from the leading edge within the lamellipodium. Table 4.7 summarizes characteristics of actin-binding proteins.

The lamellipodium is located at the leading edge of a moving cell. The forward extension of a lamellipodium is driven by actin polymerization. Lamellipodia contain extensively branched networks of actin filaments, with their plus ends oriented toward the plasma membrane (Small et al., 2002). Several proteins already discussed participate with actin in generating the forward action of the lamellipodium in the so-called ameboid movement.

Profilin, which promotes ADP/ATP exchanges by G actin to yield an ATP-bound form competent to polymerize, is at the leading edge of an advancing cell. Arp2/3, which binds to the sides of actin filaments and nucleates growth of new filaments, is associated with actin filament networks in lamellipodia. Various cross-linking proteins stabilize actin networks in lamellipodia. Newly formed filaments are stable as advancing lamellipodia move past them, until they disassemble further back from the edge.

4.5.10 Intermediate filaments

The cytoskeletons of all eukaryotic cells contain intermediate filaments (IFs). They are thinner than MTs, thicker than actin, and more stable than both. They are microscopic ropes with protein subunits in the 10 to 15 nm range. Intermediate filaments have fairly complex hierarchical organizations in which two protein building blocks generate dimers through antiparallel association of two molecules in a coiled coil configuration.

Pairs of dimers form linear protofilaments 2 to 3 nm in width. An IF is a bundle of eight protofilaments forming a cylinder about 10 nm in diameter. IFs form by association of α helix rod domains. Parallel unstaggered two-stranded helical coiled coils of 44 to 54 nm size are used. Zipper structures of hydrophobic residues are

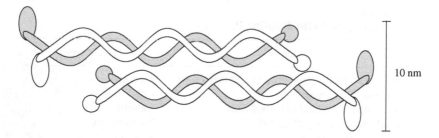

FIGURE 4.25
Intermediate filament. The filamentous protein tetramer is nonpolar and composed of two coiled coil dimers.

evident. IF aggregation is achieved when two dimers form a tetramer (see Figure 4.25). Protofilaments form longitudinal extensions of tetramers.

A number of possible topologies of filaments have been found including parallel, antiparallel, staggered, and registered. The end domains of IF structures play a significant role. Tail domains may protrude from the main body of a filament, making it accessible for interactions with other parts of the cytoplasm. The pool of unpolymerized IF subunits is insignificant since newly synthesized proteins polymerize very quickly. Filaments appear to require no cofactors or energy input for their assembly.

The IF group includes vimentin, desmins, nuclear lamins, and the keratins of epithelial cells. IFs appear to provide resilience, tensile strength, and plasticity to cells and may provide the main cytoskeleton with complex arrays. These molecules are quite unlike globular molecules, tubulin, and G actin since they form nonpolar structures. The overlap between dimers allows for the filaments to stretch and, as a result, intermediate filaments can withstand large stresses without breaking.

An extensive network of intermediate filaments surrounds the nucleus to form the nuclear envelope; it extends to desmosomes in the plasma membrane. The formation and disassembly of the nuclear lamina are regulated by phosphorylation. This allows elimination of the envelope before mitosis. The filaments also extend to the cell periphery where they maintain integrity and may connect with the membrane. Specific cells such as neurons have their own distinctive intermediate filaments, known as neurofilaments.

The tissue specificity of IFs is very high. IFs may vary substantially in sequence since at least 40 genes encode them. IF expression closely correlates to the behavior and role of the cell. IFs, although different in sequence, have the same overall structure with a long central rod domain and a coiled coil. The following are the key types of IFs.

Type I and Type II (keratins) are expressed in epithelia and sheet tissues that delimit functional components. Keratin filaments are found in the cytoplasm and loop into desmosomal plaques. They maintain physical integrity of the cell.

Type III (vimentin) can form homopolymers expressed by fibroblasts, hemopoetic cells, and endothelial cells in embryogenesis; formed in primary mesoderm and endoderm cells.

Type III (desmin) is found in smooth, striated, and cardiac muscle.

Type IV (neurofilaments) is classified into three types: NF-L, NF-M, and NF-H. They are heavily phosphorylated, especially NF-M and NF-H. Charged tails point from the filament and may space filaments and regulate axonal diameter.

Type V (lamins) form a network rather than filaments. Significant differences exist in rod and tail domain structures. Nuclear lamin proteins provide envelopes of insoluble cytoskeletal material below the nuclear membrane. Lamin disassembly at mitosis is triggered by phosphorylation.

IF control is thought to arise from phosphorylation which alters fibril assembly and solubility. Other known post-translational modifications include glycosylation of keratins and lamins. IFs generally respond to the cell cycle, movement, differentiation, etc. IF mutations cause serious pathologies, e.g., the seven keratin diseases, the most common of which is epidermolysis bullosa simplex in which basal epidermis cells detach from underlying connective tissue and the result is severe blistering.

Single-stranded DNA and RNA associate readily with many IF structures. Binding is predominantly by the H1 domain and interferes with fibril assembly. DNA–lamin interactions are thought essential for the integrity of nuclear chromatin.

IF-associated proteins are plectin (300 kDa) which associates with several cytoskeletal components and links to membranes directly, lamin receptors, and lamin-associated proteins.

4.6 Networks, stress fibers and tensegrity

To understand fully the way living systems form and function (including the mechanism of cell division), we must uncover the basic principles that guide biological organization. Many natural systems are constructed using a common form of architecture known as tensegrity, a term designating a system that stabilizes itself mechanically because of the way in which its tensional and compressive forces are distributed and balanced.

Tensegrity structures are mechanically stable because of the way the entire architecture distributes and balances mechanical stresses — not because of the strength of individual members. Since tension is continuously transmitted across all structural members, a global increase in tension is balanced by an increase in compression within members distributed throughout the structure. As described above, the interiors of living cells contain internal frameworks called cytoskeletons composed of three types of molecular protein polymers, known as MFs, IFs and MTs.

Cell shape is regulated by a complex balance of internal and external forces exerted by the extracellular matrices. This balance in terms of tensegrity was described by Ingber (1993, 1997). Cells acquire their shape from tensegrity arising from the three major types of cytoskeleton filaments and also from the extracellular matrices — the anchoring scaffolding to which cells are naturally secured. A network of contractile microfilaments in the cell exerts tension and pulls the membrane and all its internal constituents toward the nucleus at the core. Opposing this inward pull are two main types of compressive elements, one outside the cell and the other inside. The component outside the cell is the extracellular matrix; the compressive "girders" inside the cell can be MTs or large bundles of cross-linked MFs within the cytoskeleton. The third component of the cytoskeleton, the intermediate filaments, interconnect MTs and contractile MFs to the surface membrane and the nucleus.

Contractile actin bundles act as molecular cables. They exert tensile force on the cell membrane and the internal constituents, pulling them all toward the nucleus. MTs act as struts that resist the compressive force of the cables. In many cases, it is important to maintain cell shape to preserve its functionality. Chen et al. (1997) showed experimentally how cells switch between genetic programs when forced to grow into specific shapes. King and Wu revealed on theoretical grounds how cell geometry changes and the susceptibility of cells to electromagnetic fields.

Although the cytoskeleton is surrounded by membranes and penetrated by viscous fluid, this hard-wired network of molecular struts and cables stabilizes cell shape. MTs are compressed, rigid elements. Actin filaments are tensile. Maniotis and Ingber (1997) demonstrated that pulling on receptors at the cell surface produces immediate structural changes deep inside the cell. Thus, cells and nuclei do not behave like viscous containers filled with "cystoplasmic soup." The tensegrity force balance provides a means to integrate mechanics and biochemistry at the molecular level.

Ingber's tensegrity model suggests that the cytoskeleton structure can be changed by altering the balance of physical forces transmitted across the surface. This finding is important because many enzymes and other substances that control protein synthesis, energy conversion, and growth in cells can be bound by the cytoskeleton. Dynamic interplay may occur between the cytoskeletal geometry and the kinetics of biochemical reactions including gene activation. Remarkably, by simply modifying their shapes, cells may switch to different genetic programs. Cells that spread flat are believed to be more likely to divide. Round cells are thought to activate a death program called apoptosis. When cells are neither too extended nor too retracted, they do not divide or die; they differentiate in a tissue-specific manner.

Mechanical restructuring of the cell and cytoskeleton apparently tells the cell whether to grow, divide, or die. Hence, mechanical forces are transmitted over specific molecular paths in living cells. Because a local force can change the shape of an entire tensegrity structure, the binding of a molecule to a protein can cause the different, stiffened helical regions to rearrange their relative positions throughout the length of the protein. Even *in vitro*, MTs exhibit traveling waves of assembly and disassembly and the formation of polygonal networks. Reversing the laboratory conditions and causing a subsequent assembly stage revealed memory effects in these pattern formation phenomena. Table 4.8 summarizes the tension forces produced by tensegrity structures.

An important force-generating family of proteins deserves special attention due to dynamic behavior and pervading presence in all living cells. This class is known as the motor proteins.

TABLE 4.8
Tension Forces within Cells

Tensegrity	Force in pN
Tension generated by mybrofil	10^5 to 10^6
Viscous drag force on *Vorticella*	8.6×10^3
Viscous drag force on chromosomes	0.1
Force needed to stop chromosomes	700

4.7 Motor proteins and their roles in cellular processes

An important method of generating force and movement within cells is via motor proteins. The assembly and disassembly of the cytoskeleton in conjunction with force generated by motor proteins is thought to be the main mechanism for mitosis and organelle transport within cells (Leibler and Huse, 1993).

Approaching this problem from a biophysical view involves examining the interactions of motors and the cytoskeletal filaments and between the motors and cytoplasm. Advances have recently been made in our understanding of the activities of myosin and kinesin (Amos and Hirose, 1997). Motor proteins share two essential functional components in a motor domain and a cargo binding domain, and most also contain central linking domains that can be extended.

The three large families of naturally occurring motor molecules are the myosins, the kinesins, and the dyneins. All of them function by undergoing shape changes, utilizing energy from ATP. Each family has members that transport vesicles through the cytoplasm along linear assemblies of molecules — actin in the case of the myosins and tubulin for the other families. The kinesins and dyneins move or "walk" along microtubules and carry their cargo. We describe these and other motor proteins in this section and provide a mathematical description of their motion in Chapter 6.

The motors are ATPases that move objects on the filament surfaces. The axonemal dyneins drive MT sliding in cilia and flagella; cytoplasmic dynein and kinesins move many objects in opposite directions in the cytoplasm. In general, kinesins move toward the plus ends of MTs (outward) and dyneins move toward the minus ends (inward). The vesicles are attached to MAPs such as kinesin that move along the MTs and function like conveyor belts (see Figure 4.26).

These proteins have head regions that bind to MTs and also bind to ATP that provides biochemical energy. The head domains are thus ATPase motors. The tail domain binds to the organelle or vesicle to be moved. It is not known at a molecular level how the energy from ATP breakdown is converted into vectorial transport. A number of mathematical models, however, accurately describe motor protein motion (see Chapter 6) in spite of making a number of simplifying assumptions.

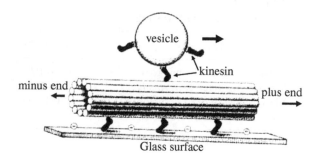

FIGURE 4.26
Vesicles are attached to MAPs that move along the MT conveyer belt via kinesin motors.

4.7.1 Myosin family

Myosins are motor proteins that move along actin filaments, while hydrolyzing ATP. Myosin is, therefore, an ATPase enzyme that converts the chemical energy stored in ATP molecules into mechanical work. The motive power for muscle activity is provided by myosin motors organized as thick filaments that interact with an array of thin actin filaments to cause shortening of elements within each myofibril. The shortening is achieved by relative sliding of the myosin and actin filament.

Myosin I has only one heavy chain, with a relatively short tail, plus two or more light chains in the neck region. While it binds to the plasma membrane, it may pull the membrane forward as it walks along a growing actin filament toward the plus end.

Myosins I and V, both of which bind to membranes, are postulated to participate in the movement of organelles along actin filaments and movements of plasma membrane relative to actin filaments. They associate with Golgi membranes that rise to secretory vesicles, including synaptic vesicles.

Myosin II, the form found in skeletal muscle, is sometimes called a conventional myosin. It includes two heavy chains, each with a globular motor domain (Ruppel et al., 1995) that includes a binding site for ATP and a domain that interacts with actin. Tail domains of the heavy chains associate in a rod-like α-helical coiled coil. Light chains associate with each heavy chain in the neck region (a total of four light chains for myosin II). Light chains of different myosin types are calmodulin or calmodulin-like regulatory proteins. They may provide stiffening in neck domains. Myosin II heads interact with actin filaments in a reaction cycle. ATP binding causes a conformational change that causes myosin to detach from actin. The active site closes, and ATP is hydrolyzed, as a conformational change (cocking of the head) results in weak binding of myosin to actin at a different place on the filament. P_i release results in a conformational change that leads to stronger myosin binding, and the power stroke. ADP release leaves the myosin head tightly bound to actin. In the absence of ATP, this produces muscle rigidity called rigor. In nonmuscle cells, myosin II is often found associated with actin filament bundles.

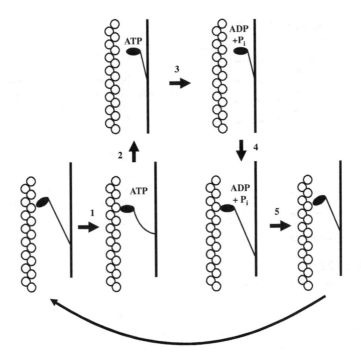

FIGURE 4.27
Scheme relating conformational changes at several stages; the largest movements are associated with binding of ATP and the subsequent release of ADP.

Myosin II is located in stress fibers, i.e., bundles of actin filaments that extend into the cell from the plasma membrane. These contract under some conditions. Myosin II has been shown by fluorescent labelling to be located predominantly at the rear ends of moving cells or in regions being retracted. Contraction at the rear of the cell probably involves sliding of actin filaments driven by bipolar myosin assemblies. The coiled coil tail domains of myosin II molecules may interact to form antiparallel bipolar complexes. These may contain many myosin molecules, as in the thick filaments of skeletal muscle or they may consist of as few as two myosin molecules.

A myosin cross-bridge acts as a rigid rod combined with an elastic element. The following scheme is based on the work of Huxley (1969) that relates conformational changes at several stages where the largest movements are associated with ATP binding and the subsequent release of ADP (see Figure 4.27): (1) the head binds to actin; (2) cross-bridge movement during the power stroke is taken up by stretching the elastic element.

Myosin V has two heavy chains like myosin II, but has a shorter coiled coil region followed by a globular domain at the end of each heavy chain tail. Calmodulin light chains are present in the neck region. Movement of myosin V along actin is processive, consistent with the role of myosin V in transporting organelles along actin filaments. Each of the two myosin V head domains dissociates from actin only when

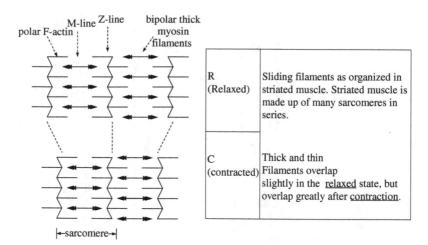

R (Relaxed)	Sliding filaments as organized in striated muscle. Striated muscle is made up of many sarcomeres in series.
C (contracted)	Thick and thin Filaments overlap slightly in the relaxed state, but overlap greatly after contraction.

FIGURE 4.28
Actin–myosin complex in muscle.

the other head domain binds to the next actin filament with the correct orientation, about 13 subunits further along the helical actin filament.

The ATP-dependent movement of myosin heads along actin filaments is accompanied by ATP hydrolysis. ATP binds to the myosin head adjacent to a seven-stranded β sheet. Loops extending from β strands interact with the adenine nucleotide. The nucleotide-binding pocket of myosin is opposite a deep cleft that bisects the actin-binding domain. Opening and closing of the cleft is proposed to cause the head to pivot about the neck region, as occupancy of the nucleotide-binding site changes and as myosin interacts with and dissociates from actin. Actin filaments may move relative to one another, as heads at the opposite ends of bipolar myosin complexes walk toward the plus ends of adjacent antiparallel actin filaments. See Figure 4.28.

Consistent with the predicted conformational cycle, different conformations of the myosin head and neck were found in crystal structures. Ca^{++} regulates actin–myosin interaction in various ways in different tissues and organisms. Some myosins are regulated by Ca^{++} binding to calmodulin-like light chains in the neck region. Some are regulated by phosphorylation of myosin light chains catalyzed by a Ca^{++}-dependent kinase or a kinase activated by a small GTP-binding protein. A complex of tropomyosin and troponin (including a calmodulin-like protein) regulates actin–myosin interaction in skeletal muscle. Caldesmon, a protein regulated by phosphorylation and Ca^{++}, controls interaction in smooth muscle.

4.7.2 Kinesin family

Kinesin is abundant in virtually all cell types at all stages of development and in all multicellular organisms. While most kinesin appears to be free in the cytoplasm, some is associated with membrane-bounded organelles, including small vesicles, the ER, and membranes that lie between the ER and Golgi. It is accepted now that

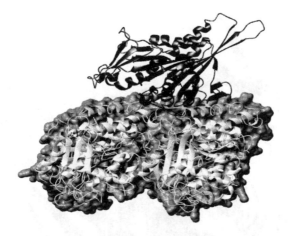

FIGURE 4.29
The kinesin motor domain binds to tubulin.

kinesin transports membrane-bounded organelles and perhaps macromolecular protein assemblies along microtubules.

Kinesin is a motor for fast axonal transport but it is not a motor for mitosis and meiosis. It is a force-generating motor protein that converts the free energy of the γ phosphate bond of ATP into mechanical work that powers the transport of intracellular organelles along MTs. The MTs always glide such that kinesin moves toward the plus or fast-growing end of the MT. This polarity and the orientation of MTs within cells show that kinesin is an anterograde motor (Case et al., 1997). Interestingly, other kinesin proteins, whose motor domains have high sequence homology and nearly identical structures, move in the opposite direction (in particular Ncd proteins; see below).

By decreasing the density of kinesin on the surface, a single kinesin molecule can be shown to move a microtubule through distances of several microns (see Figure 4.29). Interestingly, a single motor can move a microtubule as quickly — about 1 μm/s — as 10 or 100 motors.

Virtually all members of the kinesin family contain extended regions of predicted α-helical coiled coil in the main chain that likely produces dimerization. Kinesin is an $\alpha2$–$\beta2$ heterotetramer of two heavy and two light chains. Dimers of isolated head domains of kinesin possess the required ATPase properties. A large body of structural, biochemical, and biophysical evidence shows that kinesin has only one binding site per tubulin dimer (Svoboda and Block, 1994), and that the motor takes 8-nm steps from one tubulin dimer to the adjacent one in a direction parallel to the protofilaments (Howard, 2001).

Since the isolated motor domain of kinesin can hydrolyze up to 100 ATP molecules per second, it is likely that each step corresponds to one cycle of the ATPase reaction. The current model for how motor proteins generate force is that the motor contains an elastic element, a spring, that becomes strained as a result of one of the transitions between chemical states. This strain is the force that the motor puts out, and the relief of the strain is the driving force for the forward movement (Derenyi and Vicsek, 1996).

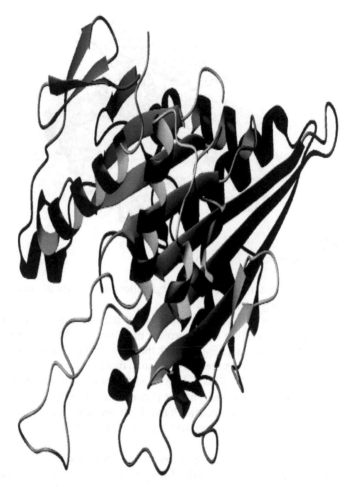

FIGURE 4.30
A ribbon diagram of the kinesin motor domain.

The kinesin superfamily contains 40 to 50 proteins known to date. Kinesins have common motor domains and variable tail domains. The structure of kinesin reveals heavy (KHC) and light (KLC) chains of 110 to 130 kDa and 60 to 80 kDa, respectively.

The motor region (Kozielski et al., 1997; Kull et al., 1996) shown in Figure 4.30 defines direction and rotational freedom of kinesin tails allowed while the tails are interpreted as acting as a spring. While the exact nature of the movement of two-headed kinesin has not been fully explained, the current opinion among biologists and biochemists leans toward the so-called hand-over-hand model of motion.

Electron microscopy of head domains bound to MTs indicates that the orientations of kinesin and Ncd heads are similar despite their different directions of movement (Fletterick, 1998). In addition, part of the kinesin head domain appears to swing in a nucleotide-dependent manner that may be related to the power stoke generation of movement. The kinesin and Ncd motor domains resemble asymmetric arrowheads with dimensions of approximately 75 by 45 by 45 Å. The bound MgADP lies in an exposed surface cleft. The core of the motor domain is an eight-stranded β sheet flanked by three α helices on each side.

Although Ncd and kinesin move in opposite directions along microtubules, the three-dimensional structures of their motor domains are remarkably similar. The eight-stranded β sheet and six α helices of the core and the three antiparallel β strands that form the small lobe are essentially identical in length, position, and orientation in both motors. The position of MgADP is virtually identical in both motors. The largest differences between kinesin and Ncd are in the surface loops near the nucleotide-binding pockets. The N and C termini of the Ncd and kinesin motor domains are positioned similarly in space and located 9 Å from one another on the side of the motor opposite to the nucleotide.

4.7.3 NCD dimer structure

The x-ray crystal structure of a dimeric form of the minus end-directed Ncd motor has been determined to 2.5 Å resolution. The model shows that the Ncd neck, the region that links the motor catalytic core to the coiled coil stalk, is helical rather than an interrupted β strand as in kinesin. Random mutagenesis of the Ncd neck resulted in a plus end-directed motor. Ncd has now been made to move like kinesin toward plus ends by fusing regions from outside the kinesin motor domain to the Ncd motor. That is, the directionality of the Ncd motor is reversed when joined to the N terminus of KHC, replacing the KHC motor domain. This suggests that direction of movement can be determined in part by a subdomain that is transferable between proteins.

Conversely, kinesin, a plus end-directed motor, has been made to move toward MT minus ends by fusing the stalk and neck of minus end-directed Ncd to the conserved KHC motor core. Mutation of the Ncd neck in the NcdKHC chimeric motor reverted the motor to slow plus end movement, showing that minus end determinants of motor directionality are present in the Ncd neck and that plus end determinants also exist in the KHC motor core.

4.7.4 An overview of KN motion models

Existing models of motor protein movement can be divided into those with diffusion and those with power strokes. Motor proteins may be directed by electrostatic interactions with the protein filaments on which they walk. In particular, the binding of kinesin to MTs has been shown to be primarily electrostatic. The motions of the motor proteins in both models are fueled through the hydrolysis of ATP since each step corresponds to one cycle of the ATPase reaction (Astumian and Bier, 1994). See Figure 4.31.

Diffusion models require an oscillating potential assumed to be driven by a conformation change of the motor. In the power stroke models by contrast, the motor protein structure changes. Ratchet potential models can be based on the Langevin

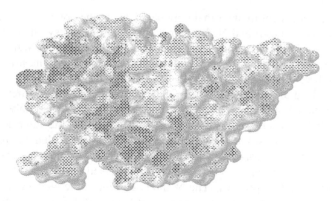

FIGURE 4.31
Electrostatic charge distribution on the kinesin motor domain.

equation with a fluctuating force (Doering et al., 1994) and/or with a Gaussian noise term. They also can be developed through Fokker-Planck formalism (Risken, 1989) for the probability distribution function $P(x, t)$ that describes the motion of the motor in a statistical manner (Jülicher et al., 1997).

While such models have been fairly successful in representing the gross features of the experimental data, the use of a hypothetical potential and a number of arbitrary parameters casts doubt on their correctness. It is also highly questionable to use a point mass approximation for kinesin since it is even larger than the tubulin dimer. The emphasis placed on stochasticity is also unwarranted for two-headed motors whose motion appears to be very deterministic. A chemomechanical cycle determines the key properties of the KN walk. Modeling motor protein motion will be addressed in Chapter 6.

4.7.5   Dyneins

Dyneins move toward the minus ends of MTs, which tend to be anchored in the centrosomes. Dynein motors also cause sliding between MTs that form the skeletons of cilia and flagella. In cilia and flagella, axonemal dynein motor molecules are attached to nine MT doublets arranged cylindrically around a pair of single microtubules. The dynein motors undergo a cycle of activity, during which they form a transient attachment to the doublet and push it toward the tip of the cilium or flagellum. At a particular ATP concentration, MT gliding velocities are found to increase with MT length. Other ciliary structures resist this sliding with the result that bends form along the length of the cilium or flagellum and propagate from base to tip or tip to base. The propagation of these bends requires coordinated actions of the several types of dynein motors present in the cilium or flagellum and the structures providing the resistance to sliding.

Dynein is also found in the cytoplasm where it is a minus end-directed motor proposed to drive fast retrograde vesicle transport in axons and other cells toward the centrosome. Cytoplasmic dynein is a rather large multisubunit protein complex composed of two identical heavy chains of about 530 kDa each, two 74-kDa intermediate chains, about four 53- to 59-kDa intermediate chains, and several light chains. Each heavy chain contains four ATP-binding sites including the ATP hydrolytic site that provides the energy for its movement along MTs and a MT-binding site. The 74-kDa intermediate chains are thought to bind dynein to its cargo, whether the cargo is a membrane-bounded vesicle in a neuron, a Golgi vesicle, a kinetochore, or a mitotic spindle astral MT. The dynein then provides the force to move the cargo along a MT toward its minus end.

We acknowledge here the use of the MolMol software in the production of the molecular structure of protein molecules shown in this chapter (Koradi, Billeter, and Wüthrich, 1996).

4.8 Centrioles, basal bodies, cilia, and flagella

Basal bodies and centrioles consist of nine-fold arrangements of triplet MTs. A molecular cartwheel fills the minus end of the cylinder and it is involved in initiating the assembly. The cylinders, called centrioles (see Figure 4.32), are always found in pairs oriented at right angles. Dense clouds of satellite material associated with the outer cylinder surfaces are responsible for the initiation of cytoplasmatic microtubules. Consequently, centrioles organize the spindle apparatus on which the chromosomes move during mitosis. They also control the directions of the movement of cilia and flagella.

4.8.1 Cilia and flagella

Cellular movement is accomplished by cilia and flagella. Cilia are hair-like structures that beat in synchrony and cause movements of unicellular paramecia. In large multicellular organisms, their role is to move fluid past cells. Cilia are also found in specialized linings in eukaryotes. Flagella are whip-like appendages that undulate to move cells. They are longer than cilia and have similar internal structures made of MTs. Cilia and flagella have the same internal structures. The major difference is length. Prokaryotic and eukaryotic flagella differ greatly. Both flagella and cilia have 9 + 2 arrangements of microtubules in which nine fused pairs of MTs are on the outside of a cylinder and two unfused MTs are in the center.

Dynein "arms" attached to the MTs serve as molecular motors. Cilia and flagella are organized from centrioles that move to the cell periphery. These are called basal bodies. Numerous cilia can project from a cell membrane. Basal bodies control the direction of movement. The difference between cilia and flagella and centrioles is that centrioles contain 9 sets of triplets and no doublet in their centers. How the triplets in the basal body turn into the cilium doublet remains a mystery. Figure 4.33 compares the cross-sections of a cilium and a centriole. Figure 4.34 compares the cross-sections of a flagellum and a MT axoneme doublet.

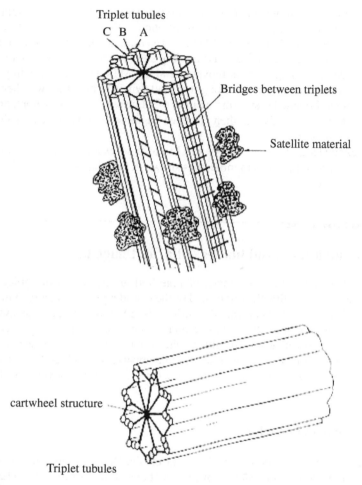

FIGURE 4.32
Structure of a centriole.

Figure 4.35 shows a longitudinal section of a paramecium cilium. it shows the basal body, transition zone, and proximal part of the axenome. The membrane MT bridges in the ciliary necklace and the plaque area are thought to be connected to the intramembranous particles seen on ciliary membranes. Figure 4.36 shows the beating motion of a cilium and its correlation with calcium waves.

Cilia and flagella move because of the interactions of a set of internal MTs (see Figure 4.37). Collectively, these are called the axoneme. Two of these MTs join to form one doublet in the cilia or flagella. Note that one tubule is incomplete and important MAPs project from one of the MT subunits. The core doublets are both complete. Extending from the doublets are sets of arms that join neighboring doublets. These are composed of dynein and are spaced at 24-nm intervals. Nexin links are spaced along the MTs to hold them together. Projecting inward are radial spokes connecting with a sheath enclosing the doublets.

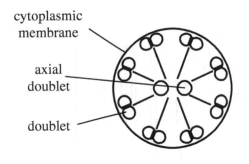

cytoplasmic
membrane

axial
doublet

doublet

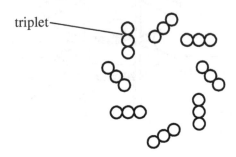

triplet

FIGURE 4.33
Comparison of structures of a cilium and a centriole.

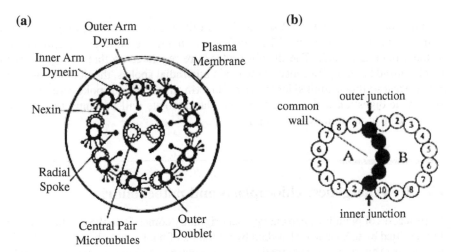

(a)

Outer Arm
Dynein

Inner Arm
Dynein

Plasma
Membrane

Nexin

Radial
Spoke

Central Pair
Microtubules

Outer
Doublet

(b)

common
wall

outer junction

inner junction

A   B

FIGURE 4.34
Cross-section through a flagellum (a) and a microtubule doublet (b).

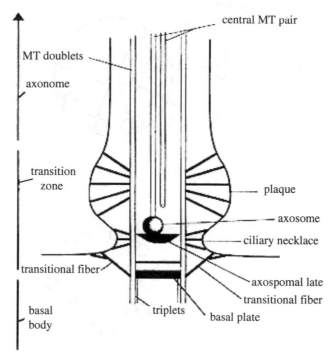

FIGURE 4.35
Longitudinal section of a cilium.

The dynein arms show ATPase activity. In the presence of ATP, they can move from one tubulin to another. They enable the tubules to slide along one another so the cilium can bend. The dynein bridges are regulated so that sliding leads to synchronized bending. Because of the nexin and radial spokes, the doublets are held in place and sliding is limited lengthwise. If nexin and the radial spokes are subjected to enzyme digestion and exposed to ATP, the doublets will continue to slide and telescope up to nine times their length.

4.9 Cell energetics: chloroplasts and mitochondria

All processes of plant life need energy that originally comes from the quanta of visible light emitted by the sun and absorbed by pigments of photosynthetic units. After this molecular excitation, the absorbed energy is accumulated and transmitted to other parts of the cell, the remainder of the plant, and finally to other organisms that cannot obtain energy by photosynthesis directly.

To accomplish this, the energy of molecular excitation is transformed into its chemical form — the so-called high-energy compounds. The most common

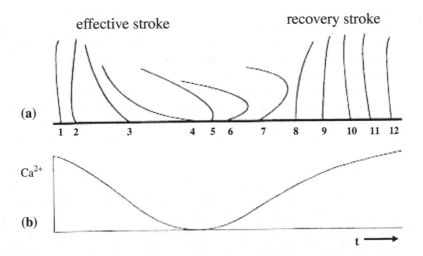

FIGURE 4.36
Different stages during the beating of a cilium (a). Oscillations of Ca^{2+} in the transition zone (b).

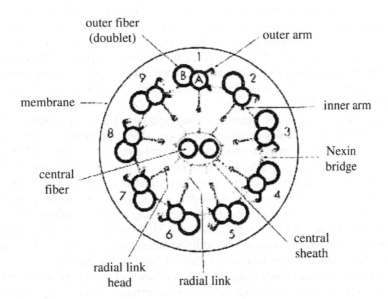

FIGURE 4.37
Electron micrograph of a cross-section of a cilium.

accumulator of chemical energy in cells is ATP, formed during photosynthesis and used in nearly all processes of energy conversion in other cells (see Chapter 6). The hydrolysis of ATP produces ADP. Catalyzed by special enzymes known as ATPases, it allows the use of this stored energy for ionic pumps, molecular synthesis, production of mechanical work, and other functions.

The amount of energy stored by the ADP → ATP reaction in cells is limited because of osmotic stability. Therefore, other molecules such as sugars and fats are used for long-term energy storage. The free energy of ATP is used to synthesize these molecules. Subsequently, in the respiratory chain, these molecules are decomposed, whereas ATP is produced again.

Chloroplasts are double-membraned, ATP-producing organelles found only in plants. Their outer membranes contain sets of thin membranes (thylakoids) organized into flattened sacks stacked up like coins. The disks contain chlorophyll pigments that absorb solar energy which is the ultimate energy source for all plant needs including manufacturing carbohydrates from carbon dioxide and water (Mader, 1996). The chloroplasts first convert solar energy into ATP-stored energy, which is then used to manufacture storage carbohydrates that can be converted back into ATP when energy is needed. Chloroplasts also possess electron transport systems for producing ATP. The electrons that enter the system are taken from water. During photosynthesis, carbon dioxide is reduced to a carbohydrate by energy obtained from ATP (Mader, 1996).

Photosynthesizing bacteria (cyanobacteria) use yet another system whereby they do not manufacture chloroplasts but use chlorophyll bound to cytoplasmic thylakoids. The two most common evolutionary theories of the origin of the mitochondria–chloroplast ATP production system are (1) endosymbiosis of mitochondria and chloroplasts from the bacterial membrane system and (2) gradual evolution of the prokaryote cell membrane system of ATP production into the mitochondria and chloroplast systems. Proponents of endosymbiosis maintain that mitochondria were once free-living bacteria, and that "early in evolution ancestral eukaryotic cells simply ate their future partners" (Vogel, 1998).

The mitochondrion is the site of aerobic respiration Most key processes of aerobic respiration occur across its inner membrane. The mitochondrion where ATP is produced functions to produce an electrochemical gradient similar to a battery (see Figure 4.38) by accumulating hydrogen ions between the inner and outer membranes. This electrochemical energy comes from the estimated 10,000 enzyme chains in the membranous sacks on the mitochondrial walls. As the charge builds up, it provides electrical potential that releases its energy by causing a flow of hydrogen ions across the inner membrane into the inner chamber. The energy causes an enzyme to be attached to ADP which catalyzes the addition of a third phosphorus to form ATP.

One theory holds that mitochondria evolved from endosymbiotic bacteria. However, the many contrasts between the prokaryotic and eukaryotic means of producing ATP provide strong evidence against the endosymbiosis theory. No intermediates to bridge these two systems have ever been found and supportive arguments are all highly speculative. In the standard picture of eukaryote evolution, the mitochondrion was a lucky accident. First, the ancestral cell, probably an archaebacterium, acquired the ability to engulf and digest complex molecules. At some point, however, this

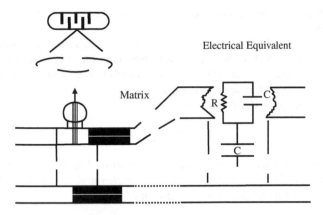

Electrical Equivalent

Matrix

FIGURE 4.38
Mitochondrial membrane and its electrical equivalent.

predatory cell failed to fully digest its prey and an even more successful cell resulted when an intended meal took up permanent residence and became the mitochondrion.

4.9.1 Cell as a thermodynamic machine

Every machine requires specific parts interconnected in an intelligent fashion to perform its desired function. In addition, a steady supply of energy is required for conversion, with some level of efficiency, into useful work. Likewise, all biological cells must have many well-engineered parts to work. Cells are constructed from smaller machines known as organelles. Organelles include mitochondria, Golgi complexes, endoplasmic reticulum, and the protein filaments of the cytoskeleton. At an even lower level, machine-like parts of the cell, such as motor proteins and enzymes, perform specific functions involving energy input and power output. These processes will be examined in Chapter 6.

ATP is a critically important macromolecule. It is a complex nanomachine that serves as the primary energy currency of cells. ATP is used to build complex molecules and provide energy to the extent that it powers virtually every activity of the cell. Nutrients contain numerous low energy covalent bonds. Most are not very useful for performing most types of work required by cells. Thus, low energy bonds must be translated into high energy bonds using ATP energy by removing a phosphate–oxygen group to turn ATP into ADP. ADP is usually immediately recycled in the mitochondria where it is recharged and reemerges as ATP.

At any instant, each cell contains about one billion ATP molecules. Because the amount of energy released in ATP hydrolysis almost meets the needs of most biological reactions, little energy is wasted (Rolfe and Brown, 1997). Generally, ATP is coupled to another nearby reaction so that the two reactions utilize the same enzymatic complex. Release of phosphate from ATP is exothermic; the coupled reaction is endothermic. The terminal phosphate group is then transferred by hydrolysis to another compound via a process called phosphorylation, yielding ADP, phosphate (P_i), and energy.

Phosphorylation often takes place in cascades and is an important signaling mechanism within cells. ATP is not excessively unstable, but it is designed for slow hydrolysis in the absence of a catalyst. This ensures that its stored energy is released only in the presence of an appropriate enzyme.

Gibbs free energy is defined as:

$$G = E + pV - TS \qquad (4.13)$$

where E is internal energy, p is pressure, V is volume, T is absolute temperature (K), and S is entropy. One of the most important biochemical reactions in the context of energy generation in animals is the oxidation of glucose:

$$C_6H_{12}O_6 + 6O_2 \rightarrow 6CO_2 + 6H_2O \qquad (4.14)$$

in which 1 mol of glucose produces $\Delta G = +686$ *kcal* of Gibbs free energy, which is the maximum work obtainable from this reaction. In total, 180 g of glucose reacts with 134 ℓ of O_2 to produce 686 kcal of energy, i.e., 5.1 kcal is produced per ℓ of O_2. Work can be accomplished indirectly by (1) burning the glucose and using the heat released or (2) using glucose as a step in series of complex reactions releasing work at the end. Animal cells utilize this process in the Krebs cycle. For every mole of glucose metabolized, 38 mol of ATP is formed from ADP in the reaction:

$$ADP + phosphate \rightarrow ATP \qquad (4.15)$$

The overall reaction can be written:

$$glucose + 6O_2 + 38ADP + 38phosphate \rightarrow 38ATP + 6CO_2 + 6H_2O \qquad (4.16)$$

which requires an input of 382 kcal of energy. However, with each ATP hydrolysis reaction:

$$ATP \rightarrow ADP + phosphate \qquad (4.17)$$

(the inverse of Equation 4.15), 8 kcal of energy is made available. Consequently, with 38 ATP molecules, one stores 304 kcal for available work, of which only about 50% is converted into useful work (as in muscle contraction); the rest is lost to heat production. The overall efficiency of biochemically based molecular "engines" in living cells is on the order of 20%.

Next we examine fat breakdown. Suppose 302 g of fat react with 414 liters of O_2 according to:

$$C_3H_5O_3(OC_4H_7)_3 + 18.5O_2 \longrightarrow 15CO_2 + 13H_2O \qquad (4.18)$$

and 1941 kilocalories of energy are produced. This corresponds to 4.7 kcal/l of O_2.

In these two examples the number of kilocalories per liter of O_2 are fairly close. For protein breakdown, a similar picture emerges. On average, 4.9 kcal are produced per liter of O_2.

4.9.2 Active transport

The discussion of molecular phenomena and biological processes has included the general categories of diffusion, osmosis, and reverse osmosis, all of which are passive on a cellular level. The driving energy for passive transport comes from molecular kinetic energy or pressure.

Another class of transport phenomena, called active transport, is based on the ability of a living membrane to supply energy to allow transport. Biological organisms must sometimes transport substances from regions of low concentration to high concentration — opposite of the direction of travel in osmosis. Of course, sufficiently large back pressure causes reverse osmosis. In certain cases, substances can move in a direction where reverse osmosis would take them even though existing pressures are insufficient to cause reverse osmosis. That means active transport must be taking place and living membranes can expend energy for transport. Active transport can also aid ordinary osmosis and explains why some transport proceeds faster than expected from osmosis or dialysis alone.

Active transport is extremely important for nerve cells. Changes in the concentrations of electrolytes across nerve cell walls are responsible for nerve impulses. After repeated nerve impulses, significant migration has occurred and active transport "pumps" electrolytes back to their original positions. For a detailed discussion of these processes, see Chapter 6.

4.10 Other organelles

Spreading throughout the cytoplasm is the endoplasmic reticulum (ER). It is a folded system of membranes that loop back and forth to provide a large surface area. The ER provides a surface area for cell reactions. It is also the site of lipid production. The two forms of ER are (1) smooth ER which has no associated ribosomes and (2) rough ER that has attached ribosomes to give it texture. The ribosomes manufacture proteins for the cell. Listed below are the remaining key organelles.

The *Golgi apparatus* is responsible for packaging proteins for cells. Once the proteins are produced by rough ER, they pass into the sack-like cisternae that constitute the main part of the Golgi body. The proteins are then squeezed into "blebs" that drift off into the cytoplasm.

Lysosomes are also called suicide sacks. They are produced by the Golgi bodies and consist of single membranes surrounding powerful digestive enzymes.

Animal cells generally contain *centrioles*, as discussed above. Cilia and flagella are found in many different life forms. Plant cells generally contain *storage vacuoles, cell walls,* and *plastids.* Vacuoles are large empty-appearing areas in the cytoplasm. They are usually found in plant cells where they store waste. As plant cells age, they grow larger and occupy most of the cytoplasm in mature cells. Cell walls are the rigid structures surrounding plant cells. They provide support.

Plastids are large organelles found in plants and some protists but not in animals or fungi. They can easily be seem through a light microscope. Chloroplasts represent one group of plastids called *chromoplasts* (colored plastids). The other class is designated *leucoplasts* (colorless plastids); they usually store food molecules. Included in this group are amyloplasts or starch plastids present, for example, in potato root cells.

Finally, *contractile vacuoles* are organelles that are critical in enabling protozoa to combat the effects of osmosis. Protozoa must constantly excrete the water that enters through their membranes.

4.11 Nucleus: nuclear chromatin, chromosomes, and nuclear lamina

The *cell nucleus* acts as the cell's headquarters since it regulates all cell activity by controlling the enzymes. It also serves as the main library by containing the blueprints and instructions based on more than a billion years of evolution. They tell the cell how to operate, how to rebuild after every cell division, and how to act and interact with other cells of the organism. The gene control system monitors signals from the nucleus and the surrounding cytoplasm. The nucleus coordinates activities such as intermediary metabolism, growth, protein synthesis, and reproduction (cell division). It seems to be structured as a hierarchy of several levels of genomic instruction. The spherical nucleus occupies about 10% of cell volume, making it the cell's most prominent feature.

The nucleus consists of a nuclear envelope (outer membrane) and nucleoplasm. The nucleoplasm contains chromatin and the nucleolus. The nuclear envelope is a double membrane composed of four phospholipid layers and large pores through which material passes. Most nuclear material consists of chromatin, the unstructured form of DNA that organizes to form chromosomes during mitosis or cell division. It consists of DNA looped around histone proteins.

The nucleolus is a knot of chromatin. It manufactures ribosomes. The chromatin is composed of DNA containing the information for the production of proteins. This information is encoded in the four DNA bases: adenine, thymine, cytosine, and guanine. The specific sequence of bases tells the cell how to order the amino acids as was discussed in Chapter 2. Three processes enable the cell to manufacture protein. Replication allows the nucleus to make exact copies of its DNA. Transcription allows the cell to make RNA working copies of its DNA. In translation, the messenger RNA lines up amino acids into a protein molecule (see Chapter 2). The code is translated on structures made in the nucleus. They are called *ribosomes* and provide structural site where mRNA sits.

The amino acids for the proteins are carried to the site by transfer RNA (tRNA). Each tRNA molecule has a nucleotide triplet that binds to the complementary sequence on the messenger RNA (mRNA). The tRNA carries the amino acid at its opposite end.

Only the cells of advanced organisms, known as *eukaryotes*, have nuclei. Generally each cell has one nucleus, but exceptions such as slime molds and the siphonales group of algae exist. Simpler single-cell organisms (*prokaryotes*), like the bacteria and cyanobacteria, have no nuclei. All their cellular information and administrative functions are dispersed throughout the cytoplasm.

A double-layered membrane, the nuclear envelope, separates contents of the nucleus from the cytoplasm. The envelope is riddled with holes called nuclear pores that allow specific types and sizes of molecules to pass between the nucleus and the cytoplasm. It is also attached to a network of tubules (the ER) where protein synthesis occurs.

4.11.1 Chromatin and chromosomes

Almost 2 m of DNA, divided into 46 individual molecules, one for each chromosome and each about 4 cm long, is packed inside the nucleus of every human cell. Containing all this material into a microscopic cell nucleus is an extraordinary feat of packaging. The DNA is combined with proteins and organized into a precise, compact structure, a dense string-like fiber called chromatin. Each DNA strand wraps around groups of small protein molecules called histones, forming a series of bead-like structures, called nucleosomes, connected by the DNA strand.

The uncondensed chromatin has a "beads on a string" appearance. The string of nucleosomes, already compacted by a factor of six, is then coiled into an even denser structure, thus compacting the DNA by a factor of 40. This compression and structuring of DNA serves several functions. The overall negative charge of the DNA is neutralized by the positive charge of the histone molecules. The DNA occupies much less space, and inactive DNA can be folded into inaccessible locations until it is needed.

Euchromatin is the genetically active portion of chromatin and is involved in transcribing RNA to produce proteins used in cell function and growth. Heterochromatin contains inactive DNA and is the portion of chromatin that is most condensed.

During the cell cycle, chromatin fibers take on different forms inside the nucleus (Mitchison, 1973). During interphase, when the cell carries out its normal functions, chromatin is dispersed throughout the nucleus in a tangle of fibers. This exposes the euchromatin and makes it available for transcription. When the cell enters metaphase and prepares to divide, the chromatin changes dramatically. First, the chromatin strands copy themselves through DNA replication, then they are compressed in a 10,000-fold compaction process into chromosomes. As the cells divide, the chromosomes separate, giving each cell a complete copy of the genetic information in its chromatin.

4.11.2 Nucleolus

The nucleolus is a membraneless organelle within the nucleus that manufactures protein-producing ribosomes. The nucleolus looks like a large dark spot within the nucleus. A nucleus may contain up to four nucleoli, but within each species the number is fixed. After a cell divides, a nucleolus is formed when chromosomes are brought together into nucleolar organizing regions. During cell division, the nucleolus disappears.

4.11.3   Nuclear envelope

The nuclear envelope is a double-layered membrane that encloses the contents of the nucleus during most of the cell lifecycle. The space between the layers is called the perinuclear space and appears to connect with the rough ER. The envelope is perforated with tiny holes called nuclear pores. The pores regulate the passage of molecules between the nucleus and cytoplasm, permitting some of them to pass through the membrane. The inner surface has a protein lining called the nuclear lamina, which binds to chromatin and other nuclear components. During mitosis, the nuclear envelope disintegrates, then reforms as the two cells complete their formation and the chromatin begins to unravel and disperse.

4.11.4   Nuclear pores

Pores in the nuclear envelope regulate the passage of molecules between the nucleus and cytoplasm, permitting some to pass through the membrane. Blocks for building DNA and RNA are allowed into the nucleus as are molecules providing energy for constructing genetic material. The pores are fully permeable to molecules up to the size of the smallest proteins, but form a barrier to keep most large molecules out of the nucleus. Some larger proteins, such as histones, are allowed passage into the nucleus. Each pore is surrounded by an elaborate protein structure called the nuclear pore complex, which probably selects large molecules for entrance into the nucleus.

4.12   Cell division

One of the most fundamental processes within the cell cycles (see Figure 4.39) of higher organisms is cell division, also known as mitosis. Mitosis ensures genetic continuity since the new daughter cells have exactly the same numbers and kinds of chromosomes as the original mother cell and the same genetic instructions are transmitted on. As shown in Figure 4.40, the process can be divided into five or six basic stages: interphase, prophase, late prophase (or prometaphase), metaphase, anaphase, and telophase.

Cells may appear inactive during interphase, but they are in fact quite active. Interphase is the longest period of the cell cycle during which DNA replicates, the centrioles divide, and proteins are actively produced (Mitchison, 1973). During interphase, two centromeres are formed by the replication of a single centrosome. Each centrosome has a pair of centrioles. MTs extend from the centrosomes to form asters. The nucleus contains one or more nucleoli. The chromosomes are still in the form of loosely packed chromatin fibers.

During prophase, the nucleoli disappear. The chromatin fibers become more tightly packed and fold into chromosomes. Each duplicated chromosome is actually two identical chromatids joined at the centromere. Also during prophase, a network of microtubules arranged between the two centrosomes forms the mitotic spindle. The centromeres move apart from each other during this stage.

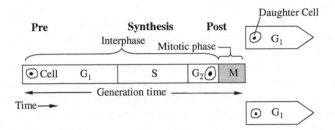

FIGURE 4.39
Eukaryotic cell cycle.

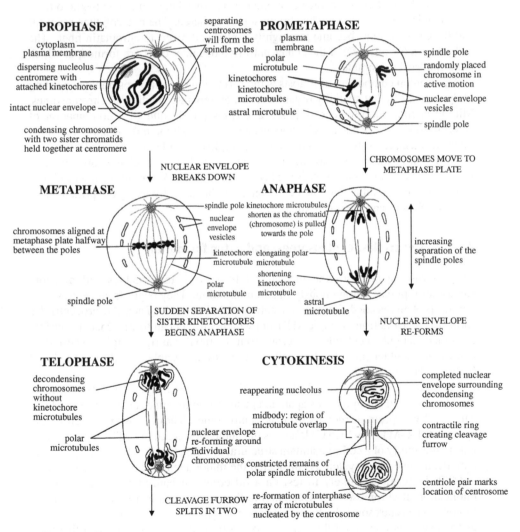

FIGURE 4.40
Stages of mitosis.

In the late prophase (or prometaphase) the nuclear envelope disappears. Spindle fibers (bundles of MTs) extend from one cell pole to the other, where the two centromeres are now located. Chromosomes then begin to attach themselves to special MTs called kinetochore MTs.

During metaphase, the chromatid pairs (chromosomes) line up along the equator of the cell. The centromeres of each chromosome are aligned with one another at what is called the metaphase plate. During anaphase, the paired centromeres of each chromosome divide, allowing the once paired chromatids to move away from each other. Each liberated chromatid is now called a chromosome. The newly formed chromosomes now move toward opposite poles of the cell. By the end of anaphase, each of the two poles of the cell has a complete collection of chromosomes.

Finally, during telophase, the cell elongates and two daughter nuclei begin to form at each pole of the cell where the chromosomes gathered. The nucleoli then reappear and the chromosomes become less tightly coiled and revert to chromatin fiber. The cell pinches (or furrows) at the center until two daughter cells are formed, at which point mitosis is complete.

An essential component of mitosis is the formation of a mitotic spindle consisting of dynamic MTs that spatially organize and then separate the divided chromosomes. Mitotic spindle morphogenesis poses a serious puzzle due to a combination of deterministic and stochastic behaviors present. Although the mitotic apparatus must function very precisely, its assembly is accomplished without a detailed blueprint; construction of the mitotic spindle is thus an example of a stochastic phenomenon where random molecular processes play a crucial role. Assembly of microtubules and the mitotic spindle can now be modeled mathematically using the principles of polymer physics and chemical kinetics.

4.12.1   Centrioles, centrosomes, and aster formation

The centrosome provides the mechanical framework that enables cell division. Centrioles replicate autonomously like mitochondria. They begin from centers containing proteins needed for their formation (tubulin, etc.). Then the procentrioles form. Each grows from a single MT from which a triplet can form. Once a centriole is formed, daughter centrioles can grow from the tubules at right angles. These then add to the daughter (in a dividing cell) or move to the periphery and form the basal body for the cilium.

Centrioles divide prior to cell division. From the centrosomes (CTRs) originate the MTs that pull chromosomes apart and push the centrosomes apart (spindle MTs). The astral MTs protrude radially from the centrosomes. The centrosome–microtubule system (CTR–MT) in which CTR acts as the nucleation centre for MTs, takes a direct part in the ability of a cell to maintain structural and functional polarity and carry out a division cycle. This system is considered the major agent of cell morphogenesis.

CTR duplication is likely to rest on a different mechanism from that used for duplication of chromosomes. Deciphering this mechanism is a major issue in cell biology. The centrosome duplication pathway might be so important for cell survival as to have been conserved throughout eukaryotic cell evolution. We can therefore speculate that a relatively small number of essential genes may be involved in this pathway. Our knowledge of the particular features of widely divergent systems and additional information from experimental approaches should ultimately allow us to understand the general principles of centrosome inheritance.

The major events that characterize CTR duplication anticipate major cell cycle processes. The temporal sequence suggests that some controls on cell division cycle may be connected with the progression of the CTR duplication cycle. The duplication of chromosomes during S phase and the duplication of centrosomes over a period encompassing most of the interphase are due to independent mechanisms. They both require coupling with the progression of the cell cycle.

The centrosome reproduction cycle requires several interconnected structural events involving duplication of both centrioles (through orthogonal budding of procentrioles in late G_1 phase, elongation during S-G_2 phase, semiconservative segregation during G_2-M, and distribution to the daughter cells as a pair of orthogonal centrioles that eventually disorient) and the maturation of the centrosome matrix, demonstrating accumulation during interphase, a profound structural remodelling during M phase, and eventually a dispersion or degradation at the outset of mitosis.

The molecular mechanisms ensuring the coupling between sequential events of the centrosome reproduction cycle and the mitotic progression have not yet been unravelled and biophysical models are still needed.

4.12.2   Chromosome segregation

The equal apportionment of daughter chromosomes to each of the two cells resulting from cell division is called chromosome segregation. It occurs during mitosis and meiosis and is regulated by DNA replication. Centromeres are the attachment points of kinetochores and are critical for segregation. Centromere structures vary among organisms. Molecular motors attached to kinetochores move chromosomes in segregation.

In biological experiments, chromosomes have been subjected to micromanipulation in order to reveal more about their movement in mitosis. By tugging on them, the forces produced by the spindle have been measured. The spindles have been chopped apart to locate the motors for chromosome movement. The current preoccupation of cell biologists like Bruce Nicklas (Nicklas and Ward, 1994; Nicklas et al., 1995) is to connect cell mechanics with molecular biology. Pulling on chromosomes alters the phosphorylation of chromosomal proteins. Different phosphorylation states signal the cell to go ahead and divide or to pause to allow time for error correction. We now wish to understand in more detail how mechanical tension produced by mitotic forces provides chemical signals that regulate the cell cycle. Models for chromosome movements based on lateral interactions of spindle microtubules are under development.

The period of DNA synthesis (S phase) ends when all nuclear DNA has been replicated and the number of chromosomes has doubled. After a cell enters S phase, it is committed to completing the cell cycle, even when environmental conditions are extremely adverse.

The separation of chromosomes is accomplished by the cytoskeleton's largest constituents, the MTs. During mitosis, MTs connect to each of the chromosomes and align them along the equatorial plate. A mysterious balance of forces prevents the MTs from separating the chromosomes until all the chromosomes are aligned and division may proceed in unison.

What then is the control mechanism in mitosis? Cells have been shown to divide without membranes. They can divide after all their motor proteins are extracted, and even without chromosomes. However, they cannot divide without MTs or centrosomes. Similarly, insufficient ATP and calcium supplies limit the ability of cells to divide.

Ted Salmon, a key figure in the field of cell biology, stated his working hypothesis (Rieder and Salmon, 1994) that mitosis will be explained by a combination of several mechanisms involving (1) the molecular and structural properties of the centrosome which organizes and nucleates polymerization of spindle microtubules, (2) the assembly of MTs which orient and participate in the generation of chromosome movements, and (3) the MT motors such as the kinesin and dynein proteins which appear to generate polarized forces along the lattice of microtubules, at kinetochores, and within the spindle fibers.

The mathematical formalism necessary for this task must encompass (1) stochastic polymerization processes, (2) nonequilibrium chemical kinetics based on structure, (3) tensegrity which reflects the mechanical properties of the networks formed, and (4) electronic and dipolar characteristics based on protein structure, bound water, and ionic solutions.

Since MTs play the key role in mitosis, new cancer therapies in development (including the use of taxol to treat ovarian and breast cancers) directly target MT assembly processes. Colchicine, colcemid, and nocadazole inhibit polymerization by binding to tubulin and preventing its addition to the plus ends. Vinblastine and vincristine aggregate tubulin and lead to MT depolymerization. Taxol stabilizes MTs by binding to a polymer. Figure 4.41 illustrates how the cytoskeletal organization can be disrupted by various chemical compounds. It is of crucial importance to our understanding of molecular level cell functioning to quantify some of the most dominant processes governing MT behavior. It is interesting to note that MTs are specialized in their functions in dividing cells as shown in Figure 4.42.

However, cell division involves more than chromosome segregation. In particular, it leads to the emergence of two daughter cells via the formation of a contractile ring and a physical separation of the two halves of the cell. This process is being modeled mathematically with reasonable success (see Figure 4.43) although the model does not exhibit the level of detail allowing full inclusion of the cytoskeletal structures (Alt and Dembo, 1995).

4.12.3 Cytokinesis

Cytokinesis involves formation of a cleavage furrow near the old metaphase plate. The furrow begins a shallow groove in the cell surface. Just below this groove is where a ring of cytoplasmic microfilaments begins to contract. The cleavage furrow deepens until the parent cell is pinched in two. The basic requirements for cytokinesis are actin, MTs, and sufficient ATP. The plane of cytokinesis is specified by astral MTs

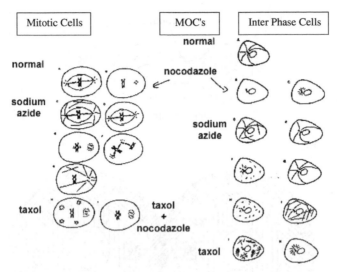

FIGURE 4.41
Disruption of the cytoskeletal organization by various chemical agents at different stages of the cell cycle.

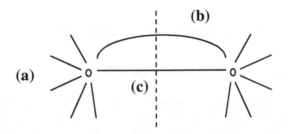

FIGURE 4.42
Specialization of microtubules: (a) astral, (b) polar, and (c) kinetochore.

and the position of the spindle. Astral MTs are sufficient to position a cleavage furrow; activation follows mitosis.

Cleavage furrow initiation requires signalling from mitotic spindle MTs. Cytokinesis is generally coupled to mitosis. Before the start of mitosis, cytoskeleton actin is distributed almost uniformly except for a concentration in the cortical layer, a situation so stable that even exposure to astral signals does not disturb it. Eventually, however, progressive stiffening occurs in the cortex area, giving rise to a zone of concentrated actin–myosin presence at the cell equator in the form of a contractile ring. The organization of the axis and the equator of the dividing cell is only a function of the mitotic spindle and asters. Computer simulations (Alt and Dembo, 1995) of the emergence and development of cleavage due to the hydrodynamics of a cross-linked polymeric fluid with a controllable viscosity successfully reproduced spatiotemporal patterns seen in experiments.

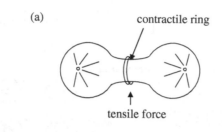

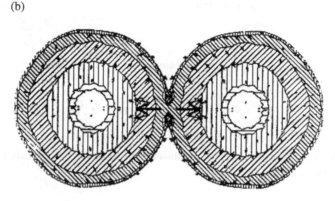

FIGURE 4.43
Development of cleavage in the cell (a) and in the simulations (b).

4.12.4   Spindle and chromosome motility

The discovery of microtubule motor proteins with a clear involvement in mitosis and meiosis has attracted great interest because motor proteins may account for many movements of the spindle and chromosomes in dividing cells. Microtubule motors have been proposed to generate the force required for spindle assembly, attachment of the chromosomes to the spindle, and movement of chromosomes toward opposite poles. Kinesin motors have been shown necessary for establishing spindle bipolarity, positioning chromosomes on the metaphase plate, and maintaining forces in the spindle.

Evidence also indicates that kinesin motors can facilitate MT depolymerization, possibly modulating microtubule dynamics during mitosis. The force generated by microtubule polymerization and depolymerization is thought to contribute to spindle dynamics and movements of chromosomes.

Many newly identified kinesin-related proteins (KRPs) localize close to the spindles in mitotically dividing cells and are implicated in spindle function by their cellular localizations and mutant effects. Remarkably, several KRPs have been demonstrated to be chromosome-associated, providing a direct link between the chromosomes and microtubules of the spindle. The demonstration of kinesin motor protein function in spindle and chromosome motility represents a major step forward in our understanding of cell division.

4.13   Cell intelligence

G. Albrecht-Buehler advocates a theory of cell functioning based on his conviction that the centriole plays the key role in orchestrating cellular activities by serving as analogs of the eyes and brain of a cell. We will present the key elements of his theory that claims the cell is an intelligent unit of life.

Cell movement is not random; it is directed and intentional. This is a crucial characteristic that distinguishes living from nonliving matter. Cells control the movement of every part of their structure. The functions of parts of cells can be likened to functions of parts of the human body. Plasma membrane and cortex correspond to skin and musculature of a cell. They consist of small, autonomously moving microplasts. Their autonomy implies that cells contain control systems preventing autonomous units from moving independently and randomly.

The bulk cytoplasm including the mitochondria, organelles, and intermediate filaments comprises the cell body and correspond to the cell's "guts" and "innards." The main cytoskeletal components are the IFs although MTs also traverse this compartment. MTs mediate between the control center (centriole) and autonomous domains. The control center detects objects and other cells by pulsating near-infrared signals. Cells have "eyes" in the form of centrioles. that can detect infrared signals and steer cell movements toward their source. Evidence indicates that signal detection is strongly localized in a narrow band of the near-infrared spectrum as shown in Figure 4.44.

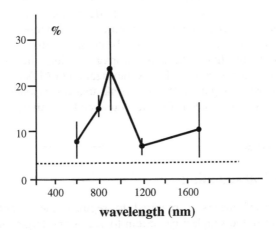

FIGURE 4.44
Percentage of cells that removed light scattering particles as a function of wavelength. The near-infrared wavelength of 800 to 900 nm is the most attractive.

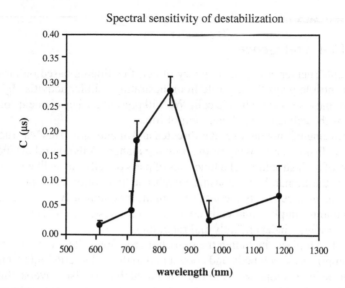

FIGURE 4.45
Sensitivity of cells to the destabilizing effects of external electromagnetic radiation.

An eye maps the directions of light sources in a one-to-one fashion. If cells can detect light sources and measure space and time variables such as angles, distances, curvatures, and durations, they must be able to derive these abstract quantities from the physical objects or signals of their environment. In response to exogenous signals, the centrosome may send destabilizing signals along its radial array of microtubules. The observed destabilization is the signal propagated along the microtubules similar to signal movement along nerves. The destabilization of MTs is strictly wavelength dependent as shown in Figure 4.45. The peak area corresponds to the same near-infrared region in which Albrecht-Buehler found cell sensitivity to other objects.

4.14   Biological signaling

Signaling by varied means is required to regulate the complex behavior of living systems from the simplest bacterium to yeasts and larger eukaryotes such as humans. The difference between monocellular and multicellular organisms is that communication must necessarily be possible among different cells of multicellular organisms. In addition to intracellular signalling, a cell must be prepared to transmit and receive extracellular signals.

The signalling mechanisms discovered to date exhibit the complexity of organic chemistry. In order to interpret signals from other cells, a cell requires special membrane receptors to detect the presence of a signaling molecules in extracellular fluid. Albrecht-Buehler proposed in his intelligent cell model that cells are able to sense light through the use of centrioles. Since the centrioles are always found with perpendicular orientations, this would allow the cell to discern directional information about a signal through latitude and longitude measurements.

That this would provide an invaluable signal receptor is clear, but the mechanism through which other cells might transmit such signals is unclear and will not be discussed further beyond noting that mitochondria may generate light signals at infrared frequencies.

A cell has several methods of communication through the use of its signaling molecules. The molecules are first packaged and then expelled. In the simplest type of messaging, the chemical signals are "dumped" outside the cell and carried diffusively. This method of communication is effective only for signaling nearby cells. Such local signaling is known as paracrine signaling. Synaptic signaling is a refined version of the paracrine model by which the signal molecule, a neurotransmitter, is released at a specifically designed interface to allow intimate contact between source and target cells. This allows quick and direct signaling but still relies on diffusion to carry the signal molecules across the narrow junction.

The last type of signaling, known as endocrine signaling, may be used when the target cells of the signal are more distant or further widespread. Molecular signals known as hormones are secreted by cells into the circulatory system. Although diffusion is used to an extent, the stream of blood (in animals) or sap (in plants) may carry the signal a long distance. Due to the dilution effect of the circulatory fluid, hormones must be effective even at low concentrations such as 10^{-8} M.

The mechanism behind the workings of these extracellular messengers is of the familiar antigen–antibody type. That means a signaling molecule may bind to specific integral membrane proteins on the target cell's exterior. The general behavior of these signaling systems is that binding of the signal molecule to the receptor induces a conformation change at the opposite end of the protein receptor within the cell interior. The conformational change may have a direct or indirect response.

In the case of a direct response, the cytoplasmic domain of the protein becomes enzymatic and catalyzes a specific chemical reaction until the signal molecule at the extracurricular end breaks down or becomes unbound. In the indirect case, the conformational change may release another signaling molecule known as a G protein. This G protein may then bind to one or more other enzymes, serving to activate or deactivate them. The indirect signal allows coordination of complementary reaction pathways and is one way in which the cell regulates its processes.

Breaking the signaling scheme into many parts also allows for magnification of the signal at each step and makes it possible to ultimately achieve a large response to a small number of signal molecules. The final response via ligand binding is the opening of ion channels. The conformational change of the receptor is such that a hydrophilic channel is opened through the cell membrane and allows passage of a specific charged molecular species such as calcium (Ca^{2+}).

The calcium ion is particularly useful because it exists outside the cell in concentrations 4 to 10 times higher than the intracellular concentration. As a result, it diffuses easily and is used as a secondary signal. The mechanism for the operation of these signals is well understood and is described in detail in textbooks such as Alberts et al. (1994).

The most interesting biophysics problems can be found within the cell where the methods of signal transduction are yet to be fully elucidated. Intracellular signaling includes mechanisms such as the action potential which is electrical in nature and driven by chemical potentials. Sensitivity of individual cells to concentration and potential gradients is necessary if the cell is to respond to gravitational or electric fields. Intracellular signaling coordinates the orchestra of cellular processes to ensure that the entire cell works in harmony.

As described earlier in this chapter, in mitosis, chromosome segregation to each pole of the mother cell is mediated by MTs. However, the simultaneity of separation must be explained and requires some kind of signal mediated by the MTs. "Treadmilling" by free MTs under conditions of dynamic instability should also be explained since the opposite ends of the MT show such coordinated behavior.

Maniotis and Ingber (1997) demonstrated how pulling on actin filaments can induce changes in the nucleus. They showed that cells are sensitive to mechanical stimulation. One hypothesis for the control of a cell over these processes is electromagnetic regulation. This form of signaling has the advantage that it is far more rapid than extracellular signaling. The cytoskeleton seems to play a key role.

Consider the action potential. The electrical signal passes along the neural membrane driven by a cascade of sodium ions flowing into the cell and is switched off by a delayed flow of potassium ions out of the cell. While the action potential moves, the signal is not attenuated. Only toxins, which can disable the functions of voltage-gated ion channels, can stop the progress of the action potential. Thus the behavior of the system appears soliton-like (Peyrard, 1995).

The cytoskeleton adopts a configuration in neurons where the orientation of MTs is parallel with uniform direction within the axon, along which the cell transmits signals to other cells. The MTs adopt an aligned configuration with a nonuniform direction in the dendrites where the signals are received from other cells. Since molecular motors such as kinesin and dynein move in opposite directions along MTs, one would suspect another reason exists for the specific structure beyond the simple ability to transport goods. MTs are known to be sensitive to electric fields (Vassilev et al., 1982; Vater et al.) and magnetic fields (Bras, 1995) and align themselves such that they are parallel to field lines.

The specific alignment in these two neuronal regions could be intended to make MTs insensitive to electric fields within dendrites and reinforce susceptibility to electric fields within the axons. In long cylindrical cells such as axons, electrical fields are able to penetrate most easily.

One of the key questions of intracellular biological signaling is whether electron transport plays a role. The transfer of an electron between proteins leads to a conformational change that exerts a physiological effect. In the cases of ion channels, the donation or acceptance of electrons changes their internal electrostatistics and affects their functions. This results in neuromodulation by changing the response characteristics of the neurons and constitutes a reprogramming of neural networks. Numerous issues in the area of cellular signaling still await proper biophysical elucidation.

References

Alberts, B. et al., *Molecular Biology of the Cell,* Garland Publishing, London, 1994.

Albrecht-Buehler, G. http://www.basic.nwu.edu/g-buehler/cellint.htm.

Alt, W. and Dembo, M., *Math. Biosci.,* 156, 207. 1999.

Amos, L.A., *Trends Cell Biol.*, 5, 48, 1995.

Amos, L.A. and Amos, W.B., *Molecules of the Cytoskeleton*, Macmillan, London, 1991.

Amos, L.A. and Cross, R.A., *Curr. Opin. Struct. Biol.*, 7,2, 1997.

Amos, L.A. and Hirose, K., *Curr. Opin. Cell Biol.*, 9, 4, 1997.

Astumian, R.D. and Bier, M., *Phys. Rev. Lett.*, 72, 1766, 1994.

Bairoch, A. and Apweiler, R., *Nucleic Acids Res.*, 26, 38, 1998.

Bayley, P.M., Schilstra, M.J., and Martin, S.R. *J. Cell Sci.*, 95, 33, 1990.

Benedek, G.B. and Villars, F.M.H., *Physics With Illustrative Examples from Medicine and Biology*, Springer, Berlin, 2000.

Boal, D. *Mechanics of the Cell*, Cambridge University Press, Cambridge, U.K., 2001.

Bras, W., Ph.D. Thesis, John Moores University, Liverpool, U.K., 1995.

Brown, J.A. and Tuszynski, J.A., *Phys. Rev. E*, 56, 5834, 1997.

Bruinsma, R., *Physics,* International Thomson Publishing, New York, 1998.

Carlier, M.F. et al., *Proc. Natl. Acad. Sci. U.S.A.*, 84, 5257, 1987.

Case, R.B. et al., *Cell*, 90, 959, 1997.

Chen, C.S. et al., *Science*, 276, 1425, 1997.

Cole, K.S., *Trans. Faraday Soc.*, 33, 966, 1937.

de Gennes, P.G., *Introduction to Polymer Dynamics,* Cambridge University Press, Cambridge, U.K., 1990.

De Robertis, E.D.P. and De Robertis, E.M.F., *Cell and Molecular Biology*, W.B. Saunders, Philadelphia, 1980.

Derenyi, I. and Vicsek, T., *Proc. Natl. Acad. Sci. U.S.A.*, 93, 6775, 1996.

Doering, C.R., Horsthemke, W., and Riordan, J., *Phys. Rev. Lett.,* 72, 2984, 1994.

Dustin, P., *Microtubules*, Springer, Berlin, 1984.

Edelstein-Keshet, L., *Eur. Biophys. J.*, 27, 521, 1998.

Fletterick, R.J., *Nature*, 395, 813, 1998.

Flory, P.J., *Statistical Mechanics of Chain Molecules*, John Wiley & Sons, New York, 1969.

Flyvbjerg, H., Holy, T.E., and Leibler, S., *Phys. Rev. Lett.* 73, 2372, 1994.

Flyvbjerg, H., Holy, T.E., and Leibler, S., *Phys. Rev. E*, 54, 5538, 1996.

Frey, E., Kroy, K., and Wilhelm, J., *Adv. Struct. Biol.*, 5, 135, 1998.

Frieden, C. and Goddette, D., *Biochemistry*, 22, 5836, 1983.

Froehlich, H., *Adv. Electr. Electron Phys.*, 53, 85, 1980.

Fygenson, D.K., Braun, E., and Libchaber, A., *Phys. Rev. D*, 50, 1579, 1994.

Geeves, M.A. and Holmes, K.C., *Annu. Rev. Biochem.*, 68, 687, 1999.

Gittes, F., Mickey, E., and Nettleton, J., *J. Cell Biol.*, 120, 923, 1993.

Hays, T.S. and Salmon, E.D., *J. Cell Biol.*, 110, 391, 1990.

Hess, B. and Mikhailov, A., *Science*, 264, 223, 1994.

Horio, T. and Hotani, H., *Nature*, 321, 605, 1986.

Houchmandzadeh, B. and Vallade, M., *Phys. Rev. E*, 6320, 53, 1996.

Howard, J., *Mechanics of Motor Proteins and the Cytoskeleton*, Sinauer Associates, Sunderland, MA, 2001.

Huxley, H. E., *Science*, 164, 1356, 1969.

Hyman, A.A. et al., *J. Cell. Biol.*, 128, 117, 1995.

Hyman, A.A. et al., *Molec. Biol. Cell*, 3, 1155, 1992.

Ingber, D.E., *Annu. Rev. Physiol.*, 59, 575, 1997.

Ingber, D.E., *J. Cell Sci.*, 104, 613, 1993.

Janmey, P.A. et al., *J. Biol. Chem.*, 269, 32503, 1994.

Jülicher, F., Adjari, A., and Prost, J., *Rev. Mod. Phys.*, 69, 1269, 1997.

King, R.W.P. and Wu, T.T., *Phys. Rev. E*, 58, 2363, 1998.

Koradi, R., Billeter, M., and Wüthrich, K., *J. Mol. Graphics*, 14, 51, 1996.

Kozielski, F. et al., *Cell*, 91, 985, 1997.

Kraulis, J., *J. Appl. Crystallogr.*, 24, 946, 1991.

Kull, F.J. et al., *Nature*, 380, 550, 1996.

Ledbetter, M.C. and Porter, K.R., *J. Cell Biol.*, 19, 239, 1963.

Leibler, S. and Huse, D.A., *J. Cell Biol.*, 121, 1357, 1993.

Lowe, J. and Amos, L.A., *Nature*, 391, 203, 1998.

Lu, Q. et al., *Adv. Struct. Biol.*, 5, 203, 1998.

Luby-Phelps, K., *Curr. Opin. Cell Biol.*, 6, 3, 1994.

Mader, S., *Biology*, 6th ed., Wm. C. Brown, Dubuque, IA, 1996.

Mandelkow, E.M. and Mandelkow, E., *Cell Motil. and Cytoskel.*, 22, 235, 1992.

Mandelkow, E.M., Mandelkow, E., and Milligan, R., *J. Cell Biol.*, 114, 977, 1991.

Maniotis, A., Chen, E., and Ingber, D.E., *Proc. Natl. Acad. Sci. USA*, 94. 849, 1997.

Marx, A. and Mandelkow, E., *Eur. Biophys. J.* 22, 405, 1994.

Mickey, B. and Howard, J., *J. Cell Biol.*, 130, 909, 1995.

Mitchison, J.M., *Biology of the Cell Cycle*, Cambridge University Press, Cambridge, U.K., 1973.

Mitchison, T. and Kirschner, M., Nature, 312, 237, 1984.

Nicklas, R.B. and Ward, S.C., *J. Cell Biol.* 126, 1241, 1994.

Nicklas, R.B., Ward, S.C., and Gorbsky, G.J., *J. Cell Biol.*, 130, 929, 1995.

Nogales, E., Wolf, S.G., and Downing, K.H., *Nature*, 391, 199, 1998.

Oosawa, F. and Asakura, S., *Thermodynamics of the Polymerization of Protein*, Academic Press, London, 1975.

Owicki, J.C., Springgate, M.W., and McConnell, H.M. *Proc. Nat. Acad. Sci. U.S.A.*, 75, 1616, 1978.

Peyrard, M., Ed., *Nonlinear Excitations in Biomolecules*, Springer, Berlin, 1995.

Pink, D.A., in *Biomembrane Structure and Function*, Chapman, D., Ed., Macmillan, London, 1984, p. 319.

Rieder, C.L. and Salmon, E.D., *J. Cell* Biol., 24, 223, 1994.

Risken, H., *The Fokker-Planck Equation*, Springer, Berlin, 1989.

Rolfe, D.F.S. and Brown, G.C., *Physiol. Rev.*, 77, 731, 1997.

Ruppel, K.M., Lorenz, M., and Spudich, J.A., *Curr. Opin. Struct. Biol.*, 5, 181, 1995.

Semënov, M.V., *J. Theor. Biol.*, 179, 91, 1996.

Sept, D., Ph.D. Thesis, University of Alberta, Edmonton, 1997.

Sept, D. et al., *J. Theor. Biol.*, 197, 77, 1999.

Singer, S.J. and Nicolson, G.L., *Science*, 175, 720, 1972.

Small, J.V et al., *Trends Cell Biol.*, 12, 112, 2002.

Stracke, R., Böhm, K.J., Wollweber, L., Tuszynski, J.A., and Unger, E., *Biochem. Biophys. Res. Commun.*, 293, 602, 2002.

Svoboda, K. and Block, S.M., *Cell*, 77, 773, 1994.

Tabony, J. and Job, D., *Nature*, 346, 448, 1990.

Tobacman, L.S. and Korn, E.D., *J. Biol. Chem.*, 258, 3207, 1983.

Tran, P.T., Walker, R.A., and Salmon, E.D., *J. Cell Biol.*, 138, 105, 1997.

Vassilev, P.M. et al., *Biosci. Rep.*, 2, 1025, 1982.

Vogel, G., *Science*, 279: 1633, 1998.

Volkenstein, M.V., *General Biophysics*, Academic Press, San Diego, CA, 1983.

Zuckermann, M.J., Georgallas, A., and Pink, D.A., *Can. J. Phys.*, 63, 1228, 1985.

5

Nonequilibrium Thermodynamics and Biochemical Reactions

5.1 Second law of thermodynamics

The entire structure of physics is based on the *principle of mechanical determinism* formulated some 300 years ago by Sir Isaac Newton. Using contemporary language, we can express it as follows:

> The law of motion and the state of the system in a given moment of time determine unambiguously and uniquely the state of this system at all moments of time, both in the future and in the past.

The great methodological role of the principle of mechanical determinism rests on the distinction between general *laws* of motion and individual *facts* in terms of the data that characterize the *state* of a given system at a chosen moment in time. The laws of physics at a microscopic level are now well understood. They are described by quantum or well approximating classical equations of motion for a system of atoms in an effective adiabatic potential that corresponds usually to the ground and, possibly, to several low-lying excited electronic states (see Section 3.2). However, an exact determination of the state of the system under consideration poses in general an insurmountable obstacle. Therefore, the methodological significance of mechanical determinism does not automatically translate into its practical power to predict states projected arbitrarily far into the future or to reconstruct states from an arbitrarily distant past.

Several reasons can be listed for difficulties related to determining the initial state. First, the system's *complexity* makes a complete characterization of its state impossible due to the sheer number of coordinates required. For example, a typical thermodynamic system, 1 cm^3 of gas under standard conditions contains 3×10^{19} molecules and their translational state should be described by a vector with more than 10^{20} components (three components of the position vector and three components of the momentum vector per molecule).

Second, the motion's *instability* leads to unpredictable consequences. An infinitesimally small uncertainty in the state coordinates over time leads to large, practically impossible to predict deviations (this is known as deterministic chaos; see Appendix C for details). Recent research results reveal that classical mechanical systems with only several degrees of freedom in their phase spaces possess only small regions filled with stable, integrable trajectories (Prigogine and Stengers, 1984, 1997).

A much more common situation is characterized either by partly stable motion (i.e., stable only for some of the degrees of freedom) or motion that is completely stochastic or unstable. Interestingly, one example is our Solar System that has been the object of precise mechanistic studies for more than 300 years. The state of an unstable system, even if it has only several degrees of freedom, must be determined with great precision if we are to predict its evolution over a reasonably long period of time. From the viewpoint of information processing, the description of a complex system with a large number of degrees of freedom and an unstable system with a small number of degrees of freedom requires an almost infinite number of bits of information.

The third reason is much more fundamental. To determine a state, one must carry out a *measurement* during which the state may be perturbed. A perturbation which is inevitable during an interaction between a macroscopic (classical) apparatus and a microscopic quantum system makes it impossible to determine the latter's wave function. Therefore, the deterministic character of Schrödinger's equation cannot be utilized. The perturbation of the state of a system by the act of observation is not restricted to quantum physics. The more detailed our experimental inquiry into the state of a biological organism, the less indifferent the latter is to such a process. Too detailed an investigation may be completely useless if the price to pay for precision is the death of the specimen during data collection.

Statistical physics (Penrose, 1979; Chandler, 1987) gives us a simple method to circumvent the problem with incomplete knowledge of a system's initial state. Instead of a single system, it concerns itself with an *ensemble* of identical copies of the same system that differ only as to their initial state. This method replaces an individual phase trajectory of classical mechanics with a phase flow. Trajectories that comprise this flow may behave in a stable manner if a small perturbation of the initial state does not lead to essential differences in their course. Conversely, they are unstable if the opposite happens, i.e., small changes in initial conditions lead to huge trajectory deformations.

For statistical mechanics, of special interest are the extremely unstable mechanical systems characterized by the property of *mixing* the phase flow exponentially (Penrose, 1979; Prigogine and Stengers, 1984, 1997). After a period called the *stochastization time*, the states of the ensemble of such systems are distributed practically uniformly over the entire available domain of phase space (see Figure 5.1). This uniform distribution of final states is completely independent of the initial state whose knowledge becomes entirely superfluous. The value of the methods of statistical mechanics in multifarious applications is a strong argument indicating that unstable behavior with exponential mixing is typical for many-body systems (*the hypothesis of molecular chaos*).

The state of an ensemble with a uniform distribution of states over the available domain in phase space of the individual systems is known as the state of *thermodynamic equilibrium*. Figure 5.1 suggests that in the process of reaching equilibrium, the volume of the occupied domain in phase space grows continuously. This is not entirely true since Liouville's theorem known from classical mechanics ensures that the volume of subsets in phase space is conserved in the process of the system's time evolution. The volume in phase space of an arbitrary subsystem is not a quantity

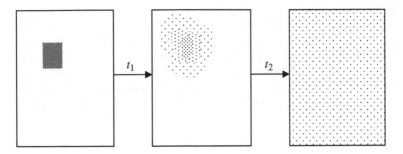

FIGURE 5.1
Time evolution of a set of initial states of a classical mechanical system charac-
terized by exponential mixing. After stochastization time, the set is distributed
uniformly over the entire available domain of the phase space. Note that time t_1
is shorter than and time t_2 longer than stochastization time.

that can be measured, however. What can be measured is the volume of a coarse-grained subsystem in which each point is replaced with a hypercube or hypersphere (pixel) whose dimensions are determined by experimental accuracy. The thus-defined volume increases during the evolution of an ensemble of systems with exponential mixing.

Giving the volume of the available domain **S** in the phase space with a measure of unity, we assign the meaning of *probability* $P(\mathbf{A})$ to the volume of an arbitrary subset **A** of **S**. More precisely, if domain **S** is not covered uniformly by its states, then we must introduce the *probability density* $\rho(s)$ determined for all states (points) s in **S**. The probability that a state belongs to a given subset **A** of the available state space **S**, is then given by a volume integral (usually in a multidimensional hyperspace), namely:

$$P(\mathbf{A}) = \int_{\mathbf{A}} ds \rho(s) \tag{5.1}$$

The probability density must be normalized to unity, so the integral over the available state space is:

$$\int_{\mathbf{S}} ds \rho(s) = 1 \tag{5.2}$$

which means that each system can be found in some state.

Assume that the probability $P(\mathbf{A})$ is known and that a system is subject to observation. As a result of finding that the state indeed belongs to the set **A** (in the theory of probability, set **A** represents an event), the observer gained a certain *amount of information* about the system under consideration and what was initially a probability $P(\mathbf{A})$ has now become a certainty.

The smaller the probability $P(\mathbf{A})$, the greater the information gain. Conversely, if $P(\mathbf{A})$ is large, the information gain is small. In other words, the information gain I is a decreasing function of P. The information about two independent events (whose probability is the product of individual probabilities) should be the sum of the information values for each event separately. The only function of P that has both of these properties takes the form:

$$I(\mathbf{A}) = -k \log_b P(\mathbf{A}) \tag{5.3}$$

where k and b are constants. The logarithm of unity is zero, so if the outcome of event $\mathbf{A}$ was certain before the observation took place, $P(\mathbf{A}) = 1$, and no new information is gained, $I = 0$. Equation (5.3) was derived in the famous paper on information theory published by Shannon in 1948. The selection of coefficients k and b determines the units of information. Assuming $k = 1$ and $b = 2$, we obtain a unit called one *bit*. One bit is the amount of information gained as a result of an experiment with two equally probable outcomes ($P = \frac{1}{2}$) which can be represented by the binary digits 0 or 1:

$$I = -\log_2 \frac{1}{2} = 1. \tag{5.4}$$

For a given probability density ρ and a given dynamical variable, i.e., a real function on a set of states:

$$x = \xi(s) \tag{5.5}$$

we can, treating the latter as a stochastic variable, determine its *mean value* as follows:

$$X = \int_S ds \rho(s) \xi(s). \tag{5.6}$$

If, in addition to the probability $P(\mathbf{A})$ that a state belongs to a certain subset $\mathbf{A}$ of the state space $\mathbf{S}$, the entire probability density $\rho(s)$ is known, according to Equation (5.6), we can determine the mean information gain achieved by the observer after the determination of state s of the system:

$$S = -k_B \int ds \rho(s) \ln \rho(s). \tag{5.7}$$

Equation (5.7) is identical to an expression first proposed by Boltzmann in 1872 and later extended to a general situation of an arbitrary statistical ensemble by Gibbs in 1902. These expressions provided a statistical interpretation of *entropy*, a quantity whose term was coined by Clausius in 1865. It represents the part of the internal energy that cannot be used for work, divided by absolute temperature T. Here, ln stands for the natural logarithm with a base $e = 2.718$, while $k_B = 1.38 \times 10^{-23}$ J/K is the gas constant R divided by Avogadro's number N_A and is called the Boltzmann constant. The unit of entropy is therefore J/K.

Equation (5.7) requires a comment. To make mathematical sense, the logarithm must be taken of a unitless quantity. However, the probability density has the dimension of the inverse volume in phase space. The unit of volume in phase space is the unit of action, i.e., the product of energy and time raised to a power equal to the number of degrees of freedom of the system. Having determined the units of action, we now move to the dimensionless function of probability density and make mathematical sense of Equation (5.7). The units of action correspond to the accuracy with which the observer can specify the microscopic state s of the system. This is, of course, somewhat arbitrary but there is a natural physical limit to this accuracy given by the quantum of action, i.e., the Planck constant $h = 6.6 \times 10^{-34}$ J·s.

Let **A** be a given domain of available states in the phase space. We calculate the entropy (Equation 5.7) for the probability density constant in this domain **A** and zero elsewhere:

$$\rho(s) = \begin{cases} \Omega^{-1} & \text{for } s \in \mathbf{A} \\ 0 & \text{for } s \notin \mathbf{A} \end{cases}. \tag{5.8}$$

For the probability to be properly normalized to unity, as in Equation (5.2), the constant Ω must represent the volume of domain **A**:

$$\Omega = \int_{\mathbf{A}} ds \tag{5.9}$$

We therefore have:

$$S = -k_B \int_{\mathbf{S}} ds\rho(s) \ln \rho(s) = k_B \int_{\mathbf{A}} ds\Omega^{-1} \ln \Omega = k_B \ln \Omega \tag{5.10}$$

Hence, the entropy is proportional to the logarithm of the volume of domain **A**, or more precisely to the number of units of volume contained in **A**. In its original formulation, Boltzmann's entropy is proportional to the logarithm of the number of microstates (states of individual systems comprising the ensemble) in a given macrostate (state of the statistical ensemble).

As stated earlier, for a proper physical description of the ensemble, we use a coarse-grained probability density. In systems that exhibit mixing, the probability density spreads over a given time scale to eventually cover the entire subset of available staxtes (see Figure 5.1). The entropy of the coarse-grained distribution in the state of thermodynamic equilibrium is greater than in its initial state. The instability of motion, i.e., the process of mixing on a microscopic scale, leads to the law of entropy increase over time. This law is commonly referred to as *the second law of thermodynamics*.

5.2 Nonequilibrium thermodynamics

The tendency of a statistical ensemble to achieve thermodynamic equilibrium with a uniform distribution of states for its constituent subsystems does not have to be monotonic in time. In general, equilibration takes place in stages and is characterized by several stochastization times with vastly different orders of magnitude. Thermodynamic equilibrium has no absolute meaning and depends on the time scale over which a given process is analyzed.

For example, consider equilibration of a hot cup of coffee to which a few drops of cream have been added. On a scale of several minutes, the process of diffusion takes place in which the concentration of cream in coffee is equilibrated. A longer time, on the order of one hour, is needed for the temperature of coffee and the surrounding air molecules to become the same. However, cold coffee in a cup is still not in complete thermodynamic equilibrium. If the surrounding air is sufficiently dry, water in the

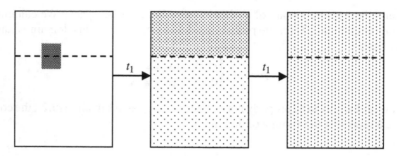

FIGURE 5.2
As a result of a bottleneck (broken line) between two subsets in the phase space, a stochastization process takes place in two stages: within a given subset (time t_1) and between the two subsets (time t_2).

cup will, after a few days, evaporate and lead to equilibrium between the liquid and gaseous phases. This is not the end of the process because porcelain is not stable forever and eventually (after a few million years perhaps) will undergo sublimation or break up and will turn into dust.

This example illustrates the *hierarchical* nature of the process of reaching thermodynamic equilibrium. In principle, for each time scale, one can identify *fast, slow* and *very slow* processes. The reason for a sharply defined hierarchy of stochastization times is the presence of bottlenecks in phase space (see Figure 5.2). The division of phase space into subsets separated by bottlenecks is due to specific organization of the system at a microscopic or macroscopic level. By definition, nonequilibrium thermodynamics is concerned with the process of reaching equilibrium by physical properties determined by intermediate slow steps. The corresponding stochastization times are called *relaxation times*. From the vantage point of the time scale of slow processes, the physical properties determined by fast processes are characterized by a so-called partial equilibrium while physical properties determined by very slow processes appear frozen.

Frozen nonequilibrium systems are everyday occurrences. For example, a certain number of carbon, hydrogen, and oxygen atoms in a ratio of 1:2:3 can exist as a diamond crystal covered with water in a pure oxygen atmosphere, or as cellulose in an identical atmosphere, or as a gas mixture of carbon dioxide and steamed water. Only the gas mixture is in thermodynamic equilibrium. The first two states are under normal conditions frozen structures.

As mentioned in Section 1.1, the atomic composition of matter in the Universe given by the ratio of neutrons to protons can also be characterized as a frozen structure. The ratio of neutrons to protons was fixed at an early stage of the evolution of the Universe when the density of neutrinos became too small to effectively control transformations of nucleons. Diamonds "remember" the enormous pressures they withstood deep below the Earth's surface and cellulose "remembers" the biosynthesis that created it in plant cells.

In certain states of matter, stochastization may not delineate well separated time scales and may produce a more or less quasicontinuous spectrum. Two states of this type, the *glassy state* and the *critical state*, have been the subjects of intensive investigation for years. Biological matter over various time scales appears to exhibit many properties of a glassy state (Frauenfelder et al., 1991).

The process of reaching complete equilibrium is treated by nonequilibrium thermodynamics as occurring over a sequence of partial equilibrium states. We are not interested in the time evolution of the system on a time scale shorter than the stochastization time for reaching partial equilibrium. Based on the ergodic theorem valid for systems with exponentially fast mixing, the ensemble average is to be replaced by the average over stochastization time. In this connection it should be stressed that during the stage of reading partial equilibrium the system under consideration need not be an actual ensemble of statistically independent subsystems. Only after the lapse of the stochastization time, due to the second important consequence of mixing, a decay of correlations (Penrose, 1979), the system is decomposed into a number of possibly identical, statistically independent subsystems. By definition, the objects of application of nonequilibrium thermodynamics are *macroscopic* systems. To see more than a collection of molecules or atoms in these systems, they must be viewed from special temporal and spatial perspectives.

Very few dynamical variables survive the averaging in Equation (5.6) in a state of complete or partial thermodynamic equilibrium and do not vanish at the end. Mutually independent nonzero averages of dynamical variables are called *thermodynamic variables* and they uniquely specify the *thermodynamic state* of the system (the state of complete or partial thermodynamic equilibrium). Only select thermodynamic variables usually have values fixed by special *constraints* imposed on the motion of the system. The values of the remaining thermodynamic variables may vary over a longer macroscopic time scale (their constraints are *imperfect*). The main task of nonequilibrium thermodynamics is the description of reaching complete equilibrium starting from a state of partial equilibrium.

Examples of thermodynamic variables are the energy of an isolated system E, its volume V, the number of molecules of a given type N_i, the net electric charge, the total electric or magnetic dipole moments (polarization or magnetization multiplied by the volume), and the shape of the system characterized by appropriate lengths and angles. If some average values of thermodynamic variables are not zero, then this is related to the property of *additivity* for thermodynamic variables. If the system is composed of two subsystems, the values of thermodynamic variables corresponding to the whole system are equal to the sums of the values of these variables for the two subsystems taken separately.

This division of the system into subsystems can be arbitrary, real or imagined, in real physical space, or in an abstract space of internal states. The only condition that must be fulfilled is the macroscopic nature of the selected subsystems. If the value of each variable characterizing an arbitrary subsystem is proportional to the size of this subsystem (i.e., is *extensive*), the thermodynamic system is called *simple* (see Figure 5.3a). Otherwise, the system is considered *complex* (see Figure 5.3b). Simple systems are internally homogeneous and only have *external constraints* defining the values of *global thermodynamic variables* for the system as a whole. Complex systems have internal structures resulting from *internal constraints* that determine the values of additional *structural (local) thermodynamic variables*.

(a)

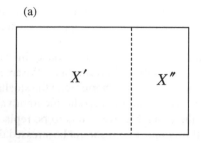

(b)

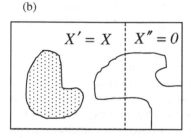

FIGURE 5.3
Division of simple (a) and complex (b) thermodynamic systems into two subsystems. In the case of a simple system, the values of a thermodynamic variable X characterizing its subsystems (X' for the first and X'' for the second) are always proportional to the size of these subsystems. However, in the case of a complex system, they may be arbitrary. Figure (b) presents a situation where the variable considered has a nonzero value only in the selected subsystem which has internal constraints.

Internal constraints can be provided by a wall separating a container of gas or liquid into two parts. It can be immobile, impenetrable to molecules, and nonconducting to energy or electric charge. In order to specify the state of a system, one must determine the volume, number of molecules of each type, energy and charge for each part of the container separately (or only for the part using the property of additivity; see Figure 5.4a). The actual wall can be less than ideal. It can even be constructed to only approximately play the role of a constraint. A wall that is not rigidly fixed becomes a movable piston; a wall incompletely impenetrable to molecules becomes a semipermeable membrane. A diathermal wall conducts a form of energy, namely heat. One made from a conducting material conducts electricity.

An electrical conductor does not necessarily require the structure of a three-dimensional wall. Likewise, a semipermeable partition does not have to be given on a macroscopic scale but may exist on a microscopic level in the internal states of molecules comprising the system. This latter type of partition is commonly found in chemical reactions. As an example, consider a simple reaction of unimolecular isomerization (see Figure 5.4b).

A molecule of a compound designated $R-C_2H_2O-R'$ (R and R' denote two arbitrary groups of atoms) may exist in two distinct chemical states (isomers) as a ketone or an alcohol. The continuum of states (the generalized coordinates and their conjugate momenta) characterize the internal degrees of freedom of the molecule, i.e., the instantaneous bond lengths, valence angles, and rotational angles. This corresponds to the contents of the box in Figure 5.4a and the two chemical states correspond to the two separate parts. A transition through the partition represents the reaction from one chemical state to another through an intermediate state (its structure is shown in the parentheses). The partition symbolizes a bottleneck (see Figure 5.2), a barrier whose character is partly entropic (a lower number of effective degrees of freedom of the intermediate transition state is due to the specific orientation of some molecular groups) and partly energetic (a temporary breakage of two covalent bonds).

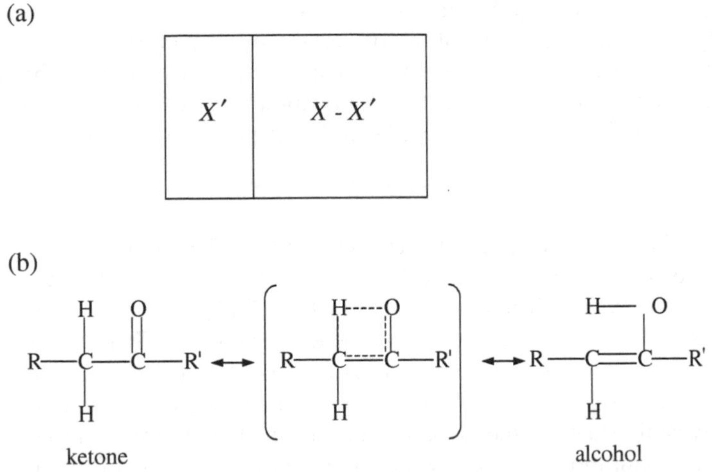

(a)

(b)

ketone   alcohol

FIGURE 5.4

(a) A complex thermodynamic system in which internal constraints are symbolically illustrated as a wall dividing the system into two parts. The thermodynamic state of the system is specified by the values of two thermodynamic variables: global X and structural X'. The structural variable X' specifies the state of one of the subsystems. The value of its counterpart for the other subsystem can be obtained using the additivity property. (b) Example of a chemical reaction of isomerization ketone–alcohol. A transition of the molecule from one chemical state to another through an intermediate state (the stucture in parentheses) corresponds to the transition of a molecule through the semipermeable partition of (a). In this case, the number of all molecules corresponds to the global variable X while the number of molecules of a given isomer corresponds to the structural variable X'. During the reaction, the value of the variable X' changes until it reaches its equilibrium value when the flux of molecules moving from left to right is balanced by the flux of molecules moving in the opposite direction.

On a short time scale, imperfect and perfect constraints behave identically by fixing the values of thermodynamic variables. Only over a longer time scale does the actual imperfect character of these constraints become apparent and determine the rates of change of these variables. Imperfect structural constraints may be more or less fictitious. Even in simple systems, in the absence of a required structure, processes of equilibration do not take place right away over the entire spatial domain of the macroscopic system.

They first take hold in small regions, then in bigger ones and finally in the biggest, as if separated by invisible partitions. In fact, no partitions inhibit the process. The reason for the slowness of the changes is the system's spatial extent. It is known from the continuum principle for physical processes that, given a constant velocity of the process, the larger the system, the slower the process. Such a hierarchical manner of describing nonequilibrium processes is provided by the thermodynamics of continuum media, in particular by hydrodynamics.

The state of partial equilibrium, i.e., a nonequilibrium thermodynamic state, is uniquely specified by the values of appropriate thermodynamic variables. In particular, they determine the probability density for the system. The entropy in this state is not only a function of probability density (see Equation 5.7), but directly a simple function of thermodynamic variables (Callen, 1985; Kondepudi and Prigogine, 1998):

$$S = S(E, X_1, \ldots, X_n). \tag{5.11}$$

Energy E has a special place in thermodynamics. The number of thermodynamic variables other than energy n is called the number of thermodynamic degrees of freedom of the system. The function in Equation (5.11) was introduced into thermodynamics by Clausius, albeit at a phenomenological level. Therefore, we call it the *Clausius entropy* in contrast to the *Boltzmann–Gibbs statistical entropy* discussed in the previous section. Note that entropy contains all the information about the thermodynamic states of the system. The irreversible progression toward the state of complete equilibrium is closely linked to the maximization of its value. Entropy S is a monotonic (strictly increasing), thus one-to-one, function of energy E and Equation (5.11) can be inverted according to:

$$E = E(S, X_1, \ldots, X_n) \tag{5.12}$$

The partial derivative of energy related to entropy is called *temperature*:

$$T \equiv \left(\frac{\partial E}{\partial S} \right)_X. \tag{5.13}$$

This derivative is to be taken when all the thermodynamic variables X_i are fixed. Absolute temperature is always positive and measured in degrees Kelvin.

In general, entropy is a monotonic function of all global variables but when taken as a function of structural (local) variables, it has a local maximum. When the constraints corresponding to one or several global or structural variables are removed (or weakened, but not enough to drive the system out of partial thermodynamic equilibrium), the values of these variables will continue evolving for as long as it takes entropy S to reach its maximum value at fixed values of the other variables, in particular the internal energy $E = $ const. Under the condition of a fixed value of entropy, $S = $ const, the internal energy evolves until it reaches its minimum value. However, the most common situation does not involve $E = $ const or $S = $ const; it involves an *isothermal* condition with a fixed value of temperature $T = $ const. This is described by a Legendre transform of internal energy (Equation 5.12) which is called *free energy*:

$$F = E - TS = F(T, X_1, \ldots, X_n). \tag{5.14}$$

Under isothermal conditions in the state of complete thermodynamic equilibrium, the free energy reaches its minimum value.

Negative derivatives of free energy with respect to thermodynamic variables at a fixed temperature:

$$A_i \equiv - \left(\frac{\partial F}{\partial X_i} \right)_T \tag{5.15}$$

are called *thermodynamic forces*. The thermodynamic force conjugate to volume is pressure, the one conjugate to the electric charge is the negative electric potential. The one conjugate to the number of molecules of one type is the corresponding negative chemical potential. Equation (5.15) relating thermodynamic variables to the corresponding forces and Equation (5.13) linking temperature with entropy or energy are referred to as *equations of state*.

As with energy and entropy, free energy is an additive quantity. If the state of a system is specified by a global variable Y and a local variable $X = Y'$, defined for a chosen (primed) subsystem (see Figure 5.3a), the free energy of the total system is the sum of the free energies of the primed and the double primed subsystems:

$$F(X, Y) = F'(X) + F''(Y - X) = F'(Y') + F''(Y'').$$ (5.16)

As a consequence, the thermodynamic force conjugate to the local variable X is equal to the difference between the global thermodynamic forces in both subsystems:

$$A \equiv -\left(\frac{\partial F}{\partial X}\right)_T = -\left(\frac{\partial F'}{\partial Y'}\right)_T + \left(\frac{\partial F''}{\partial Y''}\right)_T = A' - A''.$$ (5.17)

A thermodynamic state of a given system can be alternatively specified by the values of its global and local thermodynamic variables fixed by appropriate constraints or by the values of the conjugate forces acting on the surroundings. If the environment acts on the system with identical forces, the system is in complete thermodynamic equilibrium (see Figure 5.5). If, on the other hand, external forces do not counterbalance internal forces, the system is only in a state of partial equilibrium.

Consider an arbitrarily small change of the system's energy at a fixed temperature:

$$\Delta E = T\Delta S + \Delta F = T\Delta S - \sum_i A_i \Delta X_i.$$ (5.18)

where we have used Equations (5.13) and (5.15). According to the *first law of thermodynamics*, the energy change ΔE is achieved by providing (or removing) *heat Q* or by doing *work* on the system (or by the system) W:

$$\Delta E = Q + W.$$ (5.19)

While supplying (removing) heat is linked to a change in the value of entropy S, work involves a change in the values of thermodynamic variables X_i. However, heat Q cannot be identified directly with the quantity $T\Delta S$ and work W cannot be identified directly with a change in the free energy ΔF as Equation 5.19 can still be satisfied when something is added to and subtracted from it simultaneously. Rewriting it as:

$$\Delta E = (Q + D) + (W - D).$$ (5.20)

we obtain the general relationship:

$$Q + D = T\Delta S$$ (5.21)

and

$$W - D = -\sum_i A_i \Delta X_i.$$ (5.22)

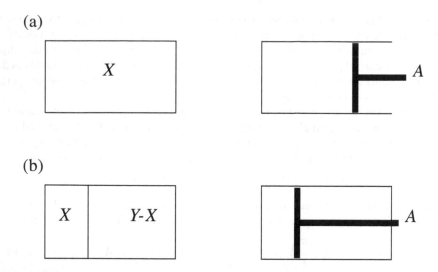

FIGURE 5.5
(a) Thermodynamic state of a simple system with a fixed value of the global thermodynamic variable X (e.g., volume) is the same as the state of this system in equilibrium with the environment that exerts a thermodynamic force A on it. This force is illustrated as a piston applying pressure on the system. (b) Thermodynamic state of a complex system specified by fixed values of global variable Y and structural variable X (represented by the volume of the subsystem separated by a partition) is the same as the state of this system when a chosen subsystem is acted upon by an external force A (represented by an internal piston) preventing it from reaching equilibrium with the rest of the system.

For processes that are simultaneously *adiabatic* (occurring without exchanging heat with the environment, $Q = 0$) and *spontaneous* (occurring without work done on or by the system, $W = 0$), the relationships in Equations (5.20) and (5.21) take the form:

$$\Delta E = 0; \quad T\Delta S = D. \tag{5.23}$$

When $E = $ const, the entropy increases to its maximum, $\Delta S \geq 0$, hence due to the positive values of temperature:

$$D \geq 0. \tag{5.24}$$

Thus:

$$T\Delta S \geq Q, \tag{5.25}$$

which constitutes a formal expression of the *second law of thermodynamics* (Kondepudi and Prigogine, 1997).

Quantity D is always positive and is called *energy dissipation*. Figure 5.6 shows the possible energy transformations that may occur under isothermal conditions. This motivates the use of the term *free energy* for the quantity F. This is the only part of the total internal energy E that can be used to perform useful work.

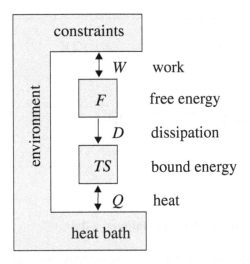

FIGURE 5.6
Energy transformations under isothermal conditions T = const.

The remaining part of the total energy is equal to TS and is *bound* because it is impossible under isothermal conditions to transfer energy to F and perform work on constraints. In spontaneous processes ($W = 0$), free energy F can only decrease or stay the same; it cannot increase. It remains constant only when dissipation D equals zero. Processes occurring without dissipation are called *reversible* since under adiabatic conditions ($Q = 0$) they do not cause entropy increases. As we will see later, they must take place infinitesimally slowly. All processes of nonequilibrium thermodynamics that proceed at finite speeds, including all biochemical processes, are associated with energy dissipation and hence are irreversible.

5.3 Rates of nonequilibrium thermodynamic processes

The energy dissipated in a nonequilibrium process going through states of partial equilibrium is related to a lack of balance between internal and external thermodynamic forces acting on the system. Positive or negative work W is performed by or against the actual external forces A_i^{ext}:

$$W = -\sum_i A_i^{\text{ext}} \Delta X_i. \tag{5.26}$$

Comparing this to Equation (5.22), the measure of dissipation is the difference between these forces and internal forces A_i determined by the equations of state (5.15):

$$D = \sum_i (A_i - A_i^{\text{ext}}) \Delta X_i. \tag{5.27}$$

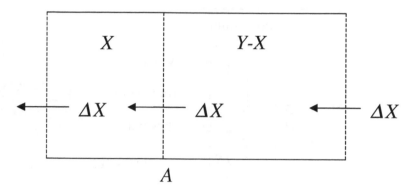

FIGURE 5.7
Open thermodynamic system in a stationary state.

Let us denote a global variable by Y and its corresponding structural variable by X (see Figure 5.7). As mentioned earlier, a state with a fixed value of the structural variable X can be maintained through completely rigid internal constraints or through an external force that exactly balances the internal force A conjugate to X. Otherwise, the system in a state of partial quilibrium will evolve to complete equilibrium.

However, if the external constraints are not completely rigid (the system as a whole is *open*), the change ΔX in the value of the structural variable X during its evolution, resulting from a transfer of some amount of X from one subsystem to another, can be compensated, providing an equal but opposite amount to the first subsystem from the outside and similarly with the second subsystem (Figure 5.7). Since the values of the variables X and Y in the system do not change, both the free energy and bound energy (entropy) remain constant:

$$F = \text{const}; \quad S = \text{const}. \tag{5.28}$$

In the absence of an external force $A^{\text{ext}} = 0$, a spontaneous change ΔX is associated with dissipation (see Equation 5.27):

$$D = A\Delta X. \tag{5.29}$$

To fulfill the conditions of Equation (5.28), D must be equal on the one hand to the work done on the system and, on the other, to the heat released to the environment (see Figure 5.6):

$$D = W = -Q. \tag{5.30}$$

The environment now performs work on the system not through external forces, but as a result of the flux of the quantity ΔX across the system (see Figure 5.7). In a nonequilibrium *stationary state* maintained due to such a flux (Kondepudi and Prigogine, 1997), the rate of dissipation, i.e., the amount of free energy change transferred into a bound energy form over a unit of time Δt, remains constant:

$$\frac{D}{\Delta t} = \frac{W}{\Delta t} = -\frac{Q}{\Delta t}. \tag{5.31}$$

In the limit of $\Delta t \to 0$, the quantity in Equation (5.31) is called the *dissipation function*. In view of the second law of thermodynamics, it is always nonzero and based on Equation (5.29), it takes the form

$$\Phi = AJ \geq 0, \tag{5.32}$$

where we have introduced *flux J*, a term equal to the rate of change (the time derivative is denoted by a dot) of the variable X:

$$J \equiv \dot{X}. \tag{5.33}$$

The direction of flux J is given by the condition that the product of its value with the force A be positive. The flux of gas in an unbiased turbine performing no work moves from an area of high pressure to one with a lower pressure but the volume of gas increases in the opposite direction. The flux of molecules in an open chemical reactor flows to an outlet with a lower chemical potential. Likewise, in a conducting material, the electric current flows from a higher potential to a lower one, and in a thermal conductor from the high temperature end to the one with a lower temperature. In general, for several structural thermodynamic variables X_i the dissipation function (Equation 5.31), according to Equation (5.27), takes the form:

$$\Phi = \sum_i A_i J_i \geq 0, \tag{5.34}$$

where

$$J_i \equiv \dot{X}_i. \tag{5.35}$$

For systems under conditions close to complete thermodynamic equilibrium, we can assume that the fluxes J_i depend linearly on the thermodynamic forces A_i (a situation designated as a *linear response to external perturbations*):

$$J_i = \sum_j L_{ij} A_j. \tag{5.36}$$

The coefficients of proportionality L_{ij} are called Onsager's *kinetic coefficients*. In this approximation, the dissipation function (Equation 5.34) is a quadratic form of thermodynamic forces:

$$\Phi = \sum_i L_{ij} A_i A_j \geq 0. \tag{5.37}$$

The sign of the form indicates that the diagonal kinetic coefficients are nonnegative:

$$L_{ii} \geq 0. \tag{5.38}$$

This is not true in general for nondiagonal coefficients. Based on statistical thermodynamics, it can be demonstrated that in the absence of magnetic fields the matrix of kinetic coefficients is symmetric:

$$L_{ij} = L_{ji}. \tag{5.39}$$

The above is an expression of the so-called *fourth law of thermodynamics* due to Onsager.

Ohm's law expresses a linear relationship between the flux of electrical charge (electric current) and a potential difference (voltage). In a diffusion process or chemical reaction, Fick's law provides a linear relationship between the flux of molecules and the chemical potential difference. Likewise, a direct proportionality exists between the heat flux and the temperature difference in a thermally conducting slab, as expressed by Fourier's law. Off-diagonal kinetic coefficients describe cross-effects. For example, due to Onsager's symmetry, it can be concluded that the thermoelectric effect (the so-called Seebeck effect) must be accompanied by an electrothermal effect (the so-called Peltier effect). In general, nonzero off-diagonal kinetic coefficients are related to *thermodynamic coupling* of various irreversible processes, e.g. chemical reactions, and with processes of *free energy transduction*. These processes play significant roles in the physiology of biological cells. However, their description in terms of a linear approximation is by and large incorrect.

The linear relationships in Equation (5.36) apply to both open and closed systems. They can be used, for example, for the simplest case of a closed system with only one nonequilibrium variable X that evolves toward its equilibrium value X^{eq}, as:

$$\dot{X} = LA. \tag{5.40}$$

The force A conjugate to X disappears in the state of complete equilibrium; hence the equation of state linking it to X takes the following form in the linear approximation:

$$X - X^{eq} = -CA. \tag{5.41}$$

Depending on the context, the coefficient C is called *capacity* or *susceptibility*. Inserting one of these equations into the other gives a linear equation that describes the system's relaxation to the state of complete equilibrium:

$$\dot{X} = -\tau^{-1}(X - X^{eq}), \tag{5.42}$$

where

$$\tau^{-1} = \frac{L}{C}. \tag{5.43}$$

Equation (5.42) is one of the simplest linear differential equations. It has the following solution:

$$X(t) - X^{eq} = (X(0) - X^{eq})e^{-\frac{t}{\tau}} \tag{5.44}$$

which describes an exponential decay of the initial value $X(0)$ to the equilibrium value X^{eq} with a characteristic relaxation time τ. The range of applicability of equations such as Equation (5.43) often exceeds greatly the range of validity of the linear approximations in Equations (5.40) and (5.41).

5.4   Single unimolecular chemical reaction

We are now interested in processes taking place not only under isothermal, but also isobaric conditions for which $P = $ const. Then the internal energy $E(S, V, N)$ should be replaced by *enthalpy* which is its Legendre transform with respect to volume V:

$$H(S, P, N) = E(S, V, N) + PV \tag{5.45}$$

and the free energy $F(T, V, N)$, should be replaced by the *free enthalpy*:

$$G(T, P, N) = F(T, V, N) + PV = H - TS. \tag{5.46}$$

The G above is sometimes called the *Gibbs free energy*; F is the *Helmholtz free energy*.

Suppose each of the N particles comprising the system can exist in two states R and P, and that in the system the following reaction can take place:

$$R \rightleftarrows P. \tag{5.47}$$

States R and P can represent different chemical states of the molecule (isomers) or different physical states, e.g., the presence of a molecule on one side of a lipid bilayer with ion channels for transport across it. In both cases, the formal description of the process is identical (see Figure 5.4). The molecule in state R will be symbolically called a *reagent* (or a *substrate*) while the molecule in state P is a *product* of the reaction. In fact, in general, the reaction can proceed in both directions. Let N_R denote the number of molecules in state R and N_P the number in state P. The total Gibbs free energy is the sum of the component free energies:

$$G = G_R(T, P, N_R) + G_P(T, P, N_P) \tag{5.48}$$

with

$$N_R + N_P = N = \text{const.} \tag{5.49}$$

Only one of N_R and N_P is an independent thermodynamic variable that character-izes a state of lack of chemical equilibrium. Let us choose $N_P = X$ as the independent variable. The thermodynamic force connected with this variable is called the *chemical affinity*, and it is equal to the difference between the chemical potential of the reagent and the product in this reaction (Hill, 1989; Kondepudi and Prigogine, 1998):

$$A = -\left(\frac{\partial G}{\partial X}\right)_{T,P,N} = -\left(\frac{\partial G_P}{\partial N_P}\right)_{T,P} + \left(\frac{\partial G_R}{\partial N_R}\right)_{T,P} = -\mu_P + \mu_R \tag{5.50}$$

For gases or perfect solutions, the chemical potential of molecules in state R takes the form (Atkins, 1998):

$$\mu_R = \mu_R^\circ + k_B T \ln \frac{N_R}{N} \tag{5.51}$$

and similarly for the molecules in state P. The fractious:

$$P_R = \frac{N_R}{N} \quad P_P = \frac{N_P}{N} \tag{5.52}$$

denote the probabilities that a given molecule is found in state R or P, respectively. The value of the *standard chemical potential* μ°_R is usually determined taking N equal to Avogadro's number (number of molecules in one mole), $N = N_A \approx 6.0 \times 10^{23}$. If the distribution of molecules is spatially homogeneous, then N_R and N_P can be expressed by *molar concentrations* denoted by the symbol of the molecule in a square bracket, i.e.:

$$[R] = \frac{N_R}{N_A V_R} \quad [P] = \frac{N_P}{N_A V_P}, \tag{5.53}$$

where V_R and V_P are the volumes taken up by the molecules of each type, respectively. They can be identical but may also differ as in the case of diffusion across a membrane. Equation (5.51) for $N = N_A$, expressed through molar concentrations [R], takes the form:

$$\mu_R = \mu_R^\circ + k_B T \ln \frac{[R]}{M}. \tag{5.54}$$

The unit of molar concentration, $1M = mol/dm^3$, has been adopted as the standard concentration that determines the standard chemical potential $\mu^\circ{}_R$ ($\mu_R = \mu^\circ{}_R$ in the case when [R] = 1M).

Equation (5.54) can be assumed to be always satisfied independently of the assumption that the gas is ideal or the solution is perfect, i.e., regardless of the statistical independence of the molecules of the system. Then this relationship should be treated as a definition of the quantity [R] which is in general called *activity*. Chemists adopted the convention that the activity of a substance in a pure phase is equal to unity (M = 1). Dissolution can be also treated as a chemical reaction and the standard chemical potential of pure solid solute is equal by definition to zero while the activity of a substance in a gas phase is expressed by pressure measured in bars (M = 1 bar = 10^5 Pa). The activity of substances in dilute liquid solutions can be approximated by their molar concentrations.

Substituting Equation (5.54) for the chemical potential into the formula in Equation (5.50), we obtain an expression for chemical affinity as:

$$A = -\frac{\Delta G^\circ}{N_A} - k_B T \ln \frac{[P]}{[R]}, \tag{5.55}$$

where

$$\Delta G^\circ = N_A (\mu_P^\circ - \mu_R^\circ) \tag{5.56}$$

is called the *free energy of the reaction* (for one mole). Equation (5.55) can be rewritten as

$$\frac{[P]}{[R]} = K e^{-\frac{A}{k_B T}}, \tag{5.57}$$

where

$$K = e^{-\frac{\Delta G^\circ}{RT}}. \tag{5.58}$$

The gas constant R is the product of Avogadro's number and the Boltzmann constant:

$$R = N_A k_B = 8.3 \times 10^{-3} kJmol^{-1}deg^{-1}. \tag{5.59}$$

In chemical equilibrium, chemical affinity vanishes, $A = 0$, since the chemical potentials of the molecules in both states are the same. Hence, the ratio of the equilibrium concentration of the molecules in state P to that in state in state R is given by the equation:

$$\frac{[P]^{eq}}{[R]^{eq}} = K. \tag{5.60}$$

The quantity K is called the *chemical equilibrium constant*. Equation (5.58) relates it to the free energy of the reaction ΔG°. For $V_R = V_P = V$ (a homogeneous

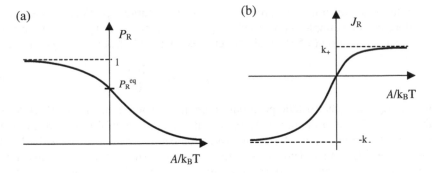

FIGURE 5.8
(a) Dependence of the molar fraction P_R on the chemical affinity (force) A for a unimolecular reaction. (b) The corresponding dependence for the flux: $J_R = dP_R/dt$.

mixture of molecules), the conservation law in Equation (5.49) can be expressed via the concentrations:

$$[R] + [P] = [R]_0, \qquad (5.61)$$

whereas the probabilities (Equation 5.52), via the *molar fractions*:

$$P_R = \frac{[R]}{[R]_0} \quad P_P = \frac{[P]}{[R]_0}. \qquad (5.62)$$

From Equation (5.57), we obtain a unique relationship between the thermodynamic variable $X = N_P = P_P N$ and its conjugate force A:

$$P_P = \frac{1}{1 + K^{-1} e^{\frac{A}{k_B T}}}. \qquad (5.63)$$

This is a chemical *equation of state* (see Figure 5.8a).

The free energy of a reaction can be divided into enthalpy and entropy components according to:

$$\Delta G^\circ = -RT \ln K = \Delta H^\circ - T \Delta S^\circ. \qquad (5.64)$$

In a single chemical reaction that takes place in a closed reactor, no work is performed. The quantity ΔH° indicates *reaction heat* (see Figure 5.6 in which energy E is replaced with enthalpy H and free energy F with the Gibbs free energy G). The heat generated during a spontaneously proceeding reaction may be given off ($\Delta H^\circ < 0$) in an *exothermic reaction* or taken up from the environment ($\Delta H^\circ > 0$) in an *endothermic* reaction. Similarly, the free energy of a reaction can be negative ($\Delta G^\circ < 0$, $K > 1$) for an *exergic reaction* or positive ($\Delta G^\circ > 0$, $K < 1$) for an *endo-ergic reaction*. If the free energy of reaction ΔG° does not depend on temperature, by differentiating Equation (5.64) and using the thermodynamic relation:

$$\left(\frac{\partial H}{\partial T}\right)_P = T\left(\frac{\partial S}{\partial T}\right)_P, \qquad (5.65)$$

it can be concluded that $\Delta S^\circ = 0$, or $\Delta G^\circ = \Delta H^\circ$. Consequently, an exergic reaction becomes automatically an exothermic reaction while an endoergic reaction becomes endothermic.

In the case of transport processes involving charged particles (ions) across membranes, for example, with the help of protein channels or ion pumps, the chemical potential in Equation (5.54) should be replaced by the *electrochemical potential*:

$$\tilde{\mu}_R = \mu_R^\circ + ze\psi_R + k_B T \ln \frac{[R]}{M}, \tag{5.66}$$

where ψ_R is the electrostatic potential of the membrane on the side of R while ze is the electric charge of the molecule expressed as a multiple of the elementary charge with an appropriate sign. A similar expression can be written for the side P of the membrane. The standard chemical potentials on both sides of the membrane are the same, $\mu^\circ_R = \mu^\circ_P$, and the equation for affinity, (5.55), takes the form of the *Nernst equation*:

$$A = -ze\Delta\psi - k_B T \ln \frac{[P]}{[R]}, \tag{5.67}$$

where

$$\Delta\psi = \psi_P - \psi_R \tag{5.68}$$

represents a *transmembrane potential*. In thermodynamic equilibrium, $A = 0$ and the transmembrane potential balances the concentration gradient of ions across the membrane:

$$\Delta\psi = -\frac{k_B T}{ze} \ln \frac{[P]^{eq}}{[R]^{eq}}. \tag{5.69}$$

The time dependence of concentrations [P] and [R] is given according to the *kinetic equation*:

$$\frac{d}{dt}[P] = -k_-[P] + k_+[R] = -\frac{d}{dt}[R] \tag{5.70}$$

which, in view of the relationship in Equation (5.61), is exactly equivalent to the equation of the type in Equation (5.42) or:

$$\frac{d}{dt}[P] = -(k_- + k_+)([P] - [P]^{eq}). \tag{5.71}$$

In spite of the fact that Equations (5.70) and (5.71) are linear, they apply to situations arbitrarily far removed from chemical equilibrium (Kurzynski, 1990). Parameters k_+ and k_- are called *forward* and *reverse reaction rate constants*, respectively. They appear in the reaction equation:

$$R \underset{k_-}{\overset{k_+}{\rightleftharpoons}} P \tag{5.72}$$

Of course, the selection of forward and reverse directions is a matter of convention, as are designations of the reagent and product. Equilibrium solutions to Equation (5.70) satisfy the relationship:

$$\frac{[P]^{eq}}{[R]^{eq}} = \frac{k_+}{k_-} = K \tag{5.73}$$

[see also Equation (5.60)] linking one of the reaction rates with the other through the equilibrium constant K.

We now define the *flux of a reaction* as the derivative below:

$$J_P = \frac{d}{dt} P_P = \frac{d}{dt} \frac{[P]}{[R]_0}. \tag{5.74}$$

Comparing it with Equation (5.33), the flux in Equation (5.74) is defined in relation to a single molecule. The kinetic equation (5.70) and the equation of state (5.63) lead to a nonlinear relationship of the flux and thermodynamic force:

$$J_P = \frac{1 - e^{-\frac{A}{k_B T}}}{k_+^{-1} + k_-^{-1} e^{-\frac{A}{k_B T}}} \tag{5.75}$$

(see Figure 5.8b). According to the second law of thermodynamics (Equation 5.33), the reaction's direction (the sign of the flux J_P) is always determined by the sign of the chemical affinity A. In view of Equation (5.55), it is given by the sign of the free energy of the reaction ΔG° and also by the reagent and product concentrations. This conclusion is correct whether the system is open or closed.

5.5 Bimolecular reactions: protolysis

The most general bimolecular reaction takes the form of an *exchange reaction*:

$$R + QP \underset{k_-}{\overset{k_+}{\rightleftarrows}} RQ + P. \tag{5.76}$$

Four thermodynamic variables undergo changes in this reaction, namely: the molecular numbers N_R, N_P, N_{QP}, and N_{RQ} that are proportional to the molecular concentrations $[R]$, $[P]$, $[QP]$, and $[RQ]$, respectively. Only one of these variables is independent which follows from the three conservation laws:

$$[R] + [RQ] = [R]_0 = \text{const},$$
$$[QP] + [P] = [P]_0 = \text{const},$$
$$[QP] + [PQ] = [Q]_0 = \text{const}. \tag{5.77}$$

The chemical affinity corresponding, for example, to the variable N_P or N_{RQ} takes the form:

$$A = -\mu_P - \mu_{RQ} + \mu_R + \mu_{QP} = -\mu_P^\circ - \mu_{RQ}^\circ + \mu_R^\circ + \mu_{QP}^\circ$$
$$- k_B T \ln \frac{[RQ][P]}{[R][QP]} \tag{5.78}$$

and denoting the standard free energy of the reaction as

$$\Delta G^\circ = N_A (\mu_P^\circ - \mu_{RQ}^\circ + \mu_R^\circ + \mu_{QP}^\circ) \tag{5.79}$$

we obtain the chemical equation of state:

$$\frac{[RQ][P]}{[R][QP]} = e^{-\frac{\Delta G^\circ}{RT}} e^{-\frac{A}{k_a T}} = K e^{-\frac{A}{k_a T}}. \tag{5.80}$$

The corresponding kinetic equation is given by:

$$\frac{d}{dt}[P] = \frac{d}{dt}[RQ] = -\frac{d}{dt}[R] = -\frac{d}{dt}[QP]$$
$$= k_+[R][QP] - k_-[RQ][P] \tag{5.81}$$

and its equilibrium solution by:

$$\frac{[RQ]^{eq}[P]^{eq}}{[R]^{eq}[QP]^{eq}} = \frac{k_+}{k_-} = K = e^{-\frac{\Delta G^\circ}{RT}}. \tag{5.82}$$

This expression is called the *law of mass action* as it means that a change in the equilibrium concentration of one substance is followed by changes in the concentrations of the other substances such that the parameter K which is only temperature dependent remains constant.

The equation of state (5.80), taking into account the relations in Equation (5.77), becomes rather complex and the kinetic equation becomes nonlinear. However, if two of the reagents are in excess of the third one:

$$[R]_0, [P]_0 \gg [Q]_0, \tag{5.83}$$

then

$$[R] \approx [R]_0, \quad [P] \approx [P]_0 \tag{5.84}$$

and the kinetic equation becomes linear. The equations of state and the relation between the flux and the force become then identical to Equations 5.63 and 5.75, with the unimolecular forward reaction rate k_+ replaced by the pseudounimolecular constant $k_+[R]$ and the unimolecular reverse reaction rate k_-, replaced by the pseudounimolecular constant $k_-[P]$.

An example of an exchange reaction taking place in an aqueous environment is the *protolysis* reaction (Atkins, 1998) with the transfer of a proton to or from a water molecule, respectively:

$$A^- + H_2O \underset{\leftarrow}{\overset{\rightarrow}{\rightleftarrows}} HA + OH^- \tag{5.85}$$

and

$$H_2O + HA \underset{\leftarrow}{\overset{\rightarrow}{\rightleftarrows}} H_3O + A^-. \tag{5.86}$$

The compound HA that is a proton donor is called an *acid*. The compound A^- that serves as a proton acceptor is called a *base* coupled to HA. We assumed here that its molecules had a negative charge but in general this does not have to be so. It can be assumed that $A^- = B$ and $HA = BH^+$.

In both reactions (Equations 5.85 and 5.86), water appears once as a base and once as an acid. A special case of the protolysis reaction is

$$H_2O + H_2O \underset{\leftarrow}{\overset{\rightarrow}{\rightleftarrows}} H_3O^+ + OH^-, \tag{5.87}$$

which is often written as an effective dissociation reaction of water into ions:

$$H_2O \rightleftarrows H^+ + OH^- \tag{5.88}$$

although it should be noted hydrogen ions never appear as unhydrated protons with diameters less than five orders of magnitude smaller than other ions. The equilibrium constant for this reaction is

$$K = \frac{[H^+]^{eq}[OH^-]^{eq}}{[H_2O]^{eq}} = 1.8 \times 10^{-16}M. \tag{5.89}$$

One mole of H_2O has a mass of 18 g; hence the molar concentration of pure water

$$[H_2O] = 55 \text{ M}. \tag{5.90}$$

In diluted solutions, this value is well approximated by $[H_2O]^{eq}$ in Equation (5.89). Using the value of the dissociation constant in Equation (5.88) we obtain,

$$[H^-]^{eq} = [OH^-]^{eq} = 10^{-7}M. \tag{5.91}$$

It is commonly accepted to use a negative decimal logarithm of the molar concentration of hydrogen ions expressed in units of M, referred to as pH:

$$pH = -\log_{10}\frac{[H^+]^{eq}}{M}. \tag{5.92}$$

For pure water we have

$$pH = 7. \tag{5.93}$$

Equilibrium constants for the reactions in Equations (5.85) and (5.86) define the ratios of the equilibrium concentrations:

$$K'_a = \frac{[H_3O^+]^{eq}[A^-]^{eq}}{[H_2O]^{eq}[HA]^{eq}} = \frac{K_a}{[H_2O]}; \quad K'_b = \frac{[HA]^{eq}[OH^-]^{eq}}{[A^-]^{eq}[H_2O]^{eq}} = \frac{K_b}{[H_2O]}. \tag{5.94}$$

Since the concentration of water is practically unchanged during the reaction, it is convenient to use constants K_a and K_b instead of K'_a and K'_b. These sets are related via:

$$K_aK_b = [H_3O^+]^{eq}[OH^-]^{eq} = 10^{-14}M^2. \tag{5.95}$$

Defining the quantities

$$pK_a = -\log_{10}\frac{K_a}{M}; \quad pK_b = -\log_{10}\frac{K_b}{M} \tag{5.96}$$

we obtain the relationship:

$$pK_a + pK_b = 14. \tag{5.97}$$

The *degree of dissociation* of acidic molecules HA is given by the ratio

$$P = \frac{[A^-]^{eq}}{[A^-]^{eq} + [HA]^{eq}}. \tag{5.98}$$

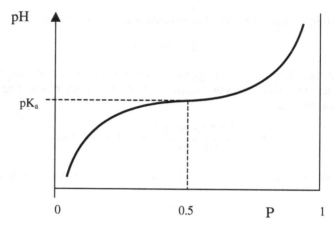

FIGURE 5.9
Titration curve corresponding to Equation (5.100).

This also leads to the relationship

$$\frac{P}{1+P} = \frac{K_a}{[H^+]^{eq}},\tag{5.99}$$

to which we apply the logarithm function on both sides and obtain the so-called *Henderson-Hasselbalch equation*:

$$pH = pK_a + \log_{10}\frac{P}{1-P}.\tag{5.100}$$

This relationship forms the quantitative basis for *titration*, i.e., an experimental method of determining the values of pK_a or pK_b from the value of pH at the inflection point (see Figure 5.9). In the neighborhood of the inflection point, the value of pH varies only slightly and the system behaves like a *buffer*. In the case of dissociation of one acid, the inflection point corresponds to $P = 0.5$ which can be changed by mixing the acid with an uncoupled base.

Changing the natural logarithms to logarithms with base 10, (using $\ln 10 = 2.3$), we can rewrite the Nernst equation (5.67) for proton channels and pumps ($ze = +|e|$) to take the form:

$$E_p = \Delta\psi - \frac{2.3RT}{F}\Delta pH,\tag{5.101}$$

where R is the gas constant and $F = N_A|e| \approx 9.65 \times 10^4$ C mol^{-1} is the Faraday constant and

$$\Delta pH = pH_P - pH_R.\tag{5.102}$$

The above expression gives the difference in pH between the membrane exterior and interior while the chemical affinity A has been replaced by the *proton-motive force*:

$$E_p = -\frac{A}{|e|}\tag{5.103}$$

(compare the sign in Equation 5.50).

Proton transfer reactions proceed quickly. Their unimolecular reaction rates are comparable to the Debye relaxation time for water that is defined by the rate of reorganization of the hydrogen bond system, and is on the order of $1 \text{ ns}^{-1} = 10^9 \text{ s}^{-1}$.

5.6 Redox reactions

The second important class of exchange reactions are *reduction-oxidation (redox) reactions* (Atkins, 1998) — chemical processes during which one or more electrons are transferred from molecule to molecule, for example:

$$Zn + Cu^{2+} \overset{\rightarrow}{\underset{\leftarrow}{}} Zn^{2+} + Cu \tag{5.104}$$

or

$$Fe + \frac{1}{2} O_2 + (H_2O) \overset{\rightarrow}{\underset{\leftarrow}{}} Fe^{2+} + 2OH^-. \tag{5.105}$$

Note that the second example exhibits an excess of water molecules. The substance that donates electrons is called a *reducer*; the one that accepts electrons is called an *oxidizer*. A reducer undergoes oxidation while an oxidizer is reduced.

Reactions of reduction and oxidation are mutually coupled and occur at the same place. They can be spatially separated in an *electrochemical cell*. Figure 5.10 illustrates the so-called Daniell cell in which the reaction of Equation (5.105) is separated into the following two reactions:

$$Zn \overset{\rightarrow}{\underset{\leftarrow}{}} Zn^{2+} + 2e^-; \quad 2e^- + Cu^{2+} \overset{\rightarrow}{\underset{\leftarrow}{}} Cu. \tag{5.106}$$

Each cell is composed of two half-cells with electrodes (electronic conductors) submerged in electrolytic solutions (ionic conductors). Electrolytes must be connected through an electrolytic key that enables the flow of ions. Electronic currents flowing between the two electrodes can be used to perform useful work. A cell functions as a chemoelectric machine. Oxidation takes place at the anode and reduction occurs at the cathode. In the case of the Daniell cell (Figure 5.10), the anode is a zinc electrode. The atoms of Zn after oxidation to ions of Zn^{2+} move through the electrolyte and leave electrons behind at the anode. The anode therefore reduces its mass and becomes negatively charged. The cathode is a copper electrode that accumulates Cu^{2+} ions after neutralization as a result of electron absorption. The cathode increases its mass and becomes positively charged.

The chemical reactions taking place spontaneously at the two electrodes continue as long as ionic concentrations in the electrolyte do not reach their equilibrium values. Later on, the reactions can be induced by applying an external source of electric current. The cell becomes an electrochemical machine called an *electrolyzer* in which electrons flow in the same direction as in the cell and the processes at the two electrodes are the same as earlier. However, the cathode must be negatively polarized in order to attract positive ions of Cu^{2+}, while the anode must be positively polarized so it can accept electrons from the Zn atoms and turn them into positively charged Zn^{2+}.

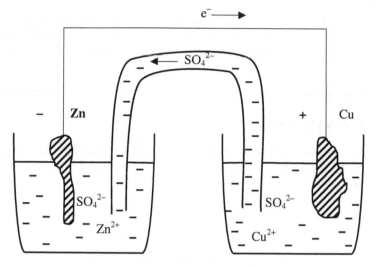

FIGURE 5.10
The Daniell electrochemical cell.

Redox processes can take place at the boundary between phases: liquid and solid or liquid and gaseous. They can also occur entirely within the liquid phase. Thus we distinguish three types of half-cells: (1) those with solid electrodes (see Figure 5.11a), (2) those with gas electrodes (see Figure 5.11b), and (3) those with neutral electrodes, also called redox half-cells (see Figure 5.11c). In their case, the active substances in the oxidized and reduced form can be found in the solution. Recall that the activity of a substance in a pure solid phase is equal to one. The activity in a gaseous phase is determined by the pressure expressed in bars and the activity in dilute liquid solution can be approximated by the value of its molar concentration in M.

In general, the redox reaction with electron transfer can be written as

$$P_{red} + R_{oxy}^{z+} \underset{\leftarrow}{\rightleftarrows} P_{oxy}^{z+} + R_{red}. \tag{5.107}$$

We need not assume that molecules R and P in the reduced state R_{red} and P_{red}, respectively, are electrically neutral. Each substance taking part in the reaction can be a positive or negative ion. In an electrochemical cell, the reaction in Equation (5.107) divides into two processes:

$$P_{red} \underset{\leftarrow}{\overset{\rightarrow}{\rightleftarrows}} P_{oxy}^{z+} + ze^- \tag{5.108}$$

and

$$R_{oxy}^{z+} + ze^- \underset{\leftarrow}{\overset{\rightarrow}{\rightleftarrows}} R_{red}. \tag{5.109}$$

The chemical equation of state for the reaction (5.107) links the activities of the substances taking part in it with the force driving it:

$$A = -\frac{\Delta G^\circ}{N_A} - k_B T \ln \frac{[R_{red}][P_{oxy}^{z+}]}{[R_{oxy}^{z+}][P_{red}]} \tag{5.110}$$

(a)   (b)   (c)

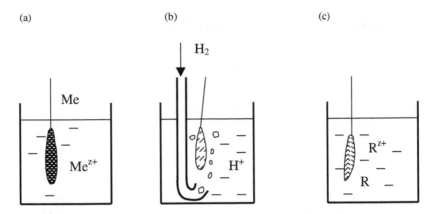

FIGURE 5.11
Three types of half-cells. (a) Half-cell with a metallic electrode. (b) Half-cell with a gas electrode. Gas at pressure P is adsorbed at the electrode made from porous nickel or covered with platinum. These substances often act as catalysts. (c) Redox half-cell. The electrode is made from a chemically neutral substance, e.g., platinum. Both oxidized and reduced forms of molecules are found in the solution.

(see Equation 5.80). In a cell serving as a chemoelectrical machine, the free chemical energy per molecule A is transformed into the free electrical energy of the charge transfer involving z electrons between electrodes at a potential difference (voltage):

$$E = E^\circ - \frac{k_B T}{z|e|} \ln \frac{[R_{red}][P_{oxy}^{z+}]}{[R_{oxy}^{z+}][P_{red}]}. \tag{5.111}$$

The relationship in Equation (5.111), similar to the relationships in Equations (5.67) and (5.101), is also called the *Nernst equation*. The voltage E, defined by the equation:

$$E = \frac{A}{z|e|} \tag{5.112}$$

(compare the sign in Equation 5.50) is called the *electromotive force* of the cell. In Equation (5.112), we have explicitly taken account of the sign of the electron charge: $e = -|e|$. E° is the electromotive force of a cell working under standard conditions where all the activities are equal to identity and the logarithm in Equation (5.111) vanishes.

The electromotive forces of different half-cells can be compared with respect to the same reference half-cell. As a convention, the *hydrogen half-cell*, which is a gas cell (Figure 5.11b), has been adopted as such a standard reference point. A layer of oxidized platinum that covers the electrode is saturated with gaseous hydrogen under a pressure of 1 bar (10^5 Pa) while the electrolyte has a standard hydrogen ion activity of $[H^+] = 1$ M, i.e., it has a pH equal to 0. For the reaction in Equation (5.108) applied to a hydrogen half-cell:

$$\frac{1}{2}H_2 \rightleftarrows H^+ + e^- \tag{5.113}$$

TABLE 5.1
Potential Energies for Key Redox Reactions

$R_{oxy}^{z+} + ze^- \rightarrow R_{red}$	$E^{\circ\prime}/V$
ferredoxin$_{oxy}^+ + e^- \rightarrow$ ferredoxin$_{red}$	-0.43
$H^+ + e^- \rightarrow \frac{1}{2}H_2$	-0.42
$(NAD^+ + 2H^+) + 2e^- \rightarrow (NADH + H^+)$	-0.32
$(NADP^+ + 2H^+) + 2e^- \rightarrow (NADPH + H^+)$	-0.32
$(FAD^+ + 2H^+) + 2e^- \rightarrow FADH_2$	-0.18
$\frac{1}{2}O_2 + (H_2O) + 2e^- \rightarrow 2OH^-$	-0.02
$(fumarate + 2H^+) + 2e^- \rightarrow$ succinate	$+0.03$
$(ubiquinone + 2H^+) + 2e^- \rightarrow$ ubiquinone H_2	$+0.10$
cytochrome $c^{3+} + e^- \rightarrow$ cytochrome c^{2+}	$+0.22$
$\frac{1}{2}O_2 + 2H^+ + 2e^- \rightarrow H_2O$	$+0.82$

while the general reaction in Equation (5.107) adopts a more specific form:

$$R_{oxy}^{z+} + \frac{z}{2}H_2 \underset{\leftarrow}{\rightarrow} R_{red} + zH^+ \tag{5.114}$$

$-E^\circ$ for this reaction is called the *standard reduction potential*. Its value varies from -3.0 V for the reduction of the lithium cation Li^+ to the metallic Li up to almost $+3.0$ V for the reduction of molecular fluorine F_2 to the fluorine anion F^-. The electromotive force of a cell working under standard conditions is equal to the difference between standard reduction potentials for the reactions taking place at the cathode and the anode, respectively. The standard reduction potential for the reaction from Cu^{2+} to Cu equals $+0.34$ V and for the reaction from Zn^{2+} to Zn, it equals -0.76 V. Therefore, the electromotive force of the Daniell cell equals $+0.34$ V $- (-0.76$ V$) = 1.10$ V.

The Nernst equation (5.111) also describes the case when both half-cells are made from identical substances (R = P) and only differ in the concentrations of the electrolyte. This cell type is called a *concentration cell*. We then have $E^\circ = 0$ and, in particular, for a cell composed of a hydrogen half-cell at an arbitrary pH and a standard hydrogen half-cell at pH = 0:

$$E = -\frac{k_B T}{|e|} \ln \frac{H^+}{1} = -\frac{k_B T}{|e|} (\ln 10) \, pH = -0.060 \text{VpH} \tag{5.115}$$

at 25°C. For pH $= 7$, $E = -0.42$ V. The value of this additional potential must be taken into account for biochemical redox reactions occurring in a buffer environment at a pH close to the neutral value. Electron transfer in such reactions is often linked to a proton transfer and hence a whole hydrogen atom is transferred. In the measurements of standard reduction potentials for biochemical reactions in an environment with pH $= 7$, one does not determine the value of E°, but instead the value of

$$E^{\circ\prime} = E^\circ - 0.42V. \tag{5.116}$$

Table 5.1 lists the values of the potential $E^{\circ\prime}$ for the most important biochemical redox reactions. The reference reaction potential in Equation (5.113) now equals just -0.42 V.

5.7 The steady state approximation: the theory of reaction rates

Consider the following set of two consecutive unimolecular reactions:

$$R \underset{k_{-1}}{\overset{k_{+1}}{\rightleftarrows}} I \underset{k_{-2}}{\overset{k_{+2}}{\rightleftarrows}} P \tag{5.117}$$

The corresponding kinetic equations take the form:

$$\frac{d}{dt}[R] = -k_{+1}[R] + k_{-1}[I],$$

$$\frac{d}{dt}[I] = -(k_{-1} + k_{-2})[I] + k_{-1}[R] + k_{-2}[P],$$

$$\frac{d}{dt}[P] = -k_{-2}[P] + k_{+2}[I]. \tag{5.118}$$

However, only two of them are independent in view of the conservation law

$$[R] + [I] + [P] = [R]_0 = \text{const} \tag{5.119}$$

that leaves only two independent variables, i.e., as many as the number of reactions. The reaction rate constants satisfy *detailed balance conditions* (in thermodynamic equilibrium the number of forward and reverse transitions is balanced for each reaction separately):

$$K_{+1}[R]^{eq} = k_{-1}[I]^{eq}; \qquad K_{+2}[I]^{eq} = k_{-2}[P]^{eq}. \tag{5.120}$$

Since only two of the three differential equations in 5.118 are linearly independent, it is easy to write their analytical solutions but they are not very transparent. In practice, one often uses the *steady-state approximation*. If the equilibrium concentration of the intermediate I is sufficiently small or if its free energy is relatively large, in view of the detailed balance conditions in (5.120) we have:

$$\frac{k_{+1}}{k_{-1}} = \frac{[I]^{eq}}{[R]^{eq}} = e^{-\frac{(G_I^\circ - G_R^\circ)}{RT}} \ll 1 \tag{5.121}$$

and

$$\frac{k_{+2}}{k_{-2}} = \frac{[P]^{eq}}{[I]^{eq}} = e^{-\frac{(G_P^\circ - G_I^\circ)}{RT}} \gg 1. \tag{5.122}$$

It follows from the above inequalities that

$$k_{-1} + k_{+2} \gg k_{+1}, k_{-2}. \tag{5.123}$$

From the second equation in (5.118), we conclude that a weakly occupied state under equilibrium conditions is also a short-lived state. Its concentration, [I], is a fast variable compared to [R] and [P] and after a short transient period

$$\tau_{tr} = (k_{-1} + k_{+2})^{-1}, \tag{5.124}$$

it reaches the steady state:

$$\frac{d}{dt}[I] = 0. \tag{5.125}$$

A steady state is constant on a short time scale. On a long time scale, however, the value of the concentration [I] follows the slow-varying concentrations [R] and [P]. From Equation (5.125) we can deduce a relationship between the stationary value of the *fast* variable [I] and the values of the *slow* variables [R] and [P]:

$$[I] = \frac{k_{+1}[R] + k_{-2}[P]}{k_{-1} + k_{+2}}. \tag{5.126}$$

Substituting it to the first or third equation in (5.118) we obtain:

$$\frac{d}{dt}[P] = -\frac{d}{dt}[R] = k_{+}[R] - k_{-}[P], \tag{5.127}$$

where

$$k_{+} = \frac{k_{+2}k_{+1}}{k_{-1} + k_{+2}} \qquad k_{-} = \frac{k_{-2}k_{-1}}{k_{-1} + k_{+2}}. \tag{5.128}$$

Equation (5.127) can be considered a kinetic equation for the effective reaction

$$R \overset{k_{+}}{\underset{k_{-}}{\rightleftarrows}} P \tag{5.129}$$

which, over a long time scale, well approximates the system of equations in (5.117). Applying the detailed balance conditions (Equation 5.120), we can rewrite the relationships in Equation (5.128) as:

$$k_{+} = \frac{[I]^{eq}}{[R]^{eq}}[(k_{-1})^{-1} + (k_{+2})^{-1}]^{-1},$$

$$k_{-} = \frac{[I]^{eq}}{[P]^{eq}}[(k_{-1})^{-1} + (k_{+2})^{-1}]^{-1}. \tag{5.130}$$

The effective reaction rates are determined by the equilibrium occupation probability for the intermediate state relative to the initial and final states, respectively, multiplied by the inverse of the sum of the average transit times from the intermediate to the initial and final states.

The steady state approximation applies to an arbitrary number of mutually coupled chemical reactions provided the fast- and slow-varying concentrations can be clearly distinguished. In particular, it can be applied to the sequence of the following reactions:

$$R \overset{k_{+1}}{\underset{k_{-1}}{\rightleftarrows}} R^{\ddagger} \overset{k_{+0}}{\underset{k_{-0}}{\rightleftarrows}} P^{\ddagger} \overset{k_{+2}}{\underset{k_{-2}}{\rightleftarrows}} P \tag{5.131}$$

with the two short-lived intermediates $R^{\ddagger}$ and $P^{\ddagger}$. The effective single reaction in Equation (5.129) is described by the kinetic equation (5.127) with the reaction constants:

$$k_+ = \frac{k_{+1}k_{+0}k_{+2}}{k_{-1}k_{-0} + k_{-1}k_{+2} + k_{+0}k_{+2}} \tag{5.132}$$

and

$$k_- = \frac{k_{-1}k_{-0}k_{-2}}{k_{-1}k_{-0} + k_{-1}k_{+2} + k_{+0}k_{+2}}. \tag{5.133}$$

Making use of three detailed balance conditions for the three reactions in Equation (5.131), we can rewrite Equations 5.132 and 5.133 as:

$$k_+ = \frac{[R^{\ddagger}]^{\mathrm{eq}}}{[R]^{\mathrm{eq}}} \left[(k_{+0})^{-1} + (k_{-1})^{-1} + \left(\frac{[P^{\ddagger}]^{\mathrm{eq}}}{[R^{\ddagger}]^{\mathrm{eq}}} k_{+2} \right)^{-1} \right]^{-1} \tag{5.134}$$

and

$$k_- = \frac{[P^{\ddagger}]^{\mathrm{eq}}}{[P]^{\mathrm{eq}}} \left[(k_{-0})^{-1} + (k_{+2})^{-1} + \left(\frac{[R^{\ddagger}]^{\mathrm{eq}}}{[P^{\ddagger}]^{\mathrm{eq}}} k_{-1} \right)^{-1} \right]^{-1}. \tag{5.135}$$

The intermediates $R^{\ddagger}$ and $P^{\ddagger}$ can be interpreted as *transition states* of the reaction (Equation 5.129; see Figures 3.6 and 5.4). Introducing, in accordance with Equations (5.58) and (5.60), the free energy differences $\Delta G_R^{\ddagger}$ and $\Delta G_P^{\ddagger}$ between the corresponding intermediate states and the initial and final states, respectively:

$$\frac{[R^{\ddagger}]^{\mathrm{eq}}}{[R]^{\mathrm{eq}}} = e^{-\frac{\Delta G_R^{\ddagger}}{k_B T}}, \quad \frac{[P^{\ddagger}]^{\mathrm{eq}}}{[P]^{\mathrm{eq}}} = e^{-\frac{\Delta G_P^{\ddagger}}{k_B T}}, \tag{5.136}$$

we can express the reaction rates (Equations 5.134 and 5.135) in terms of the *Arrhenius relation*:

$$k_+ = v_+ e^{-\frac{\Delta G_R^{\ddagger}}{RT}}, \quad k_- = v_- e^{-\frac{\Delta G_P^{\ddagger}}{RT}}. \tag{5.137}$$

where $\Delta G_R^{\ddagger}$ and $\Delta G_P^{\ddagger}$ are called *activation free energies*, while v_+ and v_-, are designated *preexponential factors*. Their reciprocals:

$$v_+^{-1} = (k_{+0})^{-1} + (k_{-1})^{-1} + \left(\frac{[P^{\ddagger}]^{\mathrm{eq}}}{[R^{\ddagger}]^{\mathrm{eq}}} k_{+2} \right)^{-1} \tag{5.138}$$

and

$$v_-^{-1} = (k_{-0})^{-1} + (k_{+2})^{-1} + \left(\frac{[R^{\ddagger}]^{\mathrm{eq}}}{[P^{\ddagger}]^{\mathrm{eq}}} k_{-1} \right)^{-1} \tag{5.139}$$

represent the sum of the three characteristic times for the microscopic mechanisms that determine the resultant reaction rate. The first contribution describes the transition between the two intermediate states assuming a local thermodynamic equilibrium between the initial and intermediate states. As a result of this transition, thermodynamic equilibrium is obviously disturbed. The second and third contributions (Equations 5.138 and 5.139) describe processes that restore local equilibrium on the side of the initial and final states, respectively.

If the limiting factor of the reaction is the transition across the barrier, i.e., the first time components in Equations (5.138) and (5.139) are the longest, the remaining contributions can be ignored, yielding:

$$v_+ = k_{+0}; \qquad v_- = k_{-0}. \tag{5.140}$$

The assumption of an infinitely fast process of reaching partial equilibrium between the initial state R and $R^{\ddagger}$ or P and $P^{\ddagger}$ forms the basis of the *transition state theory* (Atkins, 1998). Equating the transition rates k_{+0} and k_{-0} with the average frequency of thermal oscillations, the theory is limited to calculations of the activation Gibbs free energies $\Delta G_R^{\ddagger}$ and $\Delta G_P^{\ddagger}$.

Conversely, if the processes that restore partial equilibrium in the intermediate state are slower than the barrier crossing, a detailed knowledge of microscopic dynamics is required to compute the requisite expressions. This implies the knowledge of intramolecular dynamics for isomerization reactions or intermolecular dynamics for exchange reactions. In such cases, we say that a reaction is *controlled* by dynamics. Sections 3.2 to 3.4 discussed the statistical theory of reaction rates for reactions controlled by the dynamics of conformational transitions in biomolecules.

5.8 Chemical mechanisms of enzymatic catalysis

The highest rate of a chemical reaction allowed by thermodynamics is determined by the transition state theory. The expression for the rate constant (inverse mean reaction time) given by this theory takes the form:

$$k^{eq} = ve^{-\frac{\Delta G^{\ddagger}}{RT}}. \tag{5.141}$$

The exponent above represents the local equilibrium occupation probability of the transition state. $\Delta G^{\ddagger}$ denotes the free energy of activation, i.e., the difference between the free energy of the transition state and that of the initial state of the reaction. The quantity RT for the room temperature $T = 300K$ corresponds to the energy 2.5 kJ/mol (see Equation 5.59). The first factor, the average frequency of thermal oscillations, can be estimated via relationship $hv = k_B T$ to yield the value of the time period as 2×10^{-13} s. For comparison, life has existed on Earth for over 4 billion years or almost 10^{17} s, which means a factor of 10^{30} longer than this characteristic time scale. Since $10^{-30} \approx e^{-70}$, the reaction time would reach such an astronomical number for the free energy of activation that is 70 times larger than the mean thermal energy, i.e., for 175 kJ/mol.

For comparison, the bond energy in the carbon–carbon case equals approximately 350 kJ/mol, and hence is twice as large. Most reactions linked to the reorganization of covalent bonds do not go through complete bond breaking but their characteristic free energy of activation is not much lower than 175 kJ/mol. Hence, at physiological temperatures, spontaneous incidence of such reactions would require hours, years, or even millennia.

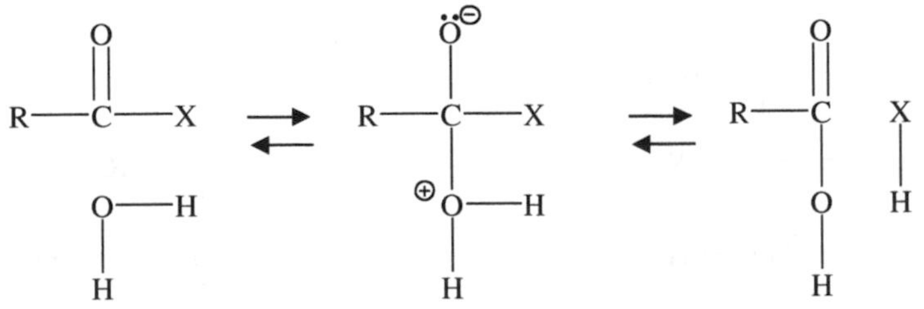

$$
\begin{array}{ccccc}
& \text{O} & & \ddot{\text{O}}{}^{\ominus} & & \text{O} \\
& \parallel & & \mid & & \parallel \\
\text{R}-\!\!&\text{C}&\!\!-\text{X} & \text{R}-\text{C}-\text{X} & \text{R}-\text{C} \quad \text{X} \\
& & \rightleftarrows & & \rightleftarrows & \\
& \text{O}-\text{H} & & {}^{\oplus}\text{O}-\text{H} & & \text{O} \quad \text{H} \\
& \mid & & \mid & & \mid \\
& \text{H} & & \text{H} & & \text{H}
\end{array}
$$

FIGURE 5.12
Uncatalyzed reaction of hydrolysis of a peptide bond ($-X = -NH-R'$) or an ester bond ($-X = -O-R'$). R and R' are arbitrary molecular groups. Two dots denote additional lone electron pair of an oxygen atom.

These time scales are too long for living organisms. Living systems accelerate the rates of almost all relevant reactions using specific *catalysts*, i.e., compounds that take part in chemical reactions and are recycled after completion. Biochemical reactions are catalyzed by protein *enzymes* or rarely by archaic RNA-based *ribozymes*. A typical time for an enzymatic reaction is 10^{-3} s, hence enzymes accelerate biochemical reactions by a factor of at least 10 million. To accelerate a reaction by a factor of ten, the activation barrier has to be lowered by 5.7 kJ/mol. Note also that transition state theory rather poorly determines the rates of reactions involving biological macromolecules for which the free energies of activation and intramolecular relaxation times appear to be important (see Sections 3.2 to 3.4).

Besides accelerating chemical reactions, enzymes fulfill two other important functions. First, they *control* reactions which means that a given reaction takes place in a cell at an appropriate moment and at the desired location. Second, enzymes *couple* reactions. To make use of a chemical reaction in a biological process of free energy transduction, it must occur simultaneously with another reaction at the same multienzymatic complex. Therefore, enzymes must be characterized by high *specificity*. Each metabolic reaction is catalyzed by its own enzyme.

Before we move onto the main topic of this section, i.e., the kinetics of enzymatic reactions and their regulatory functions, let us first investigate their specificity and the purely chemical aspect of accelerating chemical reactions by enzymes. We will limit our discussion to one well-known class of enzymes, namely *proteases* that catalyze the hydrolysis of peptide bonds. A general presentation of this topic can be found, e.g., in a book by Fersht (1999).

The reaction of hydrolysis of a peptide or an ester bond:

$$R-CO-X + H_2O \rightleftarrows R-COOH + H-X \tag{5.142}$$

($-X = -NH-R'$ or $-O-R'$ while R and R' are arbitrary molecular groups) occurs via a very short-lived tetrahedral intermediate state (see Figure 5.12). An electrostatic charge separation takes place over a reasonably long distance which is the main reason for a significant energy increase of the system.

The intermediate state is stabilized (its energy is lowered) as a result of bringing to its negative end an *electrophilic* molecule (a cation and/or an electron pair acceptor)

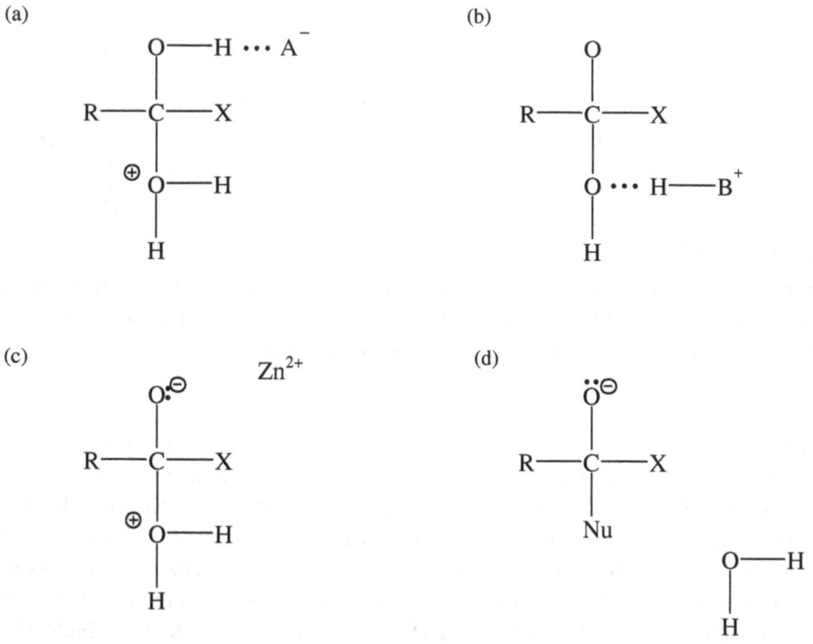

(a)

(b)

(c) Zn^{2+}

(d)

FIGURE 5.13

Four main types of chemical catalysis stabilizing a tetrahedral intermediate. (a) General acid catalysis: an acid molecule HA donates a proton to a negatively charged molecule of carboxyl oxygen and creates a hydrogen bond (dotted line). (b) General base catalysis: a base molecule B accepts a proton from a positively charged hydronic group. (c) Electrophilic hydrolysis helped by a metal cation (Zn^{2+}). (d) Nucleophilic catalysis: a nucleophile Nu^- forms a tetrahedral intermediate more stable than a molecule of water.

and to the positive end a *nucleophilic* molecule (an anion and/or an electron pair donor). An example of an electrophilic molecule is an acid molecule that detaches hydrogen in the form of a proton keeping an electron pair from the hitherto existing bond (*general acid catalysis*; see Figure 5.13a). An example of a nucleophilic molecule is a base that attaches hydrogen in the form of a proton bringing into the thus created bond its own electron pair (*general base catalysis*; see Figure 5.13b). Metal cations (Figure 5.13c) play a purely electrostatic role as electrophiles while nucleophiles often perform nucleophilic substitutions and make themselves tetrahedral intermediates. A bond hydrolyzes faster than the original bond, changing the reaction pathway (*covalent catalysis*; see Figure 5.13d).

Adding an electrophile or a nucleophile to the solution containing a hydrolyzed peptide lowers the energy contribution to the free energy of activation and simultaneously increases its entropic contribution. Stabilization of a tetrahedral intermediate requires a collision of two molecules with a proper mutual orientation which is very improbable and thus carries low entropy. The entropic contribution can be neglected when the two molecules are parts of a larger molecule within which they are already

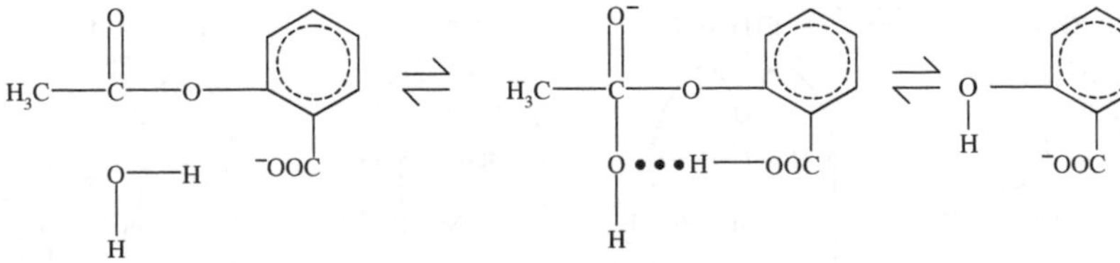

FIGURE 5.14
Example of intramolecular catalysis. The hydrolysis of acetylsalicylic acid (aspirin). The basic catalytic group $-COO^-$ belongs to the same molecule as the acetyl group CH_3-CO- which, by contact with water, is transformed into a tetrahedral intermediate and then detached.

properly oriented for bonding. A well known example of *intramolecular catalysis* is the hydrolysis of aspirin (acetylsalicylic acid) to salicylic acid (see Figure 5.14).

Usually in enzymatic catalysis, all elements of chemical catalysis are present. Consider the mechanism of action of *serine proteases* (e.g., pancreatic enzymes of mammals such as trypsin, chymotrypsin, elastase, or subtylisin of bacteria) or *cystein proteases* (e.g., papain, a plant enzyme). In both classes of enzymes, we deal with covalent catalysis which changes the reaction pathway. First, a nucleophile (in this case a hydroxyl group of one of the serins or a sulfhydryl group of one of the cysteins) covalently bonds to the substrate and forms an intermediate compound or acyl enzyme that then undergoes hydrolysis (see Fersht 1999; Figure 5.15). The nucleophile becomes active only after it donates a proton and hence covalent catalysis (nucleophilic) must be linked with a general base catalysis. In both classes of enzymes, a proton is transferred via a histidine onto an aspartate (Phillips and Fletterick, 1992; Dodson and Wlodawer, 1998; Figure 5.16). All the catalytically active molecular groups must have proper spatial orientation with respect to each other and with respect to a specifically bound substrate (see Figure 5.15). Therefore, enzymatic catalysis is to a large degree *intramolecular*.

Enzymes follow a special nomenclature (*Enzyme Nomenclature*, 1992) of which we will only list the six main classes (the names of which end with the suffix *-ase*) and give several typical examples:

Oxydoreductases catalyze electron transfer. If an electron is transferred practically only in one direction, the enzymes are called *oxidases* or *reductases*. If transfer of a proton is simultaneous (i.e., the whole hydrogen atom is transferred), then we call the enzymes *dehydrogenases*.

Transferases catalyze the transfers of radicals or molecular groups from one compound to another. An important subclass contains *phosphotransferases* or *kinaseses* that phosphorylate various substrates.

Hydrolases catalyze the hydrolysis of generalized ester bonds. Particular examples are *esterases, lipases, deoxyribonucleases*, and *proteases*. An important subclass is the group of ATP*ases* in which the hydrolysis of ATP to ADP and P_i is coupled with various processes that require free energy, e.g., the active transport across ion channels in biological membranes or the movement along cytoskeletal structures.

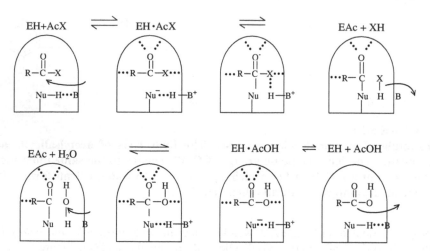

FIGURE 5.15
Hydrolysis of a peptide bond via serine and cysteine proteases proceeds in two stages. In the first stage, a covalent intermediate compound is formed, an acyl-enzyme EAc. In the second stage, a bond between the acyl group and the enzyme undergoes hydrolysis. The various functional groups in the active center are shown. Nu denotes a nucleophile, an oxygen of a side chain of serin or a sulfur of cysteine. The base B is a histidine. The bond of the radical R of the acyl group is specific while the bond of the amine group X in serin proteases is not. The latter can be readily replaced by an alcohol group. Most investigations on the mechanism of functioning of serin proteases have used esters as substrates.

Liases catalyze the nonhydrolytic breaking of various bonds. Examples are *decarboxylases* that free carbon dioxide from substrates.

Isomerases catalyze unimolecular reactions of intramolecular isomerization. When this involves relocation of groups inside molecules, the enzymes are designated *mutases* (e.g., acylmutase, phosphomutase).

Ligases (*synthetases*) catalyze the synthesis of new bonds in conjunction with the breaking (not hydrolysis) of the pyrophosphate bond in ATP to yield AMP and PP_i, e.g., DNA *polymerase, transcriptase, reverse transcriptase*.

5.9   Michaelis-Menten kinetics

At the beginning of 20th century, Adrian Brown, investigating one of the uni-molecular biochemical reactions of the form

$$R \rightleftarrows P \tag{5.143}$$

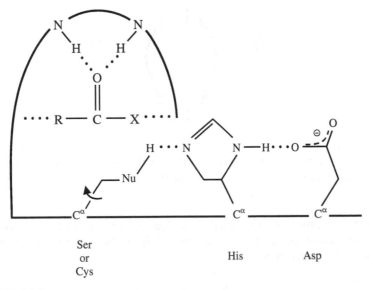

FIGURE 5.16
Pathway for the proton transfer from a nucleophile (a hydroxyl group of serine or a sulfhydryl group of cysteine) via a histidine to asparagine (the catalytic triad). After freeing from the proton, the side chain of serine or cysteine rotates around the bond $C^\alpha - C$ and the nucleophile attaches the carboxylic carbon of the substrate.

proposed a simple kinetic pathway corresponding to its catalyzed version:

$$E + R \underset{k'_-}{\overset{k'_+}{\rightleftarrows}} M \underset{k''_-}{\overset{k''_+}{\rightleftarrows}} E + P. \tag{5.144}$$

This reaction contains two well-defined steps: the binding of a reagent with an enzyme and the release of the product. E denotes a free enzyme and M denotes the *complex enzyme T-substrate*. During this reaction, the enzyme is not exhausted (see Figure 5.17). The first approximate analysis of the kinetic equations for the scheme in Equation (5.144) was carried out by Michaelis and Menten in 1913. The full analysis was performed by Haldane in his monograph on enzymes published in 1930 (Laidler and Bunting, 1973).

In the kinetic equations describing the two steps in Equation (5.144) four concentrations occur. Only two are independent due to the conservation laws:

$$[R] + [M] + [P] = [R]_0 = \text{const} \tag{5.145}$$

and

$$[E] + [M] = [E]_0 = \text{const.} \tag{5.146}$$

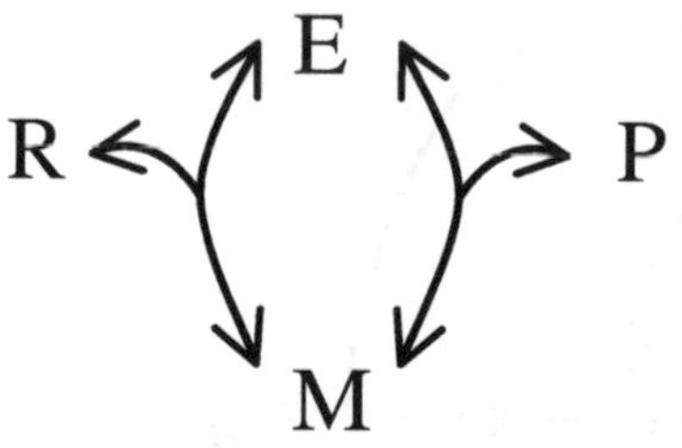

FIGURE 5.17
Enzymatic cycle with one intermediate. In the first component reaction, the enzyme is used up; in the second, it is recovered.

Figure 5.17 represents the reaction in Equation (5.144) from the viewpoint of the enzyme macromolecules. To represent it from the viewpoint of the substrate molecules, we rewrite it as:

$$R \underset{k'_-}{\overset{k'_+[E]}{\rightleftarrows}} M \underset{k''_-[E]}{\overset{k''_+}{\rightleftarrows}} P. \tag{5.147}$$

In general, the molar concentration of the enzyme is much lower than that of the substrate:

$$[E]_0 \ll [R]_0 \tag{5.148}$$

thus, the concentration [M] is much lower than the concentrations [R] and [P]. Hence, after a short prestationary stage of the reaction (see Section 5.7), we can apply the steady state approximation treating the enzyme concentration [E] as a constant. Utilizing Equation (5.128) for the scheme in Equation (5.117), in which the constants k_{+1} and k_{-2} are replaced by the coefficients $k'_+[E]$ and $k''_-[E]$, respectively, we obtain:

$$\frac{d}{dt}[P] = -\frac{d}{dt}[R] = \frac{k'_+ k''_+[R] - k'_- k''_-[P]}{k'_- + k''_+}[E]. \tag{5.149}$$

The value of the enzyme concentration [E] in its free state is determined by the total enzyme concentration $[E]_0$. An appropriate relationship is found from Equation (5.146) and the stationary condition:

$$\frac{d}{dt}[M] = -(k'_- + k''_+)[M] + k'_+[E][R] + k''_-[E][P] = 0, \tag{5.150}$$

which allows us to eliminate the fast-varying variable [M]. Thus, we obtain an equation that describes the steady state rate of product formation or reagent consumption:

$$\frac{d}{dt}[P] = -\frac{d}{dt}[R] = \frac{k_+ K_+^{-1}[R] - k_- K_-^{-1}[P]}{1 + K_+^{-1}[R] + K_-^{-1}[P]}[E]_0 \tag{5.151}$$

where

$$k_+ = k_+'', \qquad K_+ = \frac{k_-' + k_+''}{k_+'}, \qquad (5.152)$$

and

$$k_- = k_+', \qquad K_- = \frac{k_-' + k_+''}{k_-''}. \qquad (5.153)$$

It follows from Equations (5.152) and (5.153) that the parameters of the forward and reverse reactions are linked by the *Haldane equation*:

$$\frac{k_+}{K_+} \frac{K_-}{k_-} = \frac{[P]^{eq}}{[R]^{eq}} \equiv K. \qquad (5.154)$$

It is more convenient to define the enzymatic reaction rate with respect to a single molecule of the enzyme, i.e., the *reaction flux*:

$$J \equiv \frac{d}{dt}[P]/[E]_0. \qquad (5.155)$$

Using the approximate formula:

$$[R] + [P] = [R]_0 \qquad (5.156)$$

and the definition of chemical affinity for the uncatalyzed reaction in Equation (5.153):

$$\frac{[P]}{[R]} = Ke^{-\beta A}, \qquad (5.157)$$

where the equilibrium constant is:

$$K \equiv \frac{[P]^{eq}}{[R]^{eq}}, \qquad (5.158)$$

we can rewrite (5.151) in a form analogous to Equation (5.75):

$$J = \frac{1 - e^{-\beta A}}{J_+^{-1} + J_-^{-1}e^{-\beta A}}, \qquad (5.159)$$

where

$$J_\pm = \frac{k_\pm[R]_0}{K_\pm + [R]_0}. \qquad (5.160)$$

Under the condition that $[R] = [R]_0 = \text{const}$ ($[P] = 0$, $\beta A \to \infty$) or $[P] = [R]_0 = \text{const}$ ($[R] = 0$, $\beta A \to -\infty$) which corresponds to an early stage of the reaction starting from the pure R or P, respectively, the reaction flux in Equation (5.155) takes the asymptotic values $J \to J_+$ or $J \to -J_-$. The *Michaelis-Menten dependence* (Equation 5.160) of the reaction flux on the substrate concentration $[R]_0 = [R]$ or $[P]$

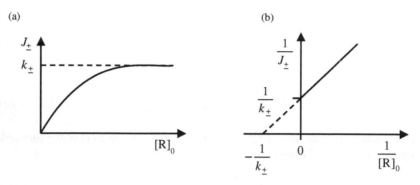

FIGURE 5.18
(a) Dependence of the asymptotic enzymatic reaction flux, $J_\pm$, on the substrate concentration $[R]_0$. (b) The Lineweaver–Burk plot for an enzyme with Michaelis–Menten kinetics.

is hyperbolic (Figure 5.18a). For low concentrations of $[R]_0$, the flux $J_\pm$ increases linearly with $[R]_0$:

$$J_\pm = \frac{k_\pm}{K_\pm}[R]_0. \tag{5.161}$$

On the other hand, for high concentrations $[R]_0$ it reaches the saturation value:

$$J = k_\pm. \tag{5.162}$$

The parameter k_+ (k_-) in the Michaelis-Menten equation can be interpreted as the *turnover number* for the enzyme, i.e., the number of product molecules P or R, respectively, produced per unit of time. The parameter K_+ (K_-) is called the *Michaelis constant*. If $k_+'' \ll k_-'$ ($k_-'' \ll k_+''$), K_+ (K_-) becomes identical with the constant of dissociation from the complex M for the substrate molecules R (or P). The conditions: $[P] = 0$, $[R] = [R]_0 = $ const or $[R] = 0$, $[P] = [R]_0 = $ const can also be satisfied in an open stationary reactor as a result of a constant removal of reaction products. This is a typical situation *in vivo* and in most experimental systems *in vitro*. The experimental results of steady state kinetics are more conveniently represented in the form of the so-called *Lineweaver–Burk plot* that gives a dependence of the inverse reaction flux versus the inverse concentration of the substrate:

$$\frac{1}{J_\pm} = \frac{1}{k_\pm} + \frac{K_\pm}{k_\pm[R]_0}. \tag{5.163}$$

It is then easy to determine both the enzyme turnover number and the value of the Michaelis constant (see Figure 5.18b).

Equations of the type in (5.159) and (5.160) for the steady state enzymatic kinetics can be derived for a much larger class of kinetic schemes than only the one in Equation (5.144) which take into account more than one intermediate state of the enzyme and even a quasi-continuous dynamic of conformational transitions (see Section 3.4; Kurzynski and Chelminiak, 2002). These types of equations usually describe experimental situations well (Fersht, 1985). The turnover numbers and the Michaelis constants, both forward and backward, constitute universal phenomeno-logical parameters for enzymatic reactions and do not have to be linked with specific kinetic schemes. The acceleration of the reaction rate by an enzyme is due to the fact that the turnover numbers $k_\pm$ are much larger than the reaction rates for noncatalyzed reactions (Equation 5.144). According to the Haldane equation (5.154), the presence of an enzyme does not affect the chemical equilibrium between R and P and the acceleration of the reverse reaction is obtained to the same degree as that for the forward reaction.

A good enzyme in the case of the kinetic scheme in Equation (5.144) should be characterized by high values of the constants k''_+ and k'_-. However, a physical limitation cannot be directly circumvented. Bimolecular reaction rates k'_+ and k''_- are controlled by diffusion and their values cannot exceed 10^9 $M^{-1}s^{-1}$. Consequently, for good enzymes not only are $k'_+[E]$ and $k''_-[E]$ small compared to the sum $k''_+ + k'_-$, which is a condition of the applicability of the stationary approximation, but also $k'_+[R]$ and $k''_-[P]$ are small compared to this sum. According to Equations (5.152) and (5.153) this implies that the inequality:

$$K_\pm \gg [R]_0 \tag{5.164}$$

is satisfied and the approximation in Equation (5.161) is valid. Indeed, most enzymes under physiological conditions do not reach saturation conditions. The condition of having the larger of $k_+/K_+ \approx k'_+$ or $k_-/K_- \approx k''_-$ close to 10^9 $M^{-1}s^{-1}$ means that the enzyme reached *kinetic perfection* (Stryer, 1995). For typical substrate concen-trations, $[R]_0 = 10^{-6}$ M reactions then take place at rates on the order of 10^3 s^{-1}.

Limitations due to long diffusion times of substrate molecules disappear in the case of enzymes that catalyze subsequent reactions of a metabolic chain in supramolecu-lar multienzymatic complexes. In the process of evolution, nature must have found this solution early on since in real biological systems this approach is very common.

5.10 Control of enzymatic reactions

The role of enzymes is to accelerate biochemical reactions and control them (Stryer, 1995; Wyman and Gill, 1990). This may include slowing them or even blocking them altogether if a need arises. The simplest method of slowing an enzymatic reaction is to reduce the effective number of enzyme molecules as a result of binding them to a molecule that resembles the substrate (*competitive inhibition*). For example, penicillin can be irreversibly bound to an enzyme that catalyzes the synthesis of the cell walls of Gram-positive bacteria. Various blockers that bind to the target receptors eliminate or significantly reduce their action.

In the presence of a competitive inhibitor I, in addition to the reaction in Equation (5.143), the reaction:

$$E + I \overset{k_{+1}}{\underset{k_{-1}}{\rightleftarrows}} E' \tag{5.165}$$

takes place. E' represents an inactive form of the enzyme. The condition for conservation of the number of enzyme molecules takes the form:

$$[E] + [M] + [E'] = [E]_0 = \text{const.} \tag{5.166}$$

We are interested only in the steady state kinetics in an open reactor to which a reagent is supplied at a constant rate and a product removed such that $[R] = \text{const}$ and $[P] = 0$. From the relationship in Equation (5.166) and two independent steady-state conditions:

$$\frac{d}{dt}[M] = -(k_-' + k_+'')[M] + k_+'[E][R] = 0 \tag{5.167}$$

and

$$\frac{d}{dt}[E'] = -k_{-1}[E'] + k_{+1}[E][I] = 0, \tag{5.168}$$

we find the stationary concentration of the enzyme-substrate complex M as:

$$[M] = \frac{[R][E]_0}{K_+(1 + K_I^{-1}[I]) + [R]}, \tag{5.169}$$

where the equilibrium constant for the inhibitor binding is

$$K_I \equiv \frac{k_{-I}}{k_{+I}} = \frac{[I]^{eq}[E]^{eq}}{[E']^{eq}} \tag{5.170}$$

and the constant K_+ is determined by the second part of Equation (5.152). Assuming that the reaction of unbinding the product is much slower than the reaction of binding to the substrate, we can approximate the value of the stationary concentration of the enzyme–substrate complex M by the equilibrium concentration value replacing the apparent dissociation constant K_+ by the actual dissociation constant:

$$K_R \equiv \frac{k_-'}{k_+'} = \frac{[R]^{eq}[E]^{eq}}{[M]^{eq}}. \tag{5.171}$$

A simplified scheme from the view of the enzyme is shown in Figure 5.19a. The value of the concentration $[M]$ is determined by the steady-state rate of production P:

$$\frac{d}{dt}[P] = k_+[M]. \tag{5.172}$$

The reaction is still of the Michaelis-Menten form but it has a changed effective dissociation constant. It is convenient to express the inverse reaction flux as a function of the inverse concentration of the reagent (using the Lineweaver–Burk plot) as:

$$\frac{1}{J_+} = \frac{1}{k_+} + (1 + K_I^{-1}[I])\frac{K_R}{k_+[R]}. \tag{5.173}$$

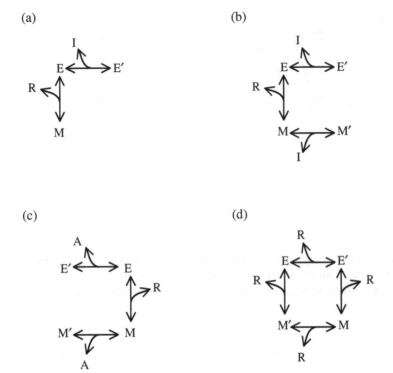

FIGURE 5.19
(a) Competitive inhibition of enzymatic catalysis. (b) Noncompetititive (all-osteric) inhibition. (c) Allosteric activation. (d) Allosteric cooperative effect.

Comparing Equations (5.173) and (5.163) we find that a competitive inhibitor does not alter the turnover number; it increases the value of the Michaelis constant (see Figure 5.20a).

In contrast to a competitive inhibitor, a *noncompetitive inhibitor* binds to both states E and M of the enzyme. In addition to the reaction of Equation (5.144), we also have two reactions below:

$$E + I \overset{K_I}{\rightleftarrows} E' \quad M + I \overset{K_I}{\rightleftarrows} M'. \tag{5.174}$$

The kinetic scheme of noncompetitive inhibition shown from the viewpoint of the enzyme is illustrated in Figure 5.19b. The condition for the conservation of the number of enzyme molecules takes the form:

$$[E] + [M] + [E'] + [M'] = [E]_0 = \text{const} \tag{5.175}$$

and three independent steady-state conditions are given by:

$$[M] = K_R^{-1}[E][R], \quad [E'] = K_I^{-1}[E][I], \quad [M'] = K_I^{-1}[M][I]. \tag{5.176}$$

As before, we assume that the steady-state condition for the catalyzed reaction is approximated by the equilibrium condition.

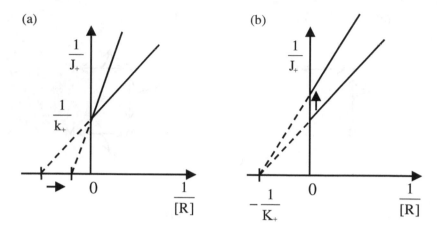

FIGURE 5.20
Lineweaver–Burk plot for an inhibited enzyme described by Michaelis–Menten kinetics. (a) Competitive inhibitor changes the value of the effective Michaelis constant. (b) Noncompetitive inhibitor alters the turnover number.

Using the four equations in (5.175) and (5.176), we find the value of the stationary (or in our approximation, equilibrium) concentration of the Michaelis complex as:

$$[M] = \frac{[R][E]_0}{(1 + K_I^{-1}[I]) + (K_R + [R])}. \tag{5.177}$$

Hence, the inverse reaction flux is given by

$$\frac{1}{J_+} = (1 + K_I^{-1}[I])\left[\frac{1}{k_+} + \frac{K_R}{k_+[R]}\right]. \tag{5.178}$$

A noncompetitive inhibitor, in contrast to the competitive one, reduces the turnover number without altering the value of the Michaelis constant (see Figure 5.20b). A noncompetitive inhibitor is an example of an *effector* (a regulatory molecule) that affects substrate binding at an active center over a certain distance (Wyman and Gill, 1990). This type of interaction, without discussing its physical nature, is known in molecular biology as *allostery* from the Greek *allos* (other) and *stereos* (space or location). The label was introduced in 1965 by Monod, Wyman, and Changeaux. An effector can act as an inhibitor and as an *activator* that will be denoted by A. An active form of an enzyme is then the form bound to an effector. Besides the reaction in Equation (5.144), we now have:

$$E' + A \overset{K_A}{\rightleftarrows} E, \quad M' + A \overset{K_A}{\rightleftarrows} M. \tag{5.179}$$

Similarly to Equation (5.74), we have only introduced equilibrium constants and assumed that they are equal. The kinetic scheme for allosteric activation from the

(a)   (b)

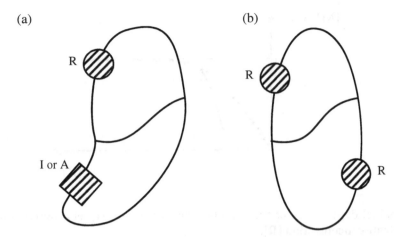

FIGURE 5.21
Allosteric interaction. (a) Allosteric heterotropic effect. A substrate and an effector (noncompetitive inhibitor or activator) bind at different locations on the enzyme. The enzyme consists of two separate entities: a catalytic and a regulatory one. (b) Allosteric homotropic effect. An enzyme is composed of at least two (usually more than two) identical catalytic subunits.

viewpoint of the enzyme is shown in Figure 5.19c. The condition of number conservation for the enzyme molecules is given in Equation (5.175), and three independent stationarity conditions take the form:

$$[M] = K_R^{-1}[E][R], \quad [E] = K_A^{-1}[E'][A], \quad [M] = K_A^{-1}[M'][A]. \qquad (5.180)$$

The stationary (and in our approximation, equilibrium) value of the concentration of the Michaelis complex is

$$[M] = \frac{[R][E]_0}{(1 + K_A[A]^{-1})(K_R + [R])}, \qquad (5.181)$$

while the inverse enzymatic reaction flux is:

$$\frac{1}{J_+} = (1 + K_A[A]^{-1})\left[\frac{1}{k_+} + \frac{K_R}{k_+[R]}\right]. \qquad (5.182)$$

An increase of the concentration [A] causes an increase of the effective turnover number with an unchanged value of the Michaelis constant that is equal, under the approximation used here, to the binding constant K_R. In the case of allosteric regulation, an enzyme is formed by two different entities: one *catalytic* and one *regulatory*. We then define an allosteric *heterotropic* effect (see Figure 5.21a) and an allosteric *homotropic* effect involving cooperative action of two or more identical catalytic entities (see Figure 5.21b) as described in the original paper of the above-mentioned scientists.

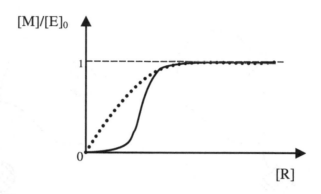

FIGURE 5.22
Sigmoidal dependence of the complex concentration [M] as a function of the substrate concentration [R].

The kinetic scheme of the cooperative interaction of two identical subunits binding the substrate is shown in Figure 5.19d. The condition of the number conservation of the enzyme molecules is given by Equation (5.175). Let us define equilibrium constants equal to

$$K_1 = \frac{[R][E]}{[E']} = \frac{[R][E]}{[M']}, \quad K_2 = \frac{[R][E']}{[M]} = \frac{[R][M']}{[M]}. \tag{5.183}$$

From (5.175) and (5.185) we readily find the concentration of the Michaelis complex as:

$$[M] = \frac{[R]^2[E]_0}{K_1 K_2 + 2K_2[R] + [R]^2}. \tag{5.184}$$

For low concentrations, Equation (5.184) represents a quadratic dependence on $[R]$:

$$[M] = \frac{[R]^2[E]_0}{K_1 K_2}, \tag{5.185}$$

but for high concentrations saturation is reached:

$$[M] = [E]_0. \tag{5.186}$$

The dependence of the reaction flux in Equation (5.172) is not hyperbolic, of the Michaelis-Menten type, but *sigmoidal* (see Figure 5.22). This type of dependence signifies the ability to effect a large change of concentration $[M]$ by changing the concentration $[R]$ only slightly. A classic example of a protein making use of this possibility is hemoglobin — a protein that transports molecular O_2 between a region with high concentration and a region of low concentration. Another example is calmodulin, a regulatory protein for many kinases. It cooperatively binds ions of Ca^{2+} (Stryer, 1995).

References

Atkins, P.W., *Physical Chemistry*, 6th ed., Oxford University, Oxford, U.K., 1998.

Callen, H. B., *Thermodynamics and Introduction to Thermostatistics*, John Wiley & Sons, New York, 1985.

Chandler, D., *Introduction to Modern Statistical Mechanics*, Oxford University Press, Oxford, U.K., 1987.

Dodson, G. and Wlodawer, A., *TIBS* 23, 347, 1998.

Enzyme Nomenclature, Academic Press, San Diego, 1992.

Fersht, A., *Structure and Function in Protein Science*, W.H. Freeman & Co., New York, 1999.

Frauenfelder, H., Sligar, S. G., and Wolynes, P. G., *Science*, 254, 1998, 1991.

Hill, T. L., *Free Energy Transduction and Biochemical Cycle Kinetics*, Springer, New York, 1989.

Kondepudi, D. and Prigogine, I., *Modern Thermodynamics*, John Wiley & Sons, Chichester, U.K., 1998.

Kurzynski, M. and Chelminiak, P., *J. Stat. Phys.*, 110, 137, 2002.

Laidler, K.L. and Bunting, P.S., *The Chemical Kinetics of Enzyme Action*, Clarendon, Oxford, U.K., 1973.

Penrose, R., *Rep. Prog. Phys.*, 42, 1937, 1979.

Phillips, M. and Fletterick, R. J., *Curr. Opin. Struct. Biol.*, 2, 713, 1992.

Prigogine, I. and Stengers, I., *Order Out of Chaos: Man's New Dialogue with Nature*, Random House, New York, 1984.

Prigogine, I. and Stengers, I., *The End of Certainty, Time, Chaos and the New Laws of Nature*, The Free Press, New York, 1997.

Stryer, L., *Biochemistry*, 4th ed., W.H. Freeman & Co., New York, 1995, chap. 8.

Wyman, J. and Gill, S.J., *Binding and Linkage. Functional Chemistry of Biological Macromolecules*, University Books, Mill Valley, CA, 1990.

6

Molecular Biological Machines

To improve a living organism by random mutation is like saying you could improve a Swiss watch by dropping it and bending one of its wheels or axis. Improving life by random mutations has the probability of zero.

Albert Szent-Gyorgi

6.1 Biological motion

All machines require specific parts such as screws, springs, cams, gears, and pulleys in order to function. Likewise, all biological machines must have many well engineered parts to work, for example, organs such as the liver, kidney, and heart. These complex life units consist of smaller parts called cells that in turn are constructed of smaller machines known as organelles that include mitochondria, Golgi complexes, microtubules, and centrioles. Below their level are parts so small they are formally classified as macromolecules (large molecules). One particularly important class of macromolecules is called motor proteins, whose chief property is the generation of forces and hence motion.

One major difference between animate and inanimate matter relates to the types of motions their systems can execute. The motion of inanimate matter is random due to thermal fluctuations or directed by external influences such as electric fields or gravity. While animate matter is subject to these influences as well, it has developed the capability of executing autonomous motion not directed by external influences. For example, bacteria seek hosts and single cellular organisms forage for food and seek mating partners. In order to do that even the simplest organisms must have the ability to harness chemical energy and transform it into mechanical work and motion. At cellular and subcellular levels, we can distinguish the following forms of molecular and supramolecular motion:

Linear diffusion of ions and molecules (entropic)
Biased diffusion waves (e.g., Ca^{2+} waves)
Motor protein motions in the form of:

Porters (cargo)
Rowers (axoplasmic transport)
Contractions (actin, myosis)

Blebbing, ruffling (microplasts)
Polymerization of protein filaments (e.g., treadmilling of actin)
Amebal swallowing (endocytosis)
Exocytosis (e.g., neurotransmitters in the synaptic connection)
Selective passage (ionic gates)
Ciliary and flagellar beating
Rotary motors (F0-ATPase)
Crawling (pseudopodia)
Action potential propagation

This list is not exhaustive. Each example given above requires a specific physical model, some of which have been developed and some still await proper analysis.

Our intention in this chapter is to focus on key elements of force generation driven by subcellular nanoscale machinery. Some topics discussed are general and theoretical and others are specific examples of biological applications. Aside from a purely curiosity-driven element, interest in these studies from a practical point of view is growing rapidly. The prospect of self-organizable, chemically driven, controllable pieces of nanomachinery and even nanofactories is very tantalizing and speculations abound as to the future applications of macromolecules such as motor proteins and protein filaments. Nature has had about 3.5 billion years to seek, test, and implement many designs. It is hoped that with the help of the emerging field of biomimetics we can adapt some of these naturally occurring designs to our specifications, for example, in new drug delivery methods.

6.2   Free energy transduction

For many historical reasons, the word *machine* has had several different meanings in most European languages. In our context, a *machine* is a physical system that enables another two systems to perform work one on each other. Undoubtedly the oldest machine used by man is the lever (see Figure 6.1a). If we place weights A_1 and A_2 at both ends, the work to be performed will involve transfer of gravitational energy from one weight to the other. Hence:

$$A_1 \Delta X_1 = -A_2 \Delta X_2. \tag{6.1}$$

The changes in height ΔX_1 and ΔX_2, respectively, are inversely proportional to the corresponding weights. Their values differ greatly which attests to the usefulness of this simple machine. Equation (6.1) can be interpreted as a condition for the conservation of gravitational potential energy, namely:

$$-\Delta G = A_1 \Delta X_1 + A_2 \Delta X_2 = 0. \tag{6.2}$$

With the help of the lever, work can be done once. The state of the lever after performing work is different from its state before performing work. However, other machines can perform work cyclically, for example, wheels and axles (Figure 6.1b).

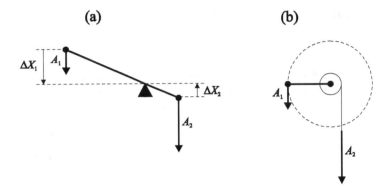

FIGURE 6.1
Lever (a) and wheel and axle (b).

Friction occurs at the pivot point of the lever and on the fulcrum (axis) of the wheel and axle. The possibility of irreversible bending of the lever at the pivot point corresponds to slippage at the axis of the wheel and axle. Accounting for friction (Equation 6.2) takes the form of the inequality below:

$$-\Delta G = A_1 \Delta X_1 + A_2 \Delta X_2 = D \geq 0, \tag{6.3}$$

where D represents irreversibly lost gravitational energy (dissipation). In the presence of friction, a cyclical machine can work in a stationary manner at a constant rate (Figure 6.2). Equation (6.3) divided by Δt corresponds to the following inequality (see Section 2.9):

$$A_1 J_1 + A_2 J_2 = \Phi \geq 0. \tag{6.4}$$

Work done per unit time is called *power*. The first term in Equation (6.4) is the *input power* and the negative of the second term expresses the *output power* while the dissipation function (rate) Φ represents the dissipated power of the machine. The ratio of the output power to the input power:

$$\eta = \frac{-A_2 J_2}{A_1 J_1} = 1 - \frac{\Phi}{A_1 J_1} \tag{6.5}$$

is called the *efficiency* of the machine.

The wheel and axle and the car (regardless of the type of engine used) are two extreme examples of so-called *mechanomechanical* machines since both variables in Equation (6.3) represent displacement, i.e., mechanical variables. Every machine that works cyclically in which the output variable X_2 is a mechanical variable is called a *motor* or an *engine*. The X_1 variable does not have to be mechanical. We also distinguish electrical engines for which the X_1 variable represents electrical charge. Its time derivative J_1 is an electrical current and the corresponding force A_1 is a voltage gradient.

Furthermore, we often deal with *thermal* engines for which X_1 represents entropy whose time derivative J_1 when multiplied by the corresponding force A_1 denoting a temperature difference expresses heat flux intensity. Finally, still in the realm of

(a)   (b)   (c)

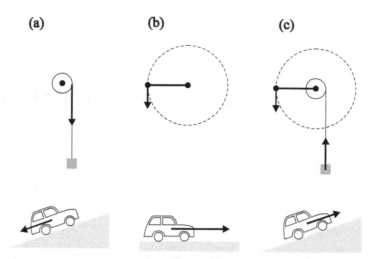

FIGURE 6.2
Simple wheel and axle and a more complicated car. (a) Weight attached to a freely unwinding rope attains a constant velocity of its downward motion due to gravity when a force of friction proportional to velocity balances the force due to gravity. A car stalled on an incline will sooner or later attain a constant velocity of its backward rolling motion. In both cases, the total gravitational energy is dissipated when this happens. (b) Rotational motion of the wheel can achieve a stationary velocity with a complete dissipation of the energy put into it. This corresponds to the motion of a car at a constant velocity on a horizontal plane. (c) Coupling of the motion of the wheel and the axle can lead to upward motion of the weight at a constant velocity and only in this case is work performed against gravity. A car engine performs work at a constant velocity up the hill as shown.

speculation, is a *chemical* motor that could directly convert chemical energy into mechanical work. In a car engine, the chemical energy obtained via internal combustion of gasoline is first converted into a heat flux and then into the motion of pistons in the cylinders. That motion in turn is converted by the axles into the rotational energy of the wheels. In the general case, variables X_1 and X_2 can have arbitrary characters. For example, an alternator is a *mechanoelectrical* machine; a battery is a *chemoelectrical* machine.

The general theory of machine action is a subfield of thermodynamics (Kondepudi and Prigogine, 1998). The origin of thermodynamics can be traced back to the early 19th century when it was necessary to investigate physical descriptions of heat engines. The French engineer Carnot has been credited with the development of a theoretical model of a heat engine that works under the condition of an alternating contact with two thermostats kept at different temperatures. The assumption was that the changes were sufficiently slow to allow the state of complete thermal equilibrium to be maintained and to ignore energy dissipation. In fact, 19th century thermodynamics was first and foremost equilibrium thermodynamics. However, the Carnot engine is a bad model for most modern machines that work under *isothermal* conditions of constant temperature and far from thermodynamic equilibrium. The inclusion of

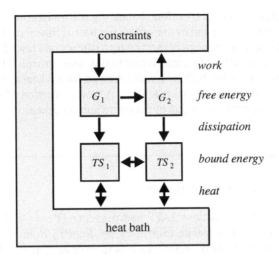

FIGURE 6.3
Pathways of energy transformation in a machine working under isothermal conditions.

dissipation in such models is very important. Also molecular biological machines that will be discussed in this chapter, such as chemomechanical motors, chemoosmotic pumps, and osmomechanical turbines work under isothermal conditions far from thermodynamic equilibrium.

Under isothermal conditions, $T = $ const, the total energy of an arbitrary thermodynamic system, regardless of whether the thermodynamic variables X_i do or do not reach their equilibrium values, can be unambiguously divided into free energy G and bound energy TS (see Section 5.2). The former changes as a result of work done either by the system or on the system. The latter changes via heat transfer. According to the second law of thermodynamics, free energy can be transformed into bound energy in a dissipative process but not the other way around. In other words, bound energy can never be used under isothermal conditions for the purpose of performing work. In our context, gravitational, chemical, and electrical energies are special cases of free energy. In the most interesting case of chemical reactions taking place under isothermal and isobaric conditions, $P = $ const, the role of free energy is played by free enthalpy (also called the Gibbs potential). Hence G is used in Equations (6.2) and (6.3).

Machines that operate under the condition of $T = $ const are *free energy transducers* (Hill, 1989; see Figure 6.3). This means that work done by the environment on one of the machine's subsystems is transformed into its free energy and this, in turn, is passed on to a second subsystem that is subsequently converted into work done by the subsystem on the environment. The free energy transfer between the two subsystems is reduced by dissipation which increases the bound energy of the two subsystems. The heat transferred to the thermostat in the course of the process can have an arbitrary sign, as can the entropy transfer between the two subsystems. At a stationary state, the values of the free and bound energy (entropy) of the two subsystems remain constant,

G_i = const and S_i = const. Therefore, balancing the transformations in Figure 6.3, the total work performed on and by the system per unit of time must be equal to the rate of dissipation that cannot be negative according to the second law of thermodynamics. The latter is equal to the heat flux transferred to the environment. This is precisely the meaning of Equation (6.4) which, originally derived for a wheel and axle, is generally correct for an arbitrary isothermal machine. A transformation of free energy takes place when, despite a positive value of the entire sum in Equation (6.4), one term has a negative sign.

6.3   Chemochemical machines

Consider two chemical reactions: ATP hydrolysis (ATP $\leftrightarrow$ ADP + P_i) and phosphorylation of a certain substrate (Sub + P_i $\leftrightarrow$ SubP). Both reactions ignore the participation of a molecule of water which is in excess. The equilibrium constants K_1 and K_2 are determined by the following quotients:

$$K_1 = \frac{[ADP]^{eq}[P_i]^{eq}}{[ADP]^{eq}}, \quad K_2 = \frac{[SubP]^{eq}}{[SubP]^{eq}[P_i]^{eq}}. \tag{6.6}$$

The first reaction is exoergic from left to right ($K_1 M^{-1} > 1, \Delta G_0 < 0$). The second is assumed to be endoergic in this direction ($K_2 M^{-1} < 1, \Delta G_0 > 0$). The second reaction can, however, take place from left to right if both reactions take place in the same reactor and if the first ensures a sufficiently high concentration of the common reagent P_i (see Figure 6.4a). It at first appears that this system acts as a chemochemical machine in which the first reaction transfers free energy to the second, but is it really so?

The thermodynamic forces acting in the two reactions are, respectively:

$$A_1 = k_B T \ \ln K_1 \frac{[ATP]}{[ADP][P_i]}, \quad A_2 = k_B T \ \ln K_2 \frac{[Sub][P_i]}{[SubP]}. \tag{6.7}$$

Under stationary conditions, the corresponding fluxes are

$$J_1 = \dot{X}_1, \quad J_2 = \dot{X}_2, \tag{6.8}$$

where

$$X_1 \equiv [ADP], \quad X_2 \equiv [SubP] \tag{6.9}$$

and they are equal. The forces in Equation (6.7) are not, however, independent from these fluxes. In Section 4.1, we demonstrated that the flux and force for each reaction taken separately are always of the same sign and hence:

$$A_1 J_1 \geq 0, \quad A_2 J_2 \geq 0 \tag{6.10}$$

in the sum of Equation (6.4). Therefore, no free energy transformation can occur. At most, an entropy transfer is possible (see Figure 6.3).

(a)

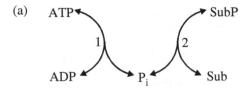

(b)

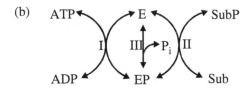

FIGURE 6.4
Coupling of two phosphorylation reactions through a common reagent P_i (a) and through a kinase enzyme E (b).

Many *in vivo* biochemical reactions take place simultaneously in the same volume of a part or the whole of a cell. However, under physiological conditions, coupling two reactions via a common reagent is not easily implemented. As an example, consider the first reaction in the glycolysis chain, i.e., glucose phosphorylation (Sub = Glu) into Glu6P which is coupled to the ATP hydrolysis. For this reaction:

$$K_2 = \frac{[Glu6P]^{eq}}{[Glu]^{eq}[P_i]^{eq}} = 6.7 \times 10^{-3}\,M^{-1}$$

Under stationary, nonequilibrium, physiological conditions, $[P_i] = 10^{-2}\,M$ and $[Glu6P] = 10^{-4}\,M$. For the reaction to move forward, the force A_2 must be positive and the concentration of glucose [Glu] must exceed $1.6\,M = 300\,g/dm^3$ which is an unrealistic value.

Nature has a different way to achieve the coupling of ATP hydrolysis and phosphorylation of the substrate, namely through an enzyme that catalyzes both reactions simultaneously (see Figure 6.4b). According to the terminology presented in Section 5.8, enzymes that catalyze the transfer of a phosphate group are called kinases. If one of the coupled reactions is ATP hydrolysis, then we refer to the enzyme as an ATPase. Figure 6.4b shows three reactions labeled I, II, and III. In addition to the transfer of the phosphate group from ATP to Sub, it is also possible to detach this group from the enzyme unproductively. This is characterized by the equilibrium constants below:

$$K_I = \frac{[ADP]^{eq}[EP]^{eq}}{[ATP]^{eq}[E]^{eq}}, \quad K_{II} = \frac{[E]^{eq}[SubP]^{eq}}{[EP]^{eq}[Sub]^{eq}}, \quad K_{III} = \frac{[E]^{eq}[P_i]^{eq}}{[EP]^{eq}}. \quad (6.11)$$

Since an enzyme cannot affect chemical equilibrium conditions, the constants in Equation (6.11) are independent of the concentrations in Equation (6.11) and the following relations are satisfied:

$$K_I K_{III} = K_1, \qquad \frac{K_{II}}{K_{III}} = K_2. \qquad (6.12)$$

As with relationships between equilibrium constants, we find corresponding relationships between thermodynamic forces:

$$A_1 = A_I + A_{III}, \qquad A_2 = A_{II} - A_{III} \qquad (6.13)$$

where the forces acting on reactions I, II, and III are defined analogously to those in Equation (6.7). Using the definitions in Equation (6.9), one can identify reaction fluxes in the two schemes shown in Figure 6.4:

$$\dot{X}_1 \equiv J_1 = J_I, \qquad \dot{X}_2 \equiv J_2 = J_{II}. \qquad (6.14)$$

Under the stationary conditions, $[E] = \text{const}$ and $[EP] = \text{const}$,

$$J_{III} = J_I - J_{II}. \qquad (6.15)$$

We can conclude from Equations (6.15) and (6.13) that the dissipation function for the system of the three reactions illustrated in Figure 6.4b can be represented as

$$\Phi = A_I J_I + A_{II} J_{II} + A_{III} J_{III}$$
$$= (A_I + A_{II})J_I + (A_{II} - A_{III})J_{II}$$
$$= A_1 J_1 + A_2 J_2. \qquad (6.16)$$

The three terms in the first equation are nonnegative but this does not mean that the two terms in the third equation must also be nonnegative. It is sufficient that for $J_1, J_2 > 0$ when $A_I, A_{II} > 0$, we have $A_{III} > A_{II}$. The other condition $A_{III} < -A_I$ is under physiological conditions of P_i concentrations impossible to satisfy. However, for $J_1, J_2 < 0$, when $A_I, A_{II} < 0$, the inequality $A_{III} > -A_I$ can take place. In both cases we deal with real transduction of the free energy. The first case, namely the transfer of free energy from subsystem 1 to subsystem 2, takes place via glucose phosphorylation at the expense of ATP hydrolysis. The second case, namely the transfer of free energy from subsystem 2 to subsystem 1, occurs in the next two stages of the glycolysis chain in which ADP phosphorylation to ATP proceeds at the expense of even higher energy substrates.

The kinetic scheme shown in Figure 6.4b can be generalized to the case of two arbitrary coupled chemical reactions:

$$R_1 \rightleftarrows P_1 \quad \text{and} \quad R_2 \rightleftarrows P_2. \qquad (6.17)$$

We assume for simplicity that both reactions are unimolecular. One reaction is a donor of free energy and the other is a free energy acceptor. Each biochemical reaction must be catalyzed by a protein enzyme. Separately, each reaction takes place in the direction determined by the second law of thermodynamics such that the amount of chemical energy dissipated is positive (see Figures 6.5a and b). Only when both reactions occur simultaneously using the same enzyme with one reaction according to the second law of thermodynamics can the second reaction be forced to take place against the second law. In this case, the latter transfers part of its free energy recovered from dissipation performing work on it (see Figure 6.5c). The mechanism of energy transfer is simple. If both reactions occur in a common cycle, they must proceed in the same direction.

(a)   (b)   (c)

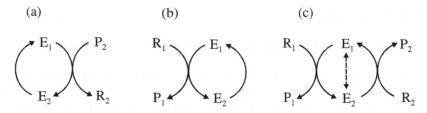

FIGURE 6.5
**(a and b) Two different chemical reactions proceeding independently from each
other. We assume that they are catalyzed by the same enzyme. The reagent for
the first reaction R_1 is coupled to the state E_1 of the enzyme, and the reagent for
the second reaction R_2 is coupled to E_2. Reactions take place under stationary
conditions as a result of keeping the values of concentrations of the reagents and
products fixed and different from those at equilibrium. Values are selected so
that the first reaction proceeds from R_1 to P_1 and the second from P_2 to R_2. (c) If
both reactions occur simultaneously using the same enzyme, the direction of the
first reaction can force a change of direction of the second. The new direction
would be opposite to the one dictated by the stationary values of the respective concentrations. The broken line denotes a possible unproductive transition
between states E_1 and E_2 of the enzyme. It is instructive to compare this diagram
with Figure 6.2.**

Figure 6.5 was intended to resemble Figure 6.2 in order to emphasize the similarities
of a chemomechanical machine and a wheel-and-axle device. In the same way turning
a wheel requires the input of work acting against the force of gravity due to the weight
attached through a rope to the wheel, the first reaction performs work against the
chemical force acting on the second reaction, forcing it to proceed in the opposite
direction. Friction associated with the motion of the wheel and the axle corresponds to energy dissipation in the common reaction cycle. Slippage of the axle with
respect to the wheel is mirrored by the possible direct reaction between states E_1 and
E_2 of the enzyme.

However, one essential difference exists. A wheel-and-axle device is a machine
characterized by macroscopic spatial organization. The enzymes enabling the operation of a chemochemical machine are microscopic or at worst mesoscopic entities.
Viewed macroscopically, the chemochemical machine is a more or less spatially
homogeneous solution of enzymes with a typical concentration of 10^{-6} M, i.e., close
to 10^{15} molecules per cubic centimeter or 10^3 molecules per cubic micrometer (the
typical size of a bacterial cell or an organellum of a eukaryotic cell).

Similarly to the suspension molecules in the solutions observed by Brown, macromolecular enzymes playing the role of biological machines move about and change
their chemical states due to thermal fluctuations. On a short time scale, energy is
borrowed from and returned to the heat bath. That the stochastic motion of biological
machines is not purely random results from their highly organized structures and the
constant input of free energy, mainly due to the hydrolysis of ATP. This is clearly
seen in experiments with single biomolecules that employ an ever more precise and
powerful arsenal of techniques (Ishijama and Yanagida, 2001a and b).

6.4   Biological machines as biased Maxwell's demons

In 1871, James Clerk Maxwell, pondering the foundations of thermodynamics, contemplated the functioning of a hypothetical entity that could observe the velocities of individual gas molecules moving about in a container. The special feature of the container would be a partition with an opening that could be covered by a latch (see Figure 6.6a). This "Gedenken (or thought) experiment" is cited in the literature as "Maxwell's demon" because of the image that a demon would be in charge of closing and opening the hole in the partition, allowing only sufficiently fast particles to move from right to left and sufficiently slow ones to pass from left to right. This would, of course, over time result in a temperature increase in the left part of the container and a decrease in the right section. The thus-created temperature gradient is clearly in contradiction with the second law of thermodynamics due to the work performed in the process by thermal fluctuations alone in a gas at a thermodynamic equilibrium.

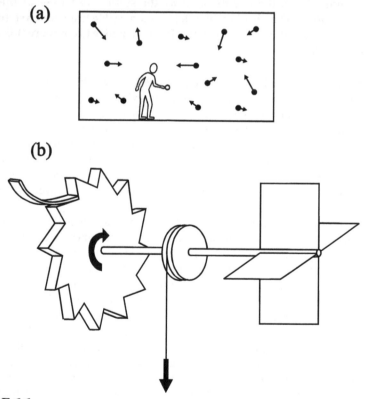

FIGURE 6.6
(a) Maxwell's demon. (b) Feynman's version of Maxwell's demon.

Fewer than 100 years later, another great physicist, Richard Feynman (1963), presented this problem in the more provocative manner shown in Figure 6.6b. A mechanical wheel (center) with a ratchet and pawl (left) can rotate around an axis. The surfaces of the attached wing-like plates (right) are bombarded with gas molecules exerting pressure on both sides of each wing. The pawl prevents the ratchet from rotating in one direction. As a result, the kinetic energy of gas fluctuations is transformed into the rotational kinetic energy of the one-directional motion of the ratchet around the axis. This device can, in principle, raise a weight against the force of gravity.

The logical error in both arguments is made when we consider direct interaction of microscopic systems (gas molecules) with macroscopic ones (hole in the partition, pawl in the wheel). However, the pawl device cannot possibly react to the collision of a single molecule unless it, too, is a microscopic object subject to the same types of thermal fluctuations. Random bending of the pawl assists with the rotation of the wheel in the opposite direction. Similarly, with the latch controlled by Maxwell's demon, to measure the speed of individual molecules, they must be in contact via a physical interaction.

To react to such an interaction, the observing device must be microscopic and will be brought rapidly to thermal equilibrium as a result of interactions with chaotically moving molecules around it. Hence, it will behave chaotically the same way as an average gas molecule does and no net macroscopic force will be generated. These unfavorable fluctuations can be reduced by lowering the temperature of the pawl or freezing the head of Maxwell's demon. In general, this can be done by reducing the entropy or supplying free energy. When this is done, the contradiction with the second law of thermodynamics is automatically removed and these mesoscopic machines will work according to the normal rules of behavior discussed earlier. The biological motors that serve as the centerpieces of this chapter provide interesting examples of such biased Maxwell's demons.

6.5 Pumps and motors as chemochemical machines

From a theoretical view, it would be convenient to treat all molecular biological machines as chemochemical machines. In this section, we will address molecular pumps and motors. In general, a chemochemical machine can be viewed as a black box entered and exited by molecules that take part in both coupled chemical reactions shown in Figure 6.7. The kinetic scheme of the reactions can be arbitrary as long as it involves at least one cycle so that the enzyme cannot be exhausted in the course of the reaction.

Four possible kinetic schemes can be devised, assuming one substrate–enzyme intermediate for each catalyzed reaction. Both reactions are coupled through a free enzyme E (see Figure 6.8a). Both are coupled through an intermediate complex M (see Figure 6.8b). The intermediate complex for one reaction appears to be the free enzyme for the second reaction (see Figure 6.8c). Both reactions proceed as alternating half-reactions (see Figure 6.8d).

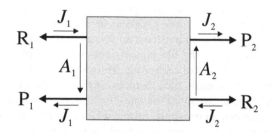

FIGURE 6.7
General scheme of a chemochemical machine coupling two reactions: $R_1 \leftrightarrow P_1$ that produces free energy and $R_2 \leftrightarrow P_2$ that consumes free energy. Both reactions can take place in either direction, as determined by the sign of the flux J_i. The forces A_i are given by the values of the relevant concentrations kept stationary.

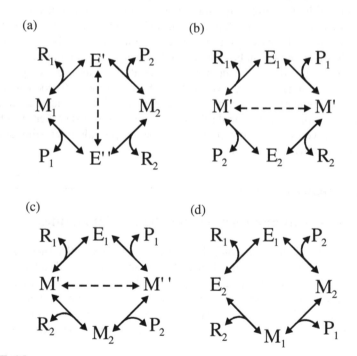

FIGURE 6.8
Four possible kinetic schemes for the system of two coupled reactions, assuming that each proceeds through one intermediate state of the enzyme. The broken lines represent unproductive (without binding to or unbinding from substrates) direct transitions between different states of the enzyme. They cause a possible mutual slippage of the two component cycles.

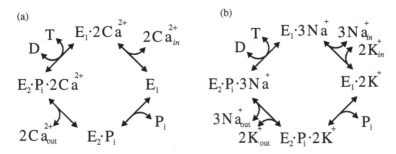

FIGURE 6.9
Simplified kinetic cycles of the calcium pump (a) and the sodium–potassium pump (b). E_1 and E_2 denote two states of an enzyme that represents a pump with a reaction center oriented to the interior and exterior compartments, respectively. T, D, and P_i represent ATP, ADP, and inorganic phosphate, respectively.

$$T \searrow \overset{D}{\nearrow} \quad A M{\cdot}D \quad \searrow \overset{P_i}{\nearrow}$$

$$A M{\cdot}T \longleftarrow - - - - \longrightarrow A M{\cdot}D{\cdot}P_i$$

$$\searrow \overset{}{A} \quad M{\cdot}T \quad \nearrow \overset{}{A}$$

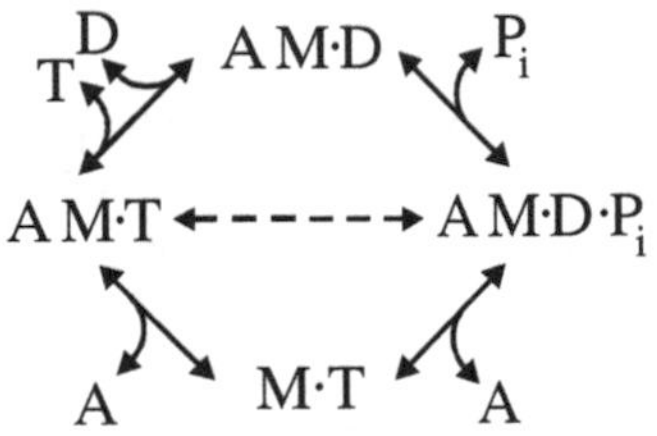

FIGURE 6.10
Simplified version of the Lymn–Taylor–Eisenberg kinetic scheme of the mechanomechanical cycle of the acto–myosin motor. M denotes the myosin head; A, the actin filament; T, D, and P_i, ATP, ADP, and an inorganic phosphate, respectively.

The scheme in Figure 6.8a applies to substrate phosphorylation. That in Figure 6.8b applies to molecular motors, and those in Figures 6.8c and d apply to molecular pumps. Treating molecular pumps as chemochemical machines poses no great problem. The molecules transported from one side of a biological membrane to the other, inside or outside the compartment, can be considered to occupy different chemical states while the transport across the membrane can be regarded as an ordinary chemical reaction (see Section 5.4). Figures 6.9a and b show simplified kinetic cycles of the calcium and sodium–potassium pumps, respectively (Stryer, 1995, Chap. 12). It is clear that both schemes are identical to the one presented in Figure 6.8d.

Molecular motors present a slightly more complicated case. Figure 6.10 shows a simplified version of the Lymn–Taylor–Eisenberg kinetic scheme of the mechano-chemical cycle of the acto–myosin motor (Howard, 2001). The scheme indicates how the ATPase cycle of myosin is related to a detached, weakly attached, and strongly attached state of the myosin head to the actin filament, respectively. Both the substrate and products of the catalyzed reaction bind to and rebind from the myosin in its

(a)   (b)

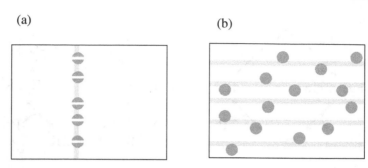

FIGURE 6.11
Functioning of biological molecular machines on a thermodynamic scale at a macroscopic level is only possible due to appropriate organization of the system. Molecular pumps are embedded into the two-dimensional structure of the membrane (a), while molecular motors (b) move along an organized system of tracks (microfilaments or microtubules).

strongly attached state, whereas the reaction takes place in the weakly attached or detached state. Only completion of the whole cycle with the ATP hydrolysis achieved in the detached state results in the directed motion of the myosin head along the actin track.

ATP hydrolysis in the weakly bound state alone is ineffective and corresponds to slippage. However, the question remains how to represent a load acting on a motor in terms of concentrations. In Figure 6.10, A (alone) denotes the actin filament before or after translation of the myosin head by a unit step. Experimental evidence indicates that an external load attached to the statistical ensemble of myosin heads (organized, in the case of myofibrils, into a system of thick filaments) influences the free energy of binding of the myosin heads to thin actin filaments (Baker et al., 1999).

The associated changes of the binding free energy can be expressed as changes of effective rather than actual concentrations of the actin filament A before and after translation. As a consequence, the acto–myosin motor can indeed be effectively treated as a usual chemochemical machine. The output flux J_2 is proportional to the mean velocity of the myosin head along the actin filament and the force A_2 is proportional to the load (Kurzynski and Chelminiak, 2002).

At a macroscopic level, the action of molecular pumps and motors manifests as a directed transport of a substance. The possible functioning of mesoscopic machines on a macroscopic scale is due to appropriate organization of the statistical ensemble. Molecular pumps are embedded in the two-dimensional structure of the membrane (see Figure 6.11a), while molecular motors move along a structurally organized system of tracks: microfilaments or microtubules (see Figure 6.11b). More will be said about these protein filaments and their self-assembly later in this chapter.

Not all biological molecular machines perform work on a macroscopic scale. Examples of such behavior are molecular turbines, e.g., the F_0 portion of ATP synthase. Since no mechanism coordinates the rotational motions of individual turbines, no macroscopic thermodynamic variable characterizes this motion. F_0 and F_1 must, therefore, be treated jointly from the macroscopic point of view, giving rise to a reversible molecular pump, the H^+ ATPase.

A large number of ingenious techniques have been designed for molecular pumps and motors that enable precise observation of the behavior of single molecules. For molecular pumps and ion channels, such a technique is the patch clamp method (Sackmann and Naher, 1995). Various motility assays apply to molecular motors (Mehta et al., 1999; Ishijima and Yanagida, 2001). All these observations reveal the stochastic nature of the behavior exhibited by molecular motors. In a particularly impressive experiment, Kitamura et al. (1999) demonstrated that a single myosin head can accidentally make two to five steps along the actin filament per single ATP molecule hydrolyzed. This undermines the prevailing hypothesis that one ATP molecule is used per step.

6.6 Flux-force dependence

To better analyze the character of flux–force relations in biological processes of free energy transduction, we now return to the simplest kinetic scheme (see Figure 6.5c) for the enzymatic coupling of a reaction supplying free energy with one that absorbs it. Figure 6.12 shows this scheme in more detail and gives constant reaction rates for the individual reactions involved. The thermodynamic forces for the three component reactions are given by the equations:

$$A_{\mathrm{I}} = k_B T \ln \frac{[E_2]^{\mathrm{eq}}[P_1]^{\mathrm{eq}}}{[E_1]^{\mathrm{eq}}[R_1]^{\mathrm{eq}}} \frac{[E_1][R_1]}{[E_2][P_1]},$$

$$A_{\mathrm{II}} = k_B T \ln \frac{[E_1]^{\mathrm{eq}}[P_2]^{\mathrm{eq}}}{[E_2]^{\mathrm{eq}}[R_2]^{\mathrm{eq}}} \frac{[E_2][R_2]}{[E_1][P_2]},$$

$$A_{\mathrm{III}} = k_B T \ln \frac{[E_1]^{\mathrm{eq}}[E_2]}{[E_2]^{\mathrm{eq}}[E_1]} \tag{6.18}$$

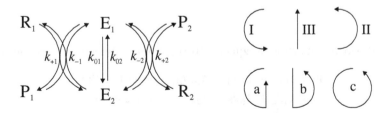

FIGURE 6.12
Enzymatic coupling of a reaction providing free energy to one absorbing free energy. Shown are constant reaction rates, conventional reaction flux directions, and cyclical fluxes.

and the corresponding reaction fluxes (see Equation 5.75) are

$$J_I = \frac{1 - e^{-\beta A_I}}{(k_{+1}[R_1])^{-1} + (k_{-1}[P_1])^{-1} e^{-\beta A_I}} [E]_0,$$

$$J_{II} = \frac{1 - e^{-\beta A_{II}}}{(k_{+2}[R_2])^{-1} + (k_{-2}[P_2])^{-1} e^{-\beta A_{II}}} [E]_0,$$

$$J_{III} = \frac{1 - e^{-\beta A_{III}}}{k_{20}^{-1} + k_{10}^{-1} e^{-\beta A_{III}}} [E]_0, \qquad (6.19)$$

where $[E]_0$ is the total enzyme concentration:

$$[E]_0 = [E_1] + [E_2] = [E_1]^{eq} + [E_2]^{eq}. \qquad (6.20)$$

Operational forces and fluxes, respectively, are given for noncatalyzed reactions as (Hill 1989)

$$A_1 = k_B T \ln \frac{[P_1]^{eq}}{[R_1]^{eq}} \frac{[R_1]}{[P_1]}, \quad J_1 = [\dot{P}_1] = -[\dot{R}_1],$$

$$A_1 = k_B T \ln \frac{[P_2]^{eq}}{[R_2]^{eq}} \frac{[R_2]}{[P_2]}, \quad J_2 = [\dot{P}_2] = -[\dot{R}_2]. \qquad (6.21)$$

Reaction fluxes J_I and J_{II} directly determine operational fluxes J_1 and J_2:

$$J_1 = J_I, \qquad J_2 = J_{II}. \qquad (6.22)$$

Comparing the forces (Equations 6.18 and 6.21), we obtain:

$$A_1 = A_I + A_{III}, \qquad A_2 = A_{II} - A_{III} \qquad (6.23)$$

and from the steady-state condition:

$$\frac{d}{dt}[E_1] = \frac{d}{dt}[E_2] = 0, \qquad (6.24)$$

we find the relationship:

$$J_1 - J_2 = J_{III}. \qquad (6.25)$$

The dissipation function in Equation (6.16) can be written three different ways:

$$
\begin{aligned}
\Phi &= A_I J_I + A_{II} J_{II} + A_{III} J_{III} && \text{(transition fluxes)} \\
&= A_a J_a + A_b J_b + A_c J_c && \text{(cycle fluxes)} \\
&= A_1 J_1 + A_2 J_2 && \text{(operational fluxes)}
\end{aligned} \qquad (6.26)
$$

where:

$$J_a + J_c = J_I, \quad J_b + J_c = J_{II}, \quad J_c - J_b = J_{III} \qquad (6.27)$$

(see Figure 6.12), thus

$$A_a = A_I + A_{III}, \quad A_b = A_{II} - A_{III}, \quad A_a = A_I + A_I \tag{6.28}$$

All terms in the first and second sum are nonnegative (Hill, 1989). Only the two terms in the third sum can have different signs and when this takes place we deal with free energy transduction. As mentioned earlier, $J_1 A_1$ is called the *power input*, whereas $-J_2 A_2$ is the *power output*. The *efficiency* of the process is defined as the ratio of power output to power input:

$$\eta = -\frac{J_2 A_2}{J_1 A_1} = \frac{J_1 A_1 - \Phi}{J_1 A_1} = 1 - \frac{\Phi}{J_1 A_1}. \tag{6.29}$$

This is often written as the product:

$$\eta = \varepsilon \rho, \tag{6.30}$$

in which the *degree of coupling* of both subsystems, ε, determines the ratio of the operational fluxes:

$$\varepsilon = \frac{J_2}{J_1}, \tag{6.31}$$

whereas ρ is the ratio of the forces:

$$\rho = -\frac{A_2}{A_1}. \tag{6.32}$$

Utilizing the relations in Equations (6.22), (6.23), and (6.25), we can eliminate from the equations in (6.19) the force A_{III}. Introducing the equilibrium constants:

$$K = \frac{[E_1]^{eq}}{[E_2]^{eq}}, \quad K_1 = \frac{[P_1]^{eq}}{[R_1]^{eq}}, \quad K_2 = \frac{[P_2]^{eq}}{[R_2]^{eq}} \tag{6.33}$$

and taking the fixed values of the total substrate concentrations:

$$[R_1] + [P_1] = [R_1]_0, \quad [R_2] + [P_2] = [R_2]_0 \tag{6.34}$$

we can obtain, after rather tedious algebra that makes use of the partial equilibrium conditions for reactions I through III, the equations that link operational fluxes with forces:

$$J_1 = \frac{[1 - e^{-\beta(A_1+A_2)} + (1 - e^{-\beta A_1})k_{20}\tau_2][E]_0}{(1 + K^{-1}e^{-\beta A_2})\tau_1 + (1 + Ke^{-\beta A_1})\tau_2 + (k_{10} + k_{20})\tau_1\tau_2},$$

$$J_2 = \frac{[1 - e^{-\beta(A_1+A_2)} + (1 - e^{-\beta A_2})k_{10}\tau_1][E]_0}{(1 + K^{-1}e^{-\beta A_2})\tau_1 + (1 + Ke^{-\beta A_1})\tau_2 + (k_{10} + k_{20})\tau_1\tau_2}, \tag{6.35}$$

where the quantities:

$$\tau_i = (k_{+i}[R_i])^{-1} = (k_{+i}[R_i]_0)^{-1}(1 + K_i e^{-\beta A_i}) \tag{6.36}$$

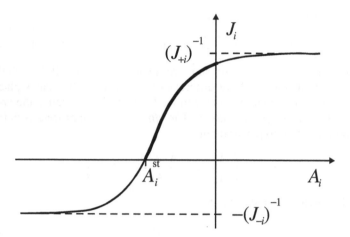

FIGURE 6.13
Character of the functional dependence of the output flux J_i versus force A_i^{st}.
Only when the stalling force A_i is negative, does the flux $J_i(A_i)$ have the opposite
sign to its conjugate force A_i and free energy transduction can take place. The
$J_i(A_i)$ dependence in this range is marked with the bold line.

$(i = 1, 2)$, have the dimension of time. In the case of no slippage, $k_{01}, k_{02} \to 0$, we
deal with complete coupling ($\varepsilon = 1$) and both fluxes become identical:

$$J_1 = J_2 = \frac{[1 - e^{-\beta(A_1 + A_2)}][E]_0}{(1 + K^{-1}e^{-\beta A_2})\tau_1 + (1 + Ke^{-\beta A_1})\tau_2}, \tag{6.37}$$

The flux–force relations (Equation 6.35) for two coupled reactions are of the same
functional form as the flux–force relation for the separate reactions in
Equation (5.159):

$$J_i = \frac{1 - e^{-\beta(A_i - A_i^{st})}}{J_{+i}^{-1} + J_{-i}^{-1}e^{-\beta(A_i - A_i^{st})}}[E]_0 \tag{6.38}$$

$(i = 1, 2)$. Now, however, the parameters J_{+i} and J_{-i} depend $[E_0]$ on another
force and their dependence on the reaction rate constants and the concentrations of
the substrates is much more complex so we do not present them here. Moreover,
additional parameters A_i^{st} determine the nonzero values of the *stalling forces*, for
which only the fluxes J_i vanish. The dependence, $J_i(A_i)$, in Equation (6.38) is
strictly increasing with an inflection point and two asymptotes (see Figure 6.13).

As noted earlier, free energy transduction takes place if one of the fluxes is of
the opposite sign to its conjugate force. From Equation (6.38), it follows that this
condition holds when the corresponding stalling force A_i^{st} is negative. The dependence
$J_i(A_i)$ in the range $A_i^{st} \leq A_i \leq 0$ can be convex, concave, or involve an inflection
point as well.

We do not discuss here the conditions for maximum efficiency of free energy
transduction as even in the linear approximation of the flux–force relations (a poor

approximation under usual physiological and laboratory conditions), the formulae for the values of forces maximizing efficiency are very complex (Westerhoff and van Dam, 1987). The conditions for maximum efficiency and maximum power output contradict each other. A machine is more efficient when its free energy dissipation is lower, i.e., it works more slowly. The slower it works, the lower is its power output.

However, maximum efficiency and maximum power output are not always at their optimum values from the point of view of a living organism. Very often the power output of biological machines is simply equal to zero, i.e., the output forces stall the machines. This can be the case with molecular motors and molecular pumps as well. Muscles of a man sustaining a big load do not perform work (physiologists use the not entirely logical "isometric contraction" phrase to describe such muscle activity) but, of course, ATP is consumed. The intracellular concentration of Ca^{2+} is kept at a very low level to avoid association with phosphate ions P_i. Potassium K^+ conversely remains at a very high level to secure an appropriate value of the plasma membrane resting potential (Stryer, 1995). Because of ATP hydrolysis by the corresponding pumps, the ions do not flow into or out of cells despite the concentration differences. All cases considered are indeed similar to that of a car that remains at the same spot on an inclined road while its wheels constantly rotate and slip (Figure 6.2).

The state of zero power output is advantageous for an organism because it can be maintained regardless of environmental changes (homeostasis). The attainable range of variability of the force stalling the machine is, however, limited. From Equation (6.35), the relations that follow are

$$-\beta A_1^{st} = \ln \frac{1 + C_1 + K_1 e^{-\beta A_2}}{1 + (C_1 + K_1)e^{-\beta A_2}},$$

and

$$-\beta A_2^{st} = \ln \frac{1 + C_2 + K_2 e^{-\beta A_1}}{1 + (C_2 + K_2)e^{-\beta A_1}}, \tag{6.39}$$

where the constants

$$C_i = \frac{k_{+i}[R_i]_0}{k_{0i}} \tag{6.40}$$

($i = 1, 2$) represent the ratios of the corresponding productive and nonproductive reaction rates (see Figure 6.12). The dependences (Equation 6.39) of the negative stalling forces A_1^{st} and A_2^{st} on A_2 and A_1, respectively, are strictly increasing but they saturate both for very large positive and negative values of the forces that determined them. For values outside the determined range, the forces can no longer stall the machine. For small values of the forces, Equation (6.39) can be linearized to yield

$$-\beta A_1^{st} = \frac{C_1}{1 + C_1 + K_1}\beta A_2, \quad -\beta A_2^{st} = \frac{C_2}{1 + C_2 + K_2}\beta A_1. \tag{6.41}$$

The proportionality coefficient is always less than or at best equal to unity. In Figure 6.14, both dependences in Equation (6.39) are drawn and domains in the (A_1, A_2) plane where free energy transduction takes place are shown.

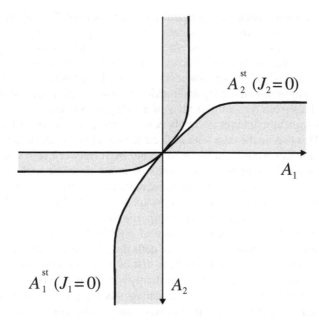

FIGURE 6.14
Areas in the (A_1, A_2) plane for which free energy transduction takes place. The boundaries of the vanishing fluxes are determined by Equation (6.39).

In terms of constants C_i in Equation (6.40), the expression for the degree of coupling is

$$\varepsilon = \frac{J_2}{J_1} = \frac{1 - e^{-\beta(A_1 + A_2)} + C_1^{-1}(1 - e^{-\beta A_2})(1 + K_1 e^{-\beta A_1})}{1 - e^{-\beta(A_1 + A_2)} + C_2^{-1}(1 - e^{-\beta A_1})(1 + K_2 e^{-\beta A_2})}. \qquad (6.42)$$

It can be readily seen that in the case when $\beta A_1 \geq 0$ and $\beta A_2 \leq 0$, the coupling coefficient is less than or equal to unity. Conversely, when $\beta A_1 \leq 0$ and $\beta A_2 \geq 0$, the coupling coefficient is greater than or equal to unity. Therefore, similarly to macroscopic machines (Figure 6.2), slippage lowers the efficiency of a chemochemical machine as can be seen in Equation (6.30). This statement, however, is not entirely general since the fluctuation mechanism of the coupling between the two reactions can lead in the case of kinetic schemes more complex than that in Figure 6.12 to increased efficiency of a chemochemical machine (Kurzynski and Chelminiak, 2002).

6.7 Overview of motor protein biophysics

Two principal motor proteins that attach to microtubules (MTs) are kinesin and dynein. While kinesin moves toward the plus ends of MTs, dynein is negative end-directed. Each of these proteins consists of a globular head region and an extended coiled coil tail section as shown schematically in Figure 6.15. The study of motor proteins including myosin, which has a similar structure, indicates that the long tail can increase

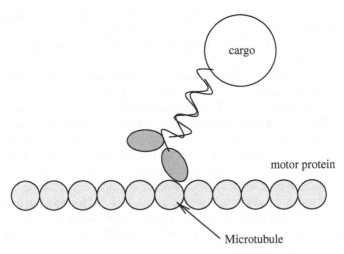

cargo

motor protein

Microtubule

FIGURE 6.15
Motor protein "walking" along a MT protofilament.

the force generated by the molecular motor and the essential components for force generation are located within the globular head.

Models of motor protein movement can essentially be divided into two different mechanisms, those with diffusion and those with a *power stroke* (Leibler and Huse, 1993). The efficient propagation of these proteins, often with pairs such as kinesin and dynein moving in opposite directions simultaneously and seemingly avoiding collisions, led to the proposal that they may be directed by electrostatic interactions with MTs (Brown and Tuszynski, 1997). It is interesting to note that the binding of kinesin to MTs has been shown to be primarily electrostatic (Woehlke et al., 1997).

The motions of the motor protein in both models are accomplished through hydrolysis of ATP. Diffusion models require an oscillating potential that is presumably driven by a conformation change of the motor-MT bond. Activation of the complex by ATP leads to a potential that is relatively flat. Diffusion occurs in this state and once the potential reverts to its asymmetric form, the geometry of the potential is such that forward propagation of the motor protein is favored. A review of such schemes may be found in Jülicher, Adjari, and Prost (1997).

In the power stroke models, by contrast, the motor protein is the entity whose structure changes. Sometimes such models are envisioned as "walking" proteins since one can visualize a protein stretching and binding at a second location before relaxing to its original conformation when the "back leg" releases its grip on the MT to start the process anew once additional ATP arrives. Both models share similar features, but the power stroke model is often preferred for the descriptions of particular features of motor proteins.

A motor protein has two or more distinct states in which at least one conformational change occurs and is driven by ATP hydrolysis as demonstrated for myosin by Ruppel, Lorenz, and Spudich (1995). Phosphorylation of the motor protein, as in the case of myosin, may lead to subsequent conformational changes. Thus, the protein may be viewed as "walking" along the MT powered by ATP. It is interesting to note that the use of GTP to control MT dynamics and ATP to control motor protein motion along MTs allows a cell to control both the cars (motor proteins) and the tracks (MTs) individually.

Buttiker (1987) and Landauer (1988) proposed a model for molecular motors that results in unidirectional motion of a molecule along a quasi-one-dimensional protein fiber. This model was essentially a generalization of the thermal ratchet model of Feynman (1969). The key mechanism is an inhomogeneous temperature distribution with the periodicity of the filamentous structure to which the motor binds. However, it is well known from the theory of thermal conductivity that temperature variations over tens of nanometers are equilibrated on time scales of microseconds. Therefore, models of this type appear unrealistic. Recent attempts at explaining motor protein behavior use isothermal ratchets in which the motor is subjected to an external potential that is periodic and asymmetric. In addition, a fluctuating force $F(t)$ is acting on the motor. A review article by Jülicher, Adjari, and Prost (1997) distinguishes three classes of such models, namely:

(1) A Langevin-based approach using the equation of motion of Doering, Horsthemke, and Riordan (1994):

$$\xi \frac{dx}{dt} = -\partial_x W(x) + F(t) \tag{6.43}$$

where x is the direction along the protofilament axis, ξ is the friction coefficient, $W(x)$ is the potential due to the fiber, and $F(t)$ is the fluctuating force whose time average is zero:

$$\langle F(t) \rangle = 0 \tag{6.44}$$

This fluctuating force is typically related to the kinetics of ATP binding to the motor protein and subsequent hydrolysis of ATP into ADP that provides the excess energy allowing the protein to unbind from the filament.

(2) An approach in which the potential fluctuates in time (Astumian and Bier, 1994):

$$\xi \frac{dx}{dt} = -\partial_x W(x, t) + f(t) \tag{6.45}$$

where the Gaussian (white) noise term $f(t)$ satisfies the fluctuation-dissipation theorem and the typical conditions are imposed:

$$\langle f(t) \rangle = 0 \tag{6.46}$$

and

$$\langle f(t) f(t') \rangle = 2kT\delta(t - t') \tag{6.47}$$

(3) A generalized model with several internal states of the particle described by the Langevin equation that depends on the state $i = 1, \dots, N$:

$$\xi_i \frac{dx}{dt} = -\partial_x W_i(x, t) + f_i(t) \tag{6.48}$$

Here, the Gaussian noise term also satisfies the fluctuation–dissipation theorem via:

$$\langle f_i(t) \rangle = 0 \tag{6.49}$$

and

$$\langle f_i(t) f_j(t') \rangle = 2kT\delta(t - t')\delta_{ij} \tag{6.50}$$

A more convenient and elegant approach is through Fokker–Planck formalism (Risken, 1989) for the probability distribution function $P(x, t)$ that describes the motion of the motor in a statistical manner. $D^{(1)}$ is the drift coefficient and $D^{(2)}$ the diffusion coefficient.

$$\frac{\partial P}{\partial t} = \left[-\frac{\partial}{\partial x} D^{(1)}(x) + \frac{\partial^2}{\partial x^2} D^{(2)}(x) \right] P(x, t) \tag{6.51}$$

Models of the above types have been fairly successful in representing the gross features of the experimental data obtained (Svoboda and Block, 1994). However, the use of a hypothetical potential and a number of arbitrary parameters casts doubt on the usefulness or even correctness of such models.

Can such models of motor protein movement be successfully applied to the problem of chromosome separation? The prevailing model of how the MTs that form the mitotic spindle separate chromosomes is that motor proteins attach to the chromosomes and the distal ends of spindle MTs. Force generation is thereby developed locally by the motor proteins and cannot depend on the length of the MT. The force attracting the chromosome to the pole has been shown to increase as the spindle shortens (Hays and Salmon, 1990). Only a repulsive force from the pole or one generated by the tubulin subunits of the MT would solve this difficulty. However, if motor proteins are responsible for force generation, we must explain why chromosome movement does not require ATP (Spurck and Pickett, 1987). It has not been shown that motor proteins are important for more than simply binding chromosomes, and the relationship between force and the lengths of spindle MTs must still be explained.

We mentioned the importance of energy generation through ATP hydrolysis utilized by motor proteins in their directed motion. We also stated that not all force generation is due to motors and that protein filaments through their polymerization may produce large amounts of force applied to particular parts of the cell. These two topics will be presented in the sections that follow.

6.8 Biochemical energy currency: ATP molecules

All life processes require energy supplies that originally come from the electromagnetic energy of visible light emitted by the sun and absorbed by pigments of photosynthetic units. The absorbed light energy is transformed, accumulated, and transmitted to other parts of cells and other parts of plants, and finally to other organisms that cannot obtain energy directly by photosynthesis. For this purpose, the energy of molecular excitation is transformed into the chemical energy of so-called high-energy compounds.

The most common storage molecule of chemical energy in cells is adenosine triphosphate (ATP), formed during photosynthesis and used in nearly all processes of energy conversion in other cells. ATP is a complex nanomachine that serves as the most common high energy compound in the human body. All living organisms produce ATP, which in turn powers virtually every activity of cells and organisms.

The hydrolysis of ATP produces adenosine diphosphate (ADP) and, when catalyzed by special enzymes known as ATPases, allows the use of stored energy for ionic pumps, molecular synthesis, production of mechanical energy, and many other applications. The amount of energy stored by the ADP–ATP reactions in cells, however, is limited because of osmotic stability. Therefore, other molecules such as sugars and fats are used for long-term energy storage. The free energy of ATP, however, is used to synthesize these molecules. Subsequently, in the respiratory chain, these molecules are decomposed, whereas ATP is produced again and again.

Approximately 7.3 kcal/mole of energy per ATP molecule is released in hydrolysis whose chemical reaction is given by the equation:

$$ATP^{4-} + H_2O \longrightarrow ADP^{3-} + HPO_4^{2-} + H^+ \tag{6.52}$$

Four negatively charged oxygen atoms are in close proximity on the ATP molecules. As they repel one another, the molecule is under considerable strain. The ADP molecule contains only three negatively charged oxygen ions and thus in the conversion, the overall electrostatic energy is reduced. Under physiological conditions, the ATP molecule is metastable. When an ATP molecule binds to certain enzymes, however, the activation barrier is lowered and the molecule moves down the potential energy curve and transforms into ADP. The ADP is usually immediately recycled in the mitochondria where it is recharged and emerges as ATP. At any instant, each cell contains about one billion ATP molecules that satisfy its needs for only a few minutes and hence must be rapidly recycled. The total human body content of ATP is only about 50 g. However, an average daily intake of 2500 food calories translates into a turnover of 180 kg of ATP.

6.8.1   Structure of ATP

ATP contains adenine, a purine base, and ribose sugar. Together they form the adenosine nucleoside. The basic building blocks in the formation of ATP are carbon, hydrogen, nitrogen, oxygen, and phosphorus which are assembled in a complex whose mass is 500 Da. One phosphate ester bond and two phosphate anhydride bonds hold the three phosphates (PO_4) and the ribose together (see Figure 6.16).

The high energy bonds of ATP are rather unstable and the energy of ATP is readily released when it is hydrolyzed in cellular reactions. While ATP is not excessively unstable, it is designed so that its hydrolysis is slow in the absence of a catalyst. This ensures that its stored energy is released rapidly only in the presence of the appropriate enzyme. ATP is an energy-coupling agent and not a fuel. When ATP is produced by one set of reactions, it is almost immediately consumed by another.

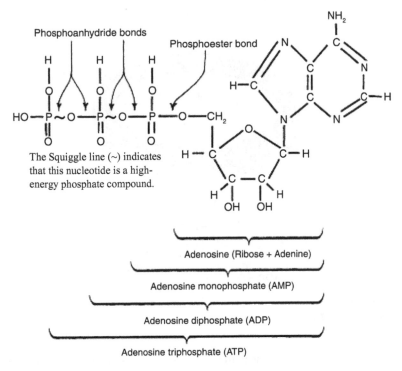

FIGURE 6.16
Chemical structure of ATP.

6.8.2 Functions of ATP

ATP is used for many cell functions including transport of substances across cell membranes, mechanical work, supplying energy needed for muscle contraction, etc. It supplies energy to the heart, skeletal muscles, chromosomes, and flagella. A major role of ATP is chemical. It supplies the energy needed to synthesize various macromolecules. ATP is also used as an on–off switch to control chemical reactions and send messages. It can bond to one part of a protein molecule, causing another part of the same molecule to change its conformation, inactivating the molecule. Subsequent removal of ATP causes the protein to return to its original shape and again become functional. The cycle can be repeated until the molecule is recycled, effectively serving as an on–off switch. Both phosphorylation and dephosphorylation can be used for this purpose.

ATP is manufactured as a result of several cellular processes including fermentation, respiration, and photosynthesis. Most commonly, cells use ADP as a precursor molecule and then add a phosphorus to it. In eukaryotes, this can occur either in the cytosol or mitochondria. Changing ADP to form ATP in the mitochondria is called chemiosmotic phosphorylation. It takes place in specially constructed chambers located in the mitochondrion's inner membranes. The mitochondrion functions to produce an electrical chemical gradient by accumulating hydrogen ions in the space between the inner and outer membranes. This energy comes from the

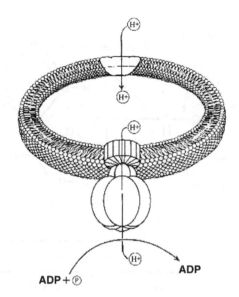

FIGURE 6.17
Functioning of ATPase.

estimated 10,000 enzyme chains in the membranous sacs on the mitochondrial walls.
See Figure 6.17.

Most food energy in cells is produced by the electron transport chain. Cellular
oxidation in the Krebs cycle causes an electron build-up that pushes $H+$ ions outward across the inner mitochondrial membrane. As the charge builds, it provides an
electrical potential that releases its energy by causing a flow of hydrogen ions across
the inner membrane into the inner chamber. The energy causes an enzyme to be
attached to ADP which catalyzes the addition of a third phosphorus to form ATP.

In the case of eukaryotic cells, the energy comes from food which is converted to
pyruvate and then to acetyl coenzyme A (acetyl CoA). Acetyl CoA then enters the
Krebs cycle which releases energy that results in the conversion of ADP back into
ATP. When the repulsion due to a high proton concentration reaches a certain level,
the hydrogen ions are forced from revolving door-like structures mounted on the inner
mitochondria membrane called ATP synthase complexes. ATP synthase functions to
reattach the phosphates to the ADP molecules, again forming ATP. Each revolution
of ATP synthase requires the energy of about nine hydrogen ions returning into the
mitochondrial inner chamber. Located on the ATP synthase are three active sites,
each of which converts ADP to ATP with every turn of the wheel. Under maximum
conditions, the wheel turns at a rate of up to 200 revolutions per second, producing
600 ATP molecules in the process.

6.8.3 Double energy packet

Although ATP contains the energy necessary for most reactions, at times more
energy is required. The solution is for ATP to release two phosphates instead of

one, producing adenosine monophosphate (AMP) plus a chain of two phosphates called a pyrophosphate. An intricate enzyme called adenylate kinase can transfer a single phosphate from ATP to AMP, producing two ADP molecules. The two ADP molecules can then enter the normal Krebs cycle designed to convert ADP into ATP.

While the main energy carrier the body uses is ATP, other energized nucleotides can also be utilized such as thymine, guanine, uracil, and cytosine for making RNA and DNA. The Krebs cycle charges only ADP, but the energy contained in ATP can be transferred to one of the other nucleosides by means of an enzyme called nucleoside diphosphate kinase. This enzyme transfers the phosphate from a nucleoside triphosphate (ATP) to a nucleoside diphosphate such as guanosine diphosphate (GDP) to form guanosine triphosphate (GTP). The nucleoside diphosphate kinase works by binding one of its six active sites to nucleoside triphosphate and releasing the phosphate which is bonded to a histidine. The nucleoside triphosphate, which is now a diphosphate, is released, and a different nucleoside diphosphate binds to the same site. As a result, the phosphate bonded to the enzyme is transferred, forming a new triphosphate. Numerous other enzymes exist in order for ATP to transfer its energy where needed. Each enzyme must be specifically designed to carry out its unique function.

6.8.4 Methods of producing ATP

A crucial difference between prokaryotes and eukaryotes is the means they use to produce ATP. All life forms produce ATP by three basic chemical methods only: oxidative phosphorylation, photophosphorylation, and substrate-level phosphorylation. In prokaryotes, ATP is produced both in cell walls and in cytosols by glycolysis. In eukaryotes, most ATP is produced in chloroplasts (plants), or mitochondria (both plants and animals). In fact, the machinery required to manufacture ATP is so intricate that viruses cannot make their own ATP.

The mitochondria produce ATP in their internal membrane systems called cristae. Since bacteria lack mitochondria and internal membrane systems, they must produce ATP in their cell membranes via two basic steps. The bacterial cell membrane contains a unique structure designed to produce ATP. There, the ATPase and the electron transport chain are located inside the cytoplasmic membranes between the hydrophobic tails of the phospholipid membrane inner and outer walls. Breakdown of sugar and other food causes the positively charged protons on the outside of the membrane to accumulate to a much higher concentration than on the membrane inside. The result of this charge difference is a dissociation of H_2O molecules into H^+ and OH^- ions. The H^+ ions produced are then transported outside the cell and the OH^- ions remain on the inside. The resultant potential energy gradient with its force, called a proton motive force, can accomplish a variety of cell tasks including converting ADP into ATP. In some bacteria such as Halobacterium, this system is modified by use of bacteriorhodopsin. Illumination causes pigment to absorb light energy, temporarily changing rhodopsin from a *trans* to a *cis* form. The *trans*-to-*cis* conversion causes deprotonation and the transfer of protons across the plasma membrane to the periplasm. The resulting proton gradient drives ATP synthesis by use of the ATPase complex. This modification allows bacteria to live in

304 Introduction to Molecular Biophysics

low-oxygen, light-rich environments. This anaerobic ATP manufacturing system unique to prokaryotes uses a chemical compound other than oxygen as a terminal electron acceptor.

6.8.5 Chloroplasts

Chloroplasts are double-membraned ATP-producing organelles found only in plants. Inside their outer membranes are thin membranes organized into flattened sacs (thylakoids) stacked up like coins. The disks contain chlorophyll pigments that absorb solar energy. The chloroplasts first convert the solar energy into ATP-stored energy, which is then used to manufacture storage carbohydrates that can be converted back into ATP when energy is needed.

Chloroplasts also possess electron transport systems for producing ATP. Electrons that enter the system are taken from water. During photosynthesis, carbon dioxide is reduced to a carbohydrate by energy obtained from ATP. Photosynthesizing bacteria (cyanobacteria) use another system involving chlorophyll bound to cytoplasmic thylakoids.

6.9 Assembly of microtubules

The assembly of cytoskeletal protein filaments (Amos and Amos, 1991) can also be used by cells to produce forces useful for motility and maintaining internal tension (Alberts et al., 1994; Luby-Phelps, 1994; Hinner et al., 1998; Frey, Kroy, and Wilhelm, 1998). The first type of protein filament whose polymerization process will be discussed is the microtubule (MT). When an individual MT is observed, it undergoes periods of almost steady growth interrupted by brief periods of rapid shortening as illustrated in Figure 6.18.

The transition from a growth period to a shortening period is known as a catastrophe event while the reverse transition is known as a rescue. The assembly dynamics at each end of the MT differ. The so-called plus end is three to five times as dynamic as the minus end. That is, both the growth and shortening of the plus end of the MT occur at rates at least three times those of the minus end but both ends are dynamic in free MTs.

Through individual MT assembly dynamics (Mitchison and Kirschner, 1984; Horio and Hotani, 1986), one observes an assembly characteristic known as *dynamic instability*. What is intriguing is how stochastic individual behavior (Cassimeris, 1993) may lead to smooth collective oscillations at high tubulin concentrations (Carlier et al., 1987; Mandelkow and Mandelkow, 1991). The answer is simply by application of statistical mechanics to ensemble averaging. A complete explanation requires application of master equation formalism (Flyvbjerg, Holy, and Leibler, 1996) or alternatively, chemical kinetics (Marx and Mandelkow, 1994; Houchmandzadeh and Vallade, 1996; Sept et al., 1999).

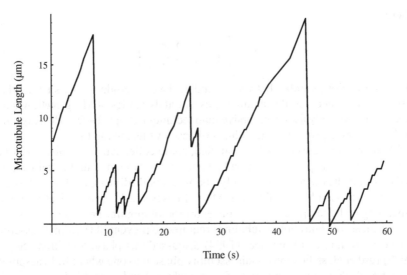

FIGURE 6.18
Growth of a single microtubule is erratic. Slow, steady growth is interrupted by large catastrophe events at about 8 s and 46 s while a rescue event occurs between two catastrophe events, around 26 s.

The assembly dynamics for an individual MT is stochastic where the rate of shortening is about 10 times the rate of growth (Mandelkow, Mandelkow, and Milligan, 1991; Tran, Walker, and Salmon, 1997). The probability of a single MT nucleating, growing, shortening, etc. depends on the local concentrations of tubulin with bound GTP and tubulin with bound GDP and other molecules found in the cytoplasm and relevant to the assembly process. Many theoretical models successfully reproduced single MT growth behavior as shown in Figure 6.18.

The challenge is to use the same model to explain ensemble dynamics, dynamic instability, and changes in dynamics when reaction conditions are altered. Suppose the length of our MT is given by the value of the variable x at an instant of time t. If we discretize the time variable, we want to write a recursion formula for the length of the MT. The simplest model allows only two possibilities: the addition of a subunit or complete collapse of the structure. Addition of a subunit may be modeled as follows (Sept et al., 1999):

$$x_{t+1} = x_t + a \tag{6.53}$$

where a is subunit length. Complete collapse of the microtubule is simply modeled using:

$$x_{t+1} = 0 \tag{6.54}$$

Suppose the probability of the addition of a tubulin subunit is p:

$$x_{t+1} = r(x_t + a) \tag{6.55}$$

where:

$$r = \left\{ \begin{array}{ll} 0 & \text{if } s > p \\ 1 & \text{if } s \le p \end{array} \right\} \tag{6.56}$$

and s is a random number between 0 and 1. Even a crude model such as that just described may successfully capture the essential dynamics of MT growth when compared by Hurst analysis or a recursive map technique (Sept, 1997). However, it lacks predictive power given that individual collapse events are random.

When an aggregation of MTs is studied, some collective properties develop. The most obvious is the phase transition behavior of the system in that certain conditions lead to overall disassembly and largely a pool of tubulin dimers, while other conditions promote self-assembly of tubulin in MTs. Phases of MT polymerization are determined largely by temperature and concentration of tubulin. For a given temperature, some critical concentration of tubulin is required to prevent MTs from disassembling. Fygenson, Braun, and Libchaber (1994) discussed this phase transition. Sept et al. (1997) further classified the nondisassembly phase into one where MTs assemble but are not nucleated and one where MTs are nucleated and assembled.

For a collection of MTs in the assembly phase, dynamic instability is the term given to the observation that MTs are growing in the immediate vicinity of other MTs that are shortening. Dynamic instability has been observed *in vivo* and *in vitro* and highlights both the nonequilibrium nature of the problem that crucially depends on energy supply via GTP hydrolysis (Hyman et al., 1992) that affects the structures of microtubules and leads to catastrophes (Hyman et al., 1995), thus creating the stochastic nature of individual MT growth. Despite these observations, ensembles of MTs show collective oscillations given suitable conditions. Specifically, when the concentration of assembled tubulin is measured, it undergoes smooth oscillations that are damped out as the energy source of GTP is depleted. Sept et al. (1997) modeled the assembly dynamics from a chemical reaction kinetics standpoint and found good agreement with the experimental data.

The principal elements in the model of Sept et al. (1997) can be summarized by the equations below. For simplicity, the MT is considered a linear polymer rather than an object of 13 protofilaments. We shall denote a MT of length n subunits by MT_n. In solution, free tubulin subunits may be bound to either GTP or GDP at their exchangeable nucleotide sites and shall be denoted T_{GTP} and T_{GDP}, respectively. Only tubulin bound to GTP is able to polymerize. The simple reaction set consists of addition, nucleation, and catastrophic collapse:

$$MT_n + T_{GTP} \underset{k_a}{\overset{k_a}{\rightleftharpoons}} MT_{n+1} \quad \text{(addition)} \tag{6.57}$$

$$nT_{GTP} \xrightarrow{k_n} MT_n \quad \text{(nucleation)} \tag{6.58}$$

and

$$MT_n \xrightarrow{k_c} nT_{GDP} \quad \text{(collapse)} \tag{6.59}$$

We again assume that all collapses are complete. To keep the model of MT assembly simple, one normally selects a specific number of dimers (n') required for nucleation. The exact value of this choice does not seem to exert a large impact on the dynamics as long as it is relatively small. The first of the preceding equations is reversible and a rate constant for the reverse reaction is considered in general and not required for this discussion. These equations can also be supplemented by the reactivation of tubulin, to make it assembly-competent, which occurs when the concentration of GTP is high:

$$T_{GDP} + GTP \xrightarrow{k_r} T_{GTP} + GDP \tag{6.60}$$

The free energy change associated with this reaction is less than the free energy change of GTP hydrolysis in solution and the difference is attributed to a structural change in the tubulin dimer. This conformational change presumably makes assembly possible. The rate constants of these reactions k_j should also be in agreement with the probabilities assigned to corresponding reactions in the individual MT model. Furthermore, temperature dependence can be built into the rate constants using empirical data on the free energy of reactants and products. When at least one autocatalytic reaction is added to the system to provide a nonlinear element, the dynamics change significantly. Consider an induced catastrophe event:

$$MT_n + T_{GDP} \xrightarrow{k_i} (n + 1)T_{GDP} \tag{6.61}$$

incorporated into the model. See Figure 6.19. This reaction has been able to reproduce oscillations observed *in vitro* caused by temperature jumps or injection of GTP into the system. It also identifies domains where tubulin is incorporated into MTs steadily and regimes where disassembly is favored. The model can reproduce the spatial pattern of MT assembly observed within cells simply by including a diffusive term in the equations of the assembly dynamics. This allows the model to be compared to the experiments performed independently by Tabony and Job (1990) and Mandelkow and Mandelkow (1992).

Agreement has been found between this model and those empirical results (Sept, 1997). Generally, the statistical dynamics of MT assembly are probably well understood except for the physical mechanism triggering catastrophes and the origins of rescue events. Again, the assumption is usually made that the energy difference is stored in the MT lattice and may be released upon disassembly or catastrophe. More recently, Hyman et al. reported on the observation of a structural change accompanying GTP hydrolysis. They discovered a length change in monomer spacing from 4.05 nm to 4.20 nm. Thus, energy may be stored locally as lattice deformation (Hyman et al., 1995). This 4% change in tubulin length results in a new moiré pattern when MTs are imaged by electron cryomicroscopy and different positions of equivalent peaks between the x-ray crystallographic diffraction patterns of GDP-MTs and GMPCPP-MTs.

Tran et al. (1997) argued that three conformational states exist with a metastable intermediary state between growing and shrinking conformations. They seem to believe that the states may be tubulin with GTP bound, with GDP. P_i bound, and finally with GDP bound at the exchangeable site. This is similar to the hypothesis of Semënov (1996) in his review of MT research. It therefore appears that in addition

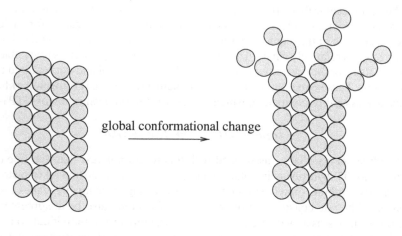

global conformational change

⟶

assembly conformation   disassembly conformation
                                                            "Ram's horns"

FIGURE 6.19
Global conformational change between assembly and disassembly. The once-straight protofilaments become curved after individual tubulin subunits undergo structural changes.

to the multitude of α and β isotypes and numerous post-translational modifications, tubulin may also exist in other conformations.

Gittes, Mickey, and Nettleton (1993) measured MT flexural rigidity by thermal fluctuations in MT shape and found that it increased when MTs were treated with MAP τ and MTs were most stabilized by preventing GTP hydrolysis. Thus, in addition to the fact that hydrolysis of GTP can destabilize MT with respect to catastrophic disassembly, it also makes the polymer more flexible (Felgner, Frank, and Schliwa, 1996). The application of taxol that reduces MT assembly dynamics increases the mechanical rigidity of MTs (Mickey and Howard, 1995).

6.10 Assembly of actin filaments

The pioneering work of Oosawa and Asakura (1975) established that spontaneous polymerization of actin monomers requires an unfavorable nucleation step followed by rapid elongation. Elongation is more accessible experimentally than nucleation, so it is much better understood, with a complete set of rate constants for association and dissociation of ATP–actin and ADP–actin subunits at both ends of the filaments (Pollard, 1986).

Nucleation has been studied by observing the complete time course of spontaneous polymerization as a function of actin monomer concentration and then finding a set of reactions and rate constants that fit these kinetic data (Tobacman and Korn, 1983). These studies concluded that actin dimers are less stable than trimers, which form nuclei for elongation.

In addition to nucleation and elongation, Oosawa et al. established that actin filaments can break and anneal end to end. Inclusion of a fragmentation reaction improved the fit of nucleation–elongation mechanisms to the observed time course of polymerization under some conditions (Wegner and Savko, 1982; Cooper et al., 1983; Buzan and Frieden, 1986). Kinetic evidence for annealing is both for (Kinosian et al., 1993; Rickard and Sheterline, 1988) and against (Carlier et al., 1984) its role in length redistribution, but the most direct evidence (electron micrographs) supports rapid annealing (Murphy et al., 1988).

The basis of the standard nucleation–elongation model is one or more unfavorable nucleation steps followed by more favorable elongation (Oosawa and Asakura, 1975). Earlier models for actin polymerization showed that the critical size for the nucleus is somewhere between a dimer and a trimer, and that the number of explicit nucleation steps does not affect the results of the model. Choosing a critical nucleus larger than three or four monomers does not affect the results of the model (Tobacman and Korn, 1983; Cooper et al., 1983; Frieden, 1983; Frieden and Goddette, 1983). With these considerations in mind, a simple five-step model was proposed by Sept et al. (1999):

$$A + A \underset{k_{-1}}{\overset{k_{+1}}{\rightleftharpoons}} A_2 \quad k_{+1} = 10 \ \mu M^{-1} s^{-1} \quad k_{-1} = 10^6 \ s^{-1} \tag{6.62}$$

$$A + A_2 \underset{k_{-2}}{\overset{k_{+2}}{\rightleftharpoons}} A_3 \quad k_{+2} = 10 \ \mu M^{-1} s^{-1} \quad k_{-2} = 10^3 \ s^{-1} \tag{6.63}$$

$$A + A_3 \underset{k_{-3}}{\overset{k_{+3}}{\rightleftharpoons}} A_4 \quad k_{+3} = 10 \ \mu M^{-1} s^{-1} \quad k_{-3} = 10 \ s^{-1} \tag{6.64}$$

$$A + A_4 \underset{k_{-4}}{\overset{k_{+4}}{\rightleftharpoons}} N \quad k_{+4} = 10 \ \mu M^{-1} s^{-1} \quad k_{-4} = 0 \ s^{-1} \tag{6.65}$$

$$A + N \underset{k_{-}}{\overset{k_{+}}{\rightleftharpoons}} N \quad k_{+} = 10 \ \mu M^{-1} s^{-1} \quad k_{-} = 1 \ s^{-1} \tag{6.66}$$

where A represents the concentrations of actin monomers, A_i is a filament with i actin monomers, and N represents the concentrations of all longer filaments. The rate constants for the last reaction were experimentally measured (Pollard, 1986) and led to the correct critical concentration ($C_c > 0.1 \mu M$). The other rate constants are only approximations from kinetic simulations, chosen to reproduce the time course of polymerization over a limited range of actin monomer concentrations.

Filaments longer than four subunits are assumed to be stable and the back-reaction rate k_{-4} is set to zero. The coupled first-order differential equations that arise from the set of reactions above are found to be "stiff" due to the large differences in forward and back reaction rates.

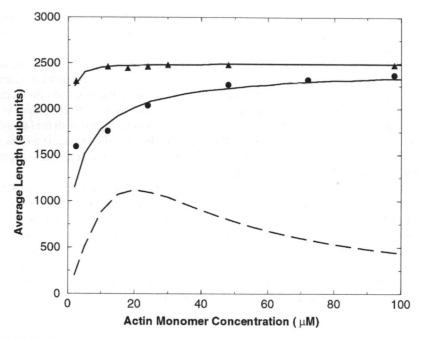

FIGURE 6.20
Dependence of the average lengths (L_n) of actin filaments on purity and monomer concentration during polymerization. The actin monomer concentration during polymerization varied as indicated. The filaments formed during each 30 ms interval were allowed to elongate over the succeeding course of the reaction. (●) Singly gel filtered actin. (▲) Doubly gel filtered actin. Theoretical lengths calculated by kinetic simulation (Sept et al., 1999) using the nucleation-elongation model and the rate constants in the text without annealing/fragmenting (dashed line) and with both annealing and fragmenting (solid line).

This set of equations produces correct polymerization curves, but the average length as a function of actin concentration is incorrect. The mean lengths of the observed filaments are almost independent of the initial concentrations of actin monomers, while this simple model predicts a mean length with quite a different behavior, especially at high concentrations of actin (see Figure 6.20). To solve this problem, additional processes are added to the model developed by Sept et al. (1999).

The average length is simply given by the total amount of polymer divided by the total number of filaments formed. Since the reaction above produces the correct time course and the extent of polymerization but not the correct average length, the simple model produces the incorrect number of filaments. The average length is too low, so the actual mechanism must produce fewer filaments. The number of filaments in the system is represented as the number of filaments formed by the addition of a monomer onto a polymer A_4. Since no back-reaction rate is involved, the equation for the change in N, the filament number concentration, is simply:

$$\dot{N} = k_+ \, AA_4 \tag{6.67}$$

Addition of monomers to existing filaments increases the concentration of polymerized actin, but does not increase N. To include filament annealing, we must also consider the break-up of a filament or fragmentation. To include filament annealing and fragmenting in the reaction scheme, the following reaction is added:

$$N + N \underset{k_f}{\overset{k_a}{\rightleftharpoons}} N \tag{6.68}$$

where k_a represents the annealing rate and k_f the fragmentation rate of a filament. Since two filaments are joined to form a new filament, the annealing rate has a quadratic dependence and the new equation for the change in N is:

$$\dot{N} = k_+ AA_4 - k_a N^2 + k_f N \tag{6.69}$$

The addition of a monomer to the fast growing barbed end of a filament is a diffusion limited process, so it is reasonable to assume that annealing of two filaments is also limited by diffusion, since similar bonds are formed and filaments diffuse more slowly than monomers. An initial rate constant for annealing was found as $k_a = 10 \mu M^{-1} s^{-1}$ for very short filaments, but as annealing progressed and the filaments became longer, the rate fell off rapidly with time (Murphy et al., 1988). Kinosian et al. (1993) found the rate of annealing to have the value of $2.2 \, \mu M^{-1} s^{-1}$.

If k_a is diffusion-limited, the diffusion we must consider is the relative diffusion of the two filaments. In fact, actin filaments above 40-nM filament concentration exhibit reptation motion (Doi, 1975; Janmey et al., 1994; Kas et al., 1996). The transverse diffusion is controlled by the mass of the polymer and the density of the polymer network, but the diffusion constant along the tube is inversely proportional to the length of the filament of the form (Kas et al., 1996):

$$D_{//} = \frac{k_B T}{\zeta L} \tag{6.70}$$

where ζ is the friction coefficient. The annealing rate constant k_a is therefore chosen to be proportional to $D_{//}$, namely:

$$k_a = k_a'/L \tag{6.71}$$

where all the constants are absorbed in the variable k_a'.

Erickson (1989) estimated the fragmentation rate to be in the neighborhood of $k_f \approx 10^{-8} s^{-1}$. This rate should be proportional to the length of the filament and the gel network formed may also affect the amount of fragmentation. Within the gel, each individual filament is constrained by its neighbors. Doi (1975) showed that the number of rods within a distance b of a given rod is $bL^2 N$ where L is the filament length and N is the filament concentration. Choosing an additional fragmentation rate proportional to this quantity results in:

$$k_f = k'_{f1}L + k'_{f2}L^2N \tag{6.72}$$

where all the constants are absorbed into the two factors k'_{f1} and k'_{f2}. Subsequently, the constant k'_{f2} is treated as a free parameter in a minimization scheme. Replacing k_a and k_f in Equation (6.68) yields:

$$\dot{N} = k_+AA_4 - k'_a\frac{N^2}{L} + k'_{f1}LN + k'_{f2}L^2N^2 \tag{6.73}$$

The amount of polymerized protein in the system is given by:

$$P = A_0 - A - 2A_2 - 3A_3 - 4A_4 \tag{6.74}$$

where A_0 represents the initial actin concentration. Using this expression, the average length of a filament is given simply by $L = P/N$ and we finally obtain:

$$\dot{N} = k_+AA_4 - k'_a\frac{N^3}{P} + k'_{f1}P + k'_{f2}P^2 \tag{6.75}$$

Note that as the actin concentration increases and more protein polymerizes, the rate of annealing decreases while the fragmentation rate increases. The values for the rate constants depend on the actin concentration and the filament density and length.

During the polymerization of actin monomers, the rates for annealing and fragmentation change with time, since they depend on filament length and density. The rate of annealing is in the range 2.2 to $10\mu M^{-1}s^{-1}$. Since $k_a = k'_a/L$ and $L > 30$ (subunits), $k'_a = 300\mu M^{-1}s^{-1}$ so that $k_a = 10\mu M^{-1}s^{-1}$ for $L = 30$. The factors k'_{f1} and k'_{f2} are treated as free parameters for fitting the two curves for mean polymer length versus actin concentration for singly and doubly filtered actin.

Singly filtered actin is estimated to contain about one part in 50,000 CapZ. Other proteins, such as severing proteins, may be present at higher concentrations in the singly gel filtered than doubly gel filtered actin (Casella et al., 1995). Severing proteins cut actin filaments. Since they act with equal probability along the length of a filament, the severing rate is proportional to L. Thus, low concentrations of severing proteins result in a larger value for k'_{f1} for singly filtered actin. It is also possible that the CapZ present in singly filtered actin increases the fragmentation rate, perhaps by changing the structure of a filament when it binds. With these points in mind, the values chosen by Sept et al. (1999) were $k'_{f2} = 1.8 \times 10^{-8}\mu M^{-1}s^{-1}$, $k'_{f1} = 2.0 \times 10^{-7}s^{-1}$ and $1.1 \times 10^{-8}s^{-1}$ for the singly and doubly gel filtered actin, respectively. In agreement with the experimental observations of Murphy et al. (1988), the annealing rate constant falls off rapidly with time. This model including annealing and fragmentation, agrees with the observed lengths over a wide range of starting actin monomer concentrations.

6.11 Muscle contraction: biophysical mechanisms and contractile proteins

Animal and human posture and motion are controlled by forces generated by muscles. A muscle is composed of a large number of fibers that can contract under direct stimulation by nerves. A muscle is connected to two bones across a joint and is attached by tendons at each end. The contraction of a muscle produces a pair of action–reaction forces between each bone and muscle. The work W performed by a contracting muscle is the product of the force of contraction, the cross-sectional area A, and the shortening distance $\Delta\ell$:

$$W = \sigma A \Delta\ell \tag{6.76}$$

where σ is the tensile stress produced by the muscle, i.e. the force of contraction per unit area. The power developed by the muscle is then:

$$P = \sigma A \frac{\Delta\ell}{\Delta t} = \sigma A v \tag{6.77}$$

and the speed of shortening $v = \dfrac{\Delta\ell}{\Delta t}$ appears to be a constant across species because, as the length of a muscle fiber increases, the time of contraction increases by the same factor. Thus, the power generated is regulated by the value of the cross-sectional area A.

Muscles in the body are not constantly generating the maximum force F_{max} they can exert. It is a matter of common experience that the smaller a weight W you hold, the faster you can lift it. Defining the contraction velocity v of a muscle as the shortening length of a muscle divided by the shortening time, one can fit the results of an isotonic measurement (tension fixed) to the relation:

$$v = b \frac{(F_{max} - W)}{(W + a)} \tag{6.78}$$

where a and b are constants, the former with the dimensions of force and the latter of velocity. This relation is known as Hill's law. According to Equation (6.77), $v = 0$ when load W is equal to the maximum force F_{max}. In Figure 6.21, we plot the contraction velocity v against load W. As the load is reduced, the muscle contracts more rapidly. In terms of the parameters in Equation (6.77), the maximum contraction velocity for zero load ($W = 0$) is:

$$V_{max} = \left(\frac{b}{a}\right) F_{max} \tag{6.79}$$

The plot in Figure 6.21 is called a force–velocity curve. When force velocity curves are measured for car engines, the force automotive engineers call the stalling force, F_{max}, will stop the engine.

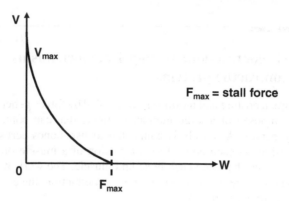

FIGURE 6.21
Force–velocity curve for a muscle.

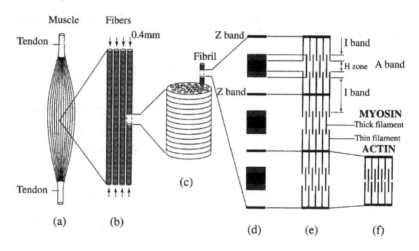

FIGURE 6.22
Composition of a muscle fiber.

In terms of their molecular structure, muscles are composed of fiber-like bundles called fascicles. The bundles consist of very long cells with diameters of approximately 0.4 mm and lengths of about 40 mm. The forces the fibers can exert are additive so that the total force generated is proportional to the number of fibers and cross-sectional area noted earlier. Fiber cells are of two types: thin filaments called actin fibers and thick filaments known as myosin fibers (see Figure 6.22). Fibers consist of fibrils that have light and dark bands (Part (b) of Figure 6.22). Myosin and actin filaments are designated (e) and (f). Each thick filament contains 200 myosin molecules. We discussed both actin and myosin proteins at some length in Chapter 4.

Fibers are arranged in alternating patterns of thick and thin filaments that may slide over each other. When we contract our muscles, motor or molecular cross-bridges pull the thin protein (see Figure 6.23) filaments over the thick ones and

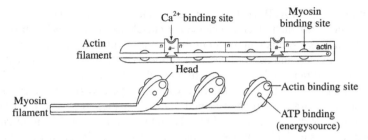

FIGURE 6.23
Principal components of the motor protein complex.

when the muscles are extended filaments do not overlap. Motor proteins are fixed to myosin fibers and move over the actin fibers, very like a collapsing telescope. The more activated the motor proteins are, the greater the muscle force generated. Force–velocity curves were measured for individual motor proteins and are similar to Hill's law (Equation 6.77), so the law has a molecular basis. The physiology of skeletal muscle contraction can be summarized as a sequence of the following molecular level events:

a) Nerve impulse reaches neuromuscular junction (acetycholine).
b) It causes a depolarization wave due to a large influx of positive ions.
c) Action potential spreads along plasma membrane, down T tubules.
d) Calcium is released from sarcoplasmic reticulum.
e) Calcium binds to troponin.
f) Conformational change in tropomyosin occurs and moves off the "hot spot" on actin.
g) The actin–myosin complex is together after splitting of ATP on the myosin head.
h) The high energy myosin head swivels, pulling actin inward in a power stroke.
i) ADP is released from the myosin head.
j) Myosin and actin remain complexed.
k) ATP binds to myosin and releases actin.

These events repeat until calcium returns to sarcoplasmic reticulum. When rigor mortis occurs, rigidity is due to the lack of ATP after death.

Rigor mortis is a state in which neither ATP nor Ca^{2+} is present, and the myosin heads are bound to actin. Relaxation state occurs when ATP is present, but Ca^{2+} is not, so myosin heads are not bound to actin. Contraction state is when both ATP and Ca^{2+} are present, and myosin heads are bound to actin. The most likely result is that when the muscle changes from rigor to relaxation, the distances between myosin heads will increase and as the muscle changes from rigor to contraction, the distances between the heads increase.

In summary, the so-called sliding filament theory maintains that muscle contraction is due to the sliding of myofilaments with no associated change in the lengths of myofilaments. A reduced width of the sarcomere which is the basic functional unit is the end result of contraction. This motion requires an energy input in the form of

ATP split by myosin ATPase. Thus during muscle contraction, the myosin and actin filaments slide past each other and cause the muscle to shorten. When broken down, myosin filaments are made up of many heads that bind to the actin and then rotate, causing the myosin and actin filaments to slide past each other. Each myosin head is made of three main helical protein strands: the regulatory light chain (RLC), the essential light chain (ELC), and the heavy chain. The myosin heads are situated in pairs, known as dimer pairs, and joined together at their bases. Most biophysical studies of muscle dynamics focused mainly on the actions of the two filaments and of the myosin heads individually, not as a pair. The purpose of the pairing of the myosin heads is still unknown.

References

Alberts, B. et al., *Molecular Biology of the Cell*, Garland Publishing, London, 1994.

Amos, L.A. and Amos, W.B. *Molecules of the Cytoskeleton*, Macmillan, London, 1991.

Astumian, R.D. and Bier, M., *Phys. Rev. Lett.*, 72, 1766, 1994.

Baker, J. et al., *Biophys. J.*, 77, 2657, 1999.

Brown, J.A. and Tuszynski, J.A., *Phys. Rev. E*, 56, 5834, 1997.

Buttiker, M., *Z. Phys. B*, 68, 161, 1987.

Buzan, J.M. and Frieden, C., *Proc. Natl. Acad. Sci. U.S.A.*, 93, 91, 1996.

Carlier, M.F. et al., *J. Biol. Chem.*, 259, 9987, 1984.

Carlier, M.F. et al., *Proc. Natl. Acad. Sci. U.S.A.*, 84, 5257, 1987.

Casella, J.F. Barron-Casella, E.A., and Torres, M.A., *Cell Motil. Cytoskel.*, 30, 164, 1995.

Cassimeris, L., *Cell. Motil. Cytoskel.*, 26, 275, 1993.

Cooper, J.A. et al., *Biochemistry*, 22, 2193, 1983.

Doering, C.R., Horsthemke, W., and Riordan, J., *Phys. Rev. Lett.*, 72, 2984, 1994.

Doi, M., *J. Physiol. Paris*, 36, 607, 1975.

Erickson, H.P., *J. Molec. Biol.*, 206, 465, 1989.

Felgner, H., Frank, R., and Schliwa, M., *J. Cell. Sci.*, 109, 509, 1996.

Feynman, R.P., Leighton, R.B., and Sands, M., *Feynman Lectures on Physics*, Vol. I., Addison-Wesley, Reading, U.K., 1963, Chap. 46.

Flyvbjerg, H., Holy, T.E., and Leibler, S., *Phys. Rev. E*, 54, 5538, 1996.

Frey, E., Kroy, K., and Wilhelm, J., *Adv. Struct. Biol.*, 5, 135, 1998.

Frieden, C., *Proc. Natl. Acad. Sci. U.S.A.*, 80, 6513, 1983.

Frieden, C. and Goddette, D., *Biochemistry*, 22, 5836, 1983.

Fygenson, D.K., Braun, E., and Libchaber, A. *Phys. Rev. D*, 50, 1579, 1994.

Gittes, F., Mickey, E., and Nettleton, J., *J. Cell Biol.*, 120, 923, 1993.

Hays, T.S. and Salmon, E.D., *J. Cell Biol.*, 110, 391, 1990.

Hill, T.L., *Free Energy Transduction and Biochemical Cycle Kinetics*, Springer, New York, 1989.

Hinner, B. et al., *Phys. Rev. Lett.*, 81, 2614, 1998.

Horio, T. and Hotani, H., *Nature*, 321, 605, 1986.

Houchmandzadeh, B. and Vallade, M., *Phys. Rev. E*, 6320, 53, 1996.

Howard, J., *Mechanics of Motor Proteins and the Cytoskeleton*, Sinauer Associates, Sunderland, MA, 2001.

Hyman, A.A. et al., *J. Cell. Biol.*, 128, 117, 1995.

Hyman, A.A. et al., *Molec. Biol. Cell*, 3, 1155, 1992.

Ishijama, M. and Yanagida, T., *Science*, 283, 1667, 2001; *Phys. Today*, Oct. 2001, p. 46.

Ishijama, M. and Yanagida, T., *TIBS*, 26, 438, 2001.

Janmey, P.A. et al., *J. Biol. Chem.*, 269, 32503, 1994.

Jüliche, F., Adjari, A., and Prost, J., *Rev. Mod. Phys.*, 69, 1269, 1997.

Käs, J. et al., *Biophys. J.*, 70, 609, 1996.

Kinosian, H.J. et al., *Biochemistry*, 32, 12353, 1993.

Kitamura, K. et al., *Nature*, 397, 129, 1999.

Kondepugi, D. and Prigogine, I., *Modern Thermodynamics*, John Wiley & Sons, Chichester, U.K., 1998.

Kurzynski, M. and Chelminiak, P., *J. Stat. Phys.*, 110, 137, 2002.

Landauer, R., *J. Stat. Phys.*, 53, 233, 1988.

Leibler, S. and Huse, D.A., *J. Cell Biol.*, 121, 1357, 1993.

Luby-Phelps, K., *Curr. Opin. Cell Biol.*, 6, 3, 1994.

Mandelkow, E.M. and Mandelkow, E., *Cell Motil. Cytoskel.*, 22, 235, 1992.

Mandelkow, E.M., Mandelkow, E., and Milligan, R., *J. Cell Biol.*, 114, 977, 1991.

Marx, A. and Mandelkow, E., *Eur. Biophys. J.*, 22, 405, 1994.

Mehta, A.D. et al., *Science*, 283, 1689, 1999.

Mickey, B. and Howard, J., *J. Cell Biol.*, 130, 909, 1995.

Mitchison, T. and Kirschner, M., *Nature*, 312, 237, 1984.

Murphy, D.B. et al., *J. Cell Biol.*, 106, 1947, 1988.

Oosawa, F. and Asakura, S., *Thermodynamics of the Polymerization of Protein*, Academic Press, London, 1975.

Pollard, T.D., *J. Cell Biol.*, 103, 2747, 1986.

Rickard, J.E. and Sheterline, P., *J. Mol. Biol.*, 201, 675, 1988.

Risken, H., *The Fokker–Planck Equation*, Springer, Berlin, 1989.

Ruppel, K.M., Lorenz, M., and Spudich, J.A., *Curr. Opin. Struct. Biol.*, 5, 181, 1995.

Sackmann, B. and Naher, E., *Single–Channel Recording*, 2nd ed., Plenum, New York, 1995.

Schafer, D., Jennings, P., and Cooper, J., *J. Cell Biol.*, 135, 169, 1996.

Semënov, M.V., *J. Theor. Biol.*, 179, 91, 1996.

Sept, D., *Phys. Rev. E*, 60, 838, 1999.

Sept, D., Elcock, A.H., and McCammon, J.A., *J. Mol. Biol.*, 294, 1181, 1999.

Sept, D. et al., *J. Theor. Biol.*, 197, 77, 1999.

Sept, D., Ph.D. Thesis, University of Alberta, Edmonton, 1997.

Sept, D., Pollard, T.D., and McCammon, J., *Biophys. J.*, 77, 2911, 1999.

Spurck, T.P. and Pickett, H.J., *J. Cell Biol.*, 105, 1691, 1987.

Stossel, J., *J. Biol. Chem.*, 269, 32503, 1994.

Stryer, L., *Biochemistry*, 4th ed., W.H. Freeman & Co., New York, 1995.

Svoboda, K. and Block, S.M., *Cell*, 77, 773, 1994.

Tabony, J. and Job, D., *Nature*, 346, 448, 1990.

Tobacman, L.S. and Korn, E.D., *J. Biol. Chem.*, 258, 3207, 1983.

Tran, P.T., Walker, R.A., and Salmon, E.D., *J. Cell Biol.*, 138, 105, 1997.

Wegner, A. and P. Savko., *Biochemistry* 21:1909–1913, 1982.

Westerhoff, H.V. and van Dam, K., *Thermodynamics and Control of Biological Free Energy Transduction*, Elsevier, Amsterdam, 1987.

Woehlke G. et al., *Cell*, 90, 207, 1997.

7

Nerve Cells

7.1 Anatomy of a nerve cell

The nervous system is the information highway of the body. The human nervous system is composed of approximately 10^{11} neurons (nerve cells) forming a network in which individual nerve cells use biochemical reactions to perform specialized functions (Nicholls et al., 2001) such as:

- Collecting information from sensory cells
- Relaying signals from one neuron to the next
- Delivering signals to muscles and other cells with instructions for action

The network consists of the central and peripheral nervous systems. The central nervous system (CNS) is composed of the brain and the spinal cord. The brain is a collection of about 10 billion interconnected neurons. The peripheral nervous system (PNS) includes neurons and nerve endings. The neuron is the basic unit and messenger of the peripheral nervous system. Excitable cells can be stimulated to create tiny electric currents. They are categorized as muscle cells and nerve cells. The electric current in neurons is used to rapidly transmit signals through the body. In muscles, it initiates contractions.

The human body can interpret, receive, and respond to sensory stimuli by the transmission of electric signals through the nervous system along fibers of networks of individual neurons. Figure 7.1 illustrates the structure of a neuron. It consists of a cell body (soma), with a nucleus connected to dendrites that serve as the inputs for signals, an axon, and axon terminals.

When a nerve signal is sent by the nervous system, the dendrites receive it. The axon then transmits the nerve signal to the axon terminals that interact through synapses with dendrites and other tissues such as muscles. The two types of axons are myelinated and unmyelinated. Unlike unmyelinated axons, myelinated axons are surrounded by sheaths of fatty tissue called myelin. Breaks in the myelin called Ranviers nodes allow nerve signals to jump from node to node. This causes faster transmission. Later in this chapter, we discuss in detail the nature of signal propagation in axons and across synapses.

Although neurons come in various morphologies (see Figure 7.2), they share a common structural organization. The central part of each neuron, called the soma, has a diameter of 10 to 80 μm and contains a nucleus. The soma is the principal site

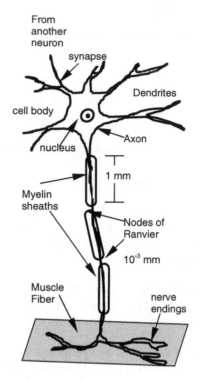

From
another
neuron
synapse

Dendrites

cell body

nucleus

Axon

1 mm

Myelin
sheaths

Nodes of
Ranvier

10^{-3} mm

Muscle
Fiber

nerve
endings

FIGURE 7.1
A neuron and its structural elements.

of protein synthesis. Emanating radially from the soma are protrusions called dendrites that collect signals from sensory cells and other neurons. The electrical impulses that carry signals travel along the plasma membrane. Each neuron is capable of receiving up to 10^5 inputs from its neighbors. A long extension of a neuron is called an axon and its length ranges from several millimeters up to a meter for motor neurons. The axon diameter varies from 0.1 to 20 μm. The giant axon of the squid has the largest diameter — up to 1 m.

The segments of the axon are encased in a myelin sheath that acts as a good insulator. The more primitive unmyelinated axons such as those found in invertebrates are cylindrical pipes whose walls consist of cell membranes. When electrical signals travel through the myelinated fibers of the axon, their propagation velocity is about 100 m/s. The average speeds range from 0.5 to 2 m/s in the brain and up to 100 m/s in motor neurons leading to muscle cells.

Each nerve can accommodate many signals simultaneously much as a number of telephone calls travel along a single cable. At the far end, each axon branches out into axon terminals that release neurotransmitter molecules into the so-called synaptic clefts or, in some cases, into gap junctions. The material released at a synapse must be recycled back into the neurotransmitter vesicles for release. This is achieved by

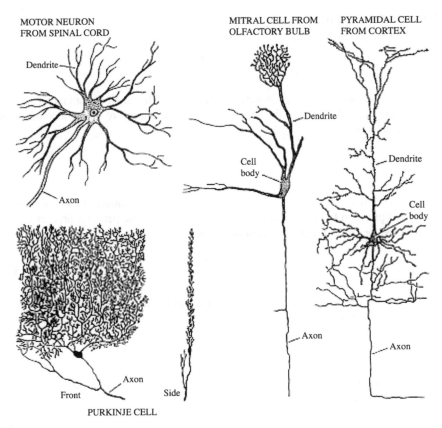

FIGURE 7.2
Specialization of neurons: shapes and sizes.

transport through axonal microtubules at speeds reaching 400 mm/day. Transport along the return direction is called retrograde transport and it occurs at roughly half the speed of anterograde transport for delivery. Figure 7.2 shows how neuron morphology and size are specialized to perform different functions.

7.2 Conducting properties of neurons

Nerve cells can be viewed as conducting cylinders surrounded by insulating sheets of biomembrane. From an electrostatic view, a sheet behaves like a capacitor with a capacitance of about 10^{-2} F/m^2. However, a biomembrane sheet, compared with rubber or glass, is not much of an insulator and it will leak current. To discuss this situation physically, we must assign conductance and resistance to a sheet of biomembrane. Suppose A is the area of a sheet of thickness D; the resistance R of the sheet is:

$$R = \rho \frac{D}{A},$$ (7.1)

where ρ is the resistivity of the membrane material. Thus the conductance $G = 1/R$ is proportional to the area of the membrane. We can therefore define a conductance per unit area by:

$$g_m = \frac{G}{A},$$ (7.2)

which has units of $1/(\Omega)m^2$, e.g., the conductance per square meter of a typical phospholipid membrane is about $91/(\Omega)m^2$. The current per unit area crossing a membrane, or current density J, is denoted by:

$$J = \frac{I}{A},$$ (7.3)

where I is the current flowing through the membrane in the direction perpendicular to it. From Ohm's law we find that:

$$I = \frac{\Delta V}{R} \text{ or } I = \Delta V \cdot G$$ (7.4)

where ΔV is the voltage difference across the membrane. From the above equations, the current density is:

$$J = g_m \Delta V.$$ (7.5)

Typical voltage differences across membranes are about 10 to 100 mV, so we expect the current density for a conductance of $91/(\Omega)m^2$ to be 0.1 to 1.0 A/m^2 (see Figure 7.3).

Suppose the concentrations of a particular ion inside and outside the cell membrane are c(in) and c(out), respectively. The equilibrium Nernst potential difference (see Chapter 3), for the ion is:

$$V_c = V(\text{in}) - V(\text{out}) = \frac{k_B T}{e} \ln \frac{c(\text{in})}{c(\text{out})}$$ (7.6)

The effective potential difference across the membrane, therefore, becomes $\Delta V - \Delta V_c$ and the true current density is given by:

$$J = g_m (\Delta V - \Delta V_c)$$ (7.7)

We may now set up an equivalent circuit (see Figure 7.4) where the resistor plays the role of leakage of current through the membrane. We imagine that the sides of the capacitor represent the axoplasm and the other the exoplasm of a nerve cell. The exoplasm is the layer of cytoplasm inside the cell and nearest the cell wall. The interior

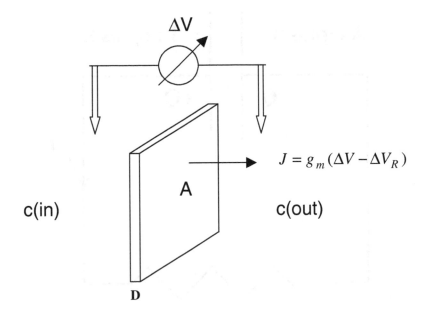

1) ΔV: potential difference
2) g_m: conductance / unit area
3) ΔV_c: Nernst potential
4a) c(in): Ion concentration inside
4b) c(out): Ion concentration outside

5) $\Delta V_c = \dfrac{k_B}{e} T ln \dfrac{c(out)}{c(in)}$

FIGURE 7.3
Electrical current density (J) across a membrane.

fluid is the axoplasm whose ionic composition differs from that of the extracellular fluid surrounding the cable. Both these fluids are electric conductors whose resistivity is typically of the order of $\rho \cong 60\,\Omega \cdot$ cm. If the section of membrane has an area A, the capacitance in the diagram is A_c where c is the capacitance per unit area of the membrane. The conductance G is therefore gA where g is the conductance per unit area. The resistance $R = 1/G = 1/gA$. For a typical biomembrane, c is about 10^{-2} F/m^2 and $g = (\Omega)^{-1}m^{-2}$.

The characteristic time for discharge of a capacitor in an RC circuit is:

$$\tau = RC = \frac{cA}{gA} = \frac{10^{-2}}{9} \cong 10^{-3}s. \tag{7.8}$$

Left alone, a membrane of a nerve cell would thus discharge in about a millisecond because of leakage. Thus, ion pumps must operate constantly to maintain membrane voltage to compensate for the leakage. The RC time calculated above, was shown by Nobel Prize winners Hodgkin and Huxley (1939), to be the characteristic switching time for nerve signals.

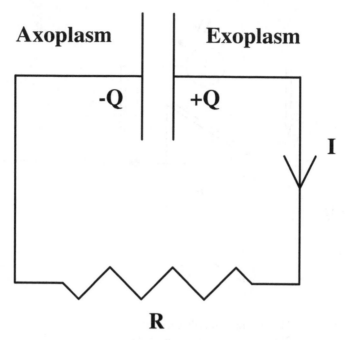

FIGURE 7.4
Equivalent circuit for current passing through a cell membrane.

The membrane wall forms a leaky capacitor with a typical value of its capacitance $C = 10^{-6}$ F/cm^2 and a finite conductivity $\sigma_M = \rho_M^{-1}$ where ρ_M is membrane resistivity; typically, $\rho_M = 2.10^3 \Omega \cdot$ cm^2. The ability of ions to pass through a membrane depends, of course, on the size, length, shape, and chemical properties of pores present in the membrane. All of these factors are included in a single parameter, average conductance of a pore or g_P, such that the membrane conductivity σ_M is given by:

$$\sigma_M = n g_P \qquad (7.9)$$

where n is the number of pores per unit area. For example, taking the radius of a squid axon as $r = 0.25 \cdot 10^{-3}$m and the typical conductivity of $\sigma_M = 0.5 \cdot 10^{-3} \Omega^{-1}$ cm^{-2}, we obtain the value of membrane conductance as σ_M times the surface area of 1 cm of the axon's length, i.e., $A = 2\pi r \ell = 1.6 \times 10^{-5}$ m^2, and hence:

$$g_M = \sigma_M A = 0.8 \cdot 10^{-4} \Omega^{-1} \qquad (7.10)$$

Conversely, the *resistance* of 1 cm of a squid axon is found to be:

$$R_M = g_M^{-1} = 12.5 \, k\Omega \qquad (7.11)$$

With a potential difference of $V = 80$ mV, we find the current flow through 1 cm of the length of an axon to be:

$$I = g_M V = 0.64 \cdot 10^{-5} A \qquad (7.12)$$

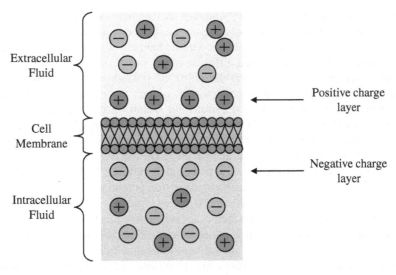

FIGURE 7.5
Formation of positive and negative charge layers on the outside and inside surfaces, respectively, of a membrane during its resting state.

which translates into a flow of about $0.4 \cdot 10^{11}$ elementary charges during the 1-ms duration of an *action potential* that travels at speeds on the order of 10 m/s. In the next section, we discuss the main features of the action potential propagation model proposed by Hodgkin and Huxley (1952) in their pioneering work.

7.3 Action potential generation: Hodgkin–Huxley equations

While extracellular fluid is rich in sodium (Na^+), and chloride (Cl^-) ions, intracellular fluid is rich in potassium (K^+) ions. The fluid inside cells also contains negatively charged proteins. Since a living cell has a selectively permeable membrane, a small build-up of negative charges occurs just inside the membrane and an equal number of positive charges occurs on the outside. See Figure 7.5. This gives rise to a *resting potential*, meaning an electric charge separation exists across the plasma membrane.

The size of the resting potential varies, but in excitable cells runs typically around −70 mV. At a molecular level, the resting potential arises from two key activities. The sodium–potassium ATPase pump pushes two K^+ ions into the cell for every three Na^+ ions it pumps out of the cell. The activity results in a net loss of positive charges within the cell. The sodium–potassium ATPase produces a concentration of Na^+ outside the cell that is some 10 times greater than that inside the cell and a concentration of K^+ inside the cell some 20 times greater than that outside the cell.

The concentrations of Cl^- and Ca^{2+} ions are also maintained at greater levels outside the cell except that some intracellular membrane-bound compartments may also have high concentrations of Ca^{2+}. Some potassium channels in the plasma membrane are "leaky" and allow slow facilitated diffusion of K^+ out of the cell.

The influx of sodium ions and the efflux of potassium ions occurs by biased diffusion. However, certain external stimuli reduce the charge across the plasma membrane. Mechanical stimuli (e.g., stretching, sound waves) activate mechanically-gated sodium channels. Certain neurotransmitters (e.g., acetylcholine) open ligand-gated sodium channels. In each case, the facilitated diffusion of sodium into the cell reduces the resting potential at that place on the cell, creating an excitatory postsynaptic potential (EPSP). If the potential is reduced to the threshold voltage (about -50 mV in mammalian neurons), an action potential is generated in the cell.

In neurons, this ranges from -40 to -90 mV — typically -70 mV — and the minus side indicates that the inside of the membrane is negative relative to the outside. When a neuron is in its resting state, it is not conducting electrical signals. The change in the resting potential is critical in the initiation and conduction of a signal. "Gates" in the membrane open when a sufficiently strong stimulus is applied to a given point on the neuron, and sodium ions rush into the cell. This rush of Na^+ ions is illustrated in Figure 7.6.

The Na^+ ions are driven into the cell by attraction to the negative ions on the inside of the cell and by the relatively high concentration of positive sodium ions outside the cell. This large influx of Na^+ ions first neutralizes the negative ions on the interior of the membrane and then causes it to become positively charged. For a very short time, as a result, the membrane potential in this localized region goes from about -70 mV (resting potential) to about $+30$ mV.

If depolarization at a location on the cell membrane reaches the threshold voltage, this leads to the opening of as many as hundreds of voltage-gated sodium channels in that portion of the plasma membrane. During the millisecond that the channels remain open, some 7000 Na^+ ions rush into the cell. The sudden complete depolarization of the membrane opens up more voltage-gated sodium channels in adjacent portions of the membrane. Consequently, a wave of depolarization sweeps along the

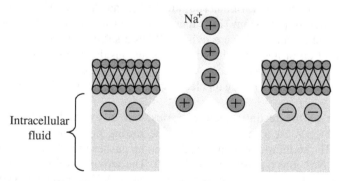

FIGURE 7.6
As a stimulus is applied, sodium ions rush into the cell.

cell membrane, giving rise to the *action potential* lasting a few milliseconds. This electrical signal propagates down the axon at speeds between 0.5 and 130 m/s, to the next neuron or to a muscle cell.

A second stimulus applied to a neuron less than 0.001 s after the first will not trigger another impulse. The membrane is depolarized, and the neuron is in its refractory period. Not until the −70 mV polarity is reestablished will the neuron be ready to fire again. Repolarization is first established by the facilitated diffusion of potassium ions out of the cell. Only when the neuron is finally rested are the sodium ions that came in at each impulse actively transported back out of the cell. In some human neurons, the refractory period lasts only 0.001 to 0.002 seconds. This means that the neuron can transmit 500 to 1000 impulses per second. It is also important that the action potential is all or none. This means that an action potential will not occur unless the depolarization threshold is met. As a nerve action potential propagates along a neuron it goes through several phases:

1. Resting state at which the membrane potential is at rest
2. Depolarization which occurs after a drastic reversal in membrane potential
3. Repolarization which occurs when the membrane potential is returning to the resting state
4. Undershoot, a situation arising when the potassium gate stays open too long

The strength of the action potential is an intrinsic property of the cell. So long as they can reach the threshold of the cell, strong stimuli produce no stronger action potentials than weak ones. However, the strength of the stimulus is given by the frequency of the action potentials that it generates.

The Hodgkin–Huxley model (Hodgkin and Huxley, 1952) describes mathematically the ionic mechanisms underlying the initiation and propagation of action potentials in the squid giant axon. Their model was formulated some 50 years ago and ranks among the most significant conceptual breakthroughs in the life sciences. It provided the foundation for modern computational neuroscience. The theory was the culmination of collaboration of Hodgkin and Huxley from 1938 to 1952. Their achievement was facilitated by four key developments.

First, Cole and Curtis (1939) demonstrated that action potential is associated with a large increase in membrane conductance. Hodgkin and Huxley then made the first intracellular recording of an action potential that demonstrated directly that it exceeded zero mV, rejecting Bernstein's hypothesis (Bernstein, 1902) that the underlying increase in membrane permeability is nonselective. Hodgkin and Katz (1949) explained the overshooting action potential by showing that it results from an increase in sodium permeability, validating the neglected work of Overton (1902).

Finally, Hodgkin, Huxley, and Katz, following Cole and Marmont, developed a voltage clamp circuit to enable measurements of ionic currents from squid axon. Hodgkin and Huxley then showed that step depolarizations of the squid axon trigger an inward current followed by an outward current. Using ionic substitution, they demonstrated that this net current could be separated into two distinct components, a rapid inward current carried by Na^+ ions, and a more slowly activating outward current carried by K^+ ions. From experiments using voltage clamp protocols, they concluded, in what was later named the "ionic hypothesis," that these two currents resulted from independent permeability mechanisms for Na^+ and K^+ with conductances changing as functions of time and membrane potential.

They modeled the observed smooth current changes in terms of pores or channels that were open or closed and by using a statistical approach that generated predictions for the probability that the channels would be open. They represented total ionic current as the sum of separate Na^+, K^+, and leak currents and set up separate equations for the gating variables m and h for the activation and inactivation of g_{Na}, respectively, or n for the activation of g_K.

The model Hodgkin and Huxley developed reproduced and explained a remarkable range of data from squid axon, including the shape and propagation of the action potential, its sharp threshold, refractory period, anode-break excitation, accommodation, and subthreshold oscillations. With minor parameter changes, the model describes many channel types, due to the generality and correctness of the approach.

As noted earlier, two ions are essential to a normal action potential: Na^+ and K^+. The depolarization phase is controlled by sodium. An external stimulus causes an influx of sodium in the nerve cell and it continues down the neural pathway until it reaches its destination. After the cell depolarizes, it must repolarize to its resting potential before it can depolarize again. The repolarization phase is controlled by potassium ions whose efflux causes the potential to return to its resting state. The two variable conductances g_K and g_{Na} used in the mathematical model represent the average effect of the binary gating of many potassium and sodium channels. The constant leakage conductance g_L represents the effects of other channels, primarily chloride, that always stay open.

The first step in calculating the total nerve action potential involves calculating the potential of the nerve cell membrane while the membrane is at rest:

$$V_m = \left(\frac{RT}{F}\right) \ln \frac{(P_{K^+}[K^+]_e + P_{Na^+}[Na^+]_e)}{(P_{K^+}[K^+]_i + P_{Na^+}[Na^+]_i)} \qquad (7.13)$$

where R is the gas constant, T is temperature, F is Faraday's constant, P_{K^+}, Na^+ are sodium and potassium permeabilities at rest, respectively, K_e^+/Na_e^+ are extracellular concentrations of sodium and potassium, and K_i^+/Na_i^+ are intracellular concentrations of sodium and potassium, respectively. The net current flowing into the cell through the channels has the effect of charging the membrane capacitance, giving the interior of the cell a membrane potential V_m relative to the exterior. The current I per unit area can be expressed in terms of the potential difference V as a sum of four terms, namely:

$$I = I_K + I_{Na} + I_L + C_m \frac{dV}{dt} \qquad (7.14)$$

where:

$$I_K = g_k(V - V_K) \qquad (7.15)$$

$$I_{Na} = g_{Na}(V - V_{Na}) \qquad (7.16)$$

$$I_L = g_L(V - V_L) \qquad (7.17)$$

which represent potassium current, sodium current, and leakage current, respectively. C_m is the capacity of the membrane surface per unit area. The variables denoted by V with various subscripts are the corresponding voltages.

Next, the conductance of sodium and potassium is modeled using the following equations:

$$g_{Na} = 120(m^3)h \qquad (7.18)$$

and:

$$g_K = 36(n^4) \qquad (7.19)$$

where m, n, and h represent the values of gates that provide access through the sodium and potassium channels. The m, n, and h gates regulate ion flow into and out of the cell and hence are associated with the action potential. The m and h gates control sodium flow, while the n gates control potassium flow. In the resting phase of the action potential, the m gate is closed and the h gate is open. Therefore, sodium is neither leaving nor entering the cell. The n gate is also closed, so potassium can neither leave nor enter the cell. During depolarization, the m gate opens to allow sodium to diffuse down its gradient, while the n gate is still closed. During repolarization, the h gate closes, preventing sodium from entering the cell. The n gate is open during this phase so potassium moves out. In the undershoot phase, the m gate closes, the h gate stays closed, and the n gate stays open. Finally, the h gate opens, the n gate closes, and resting state is once again achieved.

The three conductances of the cell membrane are: a passive (voltage-independent) leak conductance, an active (voltage-dependent) sodium conductance, and an active potassium conductance. The conductances are connected in series and described by the Nernst equation for a given ion that is in a so-called equilibrium potential. This means that the diffusion forces occurring because of different concentrations of ions outside and inside the membrane are balanced with the charge distribution built up by electrical drift. The probabilities of the gating particles h, m and n that describe the kinetics of the ionic channels are modeled by the following nonlinear, ordinary differential equations:

$$\frac{dm}{dt} = \alpha_m(V_m)(1-m) - \beta_m(V_m)(m) \qquad (7.20)$$

$$\frac{dn}{dt} = \alpha_n(V_n)(1-n) - \beta_n(V_n)(n) \qquad (7.21)$$

$$\frac{dh}{dt} = \alpha_h(V_h)(1-h) - \beta_h(V_h)(h) \qquad (7.22)$$

where n is called the activation state variable and has a simple exponential dependence governed by a single time constant, m is called the steady state activation variable for Na, i.e., the value reached by n when it is held at the potential V for a long period, and h is called the inactivation state variable, since it becomes smaller when m becomes larger. See Figure 7.7.

Note that m and h obey equations just like the one for n, but with different voltage dependences for their steady state values and time constants. Each particle follows

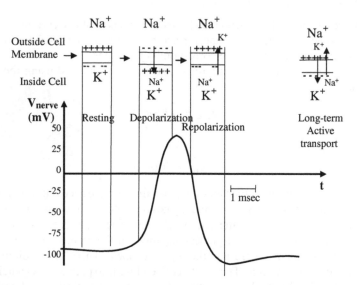

FIGURE 7.7
Profile of an action potential at a fixed location on a nerve as a function of time.

a two-state Markov model (channel closed or open), governed by the transition rates from the closed to the open state and from the open to the closed state. Thus the variables h, m, and n follow first-order kinetics, relaxing exponentially toward a steady state value with time constant. By inserting the steady-state solutions, these equations yield:

$$\alpha_m = \frac{(V_m + 25)/10}{e^{(V_m+25)/10} - 1} \tag{7.23}$$

$$\alpha_n = \frac{(V_m + 10)/100}{e^{(V_m+10)/10} - 1} \tag{7.24}$$

$$\alpha_h = 0.07e^{V_m/20} \tag{7.25}$$

$$\beta_m = 4e^{V_m/18} \tag{7.26}$$

$$\beta_n = \frac{1}{e^{(V_m+30)/10} + 1} \tag{7.27}$$

$$\beta_h = 0.125e^{V_m/80} \tag{7.28}$$

After obtaining conductance solutions, the membrane current can be calculated using the following equation:

$$I = g_{Na}(m^3)(h)(V_m - V_{Na}) + g_K(n^4)(V_m - V_K) \tag{7.29}$$

The key equation used by Hodgkin and Huxley to mathematically model nerve action potential is given below for voltage dependence on time. Integration of this equation gives the membrane potential at any given time.

$$\frac{dV}{dt} = -\frac{1}{C_m}[g_{Na}(V_m)(V_m - V_{Na}) + g_K(V_m)(V_m - V_K)] \tag{7.30}$$

where dV/dt is the change in membrane potential with respect to time and C_m is the membrane capacitance.

The difficult part of developing this model was parameterizing the time and voltage dependence of Na and K conductances. The solution was obtained by performing a series of voltage clamp experiments measuring both the total current and the current when the Na conductance was disabled. This enabled the neuroscientists to calculate the K current and, from that and the known voltages, calculate the values of the two conductances. By performing the experiments with different values of clamp voltage, Hodgkin and Huxley determined the time dependence and equilibrium values of the conductances at different voltages. The reader is referred to a recent book by Scott (2002) for a comprehensive exposition of nonlinear mathematics models in the area of neuroscience.

7.4 Structure and function of synapse

Coordination of cellular activities in animals is usually considered to involve (1) an endocrine system whose response is to hormones, i.e., chemicals secreted into the blood by endocrine glands and carried by blood to responding cells and (2) a nervous system whose response is to electrical impulses passing from the central nervous system to muscles and glands. Coordination by the nervous system is also chemical. Most neurons achieve their effects by releasing chemicals celled neurotransmitters sent to receiving cells such as postsynaptic neurons, muscle cells, and gland cells. The real distinction between nervous and endocrine coordination is that nervous coordination is faster and more localized.

The junction between the axon terminals of a neuron and the receiving cell is called a synapse. Synapses at muscle fibers are called neuromuscular junctions or myoneural junctions. The sequence of events leading to the activation of a synapse can be briefly summarized as follows:

1. Action potentials travel down the axon of the neuron to its end called the axon terminal.
2. Each axon terminal swells to form a synaptic knob.
3. The synaptic knob is filled with membrane-bound vesicles containing neurotransmitters.
4. Arrival of an action potential at the synaptic knob opens Ca^{2+} channels in the plasma membrane.
5. The influx of Ca^{2+} triggers exocytosis of some of the vesicles.
6. Neurotransmitter molecules are released into the synaptic cleft.
7. Neurotransmitter molecules bind to receptors on postsynaptic membranes.

The receptors are ligand-gated ion channels (Hille, 1982). Synaptic transmission is the propagation of nerve impulses from one nerve cell to another. The end of a presynaptic axon, where it is juxtaposed to the postsynaptic neuron, is enlarged and forms a structure known as the terminal button. An axon can make contact anywhere along the second neuron: on the dendrites (axodendritic synapse), cell body (axosomatic synapse), or axons (axoaxonal synapse). Each dendritic tree is connected to a thousand or more neighboring neurons. When one of those neurons fires, a positive or negative charge is received by one of the dendrites. The strengths of all the received charges are added together through spatial and temporal summation.

Spatial summation occurs when several weak signals are converted into a single large one, while temporal summation converts a rapid series of weak pulses from one source into one large signal. The aggregate input is then passed to the soma (cell body). However, the soma and the enclosed nucleus do not play significant roles in processing incoming and outgoing data. The part of the soma concerned with the signal is the axon hillock. If the aggregate input is greater than the axon hillock's threshold value, the neuron fires and an output signal is transmitted down the axon.

The strength of the output is constant, whether the input was just above the threshold or many times greater. The output strength is unaffected by the many divisions in the axon. It reaches each terminal button with the same intensity it had at the axon hillock. This uniformity is critical for the proper functioning of the brain where small errors can propagate and become magnified.

Each terminal button is connected to other neurons across a synapse. A synapse can be a gap between one neuron and another or between a nerve terminal and a muscle fiber across which electrical signals pass in only one direction. Conduction across a synapse therefore cannot be the same as that along the neuron, in which the spike potential travels in both directions from the point of stimulation. In the case of crayfish, synaptic conduction is purely electrical. In most other cases of synaptic conduction, the transmission is chemical and involves neurotransmitter molecules. The neurotransmitters constitute a diverse group of chemicals ranging from simple amines such as dopamine and amino acids such as γ-aminobutyric acid (GABA), to polypeptides such as the enkephalins.

The end of each presynaptic fiber contains a dense accumulation of vesicles, each about 50 nm in diameter. The nerve impulse, on arriving at the synapse, causes the vesicles to move to the presynaptic nerve membrane, fuse with it, and eject the neurotransmitters they contain into the gap. About 10^7 molecules are released; they cross the gap by reaching the end of the postsynaptic receptor in about 1 ms. The gap between neurons is about 20 nm wide and is perhaps four times larger between a neuron and a muscle fiber. Assuming that a given neurotransmitter has a net charge of several elementary charges, the current flow across the synapse is of the order of 10^{-8} A.

Neurotransmitter molecules are adsorbed once they cross the synapse and change the ion permeability of the membrane in the second fiber, starting a spike potential. The adsorbed molecules are then destroyed by an enzyme. In the case of ACh, the enzyme is acetylcholinesterase. If the enzyme is not present, the transmitter substance keeps on working, and the second fiber cannot receive a second impulse. Other important neurotransmitters are discussed below.

Acetylcholine (ACh) is widely used at synapses in the peripheral nervous system. It is released at the terminals of all motor neurons and activates skeletal muscle, preganglionic neurons of the autonomic nervous system (ANS), and postganglionic neurons of the parasympathetic branch of the ANS. It also mediates transmission at some synapses in the brain including those involved in acquisition of short term memory.

Amino acids also act in the nervous system. Glutamic acid (Glu) is used at excitatory synapses in the central nervous system. It is essential for long term potentiation (LTP), a form of memory. Like GABA, Glu acts on two types of CNS synapses: (1) fast ($\sim$ 1 msec) by opening ligand-gated Na^+ channels and (2) slow ($\sim$ 1 sec) by binding to receptors that turn on a "second messenger" cascade of biochemical changes that open channels allowing Na^+ into cells. Glycine (Gly) is used at inhibitory synapses in the CNS. GABA is used at other inhibitory synapses in the CNS.

Catecholamines are synthesized from tyrosine (Tyr). Noradrenaline (also called norepinephrine) is released by postganglionic neurons of the sympathetic branch of the ANS. Dopamine is used at certain synapses in the CNS.

Other monoamines such as serotonin, which is synthesized from tryptophan (Trp), and histamine are confined to synapses in the brain.

Peptides: Eight of the forty or more peptides are suspected to serve as neurotransmitters in the brain. They are vasopressin (ADH), oxytocin, gonadotropin-releasing hormone (GnRH), angiotensin II, cholecystokinin (CCK), substance P, and two enkephalins: Met-enkephalin (Tyr-Gly-Gly-Phe-Met) and Leu-enkephalin (Tyr-Gly-Gly-Phe-Leu). ADH, oxytocin, GnRH, angiotensin II, and CCK also act as hormones.

After its job is done, a neurotransmitter must be removed from the synaptic cleft to prepare the synapse for the arrival of the next action potential. Two methods are used, the first of which is reuptake. The neurotransmitter is taken back into the synaptic knob of the presynaptic neuron by active transport. All the neurotransmitters except acetylcholine use this method. Acetylcholine is removed from the synapse by enzymatic breakdown into inactive fragments using acetylcholinesterase.

A single neuron, especially one in the CNS, may have thousands of other neurons synapsing on it, some of which release activating (depolarizing) neurotransmitters; others release inhibitory (hyperpolarizing) neurotransmitters. The receiving cell can integrate these signals. The two types of synapses are excitatory and inhibitory.

Neurotransmitters at excitatory synapses depolarize postsynaptic membranes. For example, ACh binding to receptors on postsynaptic cells opens ligand-gated sodium channels. The channels allow an influx of Na^+ ions, reducing the membrane potential. The reduced membrane potential is called an excitatory postsynaptic potential or EPSP. If depolarization of the postsynaptic membrane reaches threshold, an action potential is generated in the postsynaptic cell.

Neurotransmitters at inhibitory synapses hyperpolarize postsynaptic membranes. For example, GABA binding to GABAA receptors on postsynaptic neurons opens ligand-gated chloride ($Cl-$) channels to GABAB receptors and activates an internal G protein and a "second messenger" that leads to the opening of nearby K^+ channels. This is a fast response that takes about 1 ms.

In both cases, the resulting facilitated diffusion of ions increases the membrane potential to as much as $-80\,\text{mV}$. The increased membrane potential is called inhibitory postsynaptic potential (IPSP) because it counteracts any excitatory signals that arrive at that neuron.

Once the molecules of neurotransmitters are released from a cell as the result of the firing of an action potential, they bind to specific receptors on the surfaces of postsynaptic cells. Every neurotransmitter has numerous subtypes of receptors. As well as on the surfaces of postsynaptic neurons, neurotransmitter receptors are found on presynaptic neurons. In general, presynaptic neuron receptors act to inhibit further release of neurotransmitter molecules.

The mechanisms by which neurotransmitters elicit responses in presynaptic and postsynaptic neurons are diverse. The physical and neurochemical characteristics of each synapse determine the strength and polarity of the new input signal. Changing the constitutions of neurotransmitter chemicals can increase or decrease the amount of stimulation that the firing axon imparts on a neighboring dendrite. Altering neurotransmitters can also change the type of stimulation from excitatory to inhibitory and vice versa. Many drugs such as alcohol and LSD dramatically affect the production or destruction of these critical chemicals.

A synapse may act in the simple manner described above or perform more complicated logical operations such as adding, subtracting, dividing, or multiplying. Addition is performed if a synapse transmits only subthreshold responses; several are necessary to produce a spike potential. Although ACh produces a spike potential at motor nerve endings, it inhibits heart muscle and in that case it subtracts. At other synapses, several arriving potentials may be necessary to produce a single response, and the synapse may divide. An impulse in one fiber may give rise to several impulses in a further fiber, and the original spike potential may be considered to multiply.

The EPSP created by a single excitatory synapse is insufficient to reach the threshold of the neuron. EPSPs created in quick succession, however, add together (summation). If they reach threshold, an action potential is generated. The EPSPs created by separate excitatory synapses $(A + B)$ can also be added together to reach threshold. Activation of inhibitory synapses (C) makes the resting potential of a neuron more negative. The resulting IPSP may also prevent what would otherwise have been effective EPSPs from triggering an action potential.

Only if, over a brief interval, the sum of depolarizing signals minus the sum of the hyperpolarizing signals exceeds the threshold of the axon hillock will an action potential be generated. Since the response at a synapse may show complex time dependence, processes analogous to other mathematical operations can also take place. Indeed, the processes at synapses are essentially nonlinear and highly complex, and display the logic circuitry of a modern computer.

Most neurotransmitter receptors belong to a class of proteins known as serpentine receptors. The class exhibits a characteristic transmembrane structure as it spans the cell membrane seven times. The link between neurotransmitters and intracellular signaling is carried out by association with G proteins (small GTP-binding and hydrolyzing proteins), protein kinases, or by the receptors in the forms of ligand-gated ion channels. One additional characteristic of neurotransmitter receptors is that they can become unresponsive upon prolonged exposure to their neurotransmitters.

Some neurotransmitters inhibit the transmission of nerve impulses by opening chloride and/or potassium channels in the plasma membrane. In each case, opening of the channels increases the membrane potential by letting negatively charged Cl^- ions in and positively charged K^+ ions out. This hyperpolarization is called inhibitory postsynaptic potential (IPSP). Although the threshold voltage of the cell is unchanged, it requires a stronger excitatory stimulus to reach threshold. A good example is GABA, which is found in the brain and inhibits nerve transmission by both mechanisms: (1) binding to GABAA receptors opens chloride channels in the neuron and (2) binding to GABAB receptors opens potassium channels.

The axons of many neurons are encased in fatty structures called myelin sheaths. They are greatly expanded plasma membranes of accessory cells called Schwann cells. Where the sheath of one Schwann cell meets the next, the axon is unprotected. The voltage-gated sodium channels of myelinated neurons are confined to these sites (nodes of Ranviers). The in-rush of sodium ions at one node creates just enough depolarization to reach the threshold of the next. In this way, the action potential jumps from one node to the next. This results in much faster propagation of the nerve impulse than is possible in nonmyelinated neurons. Multiple sclerosis is an autoimmune disorder that results in the gradual destruction of myelin sheaths.

7.5 Neural network models

As complicated as the biological neuron is, it may be simulated by a simple computer model. In such a model, the active inputs have weights that determine how much they contribute to neurons. Each neuron has a threshold value for firing. If the sum of all weights of all active inputs is greater than the threshold, the neuron is active.

The nodes in the network are vast simplifications of real neurons: they can only exist in one of two possible states (firing or not firing). Every node is connected to every other node with some strength. At any instant, a node can start or stop firing, depending on the inputs it receives from the other nodes.

A single neuron and its input weighting perform the logical expression (A and not B) when only two inputs are present. However, a biological neuron may have as many as 10,000 different inputs and may send its output (presence or absence of a short-duration spike) to many other neurons. Indeed, neurons are wired in three-dimensional patterns.

von Neumann computing machines are based on human information processing and memory abstraction; neural networks present a different paradigm for computing and are based on parallel architectures of animal brains (Gurney, 1997). Real brains, however, are orders of magnitude more complex than any artificial computer network. Neural networks are forms of multiprocessor computer systems with simple processing elements, high degrees of interconnection, simple scalar messages, and adaptive interactions of elements.

A neural network can be explicitly programmed to perform a task by creating the topology and then setting the weights of each link and threshold (Fausett, 1994).

However, this bypasses one of their unique strengths: their ability to program themselves. The most basic method of training a neural network is trial and error. If the network does not behave as it should, we can change the weighting of a random link by a random amount. If the accuracy of the network declines, we can undo the change and make another one.

Unfortunately, the number of possible weightings rises exponentially as one adds new neurons, making large general-purpose neural nets impossible to construct using trial and error methods. In the early 1980s, Rumelhart and Parker independently rediscovered an old calculus-based learning algorithm. The back-propagation algorithm compares the result obtained with the result expected and uses the information to systematically modify the weights throughout the neural network (Haykin, 1999). This training takes only a fraction of the time that trial and error takes and can train networks reliably with only a portion of the data, since it makes inferences. The resulting networks are often correctly configured to answer problems they have never been specifically trained to solve. In its simplest form, pattern recognition uses one analog neuron for each pattern to be recognized. All the neurons share the entire input space.

Training pattern recognition networks is simple. We first draw the desired pattern and select the neuron to learn the pattern. For each active pixel, we add one to the weight of the link between the pixel and the neuron in training. We subtract one from the weight of each link between an inactive pixel and the neuron in training. A more sophisticated method of pattern recognition involves several neural nets working in parallel, each looking for a particular feature of the pattern. The results of these feature detectors are then fed into another net that best matches the pattern. This method appears to be closer to the way humans recognize patterns.

One of the simplest algorithms implemented in neural network research is the so-called perceptron. The network adapts by changing the weight by an amount proportional to the difference between the desired output and the actual output. In terms of an equation, we write:

$$\Delta W_i = \eta(D - Y)I_i \tag{7.31}$$

where η is the learning rate, D is the desired output, and Y is the actual output.

The equation is called the perceptron learning rule or the delta rule. It was introduced in the early 1960s. We expose the net to the patterns shown in Table 7.1, then train the network on examples. Weights alter after each exposure to a complete set of patterns. Since $(D - Y) = 0$ for all patterns at the end of the procedure, the weights cease adapting and the network has finished learning. However, single perceptrons are limited in what they can learn, and more advanced algorithms have been designed.

Back-propagated (BP) delta rule networks (also known as multilayer perceptrons or MLPs) and radial basis function networks (RBFs) are extensions of the perceptron learning rule. Both can learn arbitrary mappings or classifications and their inputs and outputs can have real number values.

BP is a development from the simple delta rule in which hidden layers additional to the input and output layers are added. The network topology is constrained to be feed-forward, i.e., loop-free. Generally, connections are allowed from the input layer to the first hidden layer, from the first hidden layer to the second, etc., and from the last

TABLE 7.1
Pattern Recognition in Networks

I_0	I_1	Desired Output
0	0	0
0	1	1
1	0	1
1	1	1

hidden layer to the output layer. The hidden layer learns to recode the inputs. More than one hidden layer can be used. The hidden layer architecture is more powerful than single-layer networks.

The weight change rule is a development of the perceptron learning rule. Weights are changed by an amount proportional to the error at that unit multiplied by the output of the unit feeding into the weight. Running the network consists of the forward pass (outputs and the error at the output units are calculated) and the backward pass (output unit error is used to alter weights on the output units). The error at the hidden nodes is calculated by back-propagating the error at the output units through the weights, and the weights on the hidden nodes are altered using these values. A forward pass and a backward pass are performed for each data pair to be learned. This is repeated over and over until the error reaches a sufficiently low level.

RBFs are also feed-forward, but have only one hidden layer. Like BPs, RBFs can learn arbitrary mappings. The primary difference is in the hidden layer. RBF hidden layer units have receptive fields that have centers. That means they produce maximal output at particular input values. Output trails off as the input moves away from this point. Generally, the hidden unit function is Gaussian. RBFs are trained by (1) deciding how many hidden units are needed, (2) deciding on their centers and the sharpness of their Gaussians, and (3) training the output layers.

Generally, the centers and standard deviations of the Gaussians are decided first by examining the training data. The output layer weights are then trained using the delta rule. BP is the most widely applied neural network technique. RBFs have the advantage that one can add extra units with centers near parts of the input that are difficult to classify. BPs and RBFs can also be used for processing time-varying data. Simple perceptrons, BPs, and RBFs need teachers to tell them what the desired output should be, and hence they are supervised networks. An unsupervised network adapts purely in response to its inputs. Such networks can learn to pick out structures from their inputs. Although their learning process may be slow, trained networks run very fast. The Kohonen clustering algorithm clusters high dimensional input, but retains some topological ordering of the output. After training, an input causes some of the output units in some area to become active.

Note that neural networks cannot do anything that cannot be done using traditional computing techniques, but they can achieve some things that would otherwise be very difficult. In particular, they can form a model from their training data alone. This is particularly useful with sensory data that may have an algorithm that is not known a *priori* or has too many variables. It is easier to let the network learn from examples than make educated guesses about the algorithm at work. Neural networks have many applications outside neuroscience, for example, in investment analysis, signature analysis, process control, and monitoring the performance of industrial systems, among other applications.

7.5.1 Memory

One of the most important functions of the brain is the build-up and recall of memories. Due to their operational differences, we distinguish short and long term memories. Our memories function in what is called an associative or content-addressable fashion. A memory does not exist in a particular set of neurons in some isolated region of the brain. All memories are in a sense strings of memories. For example, we remember someone or something in a variety of ways, by its color, shape, sound, or smell. Thus memories are stored in association with one another.

Different sensory units lie in separate parts of the brain, so it is clear that the memory of an object is somewhow distributed throughout the brain. Positron emission tomography (PET) scans reveal a pattern of brain activity in many widely different parts of the brain during memory recall. It is possible to access the full memory by initially remembering only one or two of these characteristic features. We access the memory by its contents not by where it is stored in the neural pathways of the brain. This is a powerful and effective mechanism. It is also very different from a traditional computer in which specific facts are located in particular places in computer memory. If only partial information is available about the location in computer memory, the fact or memory cannot be recalled.

In addition to logic, neurons are also capable of storing and retrieving data from memory. A neural network can store data in two formats. Permanent data or long-term memory may be designed into the weightings of each neuron in a network. Temporary data or short-term memory can be actively circulated in a loop until needed again. Since the output of the neuron feeds back onto itself, a self-sustaining loop keeps the neuron firing even when one input is no longer active. The stored binary bit is continuously accessible by looking at the output. This configuration is called a latch. While it works perfectly in a neural network model, a biological neuron does not behave this way. After firing, a biological neuron has to rest for an ms before it can fire again. Thus one would have to link several neurons together in a duty-cycle chain to achieve the same result.

The Hopfield neural network is a simple artificial network that can store certain memories or patterns in a manner similar to a brain. The full pattern can be recovered if the network is presented with only partial information. Furthermore, the system has a degree of stability. If only a few connections between nodes (neurons) are severed, the recalled memory is not too severely corrupted and the network can respond with a "best guess."

A similar phenomenon is observed in the human brain. During a lifetime, many neurons die out without causing a catastrophic loss of individual memories. The human brain is very robust in this respect. Any general initial pattern of firing and nonfiring nodes changes over time. The crucial property of the Hopfield network that renders it useful for simulating memory recall is that it settles down after a time to some fixed pattern. Certain nodes are always on and others are always off. It is possible to arrange that these stable firing patterns of the network correspond to the memories we wish to store. In the language of memory recall, if we start the network with a pattern of firing that approximates one of the stable firing patterns (memories), it will recall the original perfect memory. The Hopfield network includes a simple way of setting up connections between nodes so that any desired set of patterns can be made into stable firing patterns. Thus, any set of memories can be burned into the network at the beginning. If we perturb the network by node activity, we are guaranteed that a memory will be recalled. It will be the one "closest" to the starting pattern. However, if the input image is sufficiently poor, it may recall an incorrect memory and the network can become confused. Nonetheless, overall, the network is reasonably robust.

A simple Hopfield neural network can perform some functions of memory recall in a manner analogous to the way we believe the brain functions. However, one major difference is that connection strengths must be set up in a certain way in an artificial network to enable it to store a predetermined set of patterns. In other words, a "smart teacher" is needed to train the network what to remember. Once that is done, the network can be left to itself to handle pattern recall. A brain has no "teacher" that tells the neurons how to link in order to store useful information. This part of the process is automatic and the system is said to be self-organizing. Chapter 8 of this book deals with other systems in the body. Chapter 9 addresses some aspects of self-organization.

References

Bernstein, J., *Pflügers Arch. Ges. Physiol.*, 92, 521, 1902.

Cole, K.S. and Curtis, H.J.J., *Gen. Physiol.*, 22, 649, 1939.

Fausett, L., *Fundamentals of Neural Networks*, Prentice-Hall, Englewood Cliffs, NJ, 1994.

Gurney, K., *An Introduction to Neural Networks*, UCL Press, London, 1997.

Haykin S., *Neural Networks*, 2nd ed., Prentice-Hall, Englewood Cliffs, NJ, 1999.

Hille, B., *Ionic Channels of Excitable Membranes*, Sinauer Associates, Sunderland, MA, 1992.

Hodgkin, A. L. and Huxley, A.F., *J. Physiol.*, 117, 500, 1952.

Hodgkin, A. L. and Huxley, A.F., *Nature*, 144, 710, 1939.

Hodgkin, A. L. and Katz, B., *J. Physiol.* 108, 37, 1949.

Overton, E., *Pflügers Arch. Ges. Physiol.*, 92, 346, 1902.

Nicholls, J.G. et al., *From Neuron to Brain*, 4th ed., Sinauer Associates, Sunderland, MA, 2001.

Scott, A.C., *Neuroscience: A Mathematical Primer*, Springer, Berlin, 2002.

8

Tissue and Organ Biophysics

8.1   Introduction

Multicellular organisms are arranged hierarchically into tissues, organs, and organ systems. Tissues are composed of colonies of cells and extracellular matrices. Some cells, e.g., white blood cells can migrate among tissues. Several different tissues may function jointly to form an organ, with one type of tissue designated to play the role of its skin. Organs may function cooperatively to form a system, e.g., the circulatory system, the nervous system, the immune system, the respiratory system, etc. (Cerdonio and Noble, 1986).

Animal tissues are classified as epithelial, connective, muscle, and nerve. Epithelial cells control the selective process of material transport. Connective tissue is composed of nerve cells, endothelial cells, and macrophages that remove debris. They also contain fibroblastic cells that are responsible for secreting extracellular matrix. Finally, connective tissue is criss-crossed by a network of collagen fibers of varying density. Osteoblasts are related to fibroblasts and inhabit the bones. Cartilage is produced by chondrocytes. Apidocytes are fat cells, distinguished by their round shapes and large sizes. Smooth muscle cells are filled with actin and myosin bundles. Muscle cells have four structurally distinct forms: skeletal, cardiac, smooth, and myoepithelial. Skeletal and cardiac muscle cells are jointly referred to as striated muscle cells due to their appearance. Smooth muscle cells are mainly present in blood vessel walls and intestines.

This chapter presents a panoramic view of the key organs and systems in human and animal bodies. Our focus is on physical phenomena and mechanisms behind the functioning of the systems, organs, and tissues. The level of sophistication employed in this chapter rarely exceeds introductory physics (Tuszynski and Dixon, 2002). Nonetheless, the combination of biological, physiological, chemical, and physical knowledge required makes this overview at times demanding. We begin by investigating the role of pressure that must be maintained by various organs in the human body for them to function properly.

8.1.1   Pressure in human organs

Since most of the human body is composed of water, one expects fluids to play a major role in functioning of the organism. Various types of fluid flowing in separate networks can be characterized by their typical pressure values P. Table 8.1 shows some of these values where pressure is stated in mm Hg.

TABLE 8.1
Typical Fluid Pressures in Human Organs

Organ	Pressure P (mm Hg)
Arterial blood pressure	
Maximum (systolic)	
Adult	100–140
Infant	60–70
Minimum (diastolic)	
Adult	60–90
Infant	30–40
Venous blood pressure	
Venules	8–15
Veins	4–8
Major veins (CVP)	4
Capillary blood pressure:	
Arteriole end	35
Venule end	15
Bladder:	
Average	0–25
During micturition	110
Brain, lying down (CSF)	5–12
Eye, aqueous humor	12–24
Gastrointestinal	10–20
Intrathoracic	−4 to −8
Middle ear	< 1

In the next several sections, we explain the origins of these pressure values in terms of the physiological processes taking place in given organs. Readers are referred to Bruinsma (1998) and Benedek and Villars (2000) for additional information.

8.2 Anatomy and physiology of human circulatory system

8.2.1 Circulation of blood

Harvey, the English physician and physiologist, determined how blood circulates in the human body. The circulation of blood supplies food and oxygen to tissues, carries waste products from the cells, and distributes heat to equalize body temperature. It also carries hormones that stimulate the activities of organs, distributes antibodies to fight infection, and handles many other functions. The circulation of the blood is both a complex process and one of great importance to overall health.

Blood is made of two main components. Its liquid basis is an intercellular fluid known as plasma. Plasma is about 90% water, 0.9% salt, and 9% proteins, sugars, and traces of other components. White and red blood cells are suspended in plasma.

Red cells have biconcave shapes, diameters of about 7.5 μm and densities around 5×10^3 kg/m^3. The average concentration of white blood cells is about 8, 000 per mm^3 and can vary from 4,500 to 11, 000 per mm^3. The viscosity of blood varies from 2.5 to 4.0 times that of water and its density is about 1, 050 kg/m^3. This translates into 5×10^{12} red blood cells per liter. While red blood cells carry oxygen and carbon dioxide to and from the lungs, the primary role of white blood cells is assistance in combatting infections. White blood cells come in several subgroups including granulocytes, lymphocytes, monocytes and natural killer (NK) cells. Leukocytes travel in the blood stream and in the connective tissues where they differentiate into macrophages to attack and remove foreign cells and debris. Blood also contains cell fragments called platelets that repair damaged blood vessels. Erythrocytes contain nuclei and other organelles only during their early development. They are expelled just before release of the cells into the blood stream. Mature blood cells have no internal structures except the membrane-bound cytoskeletons.

8.2.2 Cardiovascular system

This section discusses the mammalian circulatory system in some detail. The system is a good example of a biological transport system based on fluid flows (Zamir, 2000). Another example is air circulation through the lungs which will be discussed later. Both systems possess a number of similarities. For example, both have wide pipes that rapidly transport large volumes of liquid or gas. The pipes divide again and again into narrower pipes with slower flow velocities that form fractal networks (see Appendix C). The pipe walls become sufficiently thin to allow molecular exchange across them.

The cardiovascular system consists of the heart, arteries, and veins. The arteries carry blood to the organs, muscles, and skin and the veins return it to the heart. Oxygenated blood finally reaches the capillaries that are so small that only one red cell can only pass through them at one time.

The heart consists of two pumps connected by elastic pipes. The pressure generated in the heart forces the blood into the arteries. The right side pumps oxygen-deficient blood to the lungs via the pulmonary trunk. It then passes through pulmonary veins to the left side of the heart which then pumps it through the aorta into the arteries that branch first into arterioles and then the very narrow capillaries where the blood gives up its oxygen. The lengths of most capillaries are 50 to 100 times longer than their diameters. Capillaries are typically 1000 μm long and 5 to 20 μm in diameter.

Figure 8.1 represents the circulatory systems of mammals and birds. For blood to flow through an artery or a vein, an external force must be applied, i.e., the pressure gradient between the ends of the vessels, and obviously flows from high pressure to low pressure. This motion continues unless another force acts to prevent flow, e.g., a vessel obstruction (stenosis or atherosclerotic blockage). The velocity of blood is much greater in the major arteries (about 0.3 m/s) than in the capillaries (about 3×10^{-4} m/s). Surprisingly, the velocity increases again when the capillaries rejoin to form veins. The decrease in velocity going from the aorta to the capillaries is not

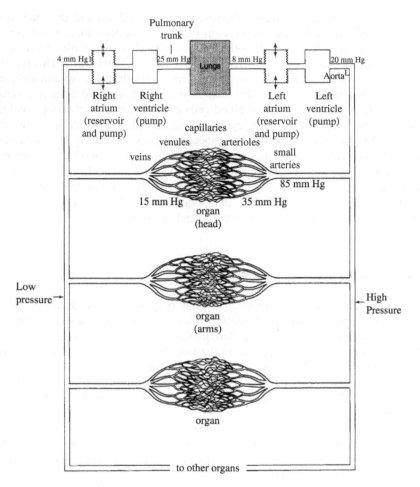

FIGURE 8.1
Major features of the circulatory system with representative blood pressures.

caused by resistance to flow. Resistance causes pressure to drop but does not affect velocity. This can be seen from the fact that blood velocity increases again in the veins while pressure continues to drop in veins leading back to the heart. Most changes in average velocity other than those produced by the heart are due to changes in the total cross-sectional area of the system during branching. For example, the aorta branches into the major arteries, each of which has a smaller cross-sectional area, but whose combined area is larger than the area of the aorta. The total flow rate in the major arteries $Q_{major\ arteries}$ is the same as in the aorta Q_{aorta} since all the blood that passes through the aorta must also pass through the major arteries. Therefore:

$$Q_{aorta} = Q_{major\ arteries}. \qquad (8.1)$$

Blood vessels resemble cylinders although in practice they may steadily decrease in size. The equation of continuity assures us that flow is equal at any point, whatever

S: Cross-section
Surface Area (cm²)

U: Flow
velocity (cm/s)

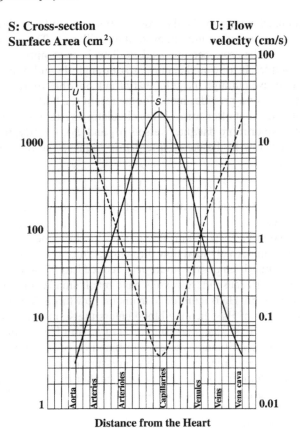

Distance from the Heart

FIGURE 8.2
Total cross-sectional area and mean flow velocity versus distance from the heart.

the degree of tapering. If the cross-sections and corresponding velocities at two points are, respectively, A_1, A_2, v_1, and v_2, from the equation of continuity, we have:

$$A_1v_1 = A_2v_2 \qquad \rightarrow \qquad v_1 = \frac{A_2}{A_1}v_2 \qquad (8.2)$$

This, of course, may be generalized to several points in a vessel. The flow rate of the blood changes rather dramatically as the blood traverses the cardiovascular system. The effective cross-sectional area as it goes through the regime with very small capillaries, i.e., cross-sectional area A, and the number of capillaries is much greater than the cross-sectional area of the aorta. The flow rate through the aorta and the capillaries must be the same so we have another form of the continuity equation applied to this case:

$$A_{aorta}V_{aorta} = A_{cap.eff}.V_{cap} \qquad (8.3)$$

Although the cross-section for each capillary is very small, the total cross-sections of all the capillaries together is about 400 times that of the aorta. Figure 8.2 shows total cross-sectional area S across a set of parallel branches, taking separately into account the cross-sections of the aorta, arteries, arterioles, capillaries, venules, and veins.

S has a pronounced maximum for the cross-sections of capillaries but the mean flow velocity U has a pronounced minimum for the capillaries. The product $Q = SU$ is the same wherever the cross-section is taken. This is an example of the generalized continuity principle. It follows that current A through the pulmonary trunk and the aorta is the same, even though the pulmonary trunk only has to transport blood to the lungs whereas the aorta provides blood for the rest of the body.

Another situation that often arises in the circulatory system is when a blood vessel bifurcates into two or more daughter vessels (d_1 and d_2). Clearly, from the equation of continuity, the flow in the parent artery p is equal to the flow in the branching daughter vessels or:

$$A_p V_p = A_{d_1} V_{d_1} + A_{d_2} V_{d_2} \tag{8.4}$$

where we labeled the velocities with vectors to account for the relative orientation of the bifurcating vessels.

The aorta is so large ($r = 9\,\text{mm}$) that a pressure difference of only 3 mm Hg is required to maintain normal blood flow through it. Thus, if the pressure of the blood is 100 mm Hg when it enters the aorta, it is reduced to 97 mm Hg when it enters the major arteries. Because these vessels have much smaller radii than the aorta, a pressure drop of 17 mm Hg is required to maintain flow through them. Therefore, the pressure is only 85 mm Hg when the blood enters the small arteries. These vessels have still smaller radii, so that a pressure drop of 55 mm Hg is required to maintain blood flow through them. Finally, there is a further drop of 20 mm Hg when the blood passes through the capillaries in which the pressure drop is less than in the small arteries even though the capillaries have much smaller radii, because the number of capillaries is very large. The blood pressure drops to only 10 mm Hg by the time it reaches the veins.

The Bernoulli equation:

$$P + \frac{\rho}{2}v^2 + y\rho g = const \tag{8.5}$$

applies to the blood flow in the aorta with its large radius. For example, we can calculate the percentage of the pressure drop in the aorta due to the presence of an atherosclerotic plaque that locally reduces the cross-sectional area to only a fifth of the normal value. We take $P_1 = 100\,\text{mm}\,Hg = 13,600$ Pa, a typical value of $v_1 = 0.12$ m/s, and use the Bernoulli equation to obtain:

$$\Delta P = P_1 - P_2 = \frac{\rho}{2}(v_2^2 - v_1^2) \tag{8.6}$$

with the continuity equation $A_1 v_1 = A_2 v_2$ and the assumption that $y_1 = y_2$, we find that $\Delta P / P = 1.2\%$ is the relative pressure decrease in the constricted area. However, the typical diameter of a small artery is only 25 μm with a blood velocity of approximately $2.8 \cdot 10^{-3}$ m/s.

When a pure liquid flows through a rigid tube, the relationship of Q, the volumetric flow rate (or volume of liquid per unit time), ΔP, the pressure gradient, L, the length of the tube, r, its radius, and fluid viscosity η is known as Poiseuille's law:

$$Q = \frac{\pi \Delta P r^4}{8\eta \ell} = \frac{\Delta P}{R} \tag{8.7}$$

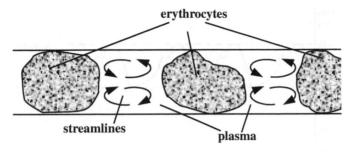

FIGURE 8.3
Bolus flow.

where R denotes resistance to flow. The applicability of Poiseuille's law to blood flow gives the correct sense of the dependence of resistance on radius and viscosity. Resistance always decreases when radius increases and it increases when viscosity increases. It is not surprising that the body adjusts blood flow by changing vessel radii, since resistance to flow is more sensitive to this parameter than to any other.

Most vessel dilation (vasodilation) and constriction (vasoconstriction) take place in the small arteries and arterioles, with some occurring in capillaries that have sphincters. Taking $\eta = 4.5 \cdot 10^{-3}$ Ns/m^2 as the typical viscosity of blood, using Equation (8.7), it is easy to calculate the pressure drop ΔP in a small artery whose length is $\ell = 5 \cdot 10^{-3}$ m as $\Delta P = 3,200\,Pa$ which, compared to the total pressure drop of $13,600\,Pa$, accounts for about 24% of the total drop in circulation.

In the larger arteries and thoracic veins, the beating action of the heart creates a pulsatile flow. The blood still advances in streamlines but the shape of the velocity profile is rather square across most of the area and no longer parabolic. If the velocity profile is examined over several cycles, the impression is that of a slug of fluid that advances four steps, backs up one, and repeats the cycle over and over.

As we proceed away from the heart, the pulses are gradually damped out due to several factors, including blood viscosity and the extensibility of the blood vessels. Many capillaries have diameters of only 5 or 6 μm, whereas erythrocytes have diameters of about 8 μm. Consequently, erythrocytes must deform to pass through capillaries. Cells occupy entire cross-sections of the interiors of vessels and pass through as a series of moving plugs with short sections of plasma trapped between them. This is known as bolus flow and is shown in Figure 8.3.

The flow in the trapped plasma sections becomes a specialized form of streamline flow. Rapidly moving streamlines down the center catch up with an erythrocyte and are deflected to the outside where their velocity, relative to the wall, becomes essentially zero. The following erythrocyte picks up these layers and forces them once more down the center. Thus, the plasma is well stirred.

The heart sends blood into the circulatory system in a periodic manner and a complete cycle lasts approximately 1.1 s. The oscillatory motion of the blood may be represented by plotting vertically the velocity of the blood flow against time (see

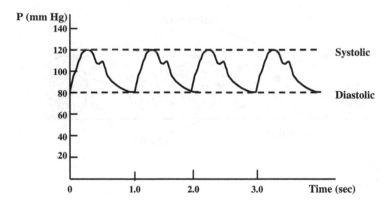

FIGURE 8.4
Blood pressure versus time in a major artery.

Figure 8.2) or the pressure as a function of time (Figure 8.4). Initially, an impulsive rise in blood flow velocity from the heart corresponding to its amplitude at the onset of systole occurs, followed by a gradual decrease in velocity while the heart is filling for its resting beat.

When flowing through a rigid vessel, blood exhibits a characteristic parabolic profile along the cross-section of the vessel where:

$$v = \frac{\Delta P}{4L\eta}(R^2 - r^2) \qquad (8.8)$$

In Equation (8.5), v denotes fluid velocity, ΔP is the pressure gradient applied to both ends of the vessel to initiate flow, L is the length of the vessel, η is blood viscosity, R is the radius of the vessel, and r is the distance from the center. Figure 8.5 clearly shows that flow velocity is largest at the center of a vessel and minimum at the edges.

Blood vessels are often curved. A good example is the carotid artery, the major vessel supplying blood to the brain. It originates from the aorta, passes through the neck, and bifurcates into two other major arteries that supply front and back circulation of the brain. The points of the velocity profile in contact with the vessel wall exhibit maximum wall stress. As the blood flows through a curved section of a blood vessel, the velocity profile bends away from the radius of vessel curvature (see Figure 8.6). A blood constituent experiences circular motion maintained by a centripetal force F_C:

$$F_C = \frac{mv^2}{r} \qquad (8.9)$$

where m is the mass of the object, v its velocity, and r the radius of the circle. As r decreases, the bend becomes sharper and the centripetal force increases.

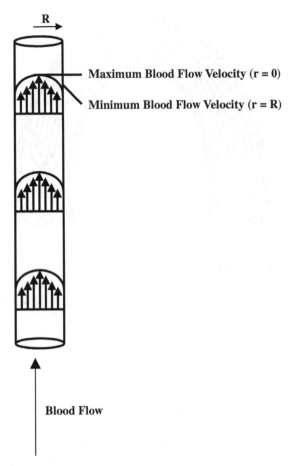

FIGURE 8.5
Velocity profile across a blood vessel's diameter.

8.2.3 Regulation of fluid between cells (interstitial fluid)

In humans and other animals, interstitial fluid is regulated by exchanges of substances with blood in capillaries. Many substances are moved across capillary walls, but the transport of water (osmosis) is of immediate interest (see Figure 8.7). Blood contains water, glucose, electrolytes (dissolved salts), gases, proteins, red and white blood cells, and waste materials such as urea. Blood enters from the left in the figure, and because the capillary is small in diameter, a significant pressure drop occurs along it.

Pressures are typically 35 mm Hg at the entrance and 15 mm Hg at the exit. Water is more concentrated in interstitial fluid than in blood, but the typical osmotic pressure difference between blood and interstitial fluid is 22 mm Hg. Reverse osmosis occurs near the entrance of the capillary. Blood pressure near the capillary exit is less than the relative osmotic pressure, and osmosis carries water from the interstitial region

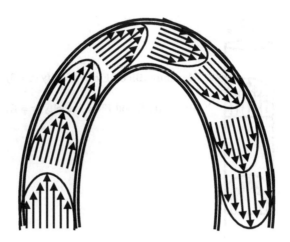

FIGURE 8.6
Blood flow in curved vessel.

into the capillary. The overall result is an exchange of water while the total amount of water in the interstitial region remains constant.

8.2.4   Gas exchange with circulatory system

Figure 8.8 represents gas exchange with the circulatory system. The partial pressure of oxygen in the lungs is only 105 mm Hg, less than that in the atmosphere, because air in the lungs is only partially replaced with each breath. Similarly, the partial pressure of carbon dioxide in the lungs is greater than in the atmosphere. Gases move in the directions of the arrows, in each case from high to low concentration (partial pressure). Oxygenated blood travels through the arteries to the capillaries, where it is transferred to cells with lower oxygen concentrations. Oxygen-poor blood travels back through the veins to the lungs, where the process begins again. A similar cycle for the removal of carbon dioxide is also shown.

Osmoregulation is very important to the human body. Unlike plant cells, eukaryotic cells cannot withstand large osmotic pressure differences. Blood has a high concentration of proteins compared with the surrounding tissues and capillaries are permeable to small molecules such as water, oxygen, carbon dioxide, and salt, but not to proteins. The capillary–tissue osmotic pressure difference that sucks water out of tissue is about 23 mm Hg. The capillaries are pressurized by blood pressure.

The average blood pressure is close to 23 mm Hg. The osmotic pressure difference is neatly offset by the hydrostatic pressure difference, but this delicate balance may be easily upset by such situations as high blood pressure or starvation. High blood pressure can lead to the accumulation of fluids in tissues, a condition known as edema. When we drink water or consume salt, we may be in danger of upsetting a delicate osmotic balance that is maintained via a regulatory mechanism provided by ADH (antidiuretic hormone). This hormone rapidly alters the water permeability of tubes connected to the urinary duct in response to changes in salt concentration.

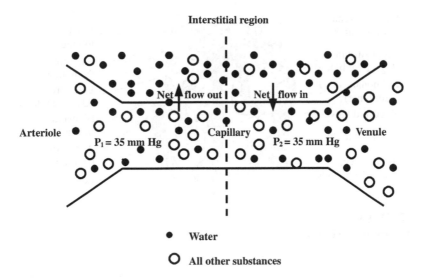

FIGURE 8.7
Water transfer between a capillary and the interstitial region. The dashed line marks the point at which the net transfer of water changes from flow out of the capillary to flow into the capillary.

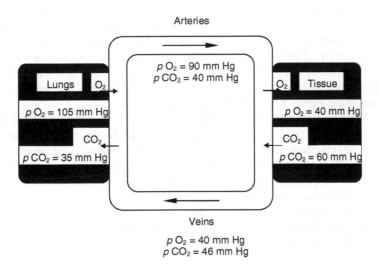

FIGURE 8.8
Interchange of gases between the lungs and blood and between the blood and tissues.

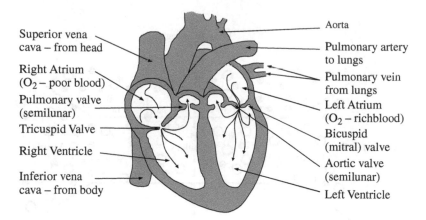

FIGURE 8.9
Structure of the heart.

8.3   Heart dynamics

Each side of the heart consists of an upper atrium and lower ventricle. Blood depleted of oxygen enters the right atrium and flows into the right ventricle which then pumps it to the lungs. It then returns through the left atrium into the left ventricle which pumps the now-oxygenated blood out through the aorta to the rest of the body. The heart has a system of valves to ensure the blood flows in the proper direction without backflow.

The two ventricles pump at the same time. The pressure in the right ventricle is quite low, about 25 mm Hg, compared to the left ventricle where the pressure can exceed 120 mm Hg at the peak (systole) of the pressure. During the resting stage of heartbeat (diastole), the pressure is typically about 80 mm Hg. Figure 8.9 shows the heart structure; Figure 8.10 shows its pumping action.

The blood volume of an average person is about 5 l. At rest, the cardiac output is approximately 5 l/min. Hence the average time it takes for the entire volume of blood to make one full circuit and return to the heart is only a minute. The mean circulation time during strenuous exercise by an athlete may be reduced to only 12 s. All this work is done by the heart against the resistance to viscous blood flow exerted by the network of interconnected blood vessels.

We define the power output P as the work done by the heart per second in pumping blood. It is equal to the average force F exerted by the heart on the blood multiplied by the the distance d over which the blood moves in 1 s, i.e.:

$$P = Fd \tag{8.10}$$

The force is equal to the pressure p exerted by the heart on the aorta multiplied by the cross-sectional area A of the aorta:

$$F = pA \tag{8.11}$$

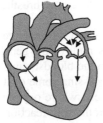 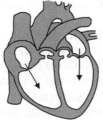 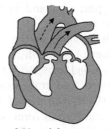

1.Blood fills both atria, some blood flows into ventricles – diastole phase of atria.

2.Atria contract, squeezing blood into ventricles – ventricular diastole.

3.Ventricles contract, squeezing blood into aorta and pulmonary arteries – ventricular systole phase.

FIGURE 8.10
Pumping action of the heart.

Blood flow Q is the volume of blood passing through the aorta in 1 s, so in 1 s, the volume of blood moves the distance:

$$d = \frac{Q}{A} \qquad (8.12)$$

In a normal adult, $Q = 0.83 \cdot 10^{-4}\,\text{m}^3/\text{s}$, and the average blood pressure is $p = 100\,\text{mm Hg} = 1.3 \cdot 10^4\,\text{N/m}^2$. Therefore, the power output of the heart is:

$$p = F \cdot d = pQ = 1.1W \qquad (8.13)$$

which amounts to only 1% of the total power dissipated by the body. It is easy to conclude that the heart of a person with abnormally high blood pressure does more work per second to maintain the same blood flow.

By measuring oxygen consumption, it can be found that a 70-kg man consumes energy at a rate of about 10 W so we infer from Equation 8.13 that the heart is about 10% efficient. However, blood pressure and the volume of blood pumped may increase by as much as a factor of five. This results in an increase of 7.5 in the power generated by the left ventricle. The power required by the right ventricle (its systolic pressure is roughly one fifth of the pressure of the left ventricle) is about a fifth the amount required by the left ventricle.

8.4 Energy management in the human body

Chemical reactions within the body are responsible for storing, releasing, absorbing, and transferring the energy humans need to move, breathe, pump blood, etc. Reactions that require energy are called endothermic reactions; those that release energy are exothermic. The energy source for endothermic reactions is the food we eat. The energy required for muscle contraction is produced by a network of chemical reactions of two types: input and output. Input ingredients come from the air we breathe and the foods we eat.

The two outputs are the carbon dioxide we exhale and the water we generate as a by-product of metabolism. To be more specific, the lungs remove oxygen from inhaled air and transport it to muscle cells via hemoglobin in the bloodstream. The body digests food in the mouth, stomach, and intestines, processing some of it into glucose ($C_6H_{12}O_6$) part of which is transported to muscle cells where oxygen and glucose combine to form water, carbon dioxide, and an exothermic reaction that produces energy E_{out}. Thus, in terms of a chemical reaction:

$$C_6H_{12}O_6 + 6O_2 \longrightarrow 6CO_2 + 6H_2O + E_{out} \qquad (8.14)$$

We can measure net oxygen inhaled and/or net carbon dioxide exhaled. The energy released can then be deduced from the glucose oxidation reaction. It takes time to breathe, transport oxygen, and deliver glucose to muscle cells, but the muscle fibers need not wait until glucose oxidation provides the energy required because the body has the ability to store energy. We have already discussed the roles of ATP, ADP, and AMP in this process (see Chapter 3).

As long as ATP is available, immediate muscle contraction can take place. When the power demand of muscles increases abruptly, the energy required cannot be supplied by oxidation because of the time required to move to a new equilibrium state. This energy gap is filled by energy stored in the muscle cells in an ATP form not requiring oxidation. After the transition time, respiration will gradually increase ATP until a new steady state is attained. During exercise, energy is supplied anaerobically (without oxygen) for the first 2 min or so, and aerobically thereafter via respiration. When the body uses energy faster than respiration can support ATP production, available energy declines — a condition called oxygen debt. Thus, at the end of vigorous exercise, when the demand for energy is returning to normal, the debt is repaid when the athlete gasps for air.

As mentioned earlier in the book, almost all metabolic energy of animals and humans comes from the conversion of ATP into ADP and AMP. The energy to convert ADP and AMP back to ATP is supplied by the oxidation of carbohydrates, fats, and proteins. Metabolism requires a constant supply of oxygen which determines the metabolic rate. Oxygen is supplied by blood pumped by the heart at a variable rate, depending on the type of activity pursued. For example, a person completely at rest consumes 15 l of oxygen per hour. Energy production is linked to the oxygen supply via the empirical relationship:

$$E = 2 \cdot 10^4 \, J/(\ell \text{ of oxygen of supplied}) \qquad (8.15)$$

Thus, at rest, the metabolic power generated is approximately:

$$P = \frac{15\ell}{h} \cdot \frac{2 \cdot 10^4 J}{\ell} \cdot \frac{1\,h}{3600\,s} = 83\,W \qquad (8.16)$$

Food energy is stored by chemical bonds and has traditionally been measured in food calories. One food calorie is equivalent to 1,000 physics calories or 4,186 joules.

To measure the energy content of food, the food is burned in oxygen and the heat produced is the energy content. The human body combines food with oxygen, i.e., it burns food to produce CO_2.

We have seen that the basic metabolic rate of an average person when resting is 80 to 100 W. Walking at 2.5 km/h consumes another 100 W. Walking faster (5 km/h) uses another 100 W. About 75% of the energy generated is converted to heat and the remainder can be used for activities like walking and running. Using 100 W as an average metabolic rate for an adult translates to a dietary requirement of 2,600 food calories daily.

8.5   Respiration biophysics

The most important function of the respiratory system is oxygenating blood and removing carbon dioxide from it. Oxygen is needed to combine chemically with food. Carbon dioxide is the primary gaseous waste product of cells. Air flows into the lungs when lung pressure is below atmospheric pressure and out of the lungs when lung pressure exceeds atmospheric pressure. The air flow obeys the familiar equation:

$$Q = \frac{(P_1 - P_2)}{R} \tag{8.17}$$

where P_1 is the pressure in the lungs and P_2 is atmospheric pressure. R is the resistance of the breathing passages to air flow. If gauge pressures are used, P_2 is zero; air flows out of the lungs when P_1 is positive (greater than atmospheric) and into the lungs when P_1 is negative (less than atmospheric).

The body changes pressure in the lungs by increasing or decreasing their volume. During inhalation, the lungs are expanded by muscle action in the diaphragm and rib cage (see Figure 8.11). The diaphragm is a sheet of muscle lying below the lungs. When relaxed, it has an upward curvature, and when contracted, it moves downward, expanding the lungs.

The interior surface of the lungs constitutes the largest surface of the human body in contact with the environment. The exchange of CO_2 and O_2 molecules between blood and the atmosphere requires about a square meter of lung surface per kilogram of body weight. This is achieved by compartmentalizing the lungs into tiny air sacs (alveoli) covered by a multitude of tiny capillaries with a very large total surface area (about $100\,m^2$). Across the surfaces of the alveoli, CO_2 and O_2 are exchanged by the blood and lungs. The surface of an alveolus is covered by a thin film of water (about $0.5\,\mu$m thick) and surrounded by lung surfactant molecules.

During inhalation the radius of an alveolus expands from about $0.5 \cdot 10^{-4}$ to $1.0 \cdot 10^{-4}$ m. Alveoli are lined with mucous tissue fluid, which normally has a surface tension of $0.050\,N/m$. As a result, the pressure difference required to inflate an alveolus is given by:

$$\Delta p = \bar{p}_i - \bar{p}_0 = \frac{2\gamma}{r} \tag{8.18}$$

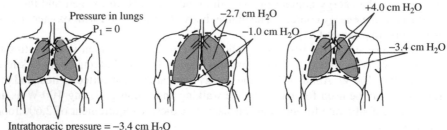

Intrathoracic pressure = −3.4 cm H₂O

FIGURE 8.11
Three stages in the breathing cycle: (a) at rest, the pressure in the lungs is zero; the lungs are open; (b) during inhalation, the chest expands, the diaphragm moves downward, and pressures drop; (c) during exhalation, the chest contracts, the diaphragm rises, and pressures rise.

where r denotes the radius and γ surface tension. Using $r = 0.5 \cdot 10^{-4}$ m and a γ typical for water, 0.070 N/m, we obtain $\Delta p = 2.8 \cdot 10^3$ N/m^2 = 21 mmHg. Thus, the gauge pressure $\bar{p}_0$ outside the alveolus would have to be -24 mm Hg since the internal pressure is $\bar{p}_i = -3$ mm Hg. The outside pressure in this case is the pressure in the space between the lungs and the pleural cavity holding the lungs. The gauge pressure in the space is negative but only about -4 mm Hg, so the actual pressure difference $\bar{p}_i - \bar{p}_0$ is only about 1 mm Hg. To overcome this difficulty, the alveoli walls secrete a surfactant that reduces surface tension by a factor of about 15.

The surfactant in the lungs is a lipoprotein. The polar head group of the lipid is hydrophilic while its long fatty acid chains are hydrophobic. Each alveolus appears to contain a fixed amount of this surfactant. Its ability to reduce surface tension depends on its concentration. Therefore, when the alveolus is deflated, the concentration of the surfactant is high and the surface tension is very low, so that the alveolus can expand without difficulty. However, as it expands, the concentration of the surfactant decreases and the surface tension increases, until a point of equilibrium is reached at maximum expansion. The importance of surface tension in the lungs is seen in victims of emphysema. Many alveoli in a person suffering from emphysema join to form fewer and larger alveoli, as shown in Figure 8.12. One effect of the larger sacs is a reduction in pressure because of their larger radii. When a person with emphysema attempts to exhale, his alveoli create a small pressure and air flow is less than normal. The linings have lost some of their elasticity.

The large area of alveoli can be reduced by other diseases. A pneumonia patient suffers from oxygen starvation because lymph and mucus in the bronchioles and alveoli reduce the gas exchange area. Supplemental oxygen may allow the patient to breathe by bringing about sufficient diffusion in the restricted area of the alveoli.

We demonstrate the general principles of gas exchange by reference to humans. Figure 8.13 illustrates the main features of the human gas exchange mechanism. Air is inhaled through the nose, passes down the trachea, and then branches into the lungs via the bronchi. Each bronchus proliferates into smaller bronchia and these lead in turn to tiny bronchioles. The structure resembles a tree with the trachea as the trunk, bronchi as the branches, and bronchioles as twigs. The bronchioles terminate in a

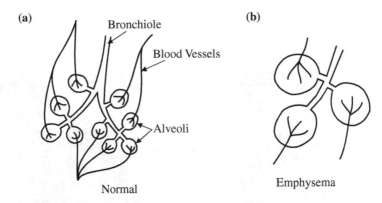

FIGURE 8.12
Alveoli and their air and blood supply in (a) normal lungs and (b) in an emphysema victim. Alveoli have joined to form fewer and larger sacs.

network of millions of alveoli (see Figure 8.14). If the alveoli were flattened, their total surface area would be about $60 \, \mathrm{m}^2$. The vital process of diffusion of oxygen and carbon dioxide takes place over this enormous area.

In general, gas transfer into and out of the blood is a primary function of the respiratory system. Diffusion of gases between air in the lungs and blood proceeds in the direction from high to low concentration, and the rate of diffusion is greatest when the difference in concentration is greatest. Diffusion obeys Fick's law, but the actual rate of exchange is greatly affected by hemoglobin in the blood.

Hemoglobin has a chemical affinity for oxygen, and its presence allows the blood to carry far more oxygen than it otherwise could. The rate of exchange also depends on blood acidity, the presence of carbon dioxide, and temperature which is cleverly designed to enhance oxygen transfer when most needed. The rate at which diffusion takes place across the walls of the alveoli is proportional to the pressure difference. This explains why breathing is difficult at high altitudes.

Blood is carried into the lungs by the pulmonary artery that terminates in a maze of tiny capillaries in very close contact with the alveoli. This allows molecules of oxygen to pass through the walls of the alveoli and capillaries and attach to hemoglobin in the blood. The metabolic combustion of oxygen in cells generates CO_2 that moves in the reverse direction and is finally exhaled. An improvement to lung ventilation is facilitated and improved by the presence of the diaphragm, a muscular partition separating the abdominal and thoracic cavities that essentially acts as a bellows to draw the air into the lungs.

Figure 8.15 shows volume flow rate versus time through the lungs. The graph rises to a fairly steady value almost immediately and then falls off toward the end of the expiration period. The expiration time is 2.5 s and the total area under the flow rate–time curve shows the total volume of gas expired. This is called the tidal volume and here it is about $450 \times 10^{-6} \, \mathrm{m}^3$. Simultaneously, nitrogen concentration can be measured as a function of time (see Figure 8.16).

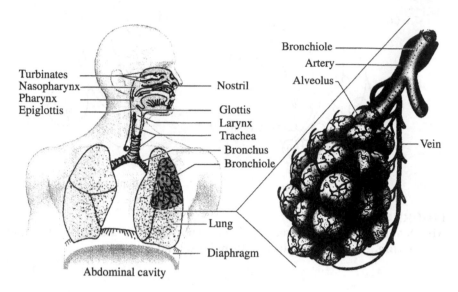

FIGURE 8.13
Human gas exchange system.

The concentration of nitrogen is zero at the beginning of expiration and then rises to a final steady value of about 72%. These observations are possible with a very simple model of the lung. We assume that the dead space is a piece of glass tubing and the volume of the alveoli is represented by a partially inflated balloon. A healthy person will have a dead space volume of $150 \times 10^{-6}\,\mathrm{m}^3$ while alveolar volume at the beginning of inspiration is about $2,550 \times 10^{-6}\,\mathrm{m}^3$.

When we breathe normally, both the dead space and alveoli are full of air at its normal concentration of 80% nitrogen. Therefore, at the beginning of inspiration, the volume of nitrogen in the lungs is $2,160 \times 10^{-6}\,\mathrm{m}^3$. When the subject breathes in $450 \times 10^{-6}\,\mathrm{m}^3$ of pure oxygen, it flushes air down from the dead space into the alveoli until, at the end of inspiration, the alveoli contain $2,160 \times 10^{-6}\,\mathrm{m}^3$ of nitrogen in a total expanded volume of $3,000 \times 10^{-6}\,\mathrm{m}^3$ The new concentration in the alveoli is 72% and the dead space is completely filled with $150 \times 10^{-6}\,\mathrm{m}^3$ of pure oxygen. Initially, only pure oxygen is breathed out and the concentration of nitrogen is zero. When the dead space has been emptied, the gas emerges from the alveoli and its concentration is 72% nitrogen.

If, at the air boundary between the dead space and the alveolar volume, no mixing took place, as soon as the gas in the dead space was expired, the nitrogen concentration would instantly rise to the final alveoli concentration of 72% (shown by the vertical line in Figure 8.16). In this ideal case, the dead space would simply be the volume corresponding to the shaded area under the flow rate–time curve. However, mixing occurs and a gradual and fairly rapid rise takes place instead of an instantaneous rise. To determine the amount of work done by breathing in clinical practice, pressure–volume curves for inspiration and expiration such as those in Figure 8.17 are drawn.

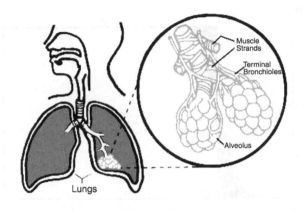

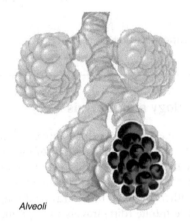

Alveoli

FIGURE 8.14
Alveoli.

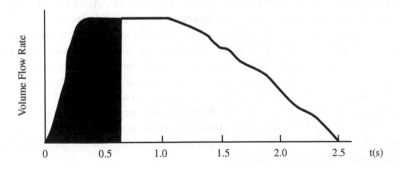

FIGURE 8.15
Variation of volume flow rate in the lungs.

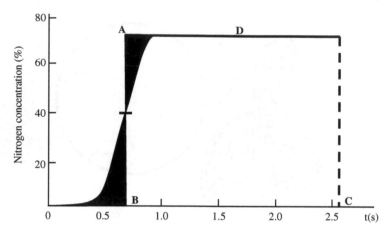

FIGURE 8.16
Variation of nitrogen concentration.

8.6 Kidney physiology and dialysis

The primary function of the kidneys is the removal of unwanted substances from blood plasma. The nephron is the functional unit of the kidney. See Figure 8.18. A natural kidney contains about 1 million nephrons. Essentially, arterial blood enters an area called Bowman's capsule, which acts as the filter in the nephron (Benedek and Villars, 2000). The filtered (clean) blood travels through a mesh, and eventually leaves the filter unit. The filtrate (dirt) travels through the tubule on its way to the ureter. See Figure 8.19.

However, before the filtrate leaves the nephron, most of it is released from the tubule into the mesh and back into the clean blood. In this rather peculiar manner, nephrons control solute concentrations in the blood. Any deviation (structure malfunction) results in accumulation of unwanted particles in the blood, eventually causing death. Figure 8.20 shows the first successful design of an artificial kidney.

As shown in the diagram, dirty blood enters from the left and is sent through a region filled with a solution called a dialysate. The dialysate is kept at a solute concentration equal to that desired in the blood plasma. In other words, the dialysate has a lower concentration of dirt than the dirty blood. Due to the difference in concentration, the dirt diffuses out of the blood and into the dialysate (see Figure 8.21). Eventually, the dirt concentration in the blood is brought to an acceptable level and returns back to the body. Using formulae derived for the diffusion equation (see Chapter 3 and Appendix A), we find the time dependence of the concentration difference and the characteristic time of equilibration as:

$$\Delta C(t) = \Delta C(0)e^{\left(\frac{t}{\tau_0}\right)} \quad \text{and} \quad \tau_0^{-1} = A\wp\frac{V_1 + V_2}{V_1 V_2} \tag{8.19}$$

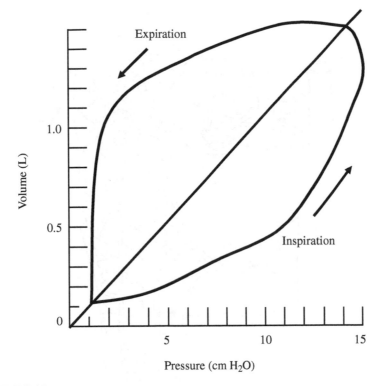

FIGURE 8.17
Plot of volume versus pressure during the inspiration and expiration components of the respiration cycle.

Volume V_1 is taken to include the volumes of the blood and the interstitial fluids (about 40 l). V_2, the volume of the dialysate, is taken to be much greater than V_1 so that the eventual solute concentration is, in fact, the desired concentration. Using $V_2 \gg V_1$, we obtain:

$$\tau_0 \cong \frac{V_1}{A\wp}$$ (8.20)

With $V_1 = 40\ l$, $\wp = 5.63 \times 10^{-4}$ cm/s (for urea), one finds that for $A = 2\,\text{m}^2$, $\tau_0 \cong 1$ h. Within a few hours, the dirty blood would be cleansed. A membranous area of $2\,\text{m}^2$ is within design limitations and is therefore not unreasonable.

8.7 Muscle biophysics

In general, a muscle is attached, via tendons, to at least two different bones. Figure 8.22 shows muscles of the human arm and Figure 8.23 shows the muscles

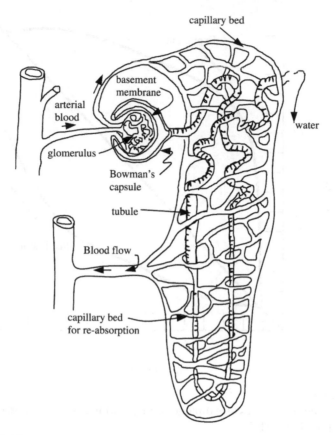

FIGURE 8.18
Nephron.

of the forefinger. Two or more bones are flexibly connected at a joint, e.g., elbow, knee, or hip. A muscle exerts a pull when its fibers contract under stimulation by a nerve but does not produce a push. Muscles that bring two limbs closer together, like the biceps in the human arm, are flexor muscles. Those that extend a limb outward, such as the triceps, are extensor muscles.

Flexor muscles are used, for example, when the upper arm lifts an object in the hand. Extensor muscles are used when throwing a ball. The human skeleton is a highly sophisticated device that transmits forces to and from various parts of the body. The muscles move the parts, generate forces, use chemical energy, and hence perform work. Thus they provide the power for movement in most multicellular organisms (Fung, 1993; Bruinsma, 1998).

Muscles generate forces by contracting after they are stimulated electrically. The tendons or rope-like attachments experience net tension after a series of these contractions, which increases with the number of electrical stimuli per second. The important function of tendons is to connect muscles to limbs. Muscles shorten the distances between the attachment points of the tendons but they cannot push these points apart. Thus pairs of muscles are necessary to operate a limb, e.g., when the

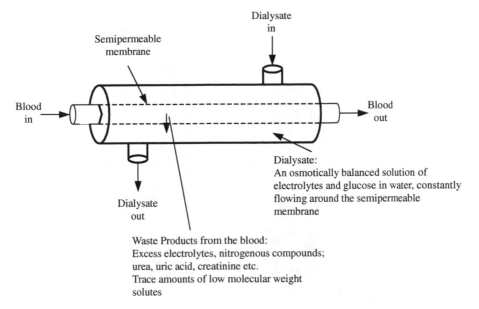

FIGURE 8.19
A schematic representation of the kidney function.

knee is bent, the hamstring muscle at the base of the thigh shortens. To straighten
the knee, the quadriceps muscle at the front of the thigh shortens. Figure 8.24 shows
counteracting muscle pairs in the human arm and leg.

Other types of muscles may join back on themselves and cause constriction of an
opening when they contract. Such muscles, called sphincters, serve several functions.
The sphincter at the lower end of the esophagus prevents the backflow of stomach
fluids. A sphincter muscle in the eye changes the curvature of the lens to allow clear
vision of near and distant objects.

Several muscles act simultaneously in the shoulder to produce total force
exerted on the arm. The addition of forces yields the expected result. The upward
and downward components of the forces cancel, leaving a large horizontal force
(see Figure 8.25).

Single muscles may be extracted from an organism and put into salt solution. They
can be electrically stimulated by applying a voltage pulse to the solution. The length
of the muscle may be kept fixed by clamping its ends. When we measure the tension
of a muscle of fixed length, we speak of isometric force measurement. Isotonic force
measurement is performed for a fixed load. The maximum isometric force F_{max} a
muscle can generate is proportional to its cross-sectional area A. Thus:

$$F_{max} = \sigma A \tag{8.21}$$

A remarkable result is that σ is approximately the same for all vertebrate muscles
and its value is $\sigma \cong 0.3 \times 10^6$ Pa. Muscles consist of fiber bundles in which each

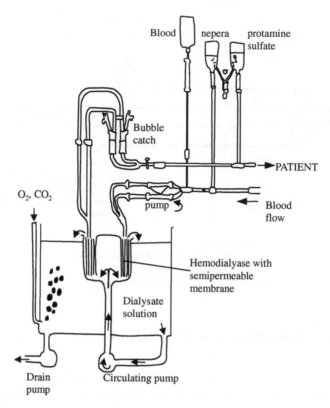

FIGURE 8.20
The first successful artificial kidney was designed by C. Colton.

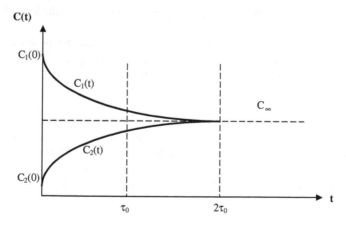

FIGURE 8.21
Time dependence of the concentration profile.

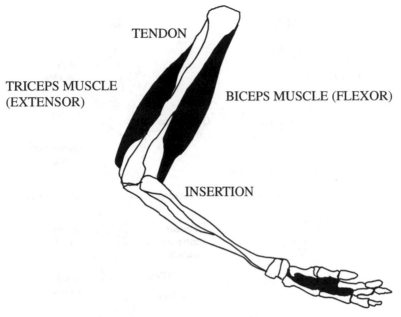

FIGURE 8.22
Triceps and biceps muscles in the human arm.

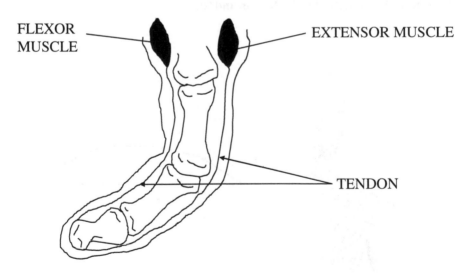

FIGURE 8.23
**The forefinger. The tendons carrying the forces exerted by the muscles extend
over joints that change the direction of the force.**

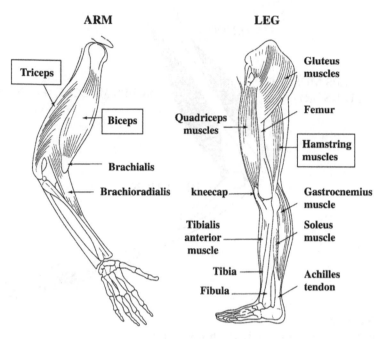

FIGURE 8.24
Counteracting muscle pairs in the arm and leg.

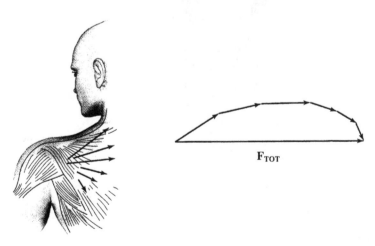

FIGURE 8.25
Principle of simultaneous actions of several muscles.

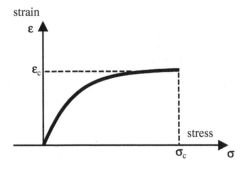

FIGURE 8.26
Stress–strain curve for bone.

fiber can contract with a given force. Thus, the total contractile force of a muscle must be proportional to its cross-sectional area.

8.8 Bone stiffness and strength

Animals develop a variety of solid materials such as bone, tooth, horn, shell, nail, and cartilage. Most of these organic compounds are complex, heterogeneous substances. The compact part of a bone, for example, consists of living cells embedded in a solid framework, a third of which is collagen and the remainder called bone salt is a mixture of inorganic materials containing calcium, phosphorus, oxygen, and hydrogen. The presence of tiny crystallites attached to the collagen fibrils increases the Young modulus of a bone.

The stress–strain curve for bone is shown in Figure 8.26 and includes both tensile and compressive applications of forces. Bone is constructed on the same principle as reinforced concrete. Concrete alone has great compressive strength but lacks tensile strength. Steel rods are embedded in reinforced concrete to provide tensile and compressive strength. Similarly, collagen adds tensile strength to the compressive strength of hydroxyapatite in bones.

Bones support loads placed on them, in particular the weight of the body. They also resist bending and twisting. Fractures are usually caused by excessive bending and twisting torques. The ability to resist bending and twisting depends on stiffness and also on the cross-sectional area and shape. Hollow bones are stiffer than solid bones of equal weight (similar to the way pipes are stiffer than rods).

Biological materials lack the high degree of order characteristic of metals and all show anisotropy in their physical properties. Collagen fibers are markedly anisotropic because they contain thread-like fibrils that lie in directions approximately parallel to the length of the fiber but their alignment with the axis of the fiber is not perfect. In certain regions, fibrils have much more random arrangements.

When tensile stress is applied to a fiber of collagen, the fibrils become more aligned parallel to the axis. This is why the value of Young's modulus for

TABLE 8.2
Tensile and Compressive Properties of Biological Materials

	Tension			Compression		
Material	Young's Modulus 10^8 N/m^2	Tensile Strength 10^7 N/m^2	ϵ_c Maximum Strain	Young's Modulus 10^8 N/m^2	Compressive Strength 10^7 N/m^2	ϵ_c Maximum Strain
Bone						
Human femur	16	12.1	0.015	9.4	16.7	0.0185
Horse femur	23	11.5	0.0075	8.3	14.2	0.024
Human vertebra	0.17	0.12	0.0058	0.088	0.19	0.025
Cartilage, human ear		0.30	0.30			
Eggshell	0.06	0.12	0.2			
Tooth, human						
Crown					14.6	0.023
Dentin				6.8	18.2	0.042
Nail, thumb	0.15	1.8	0.18			
Hair		19.6	0.40			

collagen is only 10^9N/m^2. The tensile strength of bone is due to collagen and its compressive strength is due to hydroxyapatite. As a consequence, the Young's moduli of bone and other heterogeneous substances are different for tensile and compressive stresses. Table 8.2 shows tensile and compressive properties of certain biological materials. Bone has a Young modulus nearly twice as large for tensile strength as the modulus for compressive stress, i.e., compressive stress produces twice as much strain as tensile stress of equal magnitude. Surprisingly, the mechanical properties of bones from different animals are very similar when one considers how different they are in other respects. The tensile stress of bone is 1.4 times that of steel and the compressive strength is close to that of granite. Bone is, of course, much lighter than steel or granite so it compares very favorably as a structural material.

8.9 Vision biophysics

The human eye is one of the most remarkable of all optical devices found in nature. Figure 8.27 shows the main anatomical features. An eyeball is approximately spherical with a diameter of about 25 mm. The eye is often likened to a camera because its purpose is to form an image on the retina in the same way a camera forms an image on a photographic plate.

Light enters the eye through a transparent membrane called the cornea, a membrane that covers a clear liquid region known as the aqueous humor, behind which is a

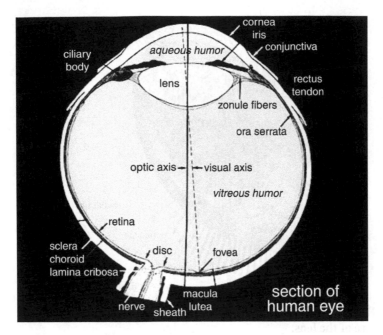

FIGURE 8.27
Cross-sectional anatomy of human eye.

diaphragm called the iris (the colored part of the eye). The iris adjusts automatically to control the amount of light entering the eye. Most of the refraction of light occurs at the cornea. The opening in the iris (pupil) through which light passes is black because no light is reflected back out from the interior of the eye. The pupil has a diameter that varies from 2 to 7 mm, decreasing in bright light and increasing in dim light. Across the pupil is the lens connected to the ciliary muscles by suspensory ligaments. The flexibility of the lens and its ability to alter shape via action of the ciliary muscle are of prime importance to the operation of the eye.

The lens is a clear, flexible, cellular structure (Figure 8.28). The cells are long and hexagonal in cross-section and are stacked in columns roughly parallel to the light path. The layers are ordered in such a way that relatively little light is scattered or absorbed. The lens consists of nested fibers that run from the front to the back (see Figure 8.27), creating an onion-like, gelatinous structure with an index of refraction of 1.40 at the center and 1.38 at the edge. The fibers are made of a protein called α-crystallin.

Cataracts are caused by the loss of structure of the crystallin proteins, leading to a clouding of the lens. The lens is suspended by tense radial filaments called the zonules of Zimm that are connected to the surrounding ciliary muscle. The lens can adjust its shape by contraction of the ciliary muscle so that, for example, when the ciliary muscle relaxes the lens is flattened because the tension in the radial filaments stretches it. Accommodation, like the adjustment of the pupil, is automatic via the autofocusing reflex. This can be overridden and it is possible to defocus when looking at an object. The change in refractive index from aqueous humor to crystallin is rather

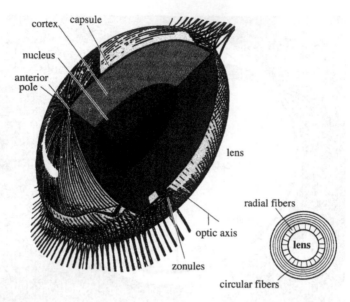

FIGURE 8.28
Structure of the lens.

small so this accommodation of the lens is modest and only fine tunes focusing. The main converging action is performed by the cornea.

Behind the lens is a region filled with a jelly-like substance called the vitreous humor. The retina that plays the role of camera film is on the curved rear surface of the eye. It is a complex array of cells called rods and cones that act to change light energy into electrical signals that then pass along the optic nerve to the brain. The region of the retina where the optic nerve enters contains no photoreceptors and constitutes a blind spot on the eye. At the center of the retina is a small area, called the fovea, about 0.25 mm in diameter, where the cones are very closely packed and the sharpest image and best color discrimination are found. Unlike a camera, the eye has no shutter. The equivalent operation is carried out by the nervous system that analyzes the signals to form images at the rate of about 30 per second. Tables 8.3 and 8.4 summarize the geometrical and refractive properties of the eye.

8.9.1 Wavelength responses

Electromagnetic radiation reaching the eye extends beyond the visible range into the infrared and the ultraviolet. However, the eye is responsive only in the range from 380 to 700 nm. The cornea absorbs most of the energy for $\lambda < 300$ mm. The risk of corneal damage in the ultraviolet range is high for people not wearing protective sunglasses. On the other end of the spectrum, water molecules in the cornea and aqueous humor absorb most of the electromagnetic radiation at $\lambda > 1200$ nm.

TABLE 8.3
Optical Constants of Helmholtz Schematic Relaxed Eye

Optical Constant	Value
Distance from front surface of cornea to front of lens	3.6 mm
Lens thickness	3.6 mm
Radii of Curvature	
Cornea	8.0 mm
Front of lens	10.0 mm
Back of lens	6.0 mm

TABLE 8.4
Refraction Indices (n) of Eye Components

Material	n
Cornea	1.38
Aqueous humor	1.33–1.34
Lens	1.41–1.45
Vitreous humor	1.34
Air	1.00
Water	1.33

Eye pigments are unresponsive to $\lambda > 800$ nm and only slightly responsive to $\lambda > 1200$ nm. In the visible region, i.e., 380 nm $< \lambda < 700$ nm, response depends rather strongly on the wavelength, i.e., color. Moreover, the response of the eye depends strongly on light intensity. If a person remains in bright surroundings, the process is called photopic vision and involves light adaptation. Conversely, in dark surroundings, eyesight is dark-adapted and the process is called scotopic vision. Changing light intensity of surroundings requires a period of readaptation during which a person is temporarily blinded.

8.9.2 Optical properties

When an object to be viewed is placed a distance d_o from the lens, the image distance always has the same value d_i which is approximately the diameter of the eyeball. Thus the power of the lens P is given by:

$$P = \frac{1}{f} = \frac{1}{d_o} + \frac{1}{d_i} \tag{8.22}$$

When the eye is fully relaxed, $d_o = \infty$ and $f = \infty$

$$P_\infty = \frac{1}{f_\infty} = \frac{1}{\infty} + \frac{1}{d_i} = \frac{1}{d_i} \tag{8.23}$$

When the eye is fully accommodated and the object is at the near point, the power P_N is:

$$P_N = \frac{1}{f_N} = \frac{1}{0.25\,\text{m}} + \frac{1}{d_i}$$

(8.24)

Thus:

$$P_N - P_\infty = \frac{1}{0.25\,\text{m}} = 4\ \text{diopters}$$

(8.25)

The power of accommodation is defined as the difference of the powers associated with the extremes of vision. For a normal eye, it is 4 diopters. Visual acuity is defined as the minimum angular separation of two equidistant points of light that can be resolved into two separate objects by the eye. If the points are closer, the eye sees light coming from a single point source. The acuity of scotopic vision (low intensity light) is highest about 20 degrees from the fovea (20% from a line through the center of the lens and the fovea) where the density of rods is greatest. For photopic vision, acuity is highest on the fovea where cones are most dense. Photopic vision is always greater than scotopic vision.

8.9.3   Light absorption and black-and-white vision

An image is projected onto the retina after light from the exterior passes through the lens. In this section, we discuss how the image is recorded and where color perception comes from. The colors of certain pigments disappear under sustained exposure to light. This process is called bleaching and can be understood by assuming that a pigment alters its structure in the excited state in a way that makes it difficult to return to the ground state. The color produced by the pigment must fade when all the pigment molecules have been altered photochemically.

This suggests that the process of vision is related to a photochemical reaction. The retina contains two types of light-sensitive cells called photoreceptor (PR) cells. The eye contains 6 to 7 million cones and about 120 million rods. The PR cells are located inside a matrix of pigment cells that acts as a nonreflecting background and are about 10^{-6} m in diameter. After light absorption, PR cells send out an electrical signal.

Cones have a relatively high threshold of light intensity before they are triggered. On firing, they send a signal to the brain and are responsible for color vision. The rods are responsible for low-intensity black-and-white vision, i.e., for night-time viewing, and have low firing thresholds.

Wilhelm Kuhne isolated the visual pigment and determined it was a light-absorbing protein called rhodopsin or visual purple. Rhodopsin is located in the rods. It strongly absorbs visible light in certain rose or purple frequency ranges. The molecules of this protein are arranged in planar structures anchored in a membrane that can transmit electrical signals in response to stimuli received from the molecules (see Figure 8.29).

The light active group is called 11-*cis*-retina. After retina absorbs a quantum of light, it isomerizes (Figure 8.29). The new structure is called 11-*trans*-retinal — a straightened-out version of *cis*-retinal (see Figure 8.30). The isomerization takes place in about 10^{-2} s. The energy of the light quantum is transferred into

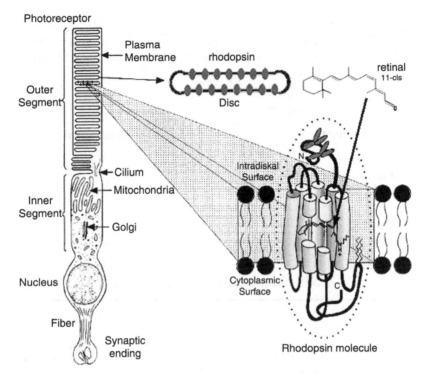

FIGURE 8.29
Planar structure of rhodopsin molecule.

restructuring of the molecule. The isomerization blocks the excited molecule from rapidly falling back into its ground state and reemitting the light quantum. Thus, rhodopsin is not a fluorescent molecule. Isomerization alters the charge of the *trans* part of the rhodopsin protein and this starts a chemical reaction and produces electrical activity by closing ion channels in the membrane. This process is the origin of the electrical signal that eventually reaches the brain as a facet of black-and-white vision.

8.9.4   Color vision

Based on the electromagnetic theory of light, we know that visible light is only a small part of the electromagnetic spectrum (see Figure 8.31).

The two major types of light-sensitive cells in the retina are rods and cones. Rods are found throughout the retina, except for a small region in the center of the fovea. Rods are more sensitive than cones and are solely responsible for peripheral vision and vision in very dark environments. Unlike cones, rods do not yield color information. Cones are concentrated in the fovea and are associated with color vision. Each of the three types of cones is sensitive to a different range of wavelengths (see Figure 8.32).

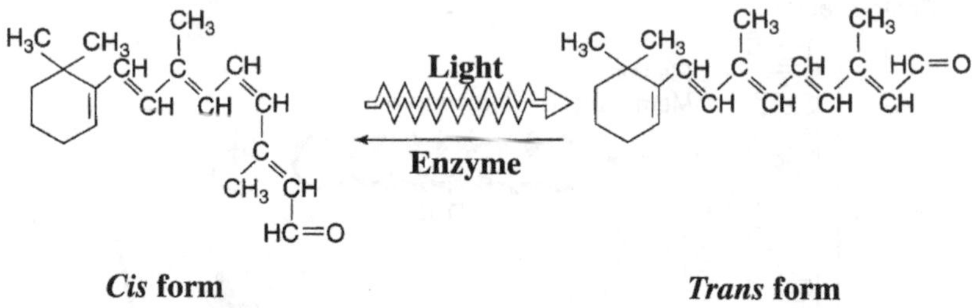

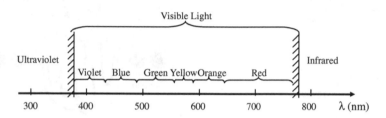

Cis **form**   *Trans* **form**

FIGURE 8.30
Retinal isomers.

FIGURE 8.31
**Small part of the electromagnetic spectrum, with the type of electromagnetic
radiation identified as a function of wavelength.**

The molecular basis of color vision was predicted by the Young–Helmholtz theory
of color vision that specifies only three color pigments: R, G, and B. Pigment R
absorbs predominantly in the red range, pigment G in the green, and pigment B in
the blue. If a light beam of many different wavelengths falls on the retina, it will
excite only a fraction of the pigment molecules in the cones. The retina will record
how many R, how many G, and how many B pigment molecules have been excited.
The information is then transmitted to the brain. While cones have three pigments
absorbing predominantly in the red, blue, and green ranges, we now know that the
Young-Helmholtz theory of color vision is far too simple.

Assuming the eye has three types of pigments, we could expect that three types of
color blindness correspond to eyes having only two of the three pigments. Research
reveals three types of dichromatic people whose sensations of color are limited to
what can be reproduced by mixing only two primary colors. Two of the three types
are very common and the third is somewhat rare. It is possible to deduce the response
curves of the three pigments assumed to be present in the human eye by comparing
the response to color of a person with normal vision to a person with each type of
dichromatic vision. Figure 8.33 illustrates these curves. However, none is similar to
curves of rhodopsin or iodopsin.

These pigments are presumably concerned merely with the gross detection of light
and are present in large quantities, particularly rhodopsin. Only if the light intensity
is great enough would color sensation be present; it would be produced by pigments
present in small amounts.

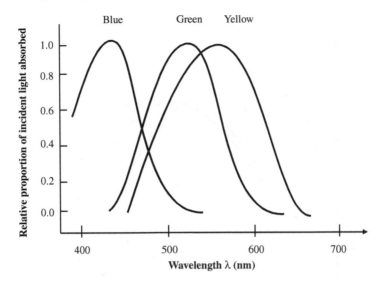

FIGURE 8.32
Relative sensitivities of the three types of cones are indicated by their relative absorption of light as a function of frequency.

8.9.5 Resolution of the human eye

The criterion for resolvability of images produced by light passing through an aperture was formulated by Lord Rayleigh and can be expressed as:

$$\alpha_{min} = 1.22\frac{\lambda}{a} \tag{8.26}$$

where α_{min} is the minimum angular separation in radians, λ is the wavelength of light in the medium, and a is the diameter of the aperture. The pupil is a variable aperture with a minimum diameter of about 2 mm. The eye is most sensitive to light of $\lambda = 500$ nm. Assuming that most refraction takes place in the cornea, we use $n = 1.33$ to find the value of α_{min} as

$$\alpha_{min} = 2.3 \cdot 10^{-4} \, \text{rad} \tag{8.27}$$

Since $\ell = 0.025$ m is the typical corneal–retinal distance, angular separation translates into linear separation according to:

$$\alpha_{min} = \frac{d}{\ell} \tag{8.28}$$

giving $d = (2.3 \cdot 10^{-4} \, \text{rad})(0.025 \, \text{m}) \cong 6 \, \mu\text{m}$. Since the average cone separation in the fovea is about 2 μm, the minimum distance of 6 μm calculated above corresponds to a span of three cones. In other words, two active cones must be separated by an inactive cone to create the sensation of an image. Note that α_{min} depends inversely on the diameter of the pupil a.

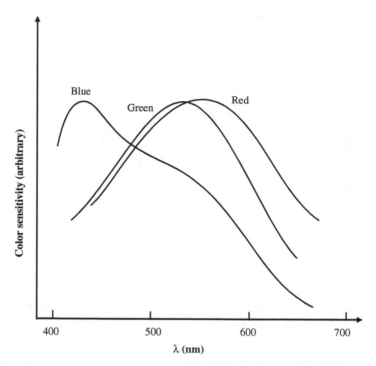

FIGURE 8.33
Theoretical response curves for three color-sensitive pigments in the human eye.

The smallest discernible detail creates an image about 6 μm in diameter, two to three times the size of a retinal cell. Outside the fovea, light-sensitive cells are not as closely spaced, and peripheral visual acuity is consequently less by a factor of about 10. The eye cannot detect detail that creates an image smaller than the spacing of light-sensitive cells. For the eye to resolve two adjacent dots, their images must fall on two nonadjacent cells. If both dots are images of small light sources and fall on adjacent cells, they are interpreted as a single larger dot.

In addition to the separation between retinal cells that limits ocular resolution, the eye also has a diffraction limit. Diffraction occurs when light passes through the pupil. The diameter of the pupil varies with light intensity, averaging about 3 mm. The outcome of the effect is such that a point source of light will not make a point image on the retina; it will produce a spot about 2 μm in diameter, very nearly the size of a retinal cell.

8.9.6 Quantum response of the eye

Materials of which the cornea, lens, and aqueous and vitreous humors are made completely absorb wavelengths below 380 nm and also absorb, to a lesser extent, light with wavelengths in the visible region. Only a few percent of quanta striking

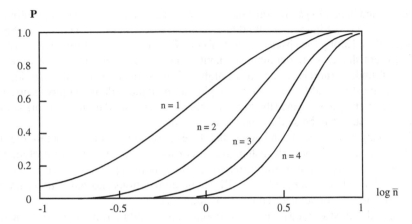

FIGURE 8.34
Frequency of visual response P of an observer (P versus log) as a function of the number of quanta n stimulating an individual receptor on the retina.

the cornea reach the retina. The rest are absorbed by materials along the path. Thus a measurement of the intensity of light I striking the cornea does not allow a direct calculation of the average number $\bar{n}$ of photons activating a single receptor on the retina. However, the two are linearly related, i.e., doubling intensity I automatically doubles the value of $\bar{n}$. Therefore:

$$\bar{n} = kI \qquad (8.29)$$

where k is a constant related to the absorbing ability of the materials of the eye and other factors are involved (Benedek and Villars, 2000).

When the eye is near the threshold of the vision process, $\bar{n}$ and I are small. Although the average number of quanta reaching each receptor on the retina is $\bar{n}$, some receptors will receive more and some less because of the random nature of the process. The statistical distribution of the frequency of occurrence of events of small probability around a mean value is well known. Thus we can work out the proportion of receptors receiving $\bar{n} - 1, \bar{n} - 2, \bar{n} - 3, \ldots$ etc. quanta and the proportion of receptors receiving $\bar{n} + 1, \bar{n} + 2, \bar{n} + 3, \ldots$ etc. We assume that a particular receptor will not pass on an electrical signal to its associated nerve fiber unless a minimum number n of quanta stimulate it.

Obviously, the smaller n and the larger $\bar{n}$, the greater the probability that any flash of light will be observed. Figure 8.34 shows the result of theoretical calculations for $n = 1, 2, 3,$ and 4 of the variation of the frequency v of response to a light flash as the number of a quanta reaching the retina is increased.

For a subject under test, the variation of frequency of response to a light flash cannot be plotted as a function of $\log \bar{n}$ as in Figure 8.34 because the value of $\bar{n}$ is not known a *priori*. However, it can be plotted as a function of $\log \bar{n}$, because as $\bar{n} = kI$ we have:

$$\log \bar{n} = \log(kI) = \log I + \log k \qquad (8.30)$$

where k and hence $\log k$ are constants. An experimental response curve, if plotted on the same scale as Figure 8.34 should be identical to the theoretical curves except that it is displaced along the abscissa axis by $\log k$. The curves for different values of n are of different shapes so the number of quanta n needed to cause a receptor to respond can be found. Parameter n may be as high as 8 or as low as 1 in human subjects.

In the most favorable cases, only one photon in an individual receptor is necessary to produce a response. Since the experiments were performed mainly on dark-adapted eyes, the results can be said to apply only to rods.

For many years it has been known that the eyes of all vertebrates contain pigments that play a part in vision. All such pigments consist of a specific type of protein called an opsin. It is present as a color group or chromophore, a particular configuration of vitamin A aldehyde called retinene. Biochemical reactions subsequently remove the retinene from the opsin and reduce it to vitamin A.

One of the most studied pigments, rhodopsin, is a combination of retinene and opsin from rods. It is also known as visual purple because of its characteristic color. When white light is passed through a cell containing rhodopsin, the absorption spectrum agrees exactly with the luminosity curve for scotopic vision uncorrected for absorption in the lens. Thus scotopic vision takes place by the absorption of light quanta by rhodopsin. Iodopsin, the pigment from cones has an absorption spectrum different from rhodopsin and when suitably corrected, different from the luminosity curve for photopic vision. This supports the idea that both rods and cones play a part in photopic vision.

When a section of a cone or rod cell is examined under an electron microscope, it consists of a stack of approximately 1,000 flattened sacs. This layered arrangement is found in all cells employing energy from light quanta, e.g., the chloroplasts responsible for photosynthesis in plants. The membranes surrounding the sacs are about 50 Å thick.

8.10   Sound perception biophysics

The sense of hearing in humans is intended to record sound and transform it into an electrical signal the nerves send to the brain. The human ear acts as a sound recording with an enormous range of intensities. Its ability to distinguish different sound frequencies is astonishing.

The ear consists of the outer or external ear, the middle ear, and the inner ear. The familiar shell-like external structure is called the pinna. It is connected to the auditory canal, a 2.5-cm long structure filled with air. The canal ends at the tympanic membrane (eardrum).

The middle ear or tympanic cavity is air-filled and situated inside the temporal bone of the head. It is closed off on one side by the tympanic membrane and the other side by two openings called the oval window and the round window. The middle ear is spanned by three small bones: the malleus or hammer, the incus or anvil, and the stapes

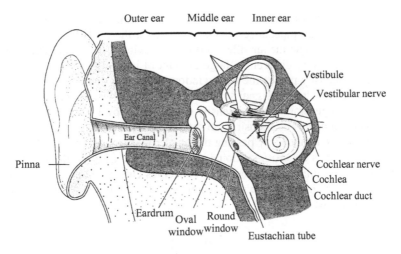

FIGURE 8.35
Anatomy of the ear.

or stirrup. The malleus is in contact with the tympanon and the incus is in contact with the oval window. Thus, the middle ear comprises the eardrum or tympanon, malleus, incus, stapes, and Eustachian tubes that transmit force exerted on the eardrum to the inner ear through the oval window. Because they form a lever system with a mechanical advantage of about two, the force delivered to the oval window is multiplied by two. The area of the oval window is about 1/20 that of the eardrum; thus the pressure created in the fluid-filled inner ear is about 40 times that exerted by sound on the eardrum. This system enables the ear to detect very low intensity sounds.

The middle ear also offers some protection against damage from very intense sounds. Muscles supporting and connecting the three small bones contract when stimulated by very intense sounds and reduce the force transmitted to the oval window by a factor of about 30. The reaction time for this defense mechanism is at least 15 ms, so it cannot protect against sudden increases in sound intensity. Figure 8.35 is a simplified representation of the anatomy of the ear. The inner ear or labyrinth is inside the temporal bone and is filled with a fluid called endolymph which is similar to the fluid inside cells.

The cochlea is the key structure of the inner ear. It has the form of a snail (*cochlea* means *snail* in Latin) and is a long spiral tube with about 2.5 turns. It starts at the oval window and ends at the round window. Inside the cochlea is the basilar membrane, a structure that is narrow and thick near the oval window and wide and thin near the round window. The organ of Corti transduces sound signals into electrical signals and rests on the basilar membrane. It consists of about 15,000 cochlea hair cells, each containing 50 to 100 hairs. The inner ear contains three semicircular canals, the vestibule, and the Eustachian tube. Sensory impulses from the inner ear pass to the brain via the vestibulo-cochlear nerve.

Sound waves impinging on the ear are pressure waves. The key purpose of the outer and middle ears is to guide the sound signal from the exterior to the oval window. Sound waves set the oval window in motion and cause motion in the endolymph fluid. An impedance mismatch exists at the oval window between the fluid-filled inner ear and the air-filled middle ear cavity. If the sound impedances of the two media are very different, a large amount of intensity may be lost. Only 2% of the signal intensity passes the oval window. However, the outer and middle ear compensate for this loss of signal strength by preamplifying sound signals as much as possible, thus generating high intensity signals at the oval window.

Amplification begins by focusing the sound wave. The power P delivered by the sound wave is the total energy passing the unit cross-section per second. Therefore, $P = IA$ where A is the cross-sectional area of the tube carrying the sound and I is the intensity. Assuming the sound wave conserves energy, we conclude that power is fixed along the channel. The sound intensity $I(A)$ then varies inversely with the cross-sectional area A. Thus:

$$I(A) = \frac{P}{A} \tag{8.31}$$

Narrowing the channel enhances sound intensity. The pinna collects sound much as a horn does and guides it to the auditory canal that amplifies the intensity. The channel has a radius of only 3 mm. The auditory canal acts like a resonant cavity for sound signals and assists the transmission of sound.

The next phase of amplification works differently in that the sound wave applies pressure oscillation to the tympanic membrane ΔP(tymp). The maximum force F on this membrane is thus equal to $F = \Delta P$(tymp)A(tymp) where A(tymp) is the surface area of the tympanic membrane, i.e., about 55 mm^2. The ear is connected mechanically to the oval window by three small bones, so the oval window also feels the oscillating force F. The area of the oval window is A(oval) = 3.2 mm^2. The force on the oval window is $F = \Delta P$(oval)A(oval). The ratio of pressure amplitudes is therefore:

$$\frac{\Delta P(oval)}{\Delta P(tymp)} = \frac{A(tymp)}{A(oval)} = 17 \tag{8.32}$$

The sound intensity increases as the square of the pressure amplitude so the middle ear magnifies the intensity by a factor of $17^2 = 289$. This more than compensates for signal loss due to impedance mismatch. The reason why the ear has three small bones instead of a single rod structure, is that a rod-like bone would deliver destructively large forces for intense sound. The mechanical structure of the ossicle is designed to damp high force levels produced on the oval window by loud sounds.

8.11   Gross features of the nervous system

The brain is probably the most complex structure in the known universe. While the brain is the product of millions of years of evolution, some structures unique to the human species appeared relatively recently. For example, only 100,000 years ago, the ancestors of modern man had brains weighing only about 0.5 kg, roughly a third of the average modern brain weight. Most of the increased weight is associated with the cortex consisting of two roughly symmetrical, corrugated, folded hemispheres astride a central core.

In evolutionary terms, all brains are extensions of the spinal cord. The distant ancestor of the human brain originated in some 500 million years ago. Life and survival were relatively simple functions so precursor brains consisted of only a few hundred nerve cells. As prehistoric creatures evolved and became more complex, so did the brain. A major change occurred when early fish crawled out of the seas onto the land. The increased difficulties of survival on land led to the development of the reptilian brain. The design is still discernible in modern reptiles and mammals and it offers a significant clue to our common evolutionary ancestry.

The next major transformation was the emergence of the mammalian brain equipped with a new structure, the cerebrum or forebrain covered by a cortex. By then, a brain consisted of hundreds of millions of nerve cells organized into separate regions associated with different tasks. About 5 million years ago, another type of cortex appeared in early humans. The surface of the cortex was organized into separate columnar regions less than 1 mm wide, each containing many millions of nerve cells. The new structure afforded much more complex processing capability. Finally, about 100,000 years ago, the new cortex underwent rapid expansion with the advent of modern humans. The cortex of the modern brain contains about two thirds of all neurons and weighs on average 1.5 kg.

The human brain consists of three separate parts. The lower section, sometimes called the brain stem, consists of the medulla oblongata that controls breathing, heart rate, and digestion and the cerebellum that coordinates senses and muscle movements. Many of their features are inherited from the reptilian brain. The second segment appears as a slight swelling in lower vertebrates and enlarges in higher primates and humans into a midbrain. The structures of the midbrain link the lower brain stem to the thalamus (for information relay) and the hypothalamus (instrumental in regulating drives and actions).

The hypothalamus is a part of the limbic system that lies above the brain stem and under the cortex and consists of a number of interconnected structures. The structures are linked to hormones, drives, temperature control, and emotions. One segment, the hippocampus, relates to memory formation. Neurons affecting the heart rate and respiration are concentrated in the hypothalamus and direct most physiological changes that accompany strong emotion. Aggressive behavior is linked to the action of the amygdala adjacent to the hippocampus which plays a crucial role in processing information as part of long-term memory. Damage to the hippocampus leads to inability to collect new stores of information. The systems of the lower and midbrain are rather simple. They are capable of registering experiences and regulating behavior largely outside of conscious awareness.

The human brain is constructed according to evolutionary development. The outer layer is the most recent development. The deeper layers contain structures inherited from reptiles and mammals. Finally, the forebrain appears as a mere bump in the brain of the frog. In higher life forms, it expands into the cerebrum and covers the brain stem in a shape resembling the head of a mushroom. It is further evolved in humans into the walnut-like configuration of left and right hemispheres. The highly convoluted surface of the hemispheres, the cortex, is about 2 mm thick and covers a total surface area of about 1.5 m^2. The structure of the cortex is extremely complicated. The cortex is where most intellectual functions are carried out. Some regions are highly specialized. For example, the occipital lobes located near the rear of the brain are associated with the visual system. The motor cortex helps coordinate voluntary muscle movements.

More neurons may be dedicated by the brain to connect to certain regions of the body. For example, fingers have many more nerve endings than toes. Symmetry between the right and left hemispheres is approximate. The brain has two occipital lobes, two parietal lobes, and two frontal lobes. For instance, the area associated with language is only in the left hemisphere. The frontal lobes occupy the front area behind the forehead and are associated with the control of responses to outside inputs. They are closely correlated with making decisions and forming judgments. In most people, the left hemisphere dominates over the right in deciding responses.

In individuals with normal hemispheric dominance, the left hemisphere that manages the right side of the body also controls language and general cognitive functions. The right hemisphere controls the left side of the body and manages nonverbal processes such as attention and pattern recognition. Although the two hemispheres are in continual communication, they act as independent parallel processors with complementary functions.

While the structural organization of the brain and the nervous system has been known for many years, the nature of the processes involved in their functioning is not completely known. We know from the investigations of generations of neurophysiologists that the basis of functioning of nerves is electrical signal propagation. Several aspects of the nervous system functioning have been modeled mathematically (Keener and Sneyd, 1998). The history of the development of neurophysiology is briefly outlined below (Selden et al., 1985).

The idea that nerves functioned by conducting "electrical powers" was first advanced by Stephen Hales. Experimental support for this concept came from Swammerdam and Glisson who showed that the muscles did not increase in volume when they contracted, concluding that the "humor" must be "ethereal" in nature. In 1786, Luigi Galvani, while dissecting the muscles of a frog, happened to touch a nerve with his scalpel while a static electrical machine was operating nearby. This caused electricity to traverse the nerve and produce muscle contraction. Based on this observation ands further experiments, he concluded that the electricity was generated within the animal's body. He designated it "animal electricity."

Alessandro Volta repeated and confirmed Galvani's observations, at first agreeing with the concept of animal electricity, then opposing it and attributing the current to a bimetal. This set off a scientific controversy that occupied the life sciences for the next century and a half. Eventually, a publication by Humboldt in 1797, just before Galvani's death, clearly established that Volta and Galvani were simultaneously right and wrong. Bimetallic electricity existed and so did animal electricity.

In the 1830s, Carlo Matteucci demonstrated that an electrical current was generated by injured tissues and that serial stacking of tissue could multiply the current in a manner similar to adding more bimetallic elements to a voltaic pile. The current existence of animal electricity was unequivocally proven. However, it was not located within the central nervous system. Du Bois-Reymond discovered that when a nerve was stimulated, an electrically measurable impulse was produced at the site of stimulation. After traveling down the nerve, it produced muscular contraction. He thus unlocked the mystery of the nerve impulse, and measured the resting potential as a steady voltage on unstimulated nerves and muscles.

Following Du Bois-Reymond's discovery, von Helmholtz measured the velocity of a nerve impulse as 30 m/s. In 1868, Julius Bernstein presented his theory of nerve action and bioelectricity, which became the cornerstone of modern neurophysiology. The Bernstein hypothesis stated that the membrane of a nerve cell could selectively pass certain kinds of ions. Situated within the membrane was a mechanism that separated negative from positive ions, permitting the positive ones to enter the cell and leaving the negative ions in the fluid outside the cell. When equilibrium was reached, an electrical (transmembrane) potential would then exist across the membrane.

The nerve impulse was simply a localized region of depolarization or loss of transmembrane potential that traveled down the nerve fiber; the membrane potential was immediately restored behind it (see Chapter 7). Bernstein also postulated that all cells possessed a transmembrane potential derived from the separation of ions, and he explained Matteucci's observation that current of injury was due to damaged cell membranes that leaked their transmembrane potentials.

In 1929, Hans Berger invented the electroencephalogram (EEG) — a standard neurological testing and diagnostic tool. In the 1930s, Burr conducted experiments on direct current potentials measurable on the surfaces of a variety of organisms and related changes in the potentials to physiological factors such as growth, development, and sleep. He formulated the concept of a bioelectric field generated by the sum total of electrical activities of all the cells of the organism.

Leao later demonstrated that depression of activity in the brain was always accompanied by the appearance of a specific type DC potential, regardless of the cause. Gerard and Libet expanded this concept in a series of experiments and concluded that the basic functions of brain excitation, depression, and integration were directed and controlled by the DC potentials. However, the sources of the DC potentials were not the transmembrane potentials. At the junction between the nerve and its end, microscopic examination revealed a gap known as a synapse.

In the 1920s, Otto Loewi experimentally showed that current transmission across the synaptic gap was due to the release of acetylcholine into the gap and it subsequently stimulated the receptor site on the end organ. Finally, the broad outlines of the Bernstein hypothesis were proven by Hodgkin, Huxley, and Eccles in the 1940s. Using microelectrodes that could penetrate nerve cell membranes in a protocol called the patch clamp technique, they demonstrated that normal transmembrane potential is produced by exclusion of sodium ions from the nerve cell interior and, when stimulated to produce an action potential, the membrane permits these ions to enter the cell.

However, it still was not clear how neurons were integrated to work together to produce coherent brain function. In the 1940s, neurophysiologists Gerard and Libet performed experiments on DC electrical potentials in the brain. Steady or slowly varying potentials oriented along the axonodendritic axis were measured. These potentials changed their magnitudes as the excitability level of the neurons was altered by chemical treatment. Gerard and Libet also found slowly oscillating potentials and "traveling waves" of potential change moving across the cortical layers of the brain at speeds of approximately 6 cm/s. These wave phenomena were interpreted as electrical currents that flowed outside the nerve cells and exerted primary controlling action on neurons.

It became apparent that brain circuitry was not a simple one-on-one arrangement. Single neurons were found to have tree-like arrangements of dendrites with input synapses from numerous neurons. Dendritic electrical potentials appeared to be additive; they did not propagate like action potentials. When a sufficient number were generated, membrane depolarization reached a critical level and an action potential was generated. Other neurons whose action potentials were inhibitory were found. Graded responses in which ion fluxes occurred across neuronal membranes were discovered. While insufficient in magnitude to produce an action potential, they produced functional changes in neurons.

The complexity of the brain structure and function is enormous. Only about 10% of the brain is composed of neurons; the remainder consists of a variety of perineural cells, mostly glia cells, that have no ability to generate action potentials. Electron microscopy revealed close associations between glia and neurons as well as between glia cells. The analog of the glia cell, the Schwann cell, was found to inhabit all peripheral nerve fibers outside of the brain and spinal cord. Schwann cells are in continuous cytoplasmic contact along the entire length of each nerve. Biochemical changes were found to occur in the glia concurrent with brain activity and strong evidence was found that glia cells were involved in memory storage. While the action potentials of neurons did not seem to influence glia cells, the reverse appeared possible. The extraneuronal currents originally described by Libet and Gerard may be associated with some electrical activity in nonneuronal glial cells.

It is feasible that the electromagnetic energy in the nervous system is a major factor exerting control over growth and development processes. For example, the nerves, acting in concert with some electrical factors of the epidermis, produce a specific sequence of electrical potential changes that cause limb regeneration in some animals.

8.12   Bioelectricity and biomagnetism

In 1941, Szent-Gyorgyi made the still controversial suggestion that semiconductivity could exist in living systems. He postulated that proteins were sufficiently well organized to function as a crystalline lattice. He proposed that fibrous proteins formed extended systems with conduction energy levels lowered to the point of permitting semiconducting current flow. This concept has proven to be of considerable value in explaining many life functions that are poorly explained when viewed solely within the framework of biochemistry.

Over the past 50 years increasing scientific data has indicated the existence of electromagnetic properties in living organisms, as proposed by Szent-Gyorgyi. For example, the perineural cells of the CNS have some properties analogous to semiconductivity and are responsible for the production and transmission of steady or slowly varying electrical currents within that tissue with many characteristics of an organized data transmission and control system. Specific solid-state electronic properties of bone matrices have been shown to produce growth in response to applied mechanical stress. This mechanism is at least partially mediated through action on the cell membrane, with subsequent activation of the internal DNA–RNA mechanism of the cell.

The only experimental work that directly applied Szent-Gyorgyi's concepts was by Huggins and Yang who showed that carcinogenic agents functioned by a combination of biochemical steric organization and biophysical mechanisms of electron transfer. These agents were shown to attach to certain areas of cell surfaces and affect electron transfers. Compounds with the same steric property but lacking electron transfer capability were believed to be noncarcinogenic (Selden et al., 1985).

Although conductivity of organic materials was reported as early as the turn of the century in solid anthracene, Szent-Gyorgyi should be credited with the idea that electron transfer may be crucial to biological processes. He suggested in 1941 that the mobile π electrons of conjugated carbon double bonds may be transferred from molecule to molecule and that this may be fundamental to the workings of biological systems. Five years later, he reported the discovery of photoconductive effects for protein films. The conduction within organic solids was then described by means of a band model as in traditional semiconductors.

Conduction is viewed as an intrinsic property of the material where thermal excitation of electrons from a valence band to the conduction band leads to a dynamic equilibrium of electron hole pairs and the result is the appearance of available charge carriers. A typical organic material has a band gap too large for charge carriers to be present. Since the number of charge carriers is temperature-dependent, conductivity is also temperature-dependent. The number density of charge carriers n is described by the following equation:

$$n_i = n_i^o \exp(-E/2kT) \tag{8.33}$$

where E is the band gap, k is Boltzmann's constant, T is the absolute temperature, and n_i^o is the density of carriers in the conduction band. The conductivity is determined by summing the contributors to conduction by holes and by electrons where the symbol μ with appropriate subscripts represents the carrier mobilities and e is the charge of an electron:

$$\Sigma = |e|(N_e \mu_e + N_h \mu_h) \tag{8.34}$$

Electronic conduction is a well studied problem in condensed matter physics. It is the phenomenon by which a current may be passed through a material. The conductivity of a material is defined experimentally by the constant σ, which relates the current density J in the material to the electric field E in the material by the following linear relationship:

$$J = \sigma E \tag{8.35}$$

A material with an electrical conductivity in excess of $10^4 \, \Omega^{-1} \, cm^{-1}$ is generally considered a conductor. Materials are classified as insulators when their electrical conductivity is below $10^{-10}\,\Omega^{-1}\,cm^{-1}$. Within the intermediate range from 10^{-9} to $10^3\,\Omega^{-1}\,cm^{-1}$, materials are classified as semiconductors. An organic semiconductor is an organic compound that exhibits properties inconsistent with electrical insulators. These organic materials may be grouped into three categories: (1) molecular crystals characterized by van der Waals bonds, (2) charge transfer complexes with covalent and coordinate bonding, and (3) polymers.

In the mid 1950s, the idea of organic semiconductors was generally ruled out due to the large activation energies in the range of several electron volts that were much higher than the energies observed in biological processes, typically less than 0.49 eV which is the free energy of ATP hydrolysis. However, several key points were overlooked. First, measurements were made only in dried systems, whereas living systems function only when wet, typically at 80% water content. It was later revealed that water causes significant changes in activation energy and may improve conductivity of proteins by a factor of up to 10^{10}!

The treatment of solvation remains a major difficulty in current studies of protein interaction. Certain proteins have unusual structures that seem specially designed for high electron conductivities and low activation energies. Essentially, these structures can be viewed in analogy to donor and acceptor impurities. Researchers measured an activation energy of about 0.3 eV for cytochrome oxidase which is known to play a key role in electron transport and is instrumental in the proton pump mechanism of energy production in mitochondria. It must be stressed, however, that most proteins have activation energies at least three times larger. Finally, conduction within individual structures may be quite different from conduction across interfaces between structures. This may be the parameter measured when an electrode is placed across a compressed mass of thousands of microscopic molecules. Hence, the conductivity may be higher within individual molecules than previously deduced from the conductivity of a molecular aggregate.

Elay et al., investigated protein conduction and found weak semiconduction in dark conditions. Conduction increased exponentially with hydration. Pethig and Szent-Gyorgyi carried out transport experiments and found that DNA and RNA had reproducible conduction with activation energy of about 1.1 eV. However, the inability of other groups to reproduce many of these experiments underlined the difficulty of working with organic matter. This difficulty is attributed to the high sensitivity of these systems to their environments.

In the nearly 60 years following Szent-Gyorgyi's prediction, kinetic evidence supports semiconduction in living cells. However, the mechanisms present in biological matter differ significantly from condensed matter behavior. Charge transport in biomolecular systems is quite different from the free flow of electrons or holes in metals and semiconductors. A single charge in a biomolecule strongly perturbs its environment and acts like a polaron. Surprisingly, environmental distortions accompanying charge transfer in proteins can, at sufficiently low temperatures, occur by quantum tunneling effects. In biomolecules, electron transfer can occur over long distances, reaching tens of angstroms. This occurs usually as hole tunneling along the covalent backbones of these molecules.

Charge transfer complexes play an important role in biology. They are responsible for generating usable energy through respiration and photosynthesis. However, not all biological electronic conduction can be described by charge transfer complexes. In some cases, electron donors and acceptors are isolated from each other. Consequently, the carriers reside in the vicinity of the same molecule for a long time before jumping to a neighboring molecule. This view gives rise to a "hopping" model because electron mobility is low and governed by activation energy. Such an interpretation is particularly useful when a material exhibits a periodic lattice or has sufficiently weak carrier concentration that interactions between carriers are negligible.

The study of the effects of electrical fields on cells goes back to 1892 when Wilhelm Roux subjected animal eggs to electric fields and observed pronounced stratification of the cytoplasm. In the decades that followed, a number of effects have been observed that implicate electric fields and/or currents in the cytoskeletal or cytoplasmic self-organization processes. For example, growing tissues and organs exhibit sensitivity to magnetic fields and electric currents. In cell division, coherent polarization waves play a key role in chromosome alignment and subsequent separation. Regulation of cellular growth and differentiation (including the growth of tumors) may be directly modulated by electromagnetic fields. Static magnetic and electric fields can alter the mitotic index and cell cycle progression of a number of cell types in various species.

Low frequency electromagnetic fields in the range of 50 to 75 Hz definitely cause perturbations in mitotic activities of plant and animal cells and a significant inhibiting effect on mitotic activity occurs early during exposure. Unfortunately, evidence indicating direct responses of microtubules to electric and magnetic fields is insufficient. Vassilev observed alignment of microtubules in parallel arrays after application of both electric and magnetic fields. This was later supported by Unger's experiments with electric fields in the range of 300 V/m and magnetic fields ranging up to 30 T as shown by Bras (1995).

Cooper (1981) further proposed that the onset of mitosis is associated with a ferroelectric phase transition that establishes an axis of oscillation for the cellular polarization wave. The mitotic spindle apparatus would delineate the polarization field with microtubules lined up along electric field lines. The poles would represent regions of highest field intensity and the equatorial plane would provide a nodal manifold. The chromosome condensation during this transformation would, in this picture, be induced by static dielectric polarization of the chromatin complex resulting from the cellular ferroelectric phase transition. Recent work by Gagliardi (2002) attempts to explain all stages of mitosis as controlled by electric fields acting on charges.

Several observations of current flows in animal cells deserve attention in this context. In the phase between fertilization and first cleavage, a steady current enters the animal pole and leaves the vegetal pole (equator). In the silk moth oocyte–nurse complex, the oocyte cytoplasm is slightly more positive than the nurse cell cytoplasm despite their connection through a broad cytoplasmic bridge passing a small electric current. A steady current enters the prospective cleavage furrows in frog and sea urchin eggs during the 10 min before cleavage initiation. About 8 min after initiation, this current reverses its direction and leaves the furrow region.

Mascarenhas demonstrated electret properties of bound water with attendant nonlinearity, hysteresis effects, and long relaxation times (on the order of 1 s). Activation energies of 7.0 to 9.0 kcal/mol have been measured, all of which indicate a long-range dipolar order leading to the formation of internal electric fields or perhaps collective oscillations of electric fields. Protonic conductivity is consistent with the above and also with GTP hydrolysis that releases an H^+ ion. The electret state of hydrated biopolymers is a general property of these systems. Bound water strongly contributes to dielectric polarization of biomolecules in solution.

Nonlinear effects in transport and polarization storage that depend on the level of hydration were discovered. A nonlinear bioelectret may stimulate ferroelectric hysteresis curves that introduce memory and irreversibility to the behavior of these systems. Conformational states of tubulin dimers in MTs are believed to be coupled to charge distribution or dipolar states (Brown and Tuszynski, 1997). This mechanism, while still requiring molecular level elucidation, may lead to the presence of piezoelectric properties.

8.12.1 Biological piezoelectricity

Piezoelectricity is the electrical response of certain materials to strain. Pyroelectricity is the electrical response of a material to a thermal gradient. Athenstadt investigated piezoelectric and pyroelectric properties of several living organisms and found that many structural tissues possess permanent electrical polarization.

The subunits comprising these tissues all have permanent electrical dipoles. In particular, Athenstaedt established that MTs are pyroelectric and hence possess electric dipoles. This is consistent with the observed alignment of MTs assembled in the presence of electromagnetic fields and with calculations presented in Chapter 4 that estimate the dipole moment of tubulin, given its structure in zinc ion-induced sheets. In the 25 years since those studies, the piezoelectric properties of many tissues have been confirmed. Tissues such as collagen and bone develop surface voltages of 10 to 150 mV in response to strain. The piezoelectric effect appears because mechanical deformation or strain displaces some electrons toward the compressed surfaces. The surfaces become negatively charged. The potential differences may stimulate bone growth.

8.12.2 Biomagnetism

Baule and McFee first detected a biomagnetic field in 1963 in association with the action of the human heart. The field intensity was five orders of magnitude lower than Earth's normal field. Later, Cohen (1967), using the same technique coupled with a signal averaging computer, presented evidence of a magnetic field of even weaker strength around a human head.

The invention of SQUID (superconducting quantum interference device) permitted detection of these and even lower intensity fields with relative ease. It was first applied by Cohen in 1970 for a more complete detection of the human magnetocardiogram. In 1972, he reported that the magnetic field in space around the human head demonstrated a wave form pattern, similar to the EEG, as measured by skin electrodes. Since then, technology improvements have permitted detection of magnetic fields associated with evoked responses. Correlations of EEG and magnetoencephalogram (MEG) results have shown that low frequency components of the EEG are well represented in the MEG, but that high frequency components (sleep spindles) are usually missing.

The retina of the eye is a direct extension of the brain and it demonstrates a number of DC electrical phenomena. The electroretinogram seems to be generated by a steady dipole field extending from the retina to the cornea. A DC magnetic field has been detected associated with it, indicating that current flow occurs. Most recently, DC magnetic fields from the brain have been detected.

8.13   Immune system and its models

The immune system (IS) of a vertebrate attacks and eventually destroys foreign invaders such as viruses and bacteria, jointly referred to as pathogens. The IS is interactively connected with the nervous system and distributed throughout the body. IS cells are highly complex living machines with specific tasks such as tagging, signaling, and binding. Antibody molecules tag and mark foreign material for removal by lymphocytes, phagocytes, and other cells that assist in the functioning of the immune system.

The key elements of the immune system function as combat troops. They include macrophages, antibodies of various specificities, T cells, B cells, helper T cells, and cytokines. A typical mammalian IS is believed to contain 10^7 to 10^8 different antibody types, each of which has a different and unique molecular composition. Understanding and modeling IS functioning is a highly complex challenge.

To identify and eliminate foreign material, the IS must first distinguish foreign molecules (antigens) from its own molecules. Failing to do so results in a host of autoimmune syndromes and diseases such as allergies, arthritis, and HIV/AIDS, to name but a few. The task requires the system to develop, learn, and remember pattern recognition. To frustrate recognition by the IS, a pathogen can alter its outermost "coat," sometimes even during the course of a single attack. The counterattack strategy involves a feedback loop. It employs genetic mechanisms to implement molecular changes of some of the IS cells on a time scale of days. This can be characterized as an adaptive process (Perelson, 1988).

The part of the antibody that identifies antigens is called a paratope (antibody-combining region). It has unique shape and molecular structure designed for specific binding. The antibody-combining region is usually formed by a groove or a pocket in the three-dimensional structure of the antibody. Paratopes are matched with epitopes of the corresponding antigen in a lock-and-key fashion. In addition to specific keys that match only certain locks, the IS has access to "master keys" that can open many locks.

The strategies developed to avoid self-attacks include (1) blocking the production of antibodies that react with the molecules of the host organism and (2) elimination or suppression of these molecules when they are produced. This aspect of the IS obviously leads to discussion of the self–not self recognition problem.

As is shown in Figure 8.36, an antibody has a paratope and an epitope. An epitope unique to a given antibody type is called an idiotope. Antibodies are produced by

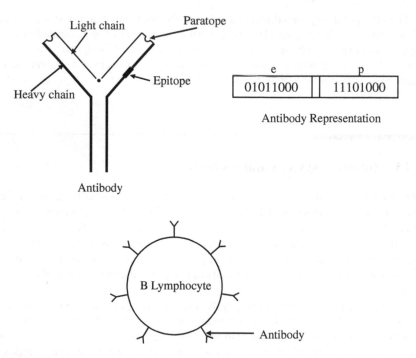

FIGURE 8.36
Structure of an antibody and its digitized representation for the purpose of modeling; sketch of a B lymphocyte.

B lymphocytes. Each lymphocyte has about 10^5 antibodies on its surface, each of which has an identical paratope acting as a sensor. When triggered by the presence of a matching epitope, B lymphocytes proliferate and secrete free antibodies. The process of amplifying only lymphocytes that produce a desired antibody type is called clonal selection. About 5% of all B lymphocytes are replaced every day with new ones generated in the bone marrow. This translates to a production level in humans of 10^6 lymphocytes per second, many of which are different.

Research indicates that antigen-antibody recognition processes extend over several levels. Type A recognizes type B, which in turn recognizes type C, etc., leading to the formation of complex reaction networks. Jerne (1973) called this type of recognition and regulation the idiotypic network model.

On a molecular level, antibodies have two polypeptide chains: one heavy and one light, each of which can exist in at least 10^4 types. The chains can combine independently leading to 10^8 possible combinations. While the heavy chain is encoded by four genes (V, D, J and C), the light chain is encoded by V, J, and C. New antibody types are generated by several processes. The key processes are cross-over, inversion, and point mutation.

Models of idiotypic networks are gross oversimplifications but they nonetheless can be easily implemented on a computer to provide a mathematical framework for the concepts described above. The major simplification is to represent the sequence of amino acids specifying the chemical properties of both the epitope and the paratope as a binary sequence (a string of zeros and ones). An antibody is represented as a pair of strings (p and e for paratope and epitope, respectively). Interactions between antibodies and antigens are modeled by matching between their complementary strings. Matches do not have to be precise. A matching threshold is defined as a minimum number of matching positions in the two strings below which no interaction takes place.

Letting $e_i(n)$ denote the nth bit of the ith epitope string, with $p_j(n)$ denoting the nth bit of the jth paratope string, and denoting the exclusive *or* operation (corresponding to complementary matching) by Λ, the matrix of matching specificities m_{ij} (Farmer et al., 1986) is:

$$m_{ij} = \sum_k G \left(\sum_n e_i(n+k) \Lambda p_j(n) - s + 1 \right) \quad (8.36)$$

where $G(x) = x$ for $x > 0$ and $G(x) = 0$ otherwise. The sum over n ranges over all possible positions on the epitope and paratope; the sum over k allows the epitope to be shifted with respect to the paratope. G measures the strength of a possible reaction between the epitope and the paratope.

For a given orientation, i.e., value of k, G is zero if less than s bits are complementary. If s or more bits are complementary, $G = 1 + \delta$, where δ is the number of complementary bits in excess of the threshold. If matches occur at more than one orientation, we sum their strength in Equation (8.36) to account for the fact that molecules may be able to interact in more than one way, and thus react more strongly because they spend more time together than molecules that can interact in only one orientation.

A subsequent simulation of the behavior of the IS can then be performed with respect to antibody and antigen concentrations and not individual molecules. Using N antibody types with concentrations $\{x_1, x_2, \ldots, x_N\}$ and n antigens with concentrations $\{y_1, \ldots, y_n\}$, the microscopic assumptions above lead to the following sets of differential equations:

$$\dot{x}_i = c \left[\sum_{j=1}^{N} m_{ji} x_i x_j - k_1 \sum_{j=1}^{N} m_{ij} x_i x_j + \sum_{j=1}^{n} m_{ji} x_i y_j \right] - k_2 x_i \quad (8.37)$$

The first term represents the stimulation of the paratope of an antibody of type i by the epitope of an antibody of type j. The second term represents the suppression of antibody of type i when its epitope is recognized by the paratope of type j. The forms of these terms are dictated by the fact that the probability of a collision between an antibody of type i and an antibody of type j is proportional to $x_i x_j$. Parameter c is a rate constant that depends on the number of collisions per unit time and the rate of antibody production stimulated by a collision.

The match specifies m_{ij} to take into account what reactions occur and how strongly. The constant k_1 represents a possible inequality between stimulation and suppression. When $m_{ij} = m_{ji}$, the interactions between paratopes and epitopes become symmetric. This system is driven by antigens, represented by the third term. To model the full immune system response, we must also introduce equations to model antigen concentrations y_i which may change as the antigens grow or are eliminated. The final term models the tendency of cells to die in the absence of any interactions at a rate determined by constant k_2.

An essential element of the model is that the list of antigen and antibody types must be dynamic, changing as new types are added or removed. Thus N and n in Equation (8.37) change with time, but on a time scale that is slow compared with the changes that occur in x_i. In the simulations, Equation (8.37) is integrated for a period, then the composition of the system is examined and updated as needed. To perform this updating, we place a minimum threshold on all concentrations, so that a variable and all of its interactions are eliminated when the concentration drops below the threshold. This simulates the elimination of an antigen or the death of the last cell expressing a given antibody type.

Much remains to be done in terms of devising realistic models of the IS. In the real immune system, antigens are sometimes remembered for long periods that may be comparable to the lifespan of the organism. The nature of the remembering mechanism is not exactly known. The diversity and complexity of the immune system still remain major challenges but modeling efforts appear to be gaining popularity and increasing their sophistication steadily (Nowak and May, 1994; May and Nowak, 2000).

References

Athenstaedt, H., *Ann. N.Y. Acad. Sci.*, 238, 68. 1974.

Bras, W., An X-ray Fibre Diffraction Study of Magnetically-Aligned Microtubules in Solution, Ph.D. thesis, Liverpool John Moores University, Liverpool, U.K., 1995.

Baule, G.M. and McFee, R., *Am. Heart, J.*, 55, 95, 1963.

Benedek, G.B. and Villars, F.M.H., *Physics with Illustrative Examples from Medicine and Biology*, Springer, Berlin, 2000.

Brown, J.A. and Tuszynski, J.A., *Phys. Rev. E*, 56, 5834, 1997.

Bruinsma, R., *Physics*, International Thomson Publishing, New York, 1998.

Cerdonio, M. and Noble, R.W. *Introductory Biophysics*, World Scientific, Singapore, 1986.

Cohen, D., Edelsack, E.A., and Zimmerman, J.E., *Appl. Phys. Lett.*, 16, 278, 1970.

Cohen, D. and Givier, E., *Appl. Phys. Lett.*, 21, 114, 1972.

Cohen, D., *Science*, 156, 652, 1967.

Cooper, M.S., *Coll. Phenom.* 3, 273, 1981.

Farmer, J.D., Packard, N.H., and Perelson, A., *Physica D*, 22, 117, 1986.

Fung, Y.C., *Biomechanics: Mechanical Properties of Living Tissues*, Springer, Berlin, 1993.

Gagliardi, L.J., *Phys. Rev. E*, 66, 11901, 2002.

Jerne, N.K., *Sci. Amer.*, 229, 52, 1973.

Keener, J.P. and Sneyd, J., *Mathematical Physiology*, Springer, Berlin, 1998.

Mascarenhas, S., *Ann. NY. Acad. Sci.*, 238, 36, 1974.

May, R.M. and Nowak, M.A. *Virus Dynamics: The Mathematical Foundations of Immunology and Virology*, Oxford University Press, Oxford, U.K., 2000.

Nowak, M.A. and May, R.M., *Proc. R. Soc. Lond. B*, 255, 81, 1994.

Perelson, A.S., Ed., *Theoretical Immunology*, Vols. 1 and 2, Addison-Wesley, Reading, MA, 1988.

Selden, G., Becker, R.O., and Guarnaschelli, M.D., Eds., *The Body Electric: Electromagnetism and the Foundation of Life*, William Morrow, New York, 1985.

Tuszynski, J.A. and Dixon, J.M., *Applications of Physics to Biology and Medicine*, John Wiley & Sons, New York, 2002.

Vassilev, P.H., Dronzin, R.T., Vassileva, M.P., and Georgiev, G.A., *Biosci. Rep.*, 2, 1025, 1982.

Zamir, M., *The Physics of Pulsatile Flow*, American Institute of Physics, New York, 2000.

9

Biological Self-Regulation
and Self-Organization

9.1   Introduction

The first issues facing a physicist trying to model biological systems are complexity and diversity. To illustrate the orders of magnitude involved, an organism such as the human body contains ~ 200 types of cells, $\sim 0.6 \times 10^{14}$ individual cells, and $\sim 10^6$ types of proteins. The entire organism is hierarchically organized and includes a number of autonomous or semiautonomous networks such as the nervous system, the immune system, the circulatory system, and others. The human nervous system is composed of 10^9 to 10^{10} neuron cells with 10^3 to 10^4 synaptic connections, each of which contains 10^5 tubulin proteins and 10^2 MAP interconnections between neighboring microtubules formed inside axons.

The dynamic functioning of the various systems relies on enzymes, motor proteins, and other macromolecules involved in intracellular trafficking and signaling. The complications do not end there. Many functional proteins exhibit conformational changes during their interactions with other molecules and post-translational modifications resulting when proteins such as actin or tubulin self-assemble into filaments.

How can we begin to understand the complexity, diversity, and interdependence of biological nanoscale machinery? This chapter addresses some of the issues contained in this rhetorical question. Our focus will be on self-organization, self-regulation, biological information, and evolutionary retention of earlier advances.

9.2   Self-organization

Ilya Prigogine in his seminal work showed that physical, chemical, and some biological systems far from thermodynamical equilibrium tend to self-organize by reducing their entropy and forming *dissipative structures*. For his contribution to chemistry, Prigogine received the Nobel Prize.

The main mathematical approach to the description of dissipative structures is via reaction–diffusion equations that are often coupled due to the chemical reactions bringing several reactant species together to form a product molecule. The nonequilibrium aspect of the theory is based on external pumping and can be modeled by external force terms in coupled reaction–diffusion equations. The new element that Prigogine explored is the emergence of limit cycles corresponding to self-sustained oscillations in chemical systems. This led to a new philosophical world view of self-organization (Prigogine and Stengers, 1984), as the arrow of time, the irreversibility of change, and the interplay between chance and necessity in real physical, chemical, and biological systems on all length and time scales (Nicolis and Prigogine, 1977, 1989; Prigogine, 1980).

Rigorous scientific models of bifurcations in nonequilibrium, nonlinear systems exhibiting the emergence of *order through fluctuations* developed by the Brussels school under Prigogine's leadership, made a huge impact across a spectrum of scientific disciplines. Inspired by Prigogine's work, Erich Jantsch (1979) made an ambitious attempt to synthesize everything known at the time about self-organizing processes from the Big Bang to the evolution of society into an all-encompassing world view. In a somewhat similar vein, Hermann Haken (1978) coined the term *synergetics* for the study of collective patterns emerging from many interacting components found in chemical reactions, crystallization patterns, lasers, slime molds, and even social upheavals.

Another Nobel laureate, biochemist Manfred Eigen (1992), focused on the origin of life, the domain where chemical self-organization and biological evolution meet. In 1979, he introduced the concept of a *hypercycle* to distinguish an autocatalytic cycle of chemical reactions containing other cycles from the less exotic limit cycle phenomena. He also introduced the idea of *quasispecies* representing the fuzzy distribution of genotypes characterizing a population of quickly mutating organisms or molecules.

The last two decades of the twentieth century witnessed a flurry of physical and mathematical modeling of nonlinear systems in physics which led to the emergence of the concept of deterministic *chaos* — a deterministic process characterized by extreme sensitivity to its initial conditions (Crutchfield et al., 1986). Chaotic behavior was studied both in abstract mathematical models and in diverse applications including weather prediction, the Solar System, and the stock exchange. Although chaotic dynamics is not strictly a form of evolution, it is an important aspect of the behaviors of complex systems. Cellular automata (see Chapter 5) which are mathematical models of distributed dynamical processes characterized by a discrete space and time evolution, have become powerful computational tools for studying complex systems and nonlinear phenomena such as chaos and pattern formation. Stephen Wolfram provided a fundamental classification of the canonical types of behavior in cellular automata models.

Catastrophe theory is another mathematical approach to strongly nonlinear systems that exhibit dramatic changes in behavior such as bifurcations. It was developed by René Thom (1975) in order to model the continuous development of discontinuous forms in organisms and other nonlinear systems, giving a mathematically precise and elegant formulation of much older work by D' Arcy Thompson (1917) on growth and form in biology.

Benoit Mandelbrot (1983) founded the field of fractal geometry which models the recurrence of similar patterns on different spatial and temporal scales characteriz-ing a huge number of natural systems. Such self-similar structures exhibit power law dependences, like the famous Zipf's law governing the frequency of words. Applications of fractality to biology and medicine are legion and include the branching structures of ferns, lungs, and blood vessels in the retina to name but a few, By studying processes such as avalanches and earthquakes, Per Bak (1988, 1991, 1996) showed that many complex systems spontaneously evolve to the critical edge between order (stability) and chaos, where the size of disturbance obeys a power law and large disturbances are less frequent than small ones. This phenomenon, which he called *self-organized criticality*, may also explain the punctuated equilibrium dynamics seen in biological evolution. The reader is referred to Appendix C for more detailed descriptions of the nonlinear concepts mentioned above.

9.3 Homeostasis

A person or animal in mortal danger prepares for action consciously and unconsciously, i.e., through voluntary and involuntary physiological changes. The body mobilizes reserves of energy and produces certain hormones such as adrenalin that prepare it for fight-or-flight response. Mobilization involves many familiar physiological reactions. In the presence of emotion, danger, or physical effort, the heart beats faster, respiration quickens, the face reddens, and perspiration appears. An individual may also experience shortness of breath, cold sweats, shivering, and trembling legs. These physiological manifestations reflect the efforts of the body to maintain its internal equilibrium. Action can be voluntary, i.e., having a drink when thirsty or eating when hungry, or involuntary, i.e., shivering, sweating, nausea, or vertigo.

The internal equilibrium of the body, the ultimate gauge of its proper functioning, involves maintenance of constant rates of concentration in the blood of certain molecules and ions that are essential to life and maintenance at specified levels of other physical parameters such as temperature despite environmental changes.

This amazing ability of the body to maintain dynamic equilibrium has intrigued many physiologists. In 1865, Claude Bernard remarked, in his *Introduction to Experimental Medicine* that the "constancy of the internal milieu was the essential condition to a free life." However, the linking of mechanisms that affect regulation of the body is credited not to Bernard, but to Walter Cannon. In 1932, Cannon, impressed by "the wisdom of the body" maintaining physiological equilibrium with such efficiency, devised the term *homeostasis*, a Greek work meaning *to remain thse same*.

Homeostasis is a remarkable property of highly complex open systems. A homeostatic entity such as a cell, a large industrial organization, or a society of ants is an open system that maintains its structure and functions using multiple dynamic equilibria closely controlled by interdependent control mechanisms. Such a system reacts to even minute environmental changes through a series of modifications of equal size and opposite direction to those that created the disturbance. The goal of these modifications is to maintain all the internal partial balances and eventually the overall system equilibrium.

Ecological, biological, and social systems are homeostatic. They oppose change using all the means at their disposal. If a system does not succeed in reestablishing its equilibria, it enters another mode of behavior, one with constraints often more severe than the previous ones. This new mode into which the system may be forced to bifurcate can lead to the destruction of the system if the disturbances persist and are not countered.

Complex systems must possess homeostasis to maintain their stability and survive. Homeostatic systems are ultrastable. Everything in their internal, structural, and functional organizations contributes to the maintenance of their organizational order. According to Jay Forrester (1961, 1973) their behavior is unpredictable, "counter-intuitive," or contravariant. When one expects a determined reaction as a result of a precise action, a completely unexpected and often contrary action occurs. For a complex system such as a school of fish or an ant colony, mere survival is not enough. It must find a way to adapt to modifications of the environment and evolve in the process. Otherwise, outside forces eventually disorganize and destroy it. However, it is a major challenge and always an open question for a stable organization whose goal is survival to find a way to adapt to change and evolve as a result of it. This concept is as relevant to an endangered species as it is to countries or even the human race as a whole. Some possible answers to this enigma can be found when we study the thermodynamics of living systems.

9.4   First law of thermodynamics and living organisms

In this section, we apply the first law of thermodynamics to the human body as an example of a homeostatic biological system. The system is understood to include all the food stored in the body. Complex chemical compounds, e.g., sugars are continually broken down by living organisms. The human body performs work on its environment and also produces heat. The first law of thermodynamics may be simply expressed as:

$$\frac{\Delta E}{\Delta t} = \frac{\Delta Q}{\Delta t} - \frac{\Delta W}{\Delta t} \tag{9.1}$$

where Δt is a short time interval. The rate at which the body does work on its environment is denoted by $\Delta W/\Delta t$ and is assumed to be positive. The rate at which the body absorbs heat from its environment is designated $\Delta Q/\Delta t$. Knowing the heat capacity of air enables us to find $\Delta Q/\Delta t$.

The rate at which chemical energy is lost by the system through the breakdown of food and production of waste products such as CO_2 and urea is represented by $\Delta E/\Delta t$. This is a negative quantity called catabolic rate, which is contributed to, for example, by the breakdown of glucose and fat. The catabolic rate in kilocalories per mole per second can be related to the oxygen consumption rate (Heusner, 1982):

$$\frac{\Delta E}{\Delta t} = 4.9 \frac{\Delta O_2 (\text{liters})}{\Delta t} \tag{9.2}$$

where $\Delta O_2(l)/\Delta t$ is the volume of oxygen consumed per second. The efficiency e of molecular activity is defined as work done per second divided by the catabolic rate:

$$e = \frac{\Delta W/\Delta t}{|\Delta E/\Delta t|} \qquad (9.3)$$

Perfectly efficient use of energy means that $e = 1$ while a complete waste of catabolic energy implies that $e = 0$. Both the numerator and denominator in Equation (9.3) can be measured. The efficiency rate depends on how many muscles are used cooperatively to produce work and on the energy used by other organs. The more muscles are used to produce useful work compared with the activity of the other organs, the more efficient the body is. A single muscle is only about 25% efficient in converting food energy to work; balance of food energy goes to thermal energy. The body can thus never be more than 25% efficient, since other organs consume energy and produce no work. Leg muscles are the largest in the body; hence cycling and climbing stairs are relatively efficient processes (20%). Shoveling uses mostly arm and shoulder muscles and its efficiency is only about 3%.

The body needs special cooling mechanisms such as sweating and vasodilation (a drug, agent, or nerve that can dilate the walls of blood vessels) to eliminate excess heat during exercise.

Table 9.1 presents catabolic rates for an average 65-kg male and associated consumption of oxygen. The catabolic rate during sleep is based on activities of the internal organs and is known as basal or resting catabolic rate. The rate of conversion of food energy to some other form is called metabolic rate. The total energy conversion rate of a person at rest is called the basal metabolic rate (BMR) and is divided among various systems in the body, as shown in Table 9.1.

Energy consumption is directly proportional to oxygen consumption, since the digestive process is the equivalent of oxidizing food. About 4.9 kcal of energy are produced for each liter of oxygen consumed, independent of the type of food and consistent with Equation (9.2). Because of this, some physiological measurements use oxygen consumption rate to determine energy production rate. The dependence of BMR on mass applies both to humans and other warm blooded animals, as shown in Figure 9.1.

The digestive process metabolizes food very effectively; only about 5% of the caloric value of foods is excreted in the feces and urine without being used. The body stores excess food energy by producing fatty tissue. Food is only one type of chemically stored energy.

9.5 Physics of animal thermoregulation

This section explains, from the point of view of thermodynamics, how the body adapts to thermal stress. The brain, for example, can be irreversibly damaged when temperature exceeds 42°C for a prolonged period. Body temperature starts to drop after submersion in ice water and hypothermia and death may follow.

TABLE 9.1
Basal Metabolic and Oxygen Consumption Rates

Organ	Power Consumed at Rest (kcal/min)	(W)	Oxygen Consumption (mL/min)	% BMR
Liver and spleen	0.33	23	67	27
Brain	0.23	16	47	19
Skeletal muscle	0.22	15	45	18
Kidney	0.13	9	26	10
Heart	0.08	6	17	7
Other	0.23	16	48	19

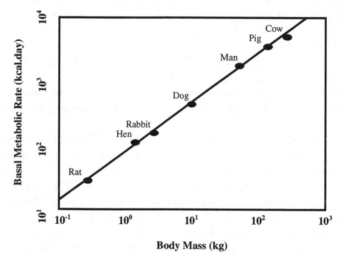

FIGURE 9.1
Plot of BMR versus mass for various animals shows a straight line relationship on a logarithmic scale indicating that BMR α $m^{3/4}$.

Animals have a variety of mechanisms to avoid such stress, e.g., sweating, shivering, cooling by flapping of wings, fluffing of feathers to improve thermal insulation, vasodilation, and vasoconstriction (narrowing of the walls of blood vessels). Feverish patients may go through cycles of shivering and sweating caused by constant switching between different thermal control strategies. Body temperatures are maintained by mammals to be, as far as possible, independent of environmental (ambient) temperature. Reptile body temperatures are usually close to ambient temperatures, but under extreme conditions reptiles too may be forced to adopt thermal control strategies.

We reexpress the first law of thermodynamics as:

$$\Delta Q = \Delta U - \Delta W \qquad (9.4)$$

relating the amount of heat lost by a system ΔQ to a decrease in internal energy ΔU of the system and to work done by the system ΔW. For a mammal, the amount of metabolic energy released by the burning of carbohydrates, etc. is ΔU. The amount of work done by the muscles and other organs is described ΔW. The heat flux I of an organism is defined as heat ΔQ lost by the mammal divided by the time it took to lose the heat. Thus:

$$I = \frac{\Delta U}{\Delta t} \tag{9.5}$$

follows from the expression of the first law of thermodynamics, provided we assume that no work is done by the mammal. The quantity on the right side of Equation (9.5) is the BMR. This quantity depends on body mass M according to an empirical formula called Kleiber's law in which:

$$\left(\frac{\Delta U}{\Delta t}\right)_{rest} = mM^{0.75} \tag{9.6}$$

The prefactor m for mammals is about 3.4 in SI units. Using Newton's law of cooling for the heat current I, the temperature of a resting organism can be expressed as:

$$T(\text{organism}) = T(\text{environment}) + \frac{1}{Ah_c}mM^{0.75} \tag{9.7}$$

where A is the body surface area and h_c is the coefficient of heat transfer. T(organism) is the temperature of the inner organs only. T(environment) is the temperature outside the skin or pelt of the animal.

We next consider the transport of heat from the inner organs across layers of fat and pelts to the outer environment. We can now use Equation (9.7) to discuss the thermal control strategies in various animals. Let us first consider seals or whales in the Arctic. The ambient temperature is that of sea water and is close to freezing. If the target temperature of the organism is 37°C, the second term in Equation (9.7) must compensate to reach that temperature. Aquatic mammals in the Arctic rely on fat layers for thermal insulation from the near-freezing environment. Heat transport through the layers of fat is by thermal conduction. For the sake of simplicity, consider a walrus to be a sphere with a radius of 1 m. The mass M is about 10^3 kg and the surface area is $4\pi\,\text{m}^2$. The outer part of the sphere is a layer of fat of thickness L. From Equation (9.7) for heat transfer we have:

$$T(\text{organism}) = T(\text{environment}) + \frac{L}{A\kappa_{\text{fat}}}mM^{0.5} \tag{9.8}$$

where $L = 8$ cm which is near the observed thickness of the fat layer of a walrus.

If a mammal has a certain body mass M and it would like its layer of fat to be as thin as possible, we must find the best value for the body surface area A. To reduce L without altering body temperature, A must also be reduced in Equation (9.8). For a given volume, the smallest surface area is a sphere. Aquatic mammals in Arctic seas, from the point of view of heat transfer, should therefore have approximately spherical shapes.

We now inquire whether size is an advantage for aquatic mammals. Body area scales with animal size L as L^2, while body mass scales as L^3. Thus, the ratio $M^{0.75}/A$ depends on body size L as:

$$\frac{M^{0.75}}{A} = \frac{(L^3)^{0.75}}{L^2} = L^{\frac{1}{4}} \tag{9.9}$$

This expression increases slowly with L and thus a larger aquatic mammal does not need as thick a layer of blubber as a smaller aquatic animal. An aquatic mammal could increase its metabolic rate above resting rate by exercising and thus increase body temperature.

Now let us consider mammals in hot climates. If the environmental temperature approaches target body temperature of 37°C, the second term of Equation (9.8) must be reduced to prevent overheating. Assume a tropical mammal has no layer of fat and many blood vessels in the skin to assist cooling so that the skin temperature equals the temperature of the organism. To reduce the second term of Equation (9.8), the surface area A must be as large as possible. Mammals living in the tropics have surprisingly large surface areas as exemplified by the large ears of the elephant and the long neck of the giraffe.

Let us take the case of an animal resting in perfectly stationary air with no convection. The animal is treated as a sphere of radius R so the coefficient of heat transfer is $h_c = \kappa(\text{air})/R$. Inserting this last expression into Equation (9.8):

$$T(\text{organism}) - T(\text{environment}) = \frac{mM^{0.75}}{4\pi R\kappa(\text{air})} \tag{9.10}$$

Assuming a density of 1000 kg/m^3 for the animal, we obtain:

$$T(\text{organism}) - T(\text{environment}) = 5637\, R^{\frac{5}{4}} \tag{9.11}$$

expressed in SI units. Again, the temperature difference between the environment and the organism grows with size. For a sphere of diameter 0.01 m in air, one obtains a temperature difference of a few degrees. For a 0.10 m diameter sphere, the difference is about 100 degrees and for a sphere with a diameter of 1 m, the difference is 1,000 degrees. However, we have neglected convection as a heat transfer mechanism.

The corresponding heat transfer coefficients in air for a sphere of about 1 m in diameter is about 1 for free convection and 10 for forced convection (in SI units). Our estimate here is 0.025 using $h_c = \kappa(\text{air})/R$. For forced convection, this is about a factor of 1,000 too small. Discounting the 1,000 factor, the temperature difference becomes a few degrees which is reasonable. Table 9.2 shows m coefficients of Kleiber's law which plays an important role in heat regulation.

Finally, let us look at thermoregulation of the human body. Heat is carried off by a combination of thermal conduction through tissue and convection through the blood stream. The human BMR is about 100 W. With intense activity, $\Delta U/\Delta t$ can rise to about 500 W. If we do not sweat, convection and radiation are the dominant heat transfer mechanisms and the heat transfer coefficient is about 15 W/m^2 K.

TABLE 9.2
Prefactor m Coefficients Based on Kleiber's Law

Organisms	m (SI units)
Mammals	3.4
Marsupials	2.3
Lizards	0.28
Salamanders	0.038

We apply Equation (9.8) using a surface area of about $1.7\,\text{m}^2$, an environmental temperature of 20°C and $\Delta U/\Delta t$ of 100 W, and obtain a body temperature of about 24°C. Thus, to reach a body temperature of 37°C, we must increase metabolic rate $\Delta U/\Delta t$ to about 425 W. At lower ambient temperatures, $\Delta U/\Delta t$ must be correspondingly higher to keep the temperature at 37°C. Humans tend to eat more in colder climates. Heat production can range from 1,000 to 5,000 if we are intensely active. In an ambient temperature of 20°C, we need a body temperature above 37°C. Overheating is therefore unavoidable if we prolong intense activity. Sweating increases the transfer coefficient from 15 to about 300 and can reduce the excess temperature by as much as 20 degrees.

Work performed during muscle contraction is accompanied by a temperature increase of the muscle fiber, i.e., energy delivered by ATP is converted partly to thermal energy. A small amount of this heat is conducted away and some is radiated. Breathing also cools the body because of the water vapor exhaled as a by-product. Convection is a very effective cooling process. When these processes are saturated, sweating is triggered to provide additional cooling. In fact, the body maintains a nearly constant temperature by eliminating excess thermal energy as rapidly as possible. Blood flow increases when blood vessels dilate; pores open so that sweating increases. Respiration rate also increases.

9.6 Allometric laws in physiology

9.6.1 Introductory scaling concepts

Human metabolic rate depends on size and level of physical activity. The normal metabolic rate under standardized conditions is proportional to the total surface area of the body as mentioned in the previous section. The proportionality of basic metabolic rate to body area is an empirical scaling law. The maximum metabolic rates of similar animals are also proportional to their total surface areas, so that metabolic rate scales as L^2 and work done by muscles scales as L^3.

Despite L^3 scaling, the speed v with which a limb moves is independent of L. The time it takes a limb to move a distance d is d/v; thus time scales as L. The power achieved scales as work divided by time or L^2. Since the oxygen required for metabolism is supplied by the blood, the metabolic rate is proportional to the volume of blood pumped per second by the heart. This is proportional to the volume V of the heart times the heart rate r (number of beats per second):

$$P \alpha V r \tag{9.12}$$

so that r scales as P/V. Since P scales as L^2 and V scales as L^3, r scales as $L^2/L^3 = 1/L$. Thus the heart of a large animal beats slower than the heart of a small animal. This indicate simple power relations governing scaling in physiology, but that is far from the truth as we try to demonstrate below.

9.6.2 Overview of allometric laws of physiology

We begin with basic ideas underlying physiological scaling laws. First, two bodies with the same proportions are isometric if the term *isometric* is used to describe geometric similarity. Bodies that are not isometric and change with size according to a particular rule are *allometric*. Many such processes can be described very simply according to so-called *allometric scaling equations*:

$$M = a \times x^k \tag{9.13}$$

where M is the body process in question (for example, metabolic rate), x is the size of the organism, and a and k are fixed constants. For example, numerous measurements of animals and humans suggest that the relationship between metabolic rate P_{met} (kcal/day) and weight m_b (kg) is given by the following equation proposed in 1932 by animal researcher Max Kleiber and later reviewed in his seminal paper published in 1947:

$$P_{met} = 73.3 \times m_b^{0.75} \tag{9.14}$$

A small amount of uncertainty is associated with the exact exponent whose average value is extremely close to 3/4. Below we explore other rules for allometric scaling, grouping them according to the values of the scaling exponents in Table 9.3. Most metabolic parameters vary with the 3/4 power of weight. It thus seems reasonable to suggest calculating certain drug doses to the 3/4 power of the weight, and not to weight or body surface area.

Kleiber reformatted the so-called Harris-Benedict equation for heat production Q by humans in a dimensionally-correct way. Namely, for men:

$$Q = 71.2 \times m_b^{3/4} \times (1 + 0.004(30 - y) + 0.010(S - 43.4)) \tag{9.15}$$

and for women as:

$$Q = 65.8 \times m_b^{3/4} \times (1 + 0.004(30 - y) + 0.010(S - 43.4)) \tag{9.16}$$

where Q is heat production in kcal/day, w is the weight in kg, y is the age in years and S called specific stature is height in $cm/m_b^{1/3}$.

TABLE 9.3
Scaling Exponents

No apparent scaling with weight can be seen

Maximum functional capillary diameter
Red cell size
Plasma protein concentrations vary little with size or species
Mean blood pressure is about 100 mm Hg more or less independent of size or species
Fractional airway dead space (V_D/V_T) is fairly constant at about a third
Body core temperature is weight-independent
Maximal tensile strength developed in muscle is scale-invariant
Maximum rate of muscle contraction appears scale invariant
Mean blood velocity is calculated to be proportional to $m_b^{-0.07}$

Almost linear scaling with weight

Heart weight (g) $= 5.8\, m_b^{0.98}$
Lung weight (g) $= 11.3\, m_b^{0.98}$
Tidal volume (ml) $= 7.69\, m_b^{1.04}$
Vital capacity (ml) $= 56.7\, m_b^{1.03}$
Lung compliance (ml/cmH$_2$O) $= 1.56\, m_b^{1.04}$
Blood volume (ml) $= 65.6\, m_b^{1.02}$
Muscle mass $= 0.40\, m_b^{1.00}$
Skeletal mass $= 0.0608\, m_b^{1.08}$

Scaling relations with quarter power

Heart rate (min^{-1}) $= 241 \times m_b^{-0.25}$
Blood circulation time (s) $= 17.4 \times m_b^{0.25}$
Respiratory rate (min$^-$1) $= 53.5 \times m_b^{-0.26}$

Scaling examples with three quarter power

P_{met} (kcal/day) $= 73.3 \times m_b^{0.75}$
V_{O2max} (ml/s) $= 1.94 \times m_b^{0.79}$
Glucose turnover (mg/min) $= 5.59 \times m_b^{0.75}$

Other allometric relations

Metabolic cost of running (ml O2 g$^-$1km$^-$1) $= 8.61\, m_b^{0.65}$
Mammal brain weight (excluding primates) $= 0.01\, m_b^{0.70}$
Monkey brain weight $= (0.02\ \text{to}\ 0.03)\, m_b^{0.66}$
Ape brain weight $= (0.03\ \text{to}\ 0.04)\, m_b^{0.66}$
Human brain weight $= (0.08\ \text{to}\ 0.09)\, m_b^{0.66}$

Source: Schmidt-Nielsen, K., *Scaling: Why Is Animal Size So Important?* Cambridge University Press, Cambridge, U.K., 1984. With permission.

Interestingly, the lifespan of an organism typically increases with its mass. For captive mammals, the following allometric equation has been found:

$$t_{\text{life}} = 11.8\, m_b^{0.20} \tag{9.17}$$

The corresponding equation for birds is:

$$t_{\text{life}} = 28.3\, m_b^{0.19} \tag{9.18}$$

Most animals except for humans seem to survive for about 800 million to a billion heartbeats. We do not know what causes aging, but it is tantalizing that the exponent in the above equations is so close to a quarter. It is not unreasonable to assume that faster metabolisms cause wear-and-tear on cellular structures such as mitochondria and this may give us fundamental insights into aging.

Many attempts have been made to explain some of the allometric equations presented above. In simple situations, we have reasonable explanations for allometric equations but serious difficulties arise in complex situations. Several attempts have been made to explain the 3/4 exponent that seems to govern metabolism related to weight. While none appears entirely convincing, below we briefly present a few of the more important ones.

McMahon (1973) argues based on engineering of the skeleton that animals cannot remain isometric with increasing size, as loads would increase more than the ability of the skeleton to withstand such loads. Furthermore, all proportional changes in bones should be similar, as different movements impose different stresses. Note that the critical length L_{cr} at which a column buckles is:

$$L_{cr} = k(E/\rho)^{1/3}d^{2/3} \tag{9.19}$$

where d is the diameter, E is Young's elastic modulus, and ρ is the density of the material. If the weight of a limb m_l is proportional to body weight m_b, m_b is proportional to $L \times d^2$. L^3 is proportional to d^2, so mass is proportional to L^4, which is another way of saying that L is proportional to $m_b^{1/4}$. Similarly, d is proportional to $m_b^{3/8}$. Because work = force × distance, the power output of a muscle is given by:

$$P_{max} = \text{maximal tensile stress} \times \text{area} \times \text{change in length/time} \tag{9.20}$$

The maximal tensile stress in a muscle is scale-invariant, as is change in length per unit time. Thus, the muscle power depends only on the cross-sectional area. If this area is proportional to d^2, then P_{max} is proportional to d^2 or to $(m_b^{3/8})^2 = m_b^{3/4}$.

West et al. (1997) used an extremely complex argument to achieve the three quarter exponent related to the energy cost of transporting essential materials. Their basic assumptions were that minimal energy was dissipated and terminal capillaries were size-invariant. Transport was through a space-filling fractal network of branching tubes and the model had fairly general applicability. They argued that the metabolic rate was proportional to the flow Q through the system, and that at the kth level of branching, $Q = N_k.\pi.r_k^2u_k$, where N_k is the number of branches at kth level and u_k is the mean flow velocity. Specifically for capillaries, $Q = N_c.\pi.r_c^2u_c$. They then assumed that capillary radius, length and mean flow r_c, l_c and u_c, respectively, are size invariant. If the allometric equation governing the relationship between Q and body mass m_b is $Q = k.m_b^a$, it follows that N_c scales in proportion to m_b^a. In other words, the number of capillaries scales in proportion to the metabolic rate.

They subsequently proved that for minimal energy dissipation, the branching of the network follows the rules for a self-similar fractal for which the number of branches increases in geometric proportion as their size decreases. If each tube at branching level k branches into n_k smaller ones, then for a self-similar fractal, $N_k = n_k$, i.e., as k increases, the number of branches increases in geometric proportion. Thus, as the mass increases, the number of capillaries increases with the logarithm of the mass. It is then shown that a, the exponent of m_b in the allometric scaling equation, is equal to $\ln(n)/\ln(\gamma \cdot \beta^2)$ where γ and β are scale factors that characterize branching of the system. Finally, they obtained the exponent $a = 3/4$.

Banavar et al. (1999) also used the concept of a branching network supplying nutrients but their argument is more general and has fewer constraints. They derived a three quarter exponent by proving that for the most efficient network, blood volume C scales with L^{D+1} where L is a parameter of linear size, and D the spatial dimension of the object (3). They further assumed that blood volume scales with mass, proving that metabolic rate B scales with L^D. Since mass is proportional to L^{D+1} and L is proportional to (metabolic rate)$^{1/D}$, they concluded that metabolic rate is proportional to mass$^{D/D+1}$.

Interestingly, the principles of symmorphosis seem to apply to the physiology of the whole organism. The concept formulated by Taylor and Weibel in 1981 maintains that biological structures are formed so as to meet but not exceed maximum requirements. For example, it would be pointless for the lungs to be able to provide many times the resting oxygen usage if the cardiovascular system could only distribute that amount. The maximum oxygen consumption of an organism appears to be usually about ten times its basal consumption (a property termed *factorial aerobic scope*). We have mentioned above the possible roles of self-similarity and fractals in human and animal physiology. Below we discuss this aspect in some detail.

9.6.3 Fractals and living systems

Examples of self-similar (fractal) objects include abstract mathematical constructions and real physical systems. In the former group we can list the Koch island, Sierpinski gasket, Cantor dust, etc. (Feder, 1988). The latter group includes snowflakes, electrical discharge patterns, ferns, corals, viscous fingering of glycerol on water surfaces, etc. Many of these examples represent biological patterns. Interestingly, fractal geometry ideas have been successfully adopted to data analysis for stock exchange postings, epidemiology, and flooding of various river systems (Mandelbrot, 1977; Feder, 1988).

In biology, we are confronted with highly complex living and evolving structures that exhibit hierarchical (and often also fractal) organization. This typically leads to a characteristic scaling of their dynamical properties and behavior, from entire organisms down to cellular, subcellular, and molecular levels. Living organisms are open, dissipative, growing, catalytic multiphase systems that try to maintain efficiency of metabolic rates by maximizing their surfaces and optimizing their transport systems for the exchange of substrates (West, 1990).

In biology, medicine and related research fields, the concept of fractality is only now beginning to take hold (Iannaccone, 1996) and its first applications have been developed (West, 1990). Fractal properties of many examples of living matter have been experimentally investigated (Sernetz et al., 1985) on several scales: individual cells (neuronal, epithelial, retinal, etc.), cell colonies and tissues, and organs (kidneys, liver, blood vessels, brain surface, trabecular bone, lungs). Experiments based on the scattering of ultrasound from the human liver yielded a value of the fractal dimension of $d_f = 2.0$. The result happens to be an integer but the whole number should not be construed to invalidate its fractal nature because the embedding dimension is 3 so the liver is still a fractal (Javanaud, 1989).

If the collagen fibers in liver form a fractal structure, two possibilities arise for the elementary constituents of the resulting fractal cluster. In the first case, they could be tropocollagen molecules about 300 nm in length. Alternatively, they could be the units spanning the cross-striations that appear every 64 nm along collagen fibrils. Cluster size is on the order of a few millimeters.

As discussed earlier, metabolic rate P_{met} scales with body size V and follows a universal relationship, the so-called allometric law of reduction of specific metabolism (Spatz, 1991), such that $P_{met} = aV^{\frac{3}{4}}$ where the exponent is clearly noninteger, implying a fractal nature of the processes. Hence, metabolic rates in living organisms are neither linearly proportional to their volume V nor to their Euclidean surface $V^{\frac{2}{3}}$. We can conclude that the fractal natures of organs appear to be reflected in metabolic rates of the processes that take place on their surfaces.

9.7   Entropy and information

One characteristic feature of living systems is their ability to reduce entropy; another is information processing and signaling capability. Living systems must communicate with the environment to survive. To cast some light on these abilities, we must first address entropy reduction and information in biological systems even those as small as single cells.

Shannon and Weaver (1963) defined information as negative entropy:

$$I = k\ell n W = -k \sum p_i \ell n(p_i) \tag{9.21}$$

Its introduction allowed a resolution of the long-standing paradox referred to as Maxwell's demon (see Chapter 6). The problem involved a theoretical creature that operated a small door between two compartments of a container filled with high and low energy gas molecules. The end result was separation of the gas into hot and cold divisions with no energy expenditure, thus contradicting the second law of thermodynamics.

Szilard's solution (1929) endowed the demon with information, i.e., Shannon's information: negative entropy balancing out the changes in the entropy of the gas. In quantitative terms, the energy cost of 1 bit of information at physiological temperature is $\ln 2 \, kT = 3 \times 0^{-21}$ J $= 18.5$ meV which should be kept in mind when discussing biological functions that are replete with information content. Information about the structure and composition to be developed, such as protein sequences and folding patterns, is very important in this context.

While statistical entropy is a probabilistic measure of uncertainty or ignorance, information is a measure of knowledge or a reduction of the uncertainty. Entropy reaches its maximum value if all microscopic states available to the system are equiprobable, that is, if we have no reason to assume one state is more probable than another. It is also clear that entropy is zero if and only if the probability of a certain state is 1 (and, of course, probabilities of all other states are zero). In that case, we have maximal certainty or complete information about state of the system.

Information was introduced by Shannon as a measure of the capacity for information transmission of a communication channel. If we obtain some information about the state of the system, it will reduce our uncertainty by excluding or reducing the probability of a number of states. The information received from an observation is equal to the degree to which uncertainty is reduced (Bennett, 1985). Although Shannon came to disavow the use of information for describing this measure because it is purely syntactic and ignores the meaning of the signal, his theory nonetheless came to be known as information theory. Entropy has been vigorously pursued as a measure for a number of higher order relational concepts including complexity and organization.

Other methods of weighting the state of a system do not adhere to the additivity condition of the probability theory requiring that the sum of probabilities must be 1. These methods involving concepts from fuzzy systems theory and possibility theory lead to alternative information theories. Together with probability theory they comprise the generalized information theory (GIT). While GIT methods are under development, the probabilistic approach to information theory still dominates applications.

9.8 Entropy reduction in living systems

Living cells are dissipative, open, far-from-equilibrium systems that lower entropy by using an influx of energy and molecular material in a multicompartment structure with specific functional characteristics. Entropy reduction was discussed early on by Schrödinger (1967). It relies on energy supply to create a metastable nonequilibrium state and electrical, pressure, and chemical potential gradients across semipermeable membranes.

Electric potential differences assist in the process. As an open system, a cell operates cyclically and exchanges material and heat with its environment. High energy molecules are absorbed through membrane pores and their energy is used to synthesize cell components and maintain ambient temperature. Heat is dissipated and waste products excreted so that excess entropy in the environment is balanced by structure and information production that lowers entropy inside the cell. This, of course, leads overall to a net entropy change close to zero when the cell fluctuates quasiperiodically. Cell death manifests in the breakdown of structures and functions leading to continuous entropy production. Overall, the entropy changes in the cell can be attributed to (1) chemical reactions, (2) mass transport in and out of the cell, (3) heat generation, and (4) information processing.

Morowitz (1995) estimated that approximately 2×10^{11} bits of information are contained in an *Escherichia coli* bacterium, the simplest and best documented organism. The number agrees with calorimetric data (Gilbert, 1966). However, the estimated information capacity in the *E. coli* genome is only 10^7 (Johnson, 1970) which is initially surprising but on closer examination, can be expected, as explained below.

All matter including living cells must obey the energy conservation principle which in this case takes the form of the first law of thermodynamics discussed earlier. A cell can be viewed as a machine, exactly the way we view a combustion engine engaged in a Carnot cycle. A cell performs work and generates heat. It requires constant supplies of energy and matter, particularly energy-giving molecules like glucose.

A more appropriate formulation of the energy balance is via Gibbs free energy that accounts for a change in the numbers of molecules and the presence of several molecular species:

$$G = U - TS + PV = \mu N \quad \text{or} \quad dG = -SdT + VdP + \mu dN \qquad (9.22)$$

Hence, the entropy differential can be written as:

$$dS = \frac{dU}{T} + P\frac{dV}{T} - \mu\frac{dN}{T} \qquad (9.23)$$

which indicates that entropy changes can be achieved through heat production, change of volume, or a flux of molecules. All these activities are relevant to living cells. Since the entropy of an ideal gas of N particles with total energy E of mass m each (Landau and Lifshitz, 1969) is:

$$\frac{S}{k} = N \left\{ \ell n \left(\frac{V}{N}\right) + \frac{3}{2}\ell n \left(\frac{mE}{3\pi \hbar^2 N}\right) + \frac{5}{2} \right\} \qquad (9.24)$$

confining molecules within space, as is the case with building a cellular structure, reduces exploration volume V and thus reduces the entropy of the system accordingly. Conversely, mixing two molecular species with numbers N_1 and N_2 in a fixed volume V by opening a partition between their compartments V_1 and V_2, increases the entropy by the amount below:

$$\Delta S = N_1 \ell n \left(\frac{N}{N_1}\right) + N_2 \ell n \left(\frac{N}{N_2}\right) \qquad (9.25)$$

Keeping various molecular species separated in individual compartments (such as mitochondria, nuclei, endoplasmic reticula, etc.) is an entropy-reducing process since our information about the system's internal distribution is enhanced.

Enzymatic catalysis against the energy barrier (see Chapter 3) typically helps achieve such a deliberate separation of molecular species. In fact, a variety of solute molecules are contained within cells. The cellular fluid (cytosol) has a chemical composition of 140 mM K^+, 12 mM Na^+, 4 mM Cl^-, and 148 mM A^- (A represents a protein; see Chapter 4). Cell walls are semipermeable membranes and permit transport of water but not of solute molecules.

We use Dalton's law to determine osmotic pressure inside a cell. A mixture of chemicals with concentrations c_1, c_2, c_3, etc. dissolved in water has a total osmotic pressure equal to the sum of the partial osmotic pressure of each chemical. The total osmotic pressure inside a cell Π_{in}, is:

$$\Pi_{in} = RT\frac{(140 + 12 + 4 + 148) \times 10^{-3}\,\text{mol}}{1\,\text{liter}} \times \frac{1\,\text{liter}}{10^{-3}\,\text{m}^3} = 7.8 \times 10^4\,\text{Pa} \quad (9.26)$$

The cell exterior is composed of 4 mM K^+, 150 mM Na^+, 120 mM Cl^-, and 34 mM A^-. As a consequence, the total osmotic pressure of the exterior Π_{out} is:

$$\Pi_{out} = RT \frac{(4 + 150 + 120 + 34) \times 10^{-3}\,\text{mol}}{1\,\text{liter}} \times \frac{1\,\text{liter}}{10^{-3}\,\text{m}^3} = 7.9 \times 10^4\,\text{Pa} \quad (9.27)$$

Because Π_{in} and Π_{out} have similar values, the osmotic pressure difference between the exterior and interior of a cell is very small; the net pressure exerted on the cell wall is the important factor. It is essential to keep interior and exterior osmotic pressures closely matched in fragile animal cells. Cells are equipped with sophisticated control mechanisms to accomplish this. The process can be seen as an entropy reduction mechanism.

Looking deeper into entropy reduction by cellular processes, in the production of macromolecules such as proteins, the assembled atoms naturally lose their degrees of freedom by joining together. In the simplest case of a peptide chain viewed as a semiflexible rod, each amino acid has three translational and three rotational degrees of freedom, in addition to internal degrees of freedom before assembly which by and large survive the assembly process.

After a peptide is assembled, only small rotations around the backbone are permitted effectively wiping out five degrees of freedom per amino acid. We can therefore view this as an entropy reduction process. This negative (structural) entropy is created in addition to the combinatorial contribution describing the probability of selecting a particular sequence of amino acids, e.g., $k \ln(20^n)$ where n is the number of amino acids in a peptide. The folding of a chain into a globular protein restricts the motion of its member groups, eliminating some rotations altogether and limiting others. This, again, can be seen as a reduction of phase space; the volume changes from Ω to Ω' with an attendant entropy reduction of $\Delta S = k \ln(\Omega/\Omega')$. For illustration purposes, we presented a somewhat simplistic approach via a microcanonical ensemble in which all states in the phase space have the same probability. In reality, due to the interactions of molecules, a canonical ensemble should be used to generate a more accurate (and more complicated) formula, namely:

$$S = kT \frac{\partial}{\partial T} \ln Z + k \ln Z \quad (9.28)$$

where $Z = \sum_i e^{-\frac{E_i}{kT}}$ is the partition function of the system (treated as an isolated one). Indeed, since the system is open (although the openness is not complete), a grand canonical ensemble technique should be used in the evaluation of the resultant entropy change.

Much has been said about the effect of the second law of thermodynamics on living systems since living systems seem to defy it. The second law of thermodynamics is valid for closed systems and it basically states that in closed systems irreversible processes such as heat generation lead to entropy increases while reversible processes involve no heat and no entropy change. No provision is made for entropy reduction, but a living cell is an open system and taken in its surroundings, the total entropy change will never be negative.

In closed systems, conditions for equilibria are expressed as minima of the appropriate thermodynamic potentials (e.g., Gibbs free energy) or maximum entropy requirements (Landau and Lifshitz, 1969). In open systems, that rule does not apply. One looks for stability conditions of a given state, i.e., whether with small perturbation, the state will evolve or retain its equilibrium value.

Another way of discussing entropy is in terms of order and disorder. The most pertinent physical transformations between states of matter that are ordered and disordered are called phase transitions (see Appendix B). Continuous (second order) phase transitions involve no entropy change at the critical points and ordering in the systems sets in gradually as seen through bifurcation of an associated order parameter. In first order phase transitions, an entropy jump is always present at the transition point. Since $Q = T \Delta S$ is the latent heat of transition, this entropy jump is proportional to the latent heat of transition. Phase transitions with both positive and negative latent heats exist, i.e., entropy creation or reduction takes place in the system on supplying or withdrawing heat, but always $\Delta G = 0$ at the transition point. This does not violate the second law of thermodynamics since the system is not isolated thermally from the environment that may receive excess heat. This example is, of course, relevant to a living cell, if one were to speculate about jump-starting a living process by physical means.

In nonequilibrium systems such as autocatalytic chemical reactions of the Brusselator type (Prigogine, 1980), order is created and sustained spontaneously by means of nonlinear interactions. Since these are open and driven systems, the second law of thermodynamics does not need to be invoked.

Another important property of nonlinear systems is the possibility of self-assembly, for example in pattern-forming crystal growth. This provides an example that has no need for an instruction-driven creation of order and structure. Sometimes, when discussing the assembly of biomatter, concern is unduly given to the need for instruction in putting the building blocks of matter together. While we have clear instructions for the amino acid sequences in the genome, the details of higher order structure formation need no special encoding. They may emerge spontaneously as an attractor in a nonlinear dynamical system we call a living cell as a result of biological self-organization (Kauffmann, 1993).

As emphasized earlier, a living cell constantly consumes energy to maintain its structure and vital functions. Most energy is in two forms: photons in plants and glucose-containing compounds in animals. Glucose is easily utilized to synthesize ATP. ATP along with its GTP analog is the common currency of biological energy as discussed in Chapter 3. An ATP molecule under standard conditions carries 7.3 kcal/mol of energy and its less common analog GTP somewhat less. Each glucose molecule gives rise to approximately $N = 30$ ATP molecules and the associated entropy production (Daut, 1987) is:

$$\frac{dS}{dt} = \Delta G(\text{glucose}) \frac{J(\text{ATP})}{NT} \qquad (9.29)$$

where $G(\text{glucose}) = 3 \times 10^6$ J/mol is the free energy of glucose oxidation and $J(\text{ATP}) = 10^{-13}$ mol/hr is the flux of resultant ATP for a single cell (Kim et al., 1991). $T = 310$ K results in roughly an entropy rate of change for a single cell in the range of 10^{-14} J/K. This can be compared to only 0.7×10^{-17} J/K of entropy reduction due to DNA-transmitted information, i.e., less than 1/1000.

This is not surprising because many other processes are at work to keep the cell in its metastable (low entropy) state. First, the membrane consisting of phospholipids comprises some 60% of cell mass and presents a highly ordered structure requiring entropy reduction. Likewise, proteins and peptides can contain up to several thousand atoms each with fairly well specified positions leading to a net entropy drop compared to a nonliving state. Finally, about 50% of the metabolic energy of the cell is utilized in pumping ions across the membrane (Rolfe and Brown, 1997), mainly as a result of transmembrane potentials and the work of ion pumps. The work of an ion pump resembles the Maxwell demon, except of course, an ion pump has no thinking function. An ion pump relies on molecular recognition mechanisms. When a pump is activated, the mechanisms lower the entropy by binding the two molecules together. The subsequent placement of an ion or a macromolecule within the confines of a membrane permanently lowers the entropy via exploration volume reduction as discussed above. Release of a waste product into the environment produces a directly opposite effect.

Chemical reactions may absorb or release heat much like first order phase transitions, i.e., they may be exothermic or endothermic. Since most living processes are in essence chemical reactions or cascades thereof, it is worth analyzing them from the entropy point of view. The most interesting information theory reactions are catalytic reactions of special types of functional proteins called enzymes. Enzymes use finely tuned selection mechanisms of molecules that have shapes complementary to recognition pockets. The device is called a lock-and-key mechanism and by forcing particular orientations of catalytically reacting molecules, enzymes increase reaction rates by several orders of magnitude. Consequently, the operation can be viewed as a type of information processing whereby the shapes of binding domains are recognized, the molecules are optimally positioned for binding, and in some cases particular bonds are broken and others created.

Some enzymes belonging to the class of allosteric proteins may adopt two or more stable conformations and act like switches — activated in one conformation and inactive in others. Following binding and catalysis, enzymes return to their original conformations thereby participating in active promotion of a particular reaction. From the point of view of information and entropy reduction, they do not overall decrease the entropy of the cell (Lowenstein, 1999). At best, they break even since all information an enzyme invests in a catalytic reaction is recollected at the end of a cycle. Furthermore, the information necessary to perform a particular function (molecular recognition) is not entirely contained in an enzyme. To be effective, an enzyme must be activated by the environment in which it resides: water, inorganic ions, etc. Thus, we can say that biological information is contained in the entire system: the cell.

9.9   Biological information

DNA production takes place even in nonreplicating cells. A typical mammalian cell polymerizes approximately 2×10^8 nucleotides of DNA a minute into hnRNA (Brand-horst and McConkey, 1974) of which only 5% ends up in the cytoplasm coding for protein synthesis (Dreyfuss et al., 1993). Because of redundancy in coding nucleotide triplets for the 20 amino acids, the original 6 bits of information in DNA translate into $\log_2(20) = 4.2$ bits in a protein.

Consequently, about 0.7×10^6 bits/s are transmitted from the nucleus to the cytoplasm. This is augmented by a small fraction of information due to mitochondrial DNA (Alberts et al., 1994). This is only a small fraction of the total information production (understood as negative entropy) of a living cell. Most information is contained in the structure of the cell and its components.

Since the Shannon information formula employs probabilities of particular states, dangers are inherent in selecting these probabilities, especially when selection is made purely combinatorially as is often the case, for example in amino acid or nucleic acid sequence determination. This is not necessarily a random choice situation akin to tossing a coin. Using $p = 1/W$ for a single element selection where W is the number of possible choices may not be correct and may yield an excessively improbable (or high information) estimate.

This would be the case if the choices of elements in a sequence are not of the same statistical weight and are biased statistically, so that a more appropriate probability value is given by the canonical ensemble Boltzmann distribution formula $p_i = p_0 e^{\frac{-E_i}{kT}}$. Of course, in order to make this estimate, one needs to know the energies E_i and hence the Hamiltonian for the system. Therefore, the apparent information estimate of $I = (k \ln N)^n$ where n is the number of members in a string may be significantly larger than the true value of $-S$ from thermodynamic estimates of a given state — a maximum entropy state for equilibrium and hence a minimum information content.

For a string of choices (e.g., amino acid sequence in a peptide or nucleic acid sequence in DNA or RNA), this may lead to "basins of attraction" favoring some combinations strongly over others. There may be evolutionary retention of favored choices and the establishment of hierarchies of order. $I = 10^{110}$ is an immense number and represents a clear computational barrier even from the standpoint of cataloging such an enormous number of objects. Immense numbers commonly appear in biology. DNA and protein sequences consist of immense numbers arising from great varieties of possible combinations from which these macromolecules may be formed. However, in view of the argument above, restricting phase space by forming basins of attraction due to intramolecular interactions may result in a hugely reduced number of combinations encountered in practice.

We must address a clear distinction between information and instruction. While information was introduced on purely statistical grounds as a measure of the number of choices possible when making a selection for a string of elements, instruction implies the existence of a message, a messenger, and a reader who will execute the message. A classic example is the synthesis of amino acids contained in the triplets of DNA and RNA base pairs. While every triplet has the same information value, namely $k \ln(4^3) = 6$ bits, some amino acids are coded uniquely by single triplets and some by two different triplets; some triplets do not encode anything. This is obvious based on the 64 possible triplets of base pairs and only 20 distinct aminoacids, hence the redundancy.

A similar difference between information and instruction can be found in the genome where in addition to the coding sequences of DNA, some of which are vital to survival, we find so-called junk DNA that has apparently no coding value but represents most of the DNA sequence. The main point to stress is that information is often confused with instruction. However, Shannon's information is a more general concept that includes purely structural aspects such as combinatorial entropy reduction and chemical entropy changes. Instruction, on the other hand, implies a message and a message reader that executes a command. In this sense, the operation is similar to the way a computer operates and follows commands.

DNA, RNA, hormones, and signaling molecules are thought to be such biological messengers. However, a vast majority of information content is not instructional in nature and is akin to simple algorithms like a logistic map or fractal recursive relations that give rise to great mathematical complexity of the results that follow. Similarly, DNA can be viewed as an algorithm that spans the awe-inspiring complexity of living cells. While coding for protein synthesis is contained in the genetic code, it is highly improbable that details of structure formation need special coding. They probably unfold because of self-organization inherent in the dynamics of the synthesized products.

In view of the discussion in this section about information content and processing in a living cell, we wish to postulate the existence of two types of information in biological systems: (1) structural information, i.e., negative entropy and (2) functional information. Structural information is simply neat and tight packing of various molecules into macromolecules and macromolecules into organelles that comprise the cell. Functional information, on the other hand, pertains to activities of a cell and the rate and number of chemical reactions generated. The two concepts are related but not identical.

An automobile presents a good analogy. It may look perfectly good but will not function if its gas line is cut. The same principle applies to a living cell. Some key reactions like synthesis of tubulin, if not properly executed, will lead to cell death. While structural information (entropy reduction) should be maximized, functional information is concerned with how rapidly information is exchanged. Therefore, a cell should tend to increase the speed of biochemical reactions and the number of molecular interactions if possible. Living processes are cyclical in nature, so one expects to maximize the rate of information (or entropy) change over time for maximum functionality. In order to optimize both structure and functionality, a living system such as a cell should strive to maximize the product of the two quantities:

$$\text{Max} \left\{ I^2(t) \left(\frac{dI(t)}{dt} \right) \right\} \tag{9.30}$$

We squared the quantities in the product due to the cyclic nature of life processes. Some aspects of this distinction between structural and functional information in biological systems were emphasized more than three decades ago by H. Froehlich (1968) who used the term *biological coherence* to describe the holistic and functional integration of information flow in living matter.

9.10   Evolutionary theories

Evolution is often seen as a trial-and-error process of variation and natural selection of systems at all levels of complexity. The phrase *natural selection* comes from the Darwinian theory of biological evolution, which distinguishes natural from artificial selection in which specific features are retained or eliminated, depending on a goal or intention.

Evolution typically leads to greater complexity although it is difficult to precisely define *complexity*. The narrow interpretation of Darwin's theory (1859) sees evolution as the result of selection by the environment acting on a population of organisms competing for limited resources. The winners of the competition, those who are most fit to gain the resources necessary for survival and reproduction, are selected and the others are eliminated. However, this view of evolution entails two strong restrictions. It assumes that a large number of configurations undergo selection and that selection is carried out by their common environment.

Some critics of Darwin's theory of evolution argue that natural selection must be augmented by self-organization in order to explain evolution. (see e.g. Jantsch, 1979; Kauffman, 1991, 1995, 1993). However, evolutionary models can be developed that allow selection or elimination of a given configuration independent of the presence of other configurations. That means a single system can pass through a sequence of configurations, some of which are retained and some eliminated. This type of selection does not presuppose the existence of an environment external to the configuration undergoing selection. It is easy to imagine configurations that are intrinsically stable or unstable. The stability of the structure functioning as a selection criterion is purely internal to the configuration and no outside forces or pressures are necessary to explain them.

The selection in such a model can be inherent in the configuration itself, and an asymmetric transition from varying to stable may be referred to as self-organization. In the present view, natural selection encompasses both external (Darwinian) selection and internal (self-organizing) selection.

9.10.1   Punctuated equilibria

Instead of a slow, continuous movement, evolution tends to be characterized by long periods of virtual standstill (equilibrium), punctuated by isolated episodes of rapid development of new forms. The punctuated equilibrium theory of Niles Eldredge and Stephen Jay Gould was postulated as a criticism of the traditional Darwinian theory of evolution.

Gould and Eldredge (1977) observed that evolution tends to happen in fits and starts, sometimes moving quickly, sometimes very slowly or not at all. On the other hand, typical variations tend to be small and that is probably why Darwin saw evolution as a slow and continuous process. Fossil evidence indicates that instead of slow, continuous progression, the evolution of life on Earth took place in a manner consistent with the Eldredge-Gould model.

To better understand the punctuated equilibria approach, consider a typical fitness landscape with valleys separated by ridges. If the evolving system has reached the bottom of a deep valley, little change will ensue because variation will fail to pull the system out of the valley. This is a negative feedback regime, in which chance fluctuations will be counteracted and return the system to its equilibrium position at the bottom of the valley. On the other hand, if only a small ridge separates the valley from a neighboring deeper valley, a chance event may be sufficient to push the system into the other valley. Such an event will become more likely when the fitness landscape changes so as to reduce the height of the ridge. Once over the ridge, the descent into the new valley will be rapid. This situation is a positive feedback regime in which deviations from the previous position are amplified. The system will evolve quickly to a new, fitter configuration. Examination of the geological records shows that many fossils correspond to the position at the bottom of the valley where organisms remained for a long time.

As emphasized throughout the book, living organisms are organized hierarchically into manifolds corresponding to their subsystems and subsubsystems. Each subsystem is described by its own set of genes. A mutation in one of the components at the lower levels generally will exert little effect on the whole. On the other hand, a mutation at the highest level, where the overall arrangement of the organism is determined, may have a spectacular impact on the lower hierarchies. For example, a single mutation may lead to inheritable diseases or deformations. Such high-level mutations are unlikely to be selected, but they can lead to revolutionary changes (Dawkins, 1983, 1986, 1989).

9.10.2 Mathematical modeling of evolution

Historically, experimental research always ignited the development of evolutionary theories. For example, the mathematical theories of population genetics of R.A. Fisher (1958), J.B.S. Haldane (1948), and S. Wright (1931) were based on experimental genetic investigations performed in the first half of the 20th century. The outstanding achievements in the field of molecular biology attained in the 1950s and 1960s constituted the underlying background for the life origin models of M. Eigen et al. (1979, 1992) and the models of regulatory genetic systems by S.A. Kauffman (1995).

Biological evolution is a complex process. Mathematical modeling can help clarify its features. In evolutionary modeling, we can distinguish the following types of approaches:

Models of molecular-genetic systems origin were constructed in attempts to determine the origins of life. The quasispecies and hypercycles of Eigen and Schuster and the sysers of V.A. Ratner and V.V. Shamin are the best known examples. They describe mathematically some hypothetical evolutionary stages of prebiological self-reproducing macromolecular systems.

General models of evolution describe informational and cybernetic aspects of the process. The neutral evolution theory of M. Kimura (1968, 1983) and S. Kauffman's automata represent sophisticated examples of such models.

Artificial life evolutionary models are aimed at understanding the formal laws of life and evolution. These models analyze the evolution of artificial organisms that live only in cyberspace.

Applied evolutionary models are computer algorithms that use evolutionary methods of optimization to solve practical problems. The genetic algorithm of J.H. Holland (1992, 1996), Holland et al. (1986), and the evolutionary programming initiated by L. Fogel et al. are well-known examples of these approaches.

Due to its significance we will elaborate on the work of Kimura (1968) who proposed the so-called neutral theory of molecular evolution. The main assumption of Kimura's theory is that mutations at the molecular level (amino acid and nucleic acid substitutions) are mostly neutral or slightly disadvantageous. While essentially disadvantageous mutations are possible, they are eliminated effectively from populations by selection. This assumption agrees well with the mutational molecular substitution rate observed experimentally and with the fact that the rates of substitutions of less biologically important parts of macromolecules are greater than for the active macromolecule centers. The mathematical models developed within the neutral theory are essentially stochastic, hence a relatively small population size plays an important role in the fixation of neutral mutations.

If molecular substitutions are neutral, why then is progressive evolution possible? To answer this, Kimura (1968, 1983) used the concept of gene duplication developed by S. Ohno (1970). According to Kimura, gene duplications create unnecessary DNA sequences that drift further because of random mutations and provide raw material for creation of new, biologically significant genes. The evolutionary concepts of the neutral theory came from the interpretation of key biological experiments. A much more abstract theory was proposed by Stuart A. Kauffman.

Current evolutionary models are strongly stimulated by the progress made in computer science, especially in areas such as artificial life and neural networks. Recent evolutionary models incorporate such notions as learning, adaptive behavior, and genetic algorithms.Mathematical models of biological evolution have developed in several distinct directions: life origin models, mathematical population genetics, models of evolution of genetic regulatory systems, and artificial life evolutionary models to name but a few (Campbell, 1974, 1987). These models provide us with a better understanding of biological evolutionary phenomena. They also offer possible generalized descriptions of biological and societal experiments (Boulding, 1978).

9.10.3   Mathematical models of population genetics

The mathematical theory of population genetics was developed by Fisher (1958), Haldane, and Wright between the 1910s and 1930s. Population genetics or the synthetic theory of evolution is based on the Darwinian concept of natural selection and Mendelian genetics. Experiments on a small fruit fly, *Drosophila*, played an important role in finding an agreement between the Darwinian assumption of gradual, continuous evolutionary improvements and the discrete character of evolution of Mendelian genetics.

According to population genetics, the main mechanism of progressive evolution is the selection of organisms with advantageous mutations. The mathematical theory of population genetics (Axelrod, 1984) describes quantitatively the gene distribution dynamics in evolving populations. The theory includes two main types of models: deterministic (implying an infinitely large population size) and stochastic (finite population size).

9.10.4 Artificial life

Artificial life or a-life is a discipline within the general area of computer science devoted to the creation and study of life-like organisms and systems built by humans (Ray, 1992). The stuff of artificial life is, in contrast to real life, inorganic and its essence is information. Computers are the breeding grounds from which these new organisms emerge. As medical scientists and biologists study life mechanisms *in vitro*, physicists, mathematicians, and computer scientists hope to create life *in silico*.

The origins of the science of artificial life can be traced to Norbert Wiener (1961) and John von Neumann (1966). In 1955 he devised an abstract mathematical machine, a cellular automaton or collection of checkerboard squares each of which could switch states. Following a simple set of rules, the automaton was capable of copying itself and the copy in turn contained the blueprint for generating another copy *ad infinitum*. von Neumann's cellular automaton could not only reproduce itself, it was also a version of the universal computer envisioned by Alan Turing. It was capable of mimicking the operations of any conceivable computing machine. In the late 1960s, a British mathematician named John Conway created a vastly simplified version of the von Neumann cellular automaton, now known as the Game of Life. To read more about cellular automata, see Wolfram (1994).

The degree to which artificial life resembles real life varies; many experimenters admit that their laboratory creations are simply simulations of some aspects of life. Practitioners of the related science of artificial intelligence (AI) distinguish between two views on the subject. Weak AI concerns computer-simulated intelligence. Strong AI is about creating intelligent computers. Similarly, we can draw a distinction between weak and strong artificial life. The goals of weak a-life practitioners are illuminating and understanding life that exists on earth (Zeleny, 1981). The boldest (strong a-life) practitioners look toward long term development of living organisms whose essence is information. These creatures may be embodied in corporeal form as robots or live in computers. Whichever form they assume, these creations are intended to be alive under every reasonable definition of life as applied to bacteria, plants, animals, and human beings.

In general, the area of artificial life can be divided into subfields such as adaptive behavior, social behavior, evolutionary biology, complex systems, neural evolution, morphogenesis, and development and learning (Waldrop, 1992).

References

Alberts, B. et al., *Molecular Biology of the Cell*, Garland Publishing, London, 1994.

Axelrod, R.M., *The Evolution of Cooperation*, Basic Books, New York, 1984.

Bak, P. and Chen, K., *Sci. Amer.*, January 1991, p. 46.

Bak, P., *How Nature Works: The Science of Self-Organized Criticality*, Springer, Berlin, 1996.

Bak, P., Tang, C., and Weisenfeld, K., *Phys. Rev. A*, 38, 364, 1988.

Banavar, J.R., Maritan, A., and Rinaldo, A., *Nature*, 299 130, 1999.

Bennett, C.H., *Dissipation, Information, Computational Complexity and the Definition of Organization: Emerging Syntheses in Science*, Pines, D., Ed., Addison-Wesley, Redwood City, CA, 1985, p. 213.

Bernard, C., *An Introduction to the Study of Experimental Medicine*, English translation, Dover Publications, London, 1957.

Boulding, K.E., *Ecodynamics: A New Theory of Societal Evolution*, Sage, London, 1978.

Brandhorst, B.P. and McConkey, E.H., *J. Mol. Biol.*, 85, 451,. 1974.

Campbell, D.T., in *Studies in the Philosophy of Biology*, Ayala, F.J. and Dobzhansky, T., Eds., Macmillan, New York, 1974.

Campbell, D.T., in *Evolutionary Epistemology, Rationality, and the Sociology of Knowledge*, Radnitzky, G. and Bartley, W.W., Eds., Open Court, La Salle, IL, 1987, p. 47.

Crutchfield, J. et al., *Sci. Amer.*, December 1986, p. 46.

Darwin, C. *The Origin of Species by Means of Natural Selection or the Preservation of Favoured Races in the Struggle for Life*, Burrow, J.W., Ed., Penguin Classics, New York, 1985 (first published by John Murray, 1859).

Daut, J., Biochem. Biophys. Acta, 895, 41, 1987.

Dawkins, R., *The Blind Watchmaker*, Longman, London, 1986.

Dawkins, R., *The Extended Phenotype*: The Gene as a Unit of Selection, Oxford University Press, Oxford, U.K., 1983.

Dawkins, R., *The Selfish Gene*, 2nd ed., Oxford University Press, Oxford, U.K., 1989.

Dreyfuss, G. et al., *Annu. Rev. Biochem.*, 62, 289, 1993.

Eigen, M. and Schuster, P., *The Hypercycle: A Principle of Natural Self Organization*, Springer, Berlin, 1979.

Eigen, M. and Winkler-Oswatitsch, R., *Steps Toward Life: A Perspective on Evolution*. Oxford University Press, New York, 1992.

Feder, J., *Fractals*, Plenum Press, New York, 1988.

Fisher, R.A., *The Genetical Theory of Natural Selection*, 2nd ed., Dover Publications, New York, 1958.

Forrester, J., *Industrial Dynamics*, MIT Press, Cambridge, MA, 1961.

Forrester, J.W., *World Dynamics*, 2nd ed., Wright-Allen Press, Cambridge, MA, 1973.

Gilbert, E.N., *Science*, 152, 320, 1966.

Gould, S.J. and Eldredge, N., *Paleobiology*, 3, 115, 1977.

Haken, H., *Synergetics*. Springer, Berlin, 1978.

Haldane, J.B.S., *J. Genet.*, 48, 277, 1948.

Heusner, A.A., Energy metabolism and body size. Is the 0.75 exponent of Kleiber's equation a statistical artifact?, *Resp. Physiol.*, 48, 1, 1982.

Heylighen, F. and Joslyn, C., in *Encyclopedia of Physical Science and Technology*, Vol. 4, 3rd ed., Meyers, R.A., Ed., Academic Press, New York, 2001, p. 155.

Holland, J.H., *Adaptation in Natural and Artificial Systems: An Introductory Analysis with Applications to Biology, Control and Artificial Intelligence*, MIT Press, Cambridge, MA, 1992.

Holland, J.H. et al., *Induction: Processes of Inference, Learning and Discovery*, MIT Press, Cambridge, MA, 1986.

Holland, J.H., *Hidden Order: How Adaptation Builds Complexity*, Addison-Wesley, Reading, MA, 1996.

Iannaccone, P.M., Ed., *Fractal Geometry in Biological Systems: An Analytical Approach*, CRC Press, Boca Raton, FL, 1996.

Jantsch, E., The Self-Organizing Universe: Scientific and Human Implications of the Emerging Paradigm of Evolution, Pergamon Press, Oxford, U.K., 1979.

Javanaud, C., J. Acoust. Soc. Am., 86 493, 1989.

Johnson, H.A., Science, 168, 1545, 1970.

Kauffman, S.A., *At Home in the Universe: The Search for Laws of Self-Organization and Complexity*, Oxford University Press, Oxford, U.K., 1995.

Kauffman, S.A., *The Origins of Order: Self-Organization and Selection in Evolution*, Oxford University Press, New York, 1993.

Kauffman, S.A., *Sci. Amer.*, August 1991, p. 78.

Kim, H.D. et al., *Blood*, 77, 3387, 1991.

Kimura, M., *The Neutral Theory of Molecular Evolution*, Cambridge University Press, Cambridge, U.K., 1983.

Kimura, M., *Nature*, 217, 624, 1968.

Kleiber, M., Body size and metabolism, *Hilgardia*, 6, 315, 1932.

Kleiber, M., *Physiol. Rev.*, 27, 511, 1947.

Landau, L.D. and Lifshitz, E.M., *Statistical Physics*. Addison-Wesley, Reading, MA, 1969.

Loewenstein, W.R., *The Touchstone of Life*. Oxford University Press, Oxford, U.K., 1999.

Mandelbrot, B., *The Fractal Geometry of Nature*, W.H. Freeman, New York, 1983.

Mandelbrot, B., *The Fractal Geometry of Nature*, W.H. Freeman, New York, 1977.

McMahon, T.A., *Science*, 179, 1201, 1973

Monod, J., *Chance and Necessity*, Collins, London, 1972.

Morowitz, H.J., *Bull. Math. Biophys.*, 17, 81, 1995.

Nicolis, G. and Prigogine, I., *Self-Organization in Non-Equilibrium Systems*, John Wiley & Sons, New York, 1977.

Nicolis, G. and Prigogine, I., *Exploring Complexity*, W.H. Freeman, New York, 1989.

Ohno, S., *Evolution by Gene Duplication*, Springer, Berlin, 1970.

Prigogine, I., and Stengers, I., *Order Out of Chaos*, Bantam Books, New York, 1984.

Prigogine, I., *From Being to Becoming: Time and Complexity in the Physical Sciences*, W.H. Freeman, San Francisco, 1980.

Ray, T.S., in *An Approach to the Synthesis of Life. Artificial Life II*, Langton, C.G. et al., Eds., Addison-Wesley, Redwood City, CA, 1991, p. 371.

Rolfe, D.F.S. and Brown, G.C., *Physiol. Rev.*, 77, 731, 1997.

Schmidt-Nielsen, K., *Scaling: Why Is Animal Size So Important?* Cambridge University Press, Cambridge, U.K., 1984.

Schrödinger, E., *What Is Life? The Physical Aspects of Living Cells*, Cambridge University Press, Cambridge, U.K., 1967.

Sernetz, M., Galleri, B., and Hofmann, J., *J. Theor. Biol.*, 117, 209, 1985.

Shannon, C.E. and Weaver, W., *The Mathematical Theory of Communication*, 5th ed., University of Illinois Press, Chicago, 1963.

Spatz, H.C., *Comp. Physiol. B*, 161, 231, 1991.

Szilard, L., *Z. Phys.*, 53, 840, 1929.

Taylor and Weibel, *Resp. Physiol.*, 44, 1, 1981.

Thom, R., *Structural Stability and Morphogenesis*, Benjamin, Reading MA, 1975.

Thompson, D., *On Growth and Form*, Cambridge University Press, Cambridge, U.K., 1917.

von Neumann, J., in *Theory of Self-Reproducing Automata*, Burks, A.W., University of Illinois Press, Champaign, 1966.

Waldrop, M.M., *Complexity: The Emerging Science at the Edge of Order and Chaos*, Simon & Schuster, New York, 1992.

West, G.B., Brown, J.H., and Enquist, B.J., *Science*, 276 122, 1997.

West, B.J., *Fractal Physiology and Chaos in Medicine*, World Scientific, Singapore, 1990.

Wiener, N., *Cybernetics: Control and Communication in the Animal and Machine*, MIT Press, Cambridge, MA, 1961.

Wolfram, S., *Cellular Automata and Complexity: Collected Papers*, Addison-Wesley, Reading MA, 1994.

Wright, S., *Genetics*, 16, 97, 1931.

Zeleny, M., Ed., *Autopoiesis: A Theory of Living Organization*, North Holland, Amsterdam, 1981.

10

Epilogue: Toward New Physics and New Biology

> Discovery consists in seeing what
> everyone else has seen and thinking
> what no one else has thought.
> A discovery is said to be an accident
> meeting a prepared mind.
>
> Albert Szent-Gyorgi

10.1 Toward new physics

The past three decades witnessed a transformation in the way we perceive physical reality. While the turn of the nineteenth century brought the quantum revolution still, maintaining the linear philosophical framework, the turn of the twentieth century undermined the reign of linearity as the prevailing scientific strategy. It is worth keeping in mind that linearity is at the core of major physical disciplines such as electromagnetism, thermodynamics, and quantum mechanics.

Many examples amassed in the physical realm and in other disciplines ranging from astrophysics to economics clearly indicate that nonlinear phenomena cannot be dismissed through perturbative corrections. Nonlinearity has many faces, as shown by several new paradigms that reflect aspects of nonlinear behavior that cannot be comprehended through a manifestly linear theory or an infinite number of perturbative corrections (Dixon, Tuszynski, and Clarkson, 1997):

- Solitons and coherent structures
- Chaos and turbulence
- Limit cycles and higher dimensional dynamical attractors
- Pattern formation and selection
- Fractals and self-similarity
- Bifurcations and catastrophes

Appendix C is an illustrative overview of these concepts for beginners.

The road to the discovery of these paradigms was paved by advances in nonlinear mathematics, some of which took more than a century to fully develop and appreciate.

In particular, nonlinear ordinary and partial differential equations were studied and classified and a wealth of exact results guided physicists to new applications, and the advent of high performance computers and numerical techniques facilitated large-scale analyses of nonlinear systems.

Many initial applications of nonlinearity focused on physical systems such as lasers, superconductors, and liquid crystals. Over time, the spillover effect was felt in other disciplines of scientific inquiry. It appears, however, that the cell biology field may be a veritable gold mine of nonlinear phenomena. In fact, it may be a cliché, but life is a strongly nonlinear and nonequilibrium phenomenon.

10.2   Toward new biology

A number of famous physicists (and probably many more less famous ones) attempted to transplant physical concepts into biology. Fröhlich (1968) introduced biological coherence, Prigogine (1980) applied limit cycles to biology and physiology. Haken (1978) included biological systems in his theory of synergetics, and finally Davydov searched for solitons in bioenergetics.

Applications of nonlinearity to life sciences proliferated rapidly in recent years, particularly in the fields of embryology, ecology, cardiology, neurology, physiology, and pharmacology, to name but a few. Nonlinear "buzz words" such as order parameters, slaved models, catastrophes, fractals, strange attractors, phase locking, and self-organization are no longer alien to life scientists.

The prevailing mind set of biologists has also undergone a gradual transformation. The old-fashioned reductionist approach has been losing ground to a more open attitude that tries to incorporate environmental effects in system behavior (see Figure 10.1). Japanese biologist M. Murase expressed it clearly by formulating a list of new emerging points of view:

- From instructionist to selectionist (pre-existing variations + adaptation)
- From genetic inheritance to epigenetic inheritance (e.g., infectious prions)
- From genetic information to epigenetic information (cells are heterogeneous; mutations occur)
- From single force to multiple force (natural selection + self organization)
- From static to dynamic (non-dividing cells, e.g., neurons undergo dramatic changes during development in response to environment)
- From reductionistic to holistic to unified (life is organized hierarchically; evolution + development)
- From elementary units to elementary processes
- From simple self-reproducing molecules to complex self-regulatory networks

Fröhlich predicted the consequences that nonlinear paradigms may hold for biology. The key ones are:

1. Stability of the system with respect to small deformations below a critical value
2. Nonlinear response to external fields

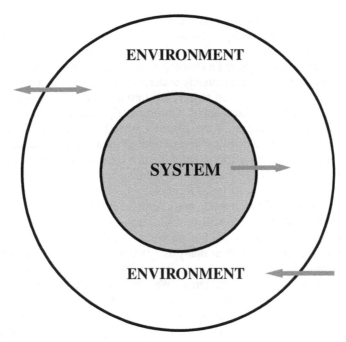

FIGURE 10.1
Interaction of a biological system with the environment.

3. Accumulation of energy in one special mode; time lag effects; frequency of incident radiation may differ from that of the coherent mode

4. Control mechanisms introduced through frequency-selective interactions, energy rate-dependence, and the relationship of size, surface stress, and frequency.

Of course, proper scientific understanding of biological systems is our ultimate goal—and probably a rather distant goal. In the meantime, the advancement of pure science via research and development intended to harness technological usefulness of biological systems is proceeding at breakneck speed. The following areas currently attract public attention.

10.2.1 Biocomputing

The current explosion of interest in the future of computing is strongly motivated by the imminent approach of the limit of classical computing as extrapolated from Moore's law (the number of transistors that can be fabricated on a commercially available silicon integrated circuit doubles every 18 to 24 months). This amazingly fast trend toward miniaturization has applied in the microelectronics sector for almost four decades.

Today, the smallest available silicon chips contain up to 100 million transistors on a few square centimeters of a wafer—translating into linear dimensions on the order of 200 nm or less. Reaching the dimensions of small clusters of several atoms, approximately 2 nm in length, will require a 10,000-fold miniaturization of micro-circuitry. Based on Moore's law, that goal will be reached between 2019 and 2028. However, heretofore unexplored technologies will have to be found before the limit is reached. Substantial effort is already under way in a field called quantum compu-tation. Because nanometer-sized objects fall into the realm of quantum mechanics, entirely different physical laws apply. Unfortunately, practical considerations such as the so-called entanglement of the system's wave function with the environment pose serious challenges to practical applications of quantum computing.

Simultaneously with the effort to build the first quantum computer, attempts have been made to use biological materials provided by nature or natural biological materi- als in combination with silicon-based technology, to devise smaller electronic devices that are structurally and functionally more flexible. It is hoped that a fast, small, and evolvable biological computer will be built very soon. We are already intimately familiar with its prototype — the human brain — a device composed of 10 billion nerve cells interconnected with as many as 1,000 neighboring neurons that commu-nicate via signals in the form of electric potential differences that travel along axons at speeds in the meters per second range. We know that waves of electric activity are correlated with cognitive functions; we do not know which brain structures and/or processes are responsible for consciousness or even what constitutes memory.

This enigma is sometimes called the mind–body problem. Without proceeding too far into this hotly debated topic, a camp of researchers galvanized by the Oxford mathematician Sir Roger Penrose believes that the fundamental nature of the mental processes lies in quantum mechanics. If true, this would, in a way, provide a neat con-ceptual link of two types of computing: nanoscale *in silico* computing and nanoscale biological or *in vivo* computing.

As described in detail in Chapter 7, nerve excitation involves passage of an elec-trical signal from one nerve cell to another in the form of synaptic transmission. Synapses are connections between nerve cells, their axons, or dendrites, and form nanometer-sized gaps crossed by neurotransmitter molecules stored in vesicles that open when stimulated. Here again is a point of contact with quantum mechanics. In addition, a very intricate structure of protein filaments fills the nerve cell body and its axons. Inside the axons lies a parallel architecture of microtubular bundles interconnected with other proteins. The structure resembles wiring of a parallel com-puter (Hameroff, 1987). This led to the hypothesis that microtubular structures may be involved in subcellular (nanoscale, possibly quantum) computation. Brown and Tuszynski (1997) demonstrated theoretically the feasibility of microtubular structures as information storing and processing devices. The concept that protein networks are strongly involved in both information processing and storage holds great promise.

Much of the factual information presented in this book suggests that living cells perform significant computational tasks. If so, can we reconcile the laws of thermo-dynamics and information theory with our knowledge of cell biology? Can we gain insights into the inner workings of the cell to enable future hybrid computers to harness the power of biological computational elements and integrate them with silicon? Although computer scientists and biochemists have not found a clear path from the test tube to the desktop, what they have found amazes and inspires them.

Without human intervention, evolution produced the smallest, most efficient computer in the world — the living cell.

The various points of reference regarding the nature of the living state undoubtedly reflect the prevailing Zeitgeist of the period in which a given theory is created. The viewpoint of representing the cell as a machine or even a factory closely mirrors the view of the industrial revolution of the 19th century. Likewise, the currently popular opinion that living cells are intensely engaged in some type of computation is closely linked with the technological revolution that ended the 20th century as a result of the proliferation of computer technology. Both points of view have merits, i.e., the cell obeys the laws of physics such as the first law of thermodynamics and hence can be viewed as a thermodynamic machine. It simultaneously acts against the second law of thermodynamics by creating structural and functional order. In other words, it creates and maintains information. Furthermore, it processes information and engages in signaling, thereby actively performing computation. It is safe to say that living cells can be viewed as both microfactories (in which nanomachines perform individual tasks) and biological computers whose nanochips are the various proteins and peptides in addition to DNA and RNA.

Most of a cell is what we might call hardware; only a small fraction is software (for example the genetic code in DNA that dictates the synthesis of proteins). Probably only a small fraction of a cell can be seen as pure information content. Do living systems contain something else that neither machines nor computers possess? They probably do. At least two properties distinguish animate matter from inanimate objects: procreative ability and autonomy expressed by free will.

On a more practical note, can biomimetics be used to enhance our computational capabilities? The answer is yes, although progress has been slow. In general terms, a chemical computer processes information by making and breaking chemical bonds and stores logic states or information in the resulting chemical (molecular) structures. A chemical nanocomputer would perform such operations selectively among a few molecules (a few nanometers wide) at a time. An alternative direction is adapting naturally occurring biochemicals for use in computing processes that do not occur in nature. Important examples are Adleman's DNA-based computer and Birge's bacteriorhodopsin-based computer memories.

Adleman (1994) first used DNA to solve a simple version of the "traveling salesman" problem where the task is to find the most efficient path through several cities. He demonstrated that the billions of molecules in a drop of DNA contained significant computational power. Digital memory can be seen in the form of DNA and proteins. Exquisitely efficient editing machines navigate through the cell, cutting and pasting molecular data into the stuff of life.

Evolution has produced the smallest, most efficient computers in the world. Even if this is an exaggeration, the innate intelligence built into DNA molecules could help fabricate tiny, complex structures using computer logic for building structures instead of crunching numbers. DNA computers would use a billion times less energy than electronic computers and store data in a trillion times less space. Moreover, computing with DNA is highly parallel. In principle trillions of DNA or RNA molecules could undergo chemical reactions and perform computations simultaneously.

Molecular biologists already have a toolbox of DNA manipulations, including enzyme cutting, ligation, sequencing, amplification, and fluorescent labeling. The idea behind DNA computing springs from a simple analogy based on the following two principles: (1) the complex structure of a living organism ultimately derives from applying sets of simple instructed operations (e.g., copying, marking, joining, inserting, deleting, etc.) to information in a DNA sequence; and (2) computation is the result of combining very basic arithmetic and logical operations.

Eric Winfree intends to create nanoscopic building blocks from DNA. The blocks will perform mathematical operations by fitting together in specific ways. DNA is not the only candidate for a biological computer chip. Birge (1994) proposed to use a light-sensitive protein dye called bacteriorhodopsin produced by bacteria. He and his collaborators have shown that bacteriorhodopsin can provide a very high density optical memory that could be integrated into an electronic computer to yield a hybrid device of much greater power than a conventional electronic computer. Electronic computers assembled with DNA and run on organic nutrients instead of electricity represent another science fiction idea that may soon become reality.

10.2.2 Biophysics: The physics of animate matter or an experimental biological tool?

For most biologists, biophysics conjures up images of complicated experimental instrumentation built by physicists. Biologists cannot deny the utility of physical equipment in their work. Progress in cytology would not have been possible without optical and electron microscopy. The deciphering of metabolic pathways is to a large degree facilitated by using radioactive isotopes as tracers. The pathways of biochemical reactions can be observed via infrared spectroscopy and nuclear magnetic resonance. X-ray crystallography and neutron diffraction techniques have enabled dramatic advances in molecular biology.

However, physics is much more than a powerful experimental tool. Theoretical physics is equally important as a conceptual tool and a language within which appropriate descriptions can be found for complex biological processes. The proportional representation of theory and experiment found in condensed matter physics is an example to be emulated by the biophysicists of the future. Elementary particle physics is involved more with theory than experimental effort. Unfortunately, biophysics, especially in the area servicing molecular biology needs, is placed at the other extreme where theoretical efforts have so far been almost negligible.

Finding new areas of application for the powerful experimental techniques of biophysics, molecular biologists adopted almost without change the original theoretical interpretations of the techniques, notwithstanding their different initial targets of investigation that were not nearly as complex as biological systems. This led to an almost universally accepted picture of biomolecules devoid of complex internal dynamics beyond rapid thermal fluctuations around well defined tertiary structures. These technically impressive images are well suited for the pages of popular science magazines and introductory texts but are not realistic representations of underlying phenomena. Biochemistry treats macromolecules as ordinary small molecules with fast internal vibrational dynamics compared to chemical time scales. This view is consistent with structural x-ray studies that assume that the difference between crystals of biological macromolecules and ordinary harmonic crystals can be distilled to numbers of atoms in the unit cell.

This simple picture is particularly seductive to enzymologists preoccupied with optimal orientations of several simultaneously catalytic molecular groups in transitional reaction states. They could conveniently avoid the dynamical properties of enzymes simply assuming that the appropriate states of the enzymes present themselves as equilibrium thermal fluctuations, as is the case with ordinary small molecules.

It is easy to understand the reason why elementary particle physics has oversubscribed theory. This is obviously dictated by the enormous costs associated with the construction of ever-larger particle accelerators. Theoretical physics has always been cheaper than experimental physics, but it is much less clear why the theoretical basis of biophysics is so underdeveloped. The blame can probably be equally shared by biologists and theoretical physicists. The former undoubtedly displayed internal resistance to learn mathematical concepts more advanced than elementary algebra and introductory probability theory. The latter may have somewhat arrogantly and nonchalantly formulated "universal" theories of biological processes using abstract language far removed from reality. Consequently, we face mutual distrust that will be difficult to overcome.

It must be overcome, rather sooner than later, as technological progress requires increasing theoretical sophistication. This is especially evident in the investigations of the dynamics of biological macromolecules such as proteins that force us to change the traditional pictures of biological processes. Their intra-molecular dynamics appears to be as slow as or slower than the associated biochemical reactions. If so, biochemical reactions should be affected to a much larger degree than predicted by the conventional theory of chemical reactions. This calls for the construction of a modern theory of enzymatic reactions that would be based on simple but adequate dynamical models. The three-dimensional spatial representation of a protein should be replaced with a four-dimensional spatiotemporal image.

It is very likely that in a near future, after construction of the essential elements of such a theoretical apparatus, we will see biophysics first and foremost as a tool of biology and of the physics of animate matter. This would be analogous to the way we view astrophysics as the physics of stellar matter. One of the tasks we set when writing this book was to prepare readers for such a change in the perception of what biophysics is and will become. We hope we have not failed in this regard.

References

Adleman, L., *Science*, 266, 1021, 1994.

Bak, P., *How Nature Works: The Science of Self-Organized Criticality*, Springer, Berlin, 1996.

Birge, R., *Am. Sci.,* July–August 1994, p. 348.

Brown, J.A. and Tuszynski, J.A., *Phys. Rev. E,* 56, 5834, 1997.

Davydov, A.S., *Quantum Mechanics and Biology,* Pergamon Press, London, 1982.

Dixon, J.M., Tuszynski, J.A., and Clarkson, P.A., *From Nonlinearity to Coherence,* Oxford University Press, Oxford, U.K., 1997.

Fröhlich, H., *Int. J. Quantum Chem.,* 2, 641, 1968.

Haken, H., *Synergetics: An Introduction to Nonequilibrium Phase Transitions and Self-Organization in Physics, Chemistry and Biology,* Springer, Berlin, 1977.

Hameroff, S., *Ultimate Computing,* Elsevier, Amsterdam, 1987.

Murase, M., *Progr. Theor. Phys.,* 95, 1, 1996.

Prigogine, I., *From Being to Becoming: Time and Complexity in the Physical Sciences,* W.H. Freeman, San Francisco, 1980.

Winfree, E., Liu, F., Wenzler, L.A., and Seeman, N.C., *Nature,* 394, 539, 1998.

Author Index

A

Athenstadt, 388

B

Bardeen, 458
Baule, 388, 392
Berger, 383
Bernard, 397, 420
Bernstein, 327, 339, 383
Birge, 429, 430, 432
Boltzmann, 34, 88, 95, 133, 136, 145, 232, 233, 238, 246, 385, 414, 484
Brown, 187, 209, 225, 227, 264, 285, 297, 316, 388, 392, 413, 422, 423, 428, 432
Burr, 383

C

Cannon, 397
Carnot, 280, 410
Changeaux, 272
Clausius, 232, 238, 459
Cohen, 388, 392
Cole, 167, 225, 327, 339
Conway, 137, 138, 157, 419
Cooper, 309, 316, 318, 387, 392, 458
Crick, 83

D

Darwin, 1, 2, 15, 416, 417, 420
Davydov, 139, 140, 141, 142, 156, 157, 158, 426, 432
Du Bois-Reymond, 382, 383

E

Eccles, 383
Ehrenfest, 459
Eigen, 2, 21, 23, 396, 417, 420, 421
Elay, 386
Eldredge, 416, 417, 421

F

Feigenbaum, 481, 507, 509, 514
Fermi, 496
Fogel, 418
Forrester, 398, 421
Frauenfelder, 112, 122, 124, 157, 275
Fricke, 167
Froehlich, 142, 143, 144, 145, 157, 169, 226, 416, 426
Fröhlich, 426, 432

G

Galvani, 382
Garcia, 125
Gerard, 383, 384
Gibbs, 210, 232, 238, 245, 247, 260, 281, 410, 412, 459, 464
Glisson, 382
Gould, 416, 417, 421
Gurwitsch, 152, 157, 158

H

Haken, 396, 421, 426, 432, 481, 514
Haldane, 265, 267, 269, 417, 418, 421
Hales, 382
Harvey, 112, 113, 157, 342

Subject Index

A

abdominal, 357
Ablowitz-Ladik equation, 497
accommodation, 328, 370, 372
acetanilide, 142, 156
acetate, 8
acetyl, 9, 302, 332
acetylcholine, 326, 333, 383
acetylcholinesterase, 333
acetylsalicylic acid, 263
acid
    acetylsalicylic, 263
    aspartic, 66
    carboxylic, 23, 24, 25, 26
    citric, 6, 8
    deoxyribonucleic, 2, 6, 82
    glutamic, 66, 89
    ribonucleic, 2, 82
    salicylic, 263
acidity, 8, 13, 20, 357
actin, 164, 176, 177, 180, 186, 187, 189, 190, 191,
    192, 194, 195, 196, 197, 198, 218,
    219, 224, 226, 228, 277, 278, 289,
    290, 291, 308, 309, 310, 311, 312,
    314, 315, 316, 317, 318, 341, 395
    F, 189
actinin, 191
action potential, 139, 169, 224, 325, 326, 327,
    328, 329, 331, 333, 334, 335, 383,
    384
activated process, 110, 116, 121
activation
    free energy of, 118, 260, 262
activation energy, 386, 387
activation free energies, 259
active transport, 99, 211, 263, 333
activity, 80, 154, 161, 164, 175, 196, 203, 206,
    209, 212, 246, 254, 255, 300, 325,
    338, 339, 342, 354, 373, 383, 384,
    387, 399, 402, 403, 428, 512
acto-myosin motor, 289, 290
acyl-enzyme, 263
acylmutase, 264

adaptive behavior, 418
adaptive process, 389
adenine, 6, 9, 12, 14, 29, 83, 84, 89, 198, 212, 300
adenosine diphosphate, 6, 28, 208, 300, 354
adenosine monophosphate, 28, 303, 354
adenosine triphosphate, 6, 28, 195, 208, 300,
    354
adenylate kinase, 303
ADH, 10, 333, 350
adherens junctions, 190
adhesion, 164, 190
adiabatic, 109, 229, 240, 241
adiabatic potential, 109, 229
ADP, 6, 7, 9, 12, 28, 39, 54, 55, 189, 191, 196,
    197, 208, 209, 210, 263, 282, 284,
    298, 300, 301, 302, 303, 308, 315,
    318, 354
aggregate, 60, 69, 161, 218, 332, 386
aggregation, 78, 192, 306, 510
air flow, 355, 356
alcohols, 23, 24, 25, 26, 53, 59
aldehyde, 24, 378
Alexander-Orbach conjecture, 130
algae
    blue-green, 160
    green, 160
algorithm
    back-propagation, 336
    Kohonen clustering, 337
aliphatic, 65, 66
allergies, 389
allometric scaling equations, 404
allosteric, 272, 273, 413
allosteric activation, 272
allosteric heterotropic effect, 273
allosteric homotropic effect, 273
allostery, 272
alloys
    binary, 491
alternator, 280
alveoli, 355, 356, 357, 358
ameboid movement, 191
amide bond, 26, 29, 67, 68
amide I bond, 140

vasoconstriction, 347, 400
vasodilation, 347, 399, 400
vein, 343, 345, 346, 350
    pulmonary, 343
    thoracic, 347
velocity, 80, 140, 145, 146, 187, 237, 290, 313, 315,
          320, 343, 344, 346, 347, 348, 383,
          405, 406, 490, 497, 498, 502, 503
ventricle
    lower, 352
venules, 345
Venus, 1
vertices, 111, 133
vesicle, 3, 4, 60, 61, 167, 181, 187, 195, 196, 198,
        202, 203, 320, 331, 332, 428
vestibule, 379
vibration, 91, 128, 140, 141, 142, 144, 169
vibrational energy, 91, 142
vibrational relaxation, 91
villi, 176
vimentin, 192
vincristine, 218
vinculin, 190
viruses, 80, 303, 389
viscosity, 80, 95, 96, 172, 219, 343, 346, 347,
        348, 458
viscous fingering, 407
vision, 139, 363, 371, 372, 373, 374, 377, 378
    photopic, 371, 372, 378
    scotopic, 371, 372, 378
visual acuity, 372, 376
vitamin A aldehyde, 378
vitamins, 89
vitreous humor, 370, 376
voltage, 79, 167, 169, 171, 224, 243, 255, 279,
        322, 323, 326, 327, 329, 331, 334,
        335, 363, 383, 504
voltage-gated, 171, 224, 326, 335
volumetric flow rate, 346
vortex, 467, 468

W

water, 1, 3, 6, 9, 11, 12, 13, 14, 17, 23, 24, 28, 33,
        48, 50, 53, 54, 55, 56, 57, 58, 59, 60,
        64, 66, 67, 69, 73, 77, 79, 80, 83, 84,
        87, 93, 95, 96, 97, 98, 99, 107, 140,
        147, 152, 159, 161, 162, 166, 167,
        172, 173, 174, 175, 208, 212, 218,
        233, 234, 250, 251, 253, 282, 304,
        341, 342, 343, 349, 350, 354, 355,
        356, 370, 386, 387, 388, 399, 401,
        403, 407, 410, 413

wave
    charge-density, 458
    solitary, 497, 498, 501
wave function, 44, 45, 428
weather prediction, 396
white blood cell, 341, 343, 349
wood, 60, 367
work, 8, 9, 41, 88, 89, 137, 159, 171, 175, 196,
        199, 209, 210, 232, 239, 240, 241,
        242, 243, 247, 253, 277, 278, 279,
        280, 281, 282, 284, 285, 286, 287,
        290, 295, 300, 301, 308, 313, 325,
        327, 338, 352, 353, 358, 362, 377,
        383, 385, 387, 395, 396, 398, 399,
        400, 401, 403, 406, 410, 413, 419,
        461, 465, 477, 481, 507
worm-like chain, 136

X

x-ray crystallography, 430
XY model, 466

Y

yeast, 8, 145, 222
Young modulus, 136, 177
Young-Helmholtz theory, 374

Z

Zimm
    zonules of, 369
zinc, 178, 253, 388, 457
zonules of Zimm, 369
zwitterionic, 161

Appendix A

Random Walks and Diffusion

Introduction

Most of the information summarized in this appendix can be found in two excellent books: Reif (1965), a volume dealing mainly with statistical physics concepts, and Benedek and Villars (2000), a treasury of applications to biology and medicine.

A statistical ensemble has N (very large) identical, similarly prepared systems. The probability of occurrence of a particular event equals the fraction of systems in the ensemble realizing it. Consider a random walk problem (Figure A.1) in which a drunk starts walking away from a lamppost. Each step has length ℓ and its direction is independent of the preceding steps. The probability of a step to the right is p; the probability of a step to the left is $q = 1 - p$. The location each time is given by $x = m\ell$ where m is an integer. The question is, after N steps, what is the probability $P(x = m\ell)$? A generalization to two or three dimensions is easy, so we will not elaborate on it.

The physical problem concerns vector addition with random direction of the individual vectors at a fixed length. The answer will provide the length and direction of the resultant vector. See Figure A.2.

Well-known physical examples of random walks are:

- Magnetism: noninteracting N spins half particles
- Diffusion of a molecule in a gas (Brownian motion)
- Light intensity due to N incoherent light sources (two-dimensional vector) (see Figure A.3)
- Growth patterns; fractals

Let us denote n_1 as the number of steps to the right and n_2 as the number of steps to the left such that:

$$N = n_1 + n_2 \quad \text{and} \quad m = n_1 - n_2 = n_1 - (N - n_1) = 2n_1 - N$$

Note that m is odd if N is odd and even if N is even. Successive steps are statistically independent of each other (no memory). First, the probability of a sequence of steps is:

$$P \propto p.p \ldots (n_1 \text{ times}) \bullet q \cdot q \cdot q \ldots (n_2 \text{ times}) = p^{n_1} q^{n_2} \qquad (A.1)$$

but $N!/(n_1! n_2!)$ combinations of such possibilities exist, so the probability is:

$$W_N(n_1) = \frac{N!}{n_1! n_2!} p^{n_1} q^{n_2} \qquad (A.2)$$

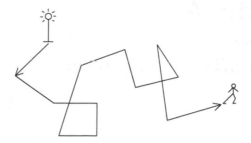

FIGURE A.1
Random path of a drunk walking away from a lamp post.

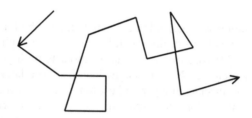

FIGURE A.2
Addition of vectors of differing direction.

FIGURE A.3
Addition of vectors of unit length.

This is an expression for a binomial distribution since:

$$(p + q)^N = \sum_{n=0}^{N} \frac{N!}{n!(N - n)!} p^n q^{N-n} \tag{A.3}$$

where n_1 determines m, so that $P_N(m) = W_N(n_1)$. Namely:

$$n_1 = 1/2(N + m) \quad \text{and} \quad n_2 = 1/2(N - m) \tag{A.4}$$

so that we obtain

$$P_N(m) = \frac{N!}{(\frac{N+m}{2})!(\frac{N-m}{2})!} p^{\frac{N+m}{2}} (1 - p)^{\frac{N-m}{2}} \tag{A.5}$$

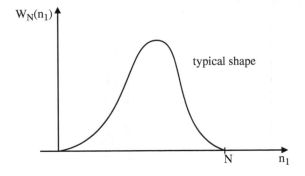

FIGURE A.4
Typical shape of the curve $W_N(n_1)$.

For $p = q = 1/2$ we have:

$$P_N(m) = \frac{N!}{(\frac{N+m}{2})!(\frac{N-m}{2})!} \left(\frac{1}{2}\right)^N \tag{A.6}$$

Summing all the possibilities we get:

$$\sum_{n_1=0}^{N} \frac{N!}{n_1!(N-n_1)!} p^{n_1} q^{N-n_1} = (p+q)^N = 1^N = 1 \tag{A.7}$$

Therefore, probability is properly normalized (see Figures A.4 and A.5). We then calculate the average number of steps to the right as:

$$\begin{aligned}
\bar{n}_1 &= \sum_{n_1=0}^{N} n_1 W(n_1) = \sum_{n_1} n_1 \frac{N!}{n_1! n_2!} p^{n_1} q^{n_2} \\
&= p \frac{\partial}{\partial p} \sum_{n_1} W(n_1) = p \frac{\partial}{\partial p} (p+q)^N \\
&= pN(p+q)^{N-1} = pN \tag{A.8}
\end{aligned}$$

To calculate this, note that:

$$n_1 p^{n_1} = p \frac{\partial}{\partial p} (p^{n_1}) \tag{A.9}$$

so that:

$$\begin{aligned}
\sum_{n_1=0}^{N} \frac{N!}{n_1!(N-n_1)!} p^{n_1} p^{N-n_1} n_1 &= \sum_{n_1=0}^{N} \frac{N!}{n_1!(N-n_1)!} \left[p \frac{\partial}{\partial p} (p^{n_1}) \right] q^{N-n_1} \\
&= p \frac{\partial}{\partial p} \left[\sum_{n=0}^{N} \frac{N!}{n_1!(N-n_1)!} p^{n_1} q^{N-n_1} \right] = p \frac{\partial}{\partial p} (p+q)^N \\
&= pN(p+q)^{N-1} = pN \tag{A.10}
\end{aligned}$$

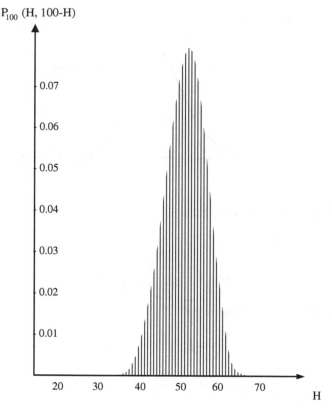

FIGURE A.5
Bernoulli distribution for $N = 100, p = q = 1/2, H = N/2 = 50, \Delta = 1/2$
$\sqrt{N} = 5$

which is valid for arbitrary p and q and $p + q = 1$. Therefore, for $p = q = 1/2$, $\bar{n}_1 = 1/2N$. In general, $\bar{n}_1 = Np$ and $\bar{n}_2 = Nq$. Thus:

$$\bar{n}_1 + \bar{n}_2 = N(p + q) = N \tag{A.11}$$

and for the displacement, $m = n_1 - n_2$, we find its average value as:

$$\overline{m} = \overline{n_1 - n_2} = \bar{n}_1 - \bar{n}_2 = N(p - q) \tag{A.12}$$

If $p = q = 1/2$, then $\overline{m} = 0$ which reflects the perfect symmetry between left and right directions.

Dispersion in the random walk is calculated as:

$$(\Delta n_1)^2 = \overline{(n_1 - \bar{n}_1)^2} = \overline{n_1^2} - (\bar{n}_1)^2 \tag{A.13}$$

We first calculate:

$$\overline{n_1^2} = \sum_{n_1=0}^{N} W(n_1)n_1^2 = \sum_{n_1=0}^{N} \frac{N!}{n_1!(N - n_1)!} p^{n_1} q^{N-n_1} n_1^2 \tag{A.14}$$

To find this, we use:

$$n_1^2 p^{n_1} = n_1 \left(p \frac{\partial}{\partial p} \right) (p^{n_1}) = \left(p \frac{\partial}{\partial p} \right)^2 (p^{n_1}) \tag{A.15}$$

Hence:

$$\sum_{n_1=0}^{N} \frac{N!}{n_1!(N-n_1)!} \left(p \frac{\partial}{\partial p} \right)^2 p^{n_1} q^{N-n_1} = \left(p \frac{\partial}{\partial p} \right)^2 \sum_{n_1=0}^{N} \frac{N!}{n_1!(N-n_1)!} p^{n_1} q^{N-n_1}$$

$$= \left(p \frac{\partial}{\partial p} \right)^2 (p+q)^N$$

$$= \left(p \frac{\partial}{\partial p} \right) [Np(p+q)^{N-1}]$$

$$= p[N(p+q)^{N-1}$$

$$+ pN(N-1)(p+q)^{N-2}] \tag{A.16}$$

but $p+q = 1$, so that:

$$\overline{n_1^2} = p[N + pN(N-1)] = Np[1 + pN - p]$$
$$= (Np)^2 + Npq = (\overline{n_1})^2 + Npq \tag{A.17}$$

Therefore:

$$\overline{(\Delta n_1)^2} = \overline{n_1^2} - (\overline{n_1})^2 = Npq \tag{A.18}$$

Hence, the rms derivation (standard deviation) is:

$$\Delta^* n_1 = \sqrt{\overline{(\Delta n_1)^2}} = \sqrt{Npq} \tag{A.19}$$

This is a linear measure of the width of the range over which n_1 is distributed. Relative width of the distribution is easily obtained as:

$$\frac{\Delta^* n_1}{\overline{n_1}} = \frac{\sqrt{Npq}}{Np} = \sqrt{\frac{q}{p}} \frac{1}{\sqrt{N}} \tag{A.20}$$

For $p = q = 1/2$, we find that $\frac{\Delta^* n_1}{\overline{n_1}} = \frac{1}{\sqrt{N}}$. As N increases $\overline{n_1} \sim N$, but $\Delta^* n_1 \sim N^{1/2}$ so the distribution becomes sharper and sharper around the mean value. The dispersion of the net displacement to the right is calculated from $m = n_1 - n_2 = 2n_1 - N$, so we find:

$$\Delta m = m - \overline{m} = 2n_1 - N - (2\overline{n_1} - N) = 2(n_1 - \overline{n_1}) = 2\Delta n_1 \tag{A.21}$$

and $(\Delta m)^2 = 4(\Delta n_1)^2$. Thus, $\overline{(\Delta m)^2} = 4\overline{(\Delta n_1)^2} = 4Npq$. In particular, for $p = q = 1/2$, $\overline{(\Delta m)^2} = N$.

Gaussian probability distributions

For large N, using Stirling's approximation, we can re-express $W(n_1)$ in terms of $P(m)$ substituting $n_1 = 1/2(N + m)$:

$$P(m) = W\left(\frac{N+m}{2}\right) = [2\pi Npq]^{-1/2} e^{\{\frac{[m-N(p-q)]^2}{8Npq}\}} \qquad (A.22)$$

Since $n_1 - Np = 1/2(N + m) - Np = 1/2[Np + Nq + m] - Np = 1/2\,[m - N(p - q)]$. Also, since $m = 2n_1 - N$, m assumes integral values separated by $\Delta m = 2$. We can also express $P(m)$ in terms of $x = m \cdot l$, the actual displacement variable. For large N, $|P(m + 2) - P(m)| << P(m)$ and $P(m)$ can be regarded as a smooth function of x. See Figure A.6.

The probability of finding the particle after N steps between x and dx is:

$$\mathcal{P}(x)\,dx = P(m)\,dx/2\ell \qquad (A.23)$$

where $\mathcal{P}(x)$ is the probability density and is equal to:

$$\mathcal{P}(x) = \frac{1}{\sqrt{2\pi}\sigma} e^{[-\frac{(x-\mu)^2}{2\sigma^2}]} \qquad (A.24)$$

where $\mu = (p - q)N\ell$ and $\sigma = 2\sqrt{Npq}\ell$.

Gaussian probability distributions occur frequently in probability theory whenever large numbers are involved. We easily check that $\mathcal{P}(x)$ is properly normalized:

$$\int_{-\infty}^{\infty} \mathcal{P}(x)\,dx = \frac{1}{\sqrt{2\pi}\sigma} \int_{-\infty}^{\infty} e^{-\frac{(x-\mu)^2}{2\sigma^2}}\,dx = \frac{1}{\sqrt{2\pi}\sigma} \int_{-\infty}^{\infty} e^{-\frac{y^2}{2\sigma^2}}\,dy$$

$$= \frac{1}{\sqrt{2\pi}\sigma}\sqrt{\pi 2\sigma^2} = 1 \qquad (A.25)$$

$y = x - \mu$, $dy = dx$ since:

$$\int_{0}^{\infty} e^{-z^2}dz = \frac{\sqrt{\pi}}{2}, \quad \text{and} \quad z = y/2^{1/2}\sigma \qquad (A.26)$$

We then calculate the mean value $\bar{x}$ as:

$$\bar{x} = \int_{-\infty}^{\infty} x\mathcal{P}(x)\,dx = \frac{1}{\sqrt{2\pi}\sigma} \int_{-\infty}^{\infty} xe^{-\frac{(x-\mu)^2}{2\sigma^2}}\,dx$$

$$= \frac{1}{\sqrt{2\pi}\sigma}\left[\int_{-\infty}^{\infty} ye^{-\frac{y^2}{2\sigma^2}}\,dy + \mu \int_{-\infty}^{\infty} e^{-\frac{y^2}{2\sigma^2}}\,dy\right] = \mu \qquad (A.27)$$

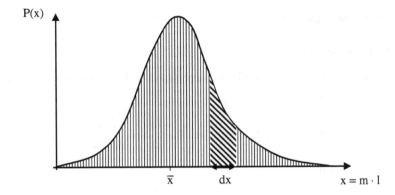

FIGURE A.6
Probability $P(n)$ as a function of x.

$P(x)$ is symmetric about $x = \mu$, the position of its maximum. The dispersion is:

$$\overline{(x - \mu)^2} = \int_{-\infty}^{\infty} (x - \mu)^2 P(x)\,dx = \frac{1}{\sqrt{2\pi}\sigma} \int_{-\infty}^{\infty} y^2 e^{-\frac{y^2}{2\sigma^2}}\,dy = \sigma^2 \qquad \text{(A.28)}$$

since:

$$\int_0^{\infty} e^{-\alpha x^2} x^2\,dx = \frac{\sqrt{\pi}}{4}\alpha^{-\frac{3}{2}} \qquad \text{(A.29)}$$

Thus:

$$\overline{(\Delta x)^2} = \overline{(x - \mu)^2} = \sigma^2 \qquad \text{(A.30)}$$

Hence, σ is the rms deviation of x from the mean of the Gaussian distribution. Thus we identify for the random walk problem the following characteristic quantities:

$$\overline{x} = \mu = (p - q)N\ell$$
$$\overline{(\Delta x)^2} = \sigma^2 = (2\sqrt{Npq}\ell)^2 = 4Npq\ell^2 \qquad \text{(A.31)}$$

which agree with those calculated earlier for the general case of arbitrary N.
Other statistical moments can be found using:

$$I(n) = \int_0^{\infty} e^{-\alpha x^2} x^n\,dx = \frac{1}{2}\Gamma\left(\frac{n+1}{2}\right)\alpha^{\frac{(n+1)}{2}} \qquad \text{(A.32)}$$

where $\Gamma(n) = (n - 1)!$

Diffusion equation

The space–time evolution of the Gaussian probability distribution is considered next (see Figure A.7). Since:

$$x = m\ell; \quad m = \frac{x}{\ell} \quad \text{and} \quad P_N(m) = P_N\left(\frac{x}{\ell}\right) = \sqrt{\frac{2}{\pi N}}e^{-\frac{x^2}{2N\ell^2}} \tag{A.33}$$

the probability density is:

$$P(x) = \sqrt{\frac{1}{2\pi N\ell^2}}e^{-\frac{x^2}{2N\ell^2}} \tag{A.34}$$

where the prefactor can be easily found from normalization. Note that:

$$\mathcal{P}(x)\Delta x = P_N(m) \tag{A.35}$$

The number of steps can be eliminated by the time variable $t = Nt_c$ where t_c is the time interval for each step or $N = t/t_c$. Hence:

$$P_N(x) = P_{\frac{t}{t_c}}(x) = P(x;t) = \sqrt{\frac{t_c}{2\pi \ell^2 t}}e^{-\frac{t_c}{2\ell^2 t}x^2} \tag{A.36}$$

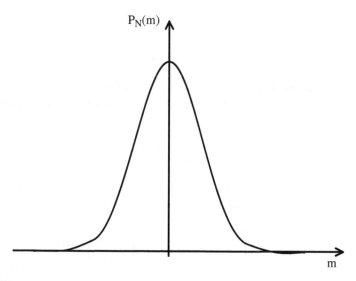

FIGURE A.7
Gaussian probability distribution.

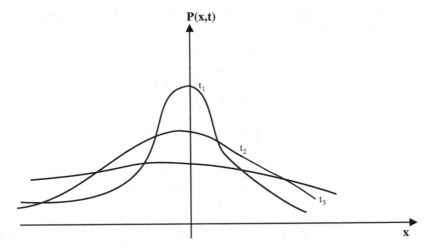

FIGURE A.8
Curves showing relation between distance variation and elapsed time.

We denote $D = \frac{\ell^2}{2t_c}$ which has the meaning of the diffusion constant so that:

$$P(x;t) = \sqrt{\frac{1}{4\pi Dt}} e^{-\frac{1}{4Dt}x^2} \qquad (A.37)$$

Also note that:

$$\overline{m_2} = N \rightarrow \frac{\overline{x^2}}{\ell^2} = \frac{t}{t_c} \qquad (A.38)$$

or equivalently:

$$\overline{x^2} = \ell^2 \frac{t}{t_c} = \left(\frac{\ell^2}{2t_c}\right) \cdot 2t = 2Dt \qquad (A.39)$$

which is a one-dimensional result relating the variance in the distance and the time elapsed during the walk. See Figure A.8.

Probability of displacement for a three-dimensional random walk

The key assumptions of the model are:

- Choice of direction (one out of three) is arbitrary
- Equal probability of plus or minus direction along each axis
- Each step is of length ℓ and takes time t_c

For a very large number of steps $N(\to \infty)$, about a third will be along x, a third along y, and a third along z:

$$P(m_x, m_y, m_z) = P_{N/3}(m_x)P_{N/3}(m_y)P_{N/3}(m_z)$$

$$= \left(\sqrt{\frac{2}{\pi N/3}}\right)^3 e^{-\frac{m_x^2}{2N/3}} \cdot e^{-\frac{m_y^2}{2N/3}} \cdot e^{-\frac{m_z^2}{2N/3}}$$

$$= \left(\frac{6}{\pi N}\right)^{3/2} e^{-\frac{3}{2N}(m_x^2+m_y^2+m_z^2)} \tag{A.40}$$

Next, write $(\frac{x}{\ell})$ for m_x, $(\frac{y}{\ell})$ for m_y, and $(\frac{z}{\ell})$ for m_z, and write $(\frac{t}{t_c})$ for N. The sum of all $\mathcal{P}_N(m_x, m_y, m_z)$ represents a displacement of the particle in the interval Δx along x, Δy along y, and Δz along z, such that:

$$\approx P_{\frac{t}{t_c}}\left(\frac{x}{\ell}, \frac{y}{\ell}, \frac{z}{\ell}\right)\left(\frac{\Delta x}{2\ell}\right)\left(\frac{\Delta y}{2\ell}\right)\left(\frac{\Delta z}{2\ell}\right) \tag{A.41}$$

properly normalize the distribution. Therefore, the probability that starting from the origin $t = 0$, the particle will end up in the cube $(x, x+\Delta x)\cdot(y, y+\Delta y)\cdot(z, z+\Delta z)$ after time t, can be written as:

$$P(x, y, z; t) = \left(\frac{3t_c}{2\pi t\ell^2}\right)^{\frac{3}{2}} e^{-(\frac{3t_c}{2t\ell^2})(x^2+y^2+z^2)} \Delta x \Delta y \Delta z \tag{A.42}$$

Introducing a three-dimensional diffusion constant $D = \frac{\ell^2}{6t_c}$, we express this probability distribution as:

$$P(x, y, z; t) = \left(\frac{1}{4\pi Dt}\right)^{\frac{3}{2}} e^{-\frac{x^2+y^2+z^2}{4Dt}} \tag{A.43}$$

and also find:

$$\overline{m_x^2} = \frac{N}{3} \qquad \frac{\overline{x^2}}{\ell^2} = \frac{1}{3}\frac{t}{t_c} \tag{A.44}$$

Alternatively:

$$\overline{x^2} = 2Dt; \quad \overline{y^2} = 2Dt; \quad \overline{z^2} = 2Dt \tag{A.45}$$

Consequently,

$$\overline{r^2} = \overline{x^2 + y^2 + z^2} = 6Dt \tag{A.46}$$

We assume statistical independence of the x, y, z displacements, i.e.:

$$P(x, y, z; t)\Delta x \Delta y \Delta z = \left(\sqrt{\frac{1}{4\pi Dt}}e^{-\frac{x^2}{4Dt}}\Delta x\right)\left(\sqrt{\frac{1}{4\pi Dt}}e^{-\frac{y^2}{4Dt}}\Delta y\right)$$

$$\left(\sqrt{\frac{1}{4\pi Dt}}e^{-\frac{z^2}{4Dt}}\Delta z\right) \tag{A.47}$$

TABLE A.1
Typical Diffusion Constants

	$D(\text{cm}^2/\text{s})$
In air	
H	0.634
H_2O	0.239
In water	
O_2	10^{-5}
DNA	10^{-8}

This is properly normalized since:

$$\sum_{\text{all}\,\Delta x} P(x;t)\Delta x = \int \sqrt{\frac{1}{4\pi Dt}} e^{-\frac{x^2}{4Dt}}\, dx = 1 \tag{A.48}$$

Equivalently, the time- and space- dependent probability distribution can be written as:

$$P(x;t) = \sqrt{\frac{1}{2\pi x^2}} e^{-\frac{x^2}{2x^2}} \tag{A.49}$$

Typical diffusion constants of small molecules in air at atmospheric pressure are listed in Table A.1. The value of D in air is typically $\sim 10^{-1}$ cm^2/s since $\ell \sim 10^{-5}$ and $t_c \sim 2 \cdot 10^{-10}$ s. Differences in the diffusion constant are mainly due to different densities (mean free paths) of the molecules of the medium. D depends on the molecules that execute the random walk, namely:

$$D\alpha \frac{\ell^2}{t_c} = \ell\left(\frac{\ell}{t_c}\right) = \ell\bar{v} \approx \ell\sqrt{\frac{3kT}{m}} \tag{A.50}$$

so D is proportional to ℓ and $\sqrt{T}$.

Importantly, in dilute media, binary collisions dominate; in dense media multiple collisions dominate (see Figure A.9).

Diffusion equation

Usually, a large number of independent particles is assumed to be engaged in a random walk. It is convenient then to introduce a local particle concentration:

$$C(x, y, z; t) = \frac{\text{\#particles}}{\text{unit volume}} = \lim_{\Delta V \to 0}\left(\frac{\Delta N}{\Delta V}\right) \tag{A.51}$$

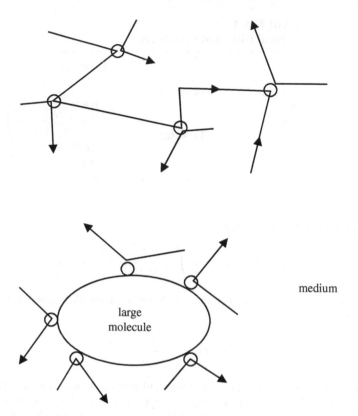

FIGURE A.9
(a) Binary collisions (dilute media). (b) Multiple collisions (dense media).

where C may be inhomogeneous and may evolve in time. Nonuniform concentrations initially may, through diffusion, eventually become uniform. With the assumption that $C(x, 0)$ at $t = 0$, what is the value of $C(x, t)$ at a later time t? The solution is found using random walk concepts (see Figure A.10).

With ΔN_o particles starting at $t = 0$, x' and ΔN particles ending at $x - x'$ after t, ΔN is proportional to $\mathcal{P}(x - x', t)\Delta x$, where $x - x'$ is the displacement so that:

$$\Delta N(x) = \sum_{\text{all intervals } \Delta x' = x - x'} \Delta N_o(x') P(x - x'; t)\Delta x \qquad (A.52)$$

which means we must collect all the contributions from initial positions through all possible intermediate paths. Note that: $\Delta N_o(x') = C_o(x')\Delta x'$ and $\Delta N(x) = C(x, t)\Delta x$. This means:

$$C(x; t)\Delta x = \sum_{\text{all } \Delta x'} \Delta x' C_o(x') P(x - x'; t)\Delta x \qquad (A.53)$$

time

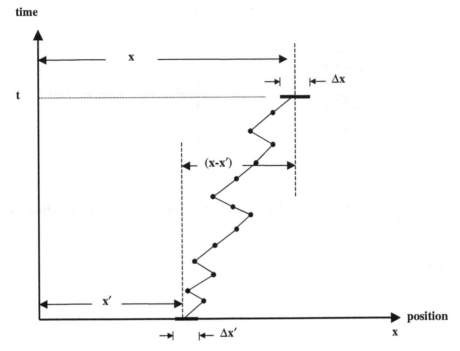

FIGURE A.10
Distance versus time plot in a typical random walk process.

Converting this sum to an integral yields:

$$C(x;t) = \int dx' C_o(x') P(x - x', t) \tag{A.54}$$

From random walk calculations, we have:

$$P(x - x'; t) = \sqrt{\frac{1}{4\pi Dt}} e^{-\frac{(x-x')^2}{4Dt}} \tag{A.55}$$

This is an integral representation of the diffusion process which can be calculated knowing $C_o(x')$ everywhere.

Example

See Figure A.11 for an illustration of the physical situation given by the initial concentration:

$$C_o(x) = \begin{cases} 0 & \text{for } x > 0 \\ C_o & \text{for } 0 \ge x \ge -L \end{cases} \tag{A.56}$$

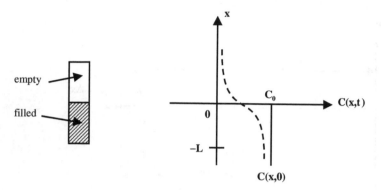

FIGURE A.11
Concentration varies with distance of diffusing particles in a cylindrical container.

Hence:

$$C(x, t) = C_o \int_{-L}^{0} dx' P(x - x', t) \qquad (A.57)$$

or:

$$C(x, t) = \frac{C_o}{\sqrt{4\pi Dt}} \int_{-L}^{0} dx' e^{-\frac{(x-x')^2}{4Dt}} \qquad (A.58)$$

Substituting: $s = \frac{(x-x')}{\sqrt{4Dt}}$ gives:

$$C(x, t) = \frac{C_o}{\sqrt{\pi}} \int_{x'=-L}^{x'=0} ds' e^{-s'^2} = \frac{C_o}{\sqrt{\pi}} \int_{\frac{x}{\sqrt{4Dt}}}^{\frac{x+L}{\sqrt{4Dt}}} e^{-s'^2} ds' \qquad (A.59)$$

Introducing the error integral (Figure A.12):

$$\Phi(s) = \frac{2}{\sqrt{\pi}} \int_{0}^{s} ds' e^{-s'^2} \qquad (A.60)$$

we get:

$$C(x, t) = \frac{C_o}{2} \left\{ \Phi\left(\frac{x+L}{\sqrt{4Dt}}\right) - \Phi\left(\frac{x}{\sqrt{4Dt}}\right) \right\} \qquad (A.61)$$

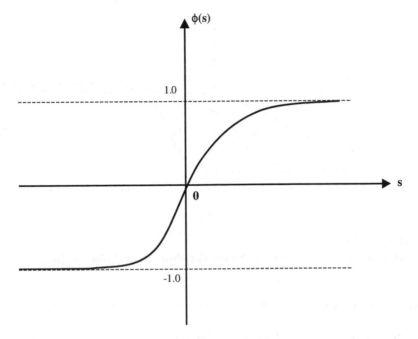

FIGURE A.12
Plot of the error integral.

For $t \ll \frac{L^2}{4D}$ and L very long, $\frac{x+L}{\sqrt{4Dt}} \gg 1$, so that:

$$C(x, t) \approx \frac{C_o}{2}\left[1 - \Phi\left(\frac{x}{\sqrt{4Dt}}\right)\right] \tag{A.62}$$

See Figures A.13 and A.14. Results have has been verified experimentally by Lam and Polson through a measurement of $\frac{\partial C}{\partial x}$ for proteins. Note that:

$$\frac{\partial C}{\partial x} = -\frac{C_o}{2}\frac{\partial \Phi(\frac{x}{\sqrt{4Dt}})}{\partial x} = -\frac{C_o}{2}\left(\frac{\partial \Phi}{\partial s}\right)\left(\frac{\partial s}{\partial x}\right) \tag{A.63}$$

$$\frac{\partial s}{\partial x} = \frac{1}{\sqrt{4Dt}} \qquad \frac{\partial \Phi}{\partial s} = \frac{2}{\sqrt{\pi}}e^{-s^2} \quad \text{hence} \tag{A.64}$$

$$\frac{\partial C}{\partial x} = -\frac{C_o}{\sqrt{4\pi Dt}}e^{-\frac{x^2}{4Dt}} \tag{A.65}$$

at $x = 0$, $(\frac{\partial C}{\partial x})_{x=0} = -\frac{C_o}{\sqrt{4\pi Dt}}$ which can be used to determine the diffusion constant D.

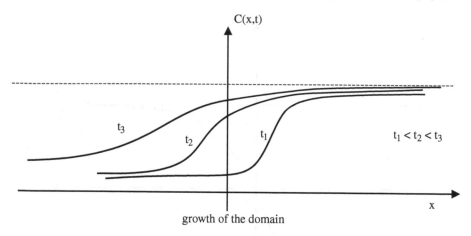

C(x,t)

t_3

t_2

t_1

$t_1 < t_2 < t_3$

x

growth of the domain

FIGURE A.13
Concentration $C(x, t)$ showing growth of a domain at different times.

Diffusion equation (differential form)

It is useful to have a local (differential) form of the evolution process. We start from the integral form:

$$C(x; t) = \int dx' C_o(x'; t_0) P(x - x', t_1 - t_0) \tag{A.66}$$

or, for close times, t and $t + \Delta t$:

$$C(x; t + t) = \int dx' C_o(x', t) P(x - x', \Delta t) \tag{A.67}$$

$$C(x; t + \Delta t) = \int ds\, P(s, \Delta t) C(x + s, t) \tag{A.68}$$

If Δt is very small, $\mathcal{P}(x, \Delta t)$ is a narrowly peaked Gaussian with its width $\sim \sqrt{D\Delta t}$ (see Figure A.15). The product of $\mathcal{P}(s, \Delta t)$ and $C(x + s, t)$ is small unless s is close to 0. Close to 0 we expand:

$$C(x + s, t) = C(x, t) + s\frac{\partial C(x, t)}{\partial x} + \frac{1}{2}s^2\frac{\partial^2 C(x, t)}{\partial x^2} + \cdots \tag{A.69}$$

Consequently,

$$C(x, t + \Delta t) = C(x, t) \int ds\, P(s, \Delta t) + \frac{\partial C(x, t)}{\partial x} \int ds\, P(s, \Delta t) s$$

$$+ \frac{1}{2}\frac{\partial^2 C(x, t)}{\partial x^2} \int ds\, P(s, \Delta t) s^2 + \cdots \tag{A.70}$$

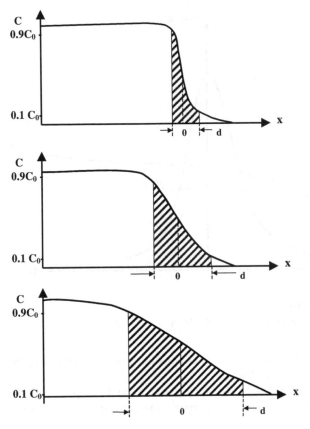

FIGURE A.14
Profile of $C(x)$ at three characteristic times showing evolution of the boundary region.

but we know that:

$$\int ds\, P(s, \Delta t) = 1$$

$$\int ds\, P(s, \Delta t)s = \bar{s} = 0$$

$$\int ds\, P(s, \Delta t)s^2 = \overline{s^2} = 2D\Delta t \qquad (A.71)$$

Hence:

$$C(x, t + \Delta t) \approx C(x, t) + D\frac{\partial C(x, t)}{\partial x^2}\Delta t \qquad (A.72)$$

or:

$$\frac{C(x, t + \Delta t) - C(x, t)}{\Delta t} = \frac{\partial C}{\partial t} = D\frac{\partial^2 C(x, t)}{\partial x^2} \qquad (A.73)$$

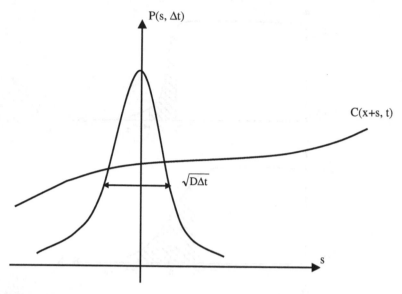

FIGURE A.15
Narrowly peaked Gaussian of width $\sim \sqrt{D\Delta t}$.

giving the meaning that diffusion equation gives a relationship between temporal variation and spatial change.

Both the integral and differential forms of the diffusion equation are equivalent. It is a linear equation, hence a superposition of two solutions is also a good solution of this equation:

$$C(x, t) = aC_1(x, t) + bC_2(x, t) \tag{A.74}$$

In particular, note that if $C(x, t) = C(x)$, then $C(x) = C_0 + C_1 x$ (Figure A.16).

Example

Suppose we seek a solution of the diffusion equation in terms of Fourier modes:

$$C(x, t = 0) = C_0 + \Delta C_0 \sin(2\pi n x / L) \tag{A.75}$$

We assume that:

$$C(x, t) = C_0 + \Delta C(t) \sin(2\pi n x / L) \tag{A.76}$$

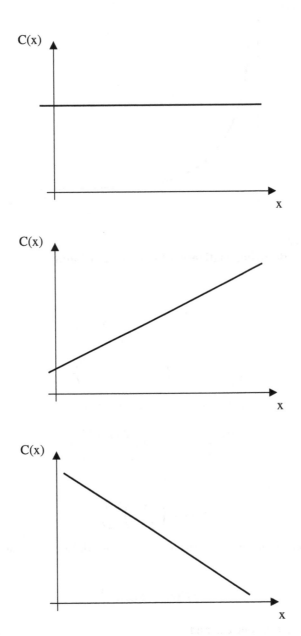

FIGURE A.16
Ways in which *C* may vary with *x*: (a) constant, (b) linearly increasing, and (c) linearly decreasing.

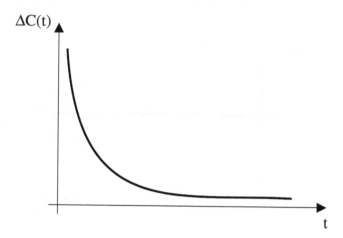

FIGURE A.17
Exponentially decaying relationship between $\Delta C(t)$ and t.

Then:

$$\frac{\partial C}{\partial t} = \frac{d}{dt}\Delta C \sin\left(\frac{2\pi n}{L}x\right)$$

$$\frac{\partial^2 C}{\partial x^2} = -\Delta C \left(\frac{2\pi n}{L}\right)^2 \sin\left(\frac{2\pi n}{L}x\right) \qquad (A.77)$$

Hence, inserting it into the diffusion equation gives:

$$\frac{d\Delta C(t)}{dt} = \left[-D\left(\frac{2\pi n}{L}\right)^2\right]\Delta C \qquad (A.78)$$

so that:

$$\tau_0 = \left(\frac{L}{2\pi n}\right)^2 \frac{1}{D} \qquad (A.79)$$

where τ_o is relaxation time such that the concentration difference decays exponentially
(Figure A.17):

$$\Delta C(t) = \Delta C_o e^{-\frac{t}{\tau_o}} \qquad (A.80)$$

Knowing L and t_o, one can find:

$$D = \frac{1}{\tau_o}\left(\frac{L}{2\pi n}\right)^2 \qquad (A.81)$$

Light scattering experiments give information about $\tau_o = \tau$ via the frequency;
$\omega = \frac{1}{2\pi \tau_o}$.

Particle conservation

In the absence of chemical reactions, the total number of molecules of a given kind remains constant (Figure A.18):

$$N = \text{number of particles in V} = \int_V C(x, y, z; t)\, dV \qquad \text{(A.82)}$$

The concept of particle flow is naturally defined (Figure A.19) from:

$$N_1(t) = A \int_{-L}^{0} dx\, C(x, t) \qquad \text{(A.83)}$$

$$N_2(t) = A \int_{0}^{L} dx\, C(x, t) \qquad \text{(A.84)}$$

as:

$$j = \frac{J}{A} = \text{current density} \qquad \text{(A.85)}$$

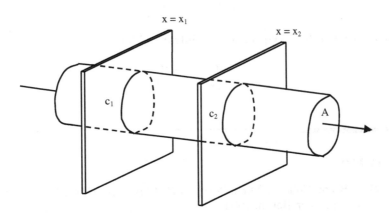

FIGURE A.18
Cylindrical tube with particle flow.

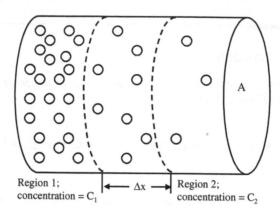

Region 1; $\leftarrow \Delta x \rightarrow$ Region 2;
concentration = C_1 concentration = C_2

FIGURE A.19
Particle flow between areas of differing concentrations.

so we have:

$$\frac{dN_{12}}{dt} = \frac{d}{dt}\int_{x_1}^{x_2} dx\, C(x,t) = \int_{x_1}^{x_2} dx\, \frac{\partial C(x,t)}{\partial t}$$
$$= j(x_1) - j(x_2) \tag{A.86}$$

Thus we get, by differentiation with respect to x, a continuity equation:

$$\frac{\partial C}{\partial t} = -\frac{\partial j}{\partial x} \tag{A.87}$$

Together with Fick's first law of diffusion:

$$j(x,t) = -D\frac{\partial C}{\partial x} \tag{A.88}$$

we arrive at the diffusion equation in its differential form:

$$\frac{\partial C}{\partial t} = D\frac{\partial^2 C}{\partial x^2} \tag{A.89}$$

References

Benedek, G.B. and Villars, F.M.H., *Physics With Illustrative Examples from Medicine and Biology*, Springer, Berlin, 2000.

Reif, F., *Fundamentals of Statistical and Thermal Physics*, McGraw Hill, New York, 1965.

Appendix B

Models of Phase Transitions and Criticality

Introduction

One of the characteristics of systems undergoing phase transitions is the emergence of long-range order in which the value of a physical quantity at an arbitrary point in the system is correlated with its value at a point located a long distance away from it. In many-body systems kept at sufficiently high temperatures, thermal energy is bound to exceed that of the interactions between particles leading to the predominance of disorder.

Long-range order can, in general, arise for two different reasons. The first is the existence of long-range forces that maintain coupling between particles and result in "rigidity" of the system (Anderson, 1984). The second is that when a symmetry present in the disordered phase is broken, uniformity is adopted on a macroscopic scale. This can be seen in many condensed matter phenomena such as superconductivity, ferromagnetism, superfluidity, liquid crystals, etc. (White and Geballe, 1979). A local application of a short-range perturbation will be felt throughout the system as a result of interactions between its parts and the generalized rigidity associated with an ordered phase.

We will briefly discuss representative examples from condensed matter physics that illustrate long-range order and related concepts. A large portion of this appendix is a condensed version of the material that can be found in Dixon, Tuszynski, and Clarkson (1997), among other excellent sources of information in this area.

Crystals

A classic example of long-range order is a perfect crystal in which the mass density distribution has infinite correlation length related to the onset of spatial periodicity. Crystal formation is shown by a regular diffraction pattern associated with Fourier components in mass density distribution (Kittel, 1956) with:

$$\rho(r) = \sum_G \rho_G e^{i\vec{G}\cdot\vec{r}} \tag{B.1}$$

where G represents vectors in the reciprocal space. The set of numbers ρ_G can be used as a quantity characterizing the low-temperature ordered (crystal) phase. Such quantities are referred to, following Landau, as order parameters (Landau and Lifshitz, 1959).

However, since several coefficients in the expansion in Equation (B.1) are nonzero, they define a multicomponent order parameter. As a result, we may obtain one of the four types of crystal lattices in two-dimensional space or one of the 14 Bravais lattice types in three spatial dimensions. These are built using rotational axes, reflection planes, and inversion centers. The addition of space translation results in several combination symmetry elements (sliding reflection planes, screw rotational axes, and improper rotational axes) and gives rise to 230 space groups. Phase transitions involving crystal symmetries will refer to crystallization from a liquid solution or via adsorption on a surface of gas molecules and to transformations from one crystal type to another.

Structural phase transitions

Distortions resulting from temperature changes or application of external stress may lead to structural phase transitions. As a consequence of structural phase transitions between various point or space groups of a crystal, new mechanical, electrical, and sometimes magnetic properties and their associated thermodynamic anomalies at transition temperature may be observed. The two main classes of structural phase transitions (Bruce and Cowley, 1981) are (1) displacive and (2) order–disorder type.

The two types can be distinguished by determining whether the local energy change between two equivalent configurations is smaller or greater than the elastic coupling energy, respectively. In displacive transitions, phonons cause the transition and the order parameter is the amplitude of the related lattice distortion. This yields a change from one lattice ordering to another. Simultaneously, the frequency of the associated phonon mode (or soft mode) tends to zero as $T \rightarrow T_c$, creating lattice instability.

Order-disorder transitions feature a transformation between randomly distributed atoms in their local double-well potential bottoms $(T > T_c)$ on the one hand and an ordered arrangement $(T < T_c)$ on the other. This is usually an abrupt transition associated with a soft diffusive mode characterizing large amplitude thermal hopping between degenerate potential well bottoms. To describe the associated behavior, displacive transitions require a continuous model of a Landau-Ginzburg type with ensuing solitary waves in the form of a traveling domain wall separating regions of particular order. Order-disorder transitions make use of Ising model calculations with effective (not real) spin variables.

Ferroelectricity

Structural phase transitions are commonly associated with changes in the distribution of electric charges giving rise to a nonzero value of the polarization vector and result in several types of ferroelectric phenomena.

Magnetic orderings

Atomic spins of many compounds and alloys tend to spontaneously form domains within which their individual spins are parallel or antiparallel. In addition to these two basic magnetic types of order (ferromagnetic and antiferromagnetic, respectively), more complicated regular spatial arrangements of localized magnetic moments are encountered in solids. Typically, lowering the temperature below the transition level T_c (called Curie temperature for ferromagnets and Néel temperature for antiferro-magnets) causes the magnetic interactions to overcome thermal fluctuations and an ordered state is adopted by the system.

In these cases the prototypical (disordered) phase is a paramagnetic one that exists for $T > T_c$ in which spins are oriented randomly. To measure the degree of order in a complex magnetic phase, as many order parameter components as there are distinct sublattices may have to be used. For ferromagnets, the order parameter the net magnetization and it is a conserved quantity. In antiferromagnets, the order parameter is the so-called staggered magnetization $\vec{M}_1 - \vec{M}_2$, where $\vec{M}_1$ and $\vec{M}_2$ are magnetization vectors for the two sublattices; it is not a conserved quantity.

Ordering transitions in binary alloys

In a brass alloy, for example, containing equal concentrations of copper and zinc, as the temperature is increased, a crystalline structure with ordered atoms undergoes a transition to a disordered structure that differs only in the atomic positions. The relevant order parameter may be defined as:

$$\eta \equiv \frac{N_1(1) - N_1(2)}{N_1(1) + N_1(2)} \tag{B.2}$$

where $N_i(\alpha)$ is the number of type i atoms in position $\alpha (i = 1, 2; \alpha = 1, 2)$.

Superconductivity

A number of metals and alloys and ceramic (high-T_c) superconductors (Phillips, 1989) exhibit new ordered states in conduction electron degrees of freedom arising below their characteristic critical temperature T_c. This state has two important features: ideal conductivity (zero resistance) and perfect diamagnetism also called the Meissner effect (expulsion of magnetic flux lines).

Bardeen, Cooper, and Schrieffer demonstrated that the ground state of a conventional low-temperature superconductor is formed by bound pairs of electrons (Cooper pairs) with wave vectors $\vec{k}$ and $-\vec{k}$ and spins s and $-s$, respectively. The order parameter, therefore, can be chosen as the wavefunction of the Cooper pair condensate $\psi(r)$ and exhibits a Hopf bifurcation at $T = T_c$.

Superfluidity

Superfluid properties are manifested by the absence of viscosity and have been detected experimentally in both ^{4}He and ^{3}He. The atoms of ^{4}He are bosons. Below a transition temperature T_λ, they undergo Bose condensation into a $k = 0$ (zero momentum) mode. The associated order parameter is the condensate's quantum wavefunction. ^{3}He atoms are fermions. Below T_λ (2.7 mK, three orders of magnitude lower than for ^{4}He) are known to form Cooper pairs, much as superconducting electrons do.

Liquid crystals

Liquid crystals comprise a large class of organic anisotropic fluids composed of strongly elongated molecules. Three basic types of liquid crystals can be distinguished: nematic, smectic, and cholesteric. The smectic group contains three additional subtypes.

The nematic phase is characterized by a direction to which most of the molecules are parallel, so that the order parameter is a second rank tensor describing correlations along a given direction. In addition to directionality, the smectic phases show layering patterns. The cholesteric phase is characterized by chirality of molecules.

Other examples of ordering and phase transition include binary fluids, the metal insulator transition, polymer transitions, spin- and charge-density waves, etc. The important point is to note the similarities and analogies manifested by (1) order parameters; (2) similar features of phase diagrams; and (3) singularities in the responses of each system to external influences at critical temperatures.

Systems undergoing phase transitions have phase diagrams that delineate the regions of stability of equilibrium phases in the space of thermodynamic coordinates (pressure p, temperature T, density ρ, etc.). Special points have been labeled on these diagrams: (1) critical (end) point; (2) triple point; (3) tricritical point.

Boundaries of the regions of a phase are drawn as continuous or broken lines. Continuous lines refer to continuous (second order) transitions; broken lines refer to discontinuous (first order) phase transitions. First order phase transitions are associated with discontinuities of the order parameter, nonzero latent heat, and hysteresis effects. Second order phase transitions have continuous order parameters, no latent heat, and singularity in specific heat (Stanley, 1972). An analytic prescription for finding equilibrium values of thermodynamic quantities is through solving the equation of state of the form:

$$f(T, p, \rho) = 0 \tag{B.3}$$

If its solution is single valued, a unique phase is obtained; on the other hand, if it is multivalued, then provided all of these solutions are stable, coexistence manifolds (points, lines, planes, etc.) are found for the phase diagram. The general equilibrium conditions for q phases are:

$$\mu_1(T, x) = \mu_2(T, x) = \cdots = \mu_q(T, x) \tag{B.4}$$

where μ_i represents the chemical potential (Gibbs free energy per particle) of the ith phase and x is the generalized thermodynamic force (e.g., pressure). Thus, Equation (B.4) determines a $(3 - q)$-dimensional manifold for single-component systems; for example, $q = 3$ gives a triple point where three phase boundaries meet. An extension of this method to multicomponent systems yields the so-called Gibbs phase rule (Landau and Lifshitz, 1959): if an s component system has q phases in equilibrium with z generalized thermodynamic forces (including temperature), the dimensionality (dim) of the coexistence manifold is:

$$\dim = s - q + z \tag{B.5}$$

The generally accepted way of classifying phase transitions was proposed by Ehrenfest and is due to the behavior of the appropriate thermodynamic potential, say, free energy F. It states that an nth order phase transition occurs when F and its derivatives up to the $(n - 1)$th are continuous at T_c while the nth derivative is discontinuous. Differentiation is taken with respect to an arbitrary independent thermodynamic variable (e.g., the order parameter). Figure B.1 shows the difference in $F(T)$ for first and second order phase transitions. Note the thermal hysteresis (irreversibility of the phase sequence on changing temperature in two opposite directions) in the case of first order transitions and the coexistence of phases a and b between T_a and T_b.

The measurable consequences for first order transitions are the so-called Clausius–Clapeyron equations:

$$\frac{dX}{dT} = \frac{\Delta S}{\Delta x} \tag{B.6}$$

where S denotes entropy (and hence latent heat $Q = T \Delta S$) and X a generalized thermodynamic coordinate (x is its conjugate), e.g., $\frac{dP}{dT} = \frac{Q}{V_A - V_B}$ for a fluid and $\frac{dH}{dT} = \frac{Q}{T(M_A - M_B)}$ for a magnet, which describes the slope of the transition line. For second order transitions, the Ehrenfest equations are found as:

$$\frac{dX}{dT} = \frac{\Delta(\frac{\partial S}{\partial x})}{\Delta \chi_T} \tag{B.7}$$

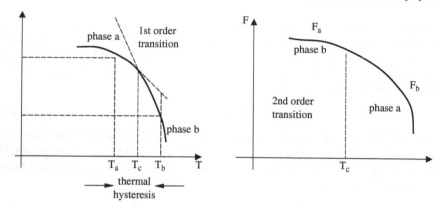

FIGURE B.1
Free energy plots for (a) first and (b) second order phase transitions.

where χ_T is the isothermal susceptibility $\chi_T = \frac{\partial X}{\partial x}$. In the two examples:

$$\frac{dP}{dT} = \frac{\Delta C_p}{T \Delta(\frac{\partial S}{\partial P})} \quad \text{and} \quad \frac{dM}{dT} = -\frac{\Delta(\frac{\partial M}{\partial T})}{\Delta \chi_T} \qquad (B.8)$$

respectively.

Broken symmetry is a situation in which the new ground state of a system does not possess the full symmetry group of the Hamiltonian (Anderson, 1984). Broken symmetry can occur spontaneously (by lowering the temperature, for example) or by application of an external field or constraint. A classic example is the ferromagnetic-to-paramagnetic phase transition at T_c where the full rotational symmetry of the paramagnetic phase is broken by the axial nature of the new ground ferromagnetic state below the Curie temperature. When a broken symmetry is continuous (e.g., translational invariance), a new excitation may appear. It is called a *massless Goldstone boson* and its frequency goes to zero at long wavelengths. The principle is sometimes called Goldstone's theorem (White and Geballe, 1979). Examples of Goldstone bosons include ferromagnetic domain walls and acoustic soft modes in structural phase transitions.

In some cases, such as the onset of ferroelectricity with a transverse optical branch or structural transitions with the softening of a collective excitation, a soft mode is responsible for the transition. The soft mode's frequency ω_k for the wave vector k tends to 0 as $T \rightarrow T_c$. The following are the types of broken symmetries:

- Translational (crystal formation, structural transition)
- Gauge symmetry (superfluidity, superconductivity)
- Time reversal (ferromagnets)
- Local rotational (liquid crystals)
- Rotational (some structural phase transitions)
- Space inversion (ferroelectricity)

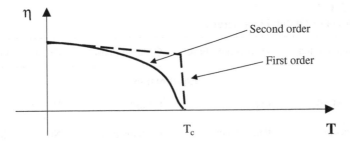

FIGURE B.2
Typical plot of the scalar order parameter as a function of temperature for first and second order phase transitions.

Gauge symmetry is of special importance since it is a universal property of Hamiltonians whenever the total number of particles or a generalized charge-like conserved quantity exists. The order parameter ψ is a complex quantity and its local density can be defined as $\rho = \psi^* \psi(r)$ so that a phase shift of ψ according to $\psi \rightarrow \psi\, e^{i\phi}$ leaves the Hamiltonian invariant. Furthermore, local gauge symmetry may exist where $\phi = \phi(r)$ but is only known to occur in the case of electric charge since it requires a gauge field that couples to the conserved charge (in this case, the electromagnetic field).

Near a second order phase transition, due to the reduction of rigidity of the system, or conversely due to its softening, critical fluctuations will dominate no matter what the dimensionality and symmetry of the disordered phase. Their amplitude diverges as $T \rightarrow T_c$. Critical fluctuations become of quasimacroscopic size near the critical point. They cannot be associated with collective modes in all cases (and are called diffusive since they have broad spectra). Those associated with collective modes are called soft modes. In addition to critical fluctuations, we also have quantum excitations called collisionless modes required by symmetries or conservation laws. In some cases, hydrodynamic modes are also present (e.g., first sound, second sound, etc. in superfluids).

An inherent property of systems undergoing phase transitions is the decrease of rigidity. Anderson (1984) defined generalized rigidity as the propensity of a system to evolve toward a stable ground state in the low symmetry phase. Consequently, work is required to change this state to another equilibrium. An associated property is the presence of small amplitude oscillations about the equilibrium state.

The concept of an order parameter was introduced by Landau (Landau and Lifshitz, 1959) and it still does not have a precise definition today. It is a thermodynamic bulk quantity that is invariant with respect to the symmetry group of the low temperature (and low symmetry) phase, zero above the transition temperature, and nonzero below transition temperature. It is a quantitative measure of the amount and type of order built up in the neighborhood of the critical point. Below T_c, it can be a degenerate quantity. To find the equation of state, a minimization procedure must be followed for an appropriate thermodynamic potential. From its original application to second order phase transitions, the idea of an order parameter, η, has been extended to first

TABLE B.1
Examples of Order Parameters

Phenomenon	Disordered Phase	Ordered Phase	Order Parameter
Equilibrium			
Condensation	Gas	Liquid	Density difference $\rho_L - \rho_G$
Spontaneous magnetization	Paramagnet	Ferromagnet	Net magnetization M
Antiferromagnetism	Paramagnet	Antiferromagnet	Staggered magnetization $M_1 - M_2$
Superconductivity	Conductor	Superconductor	Cooper pair wave function ψ
Alloy ordering	Disordered Mixture	Sublattice ordered alloy	Sublattice concentration
Ferroelectricity	Paraelectric	Ferroelectric	Polarization
Superfluidity	Fluid	Superfluid	Condensate wave function
Nonequilibrium			
Tunnel diode	Insulator	Conductor	Capacitance charge
Laser action	Lamp (incoherent)	Laser (coherent)	Electric field intensity
Super-radiant source	Noncoherent polarization	Coherent polarization	Atomic polarization
Fluid convection	Turbulent flow	Bénard cells	Amplitude of mode

order transitions (Figure B.2). It has been generalized from a scalar to a time- and space-dependent function. Examples of diverse applications of the order parameter concept to equilibrium and nonequilibrium critical phenomena are listed in Table B.1.

A useful concept in analyzing phase transitions theoretically and experimentally is that of a critical exponent (Stanley, 1972). In general, if a bulk physical quantity $Q(T)$ diverges or tends to a constant value as the temperature T tends to T_c, its behavior in the vicinity of T_c can be characterized by first defining a dimensionless small quantity ε as:

$$\varepsilon = \frac{T - T_c}{T_c} \tag{B.9}$$

which is known as the reduced temperature. Calculating the associated critical exponent called μ is done according to:

$$\mu = \lim_{\varepsilon \to 0} \frac{\ln Q(\varepsilon)}{\ln \varepsilon} \tag{B.10}$$

Figure B.3 illustrates the four typical behaviors of $Q(\varepsilon)$ following Stanley (1972).

The most important critical exponents are usually denoted as α, β, γ, δ, ν, and η and they describe, respectively, the behavior of specific heat, order parameter, isothermal

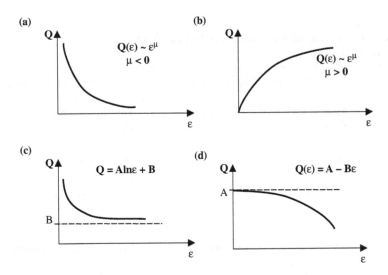

FIGURE B.3
Four generic behaviors near criticality.

susceptibility, response to an external field, correlation length, and pair correlation function.

Definitions for liquid–vapor transitions and magnetic systems are given in Table B.2 following Stanley (1972). The pair correlation function in Table B.2 is defined as:

$$\Gamma(r) \equiv \int d^3 x \langle (\psi(r) - \bar{\psi})(\psi(0) - \bar{\psi}) \rangle \qquad (B.11)$$

and $\langle \ldots \rangle$ denotes ensemble averaging; $\bar{\psi}$ is the equilibrium value of the order parameter ψ. The correlation length ξ can be defined in many ways, one of which is:

$$\xi \equiv \left[-\frac{1}{2} \Gamma^{-1}(0) \left(\frac{\partial^2 \Gamma(k)}{\partial k^2} \right)_{k=0} \right]^{-\frac{1}{2}} \qquad (B.12)$$

where $\Gamma(k)$ is the Fourier transform of $\Gamma(r)$.

Through the use of thermodynamic relations for the above quantities, it was determined that critical exponents must satisfy a number of inequalities, some of which are listed below (Stanley, 1972):

$$\alpha' + 2\beta + \gamma' \geq 2 \qquad \text{(a)}$$

$$\alpha' + \beta(\delta + 1) \geq 2 \qquad \text{(b)}$$

$$\gamma' + (\delta + 1) \geq (2 - \alpha')(\delta - 1) \qquad \text{(c)}$$

TABLE B.2
Definitions of Critical Exponents for Liquid–Vapor
and Magnetic Systems

Exponent	Definition (Liquid–Vapor)	Definition (Magnetic)
α' α	Specific heat at constant volume $C_v \sim (-\varepsilon)^{-\alpha'}$ $C_v \sim (\varepsilon)^{-\alpha}$	Specific heat at constant H $C_H \sim (-\varepsilon)^{-\alpha'}$ $C_H \sim (\varepsilon)^{-\alpha}$
β	Density difference $\rho_L - \rho_G \sim (-\varepsilon)^{\beta}$	Magnetization $M \sim (-\varepsilon)^{\beta}$
γ' γ	Isothermal compressibility $\kappa_T \sim (-\varepsilon)^{-\gamma'}$ $\kappa_T \sim \varepsilon^{-\gamma}$	Isothermal susceptibility $\chi_T \sim (-\varepsilon)^{-\gamma'}$ $\chi_T \sim \varepsilon^{-\gamma}$
δ	Pressure density critical isotherm $P - P_c \sim |\rho_L - \rho_G|^{\delta}$ $(T = T_c)$	Magnetic field magnetization $H \sim |M|^{\delta}$ $(T = T_c)$
ν' ν	Correlation length $\xi \sim (-\varepsilon)^{-\nu'}$ $\xi \sim \varepsilon^{-\nu}$	Correlation length $\xi \sim (-\varepsilon)^{-\nu'}$ $\xi \sim \varepsilon^{-\nu}$
η	Density–density pair correlation function $\Gamma(r) \sim |r|^{-(d-2+\eta)}$	Spin–spin pair correlation function $\Gamma(r) \sim |r|^{-(d-2+\eta)}$

$$\gamma' \geq \beta(\delta - 1) \qquad \text{(d)}$$

$$\frac{d(\delta - 1)}{(\delta + 1)} \geq 2 - \eta \qquad \text{(e)}$$

$$\frac{d\gamma'}{(2 - \alpha')} \geq \frac{d\gamma'}{(2\beta + \gamma')} \geq 2 - \eta \qquad \text{(f)}$$

$$(2 - \eta)\nu \geq \gamma \qquad \text{(g)}$$

$$d\nu' \geq 2 - \alpha'; \, d\nu \geq 2 - \alpha' \qquad \text{(h)} \qquad \text{(B.13)}$$

where d denotes spatial dimensionality of the system under consideration.

Early on in the development of the theory of phase transitions, it was realized that the values of the various critical exponents satisfy a number of relationships that can be derived using scaling concepts. In particular, the static scaling hypothesis asserts that the relevant thermodynamic potential, say, the Gibbs potential $G(T, H)$ for magnetic systems, is a generalized homogeneous function (Stanley, 1972) in that it satisfies:

$$G(\lambda^{a_\varepsilon}\varepsilon, \lambda^{a_H} H) = \lambda G(\varepsilon, H) \qquad \text{(B.14)}$$

for an arbitrary value of λ and specific values of a_ε, a_H that are characteristic exponents of a given phase transition. It can be demonstrated that Equation (B.14) leads to a number of similar relations for other thermodynamic quantities, giving:

$$\beta = \frac{1 - a_H}{a_\varepsilon}, \delta = \frac{a_H}{1 + a_H}, \gamma' = \frac{2a_H - 1}{a_\varepsilon}, \gamma = \frac{2a_H - 1}{a_\varepsilon}, \alpha' = 2 - \frac{1}{a_\varepsilon}. \quad \text{(B.15)}$$

Using the above, it can be shown that the following relationships are satisfied:

$$\alpha' + 2\beta + \gamma' = 2 \qquad \text{(a)}$$

$$\alpha + \beta(\delta + 1) = 2 \qquad \text{(b)}$$

$$\gamma(\delta + 1) = (2 - \alpha)(\delta - 1) \qquad \text{(c)}$$

$$\gamma' = \beta(\delta - 1) \qquad \text{(d)}$$

$$\alpha = \alpha' \qquad \text{(e)}$$

$$\gamma = \gamma' \qquad \text{(f)} \qquad \text{(B.16)}$$

Scaling is indeed an important property of the critical state. The remaining two exponents v and η should be involved in a scaling relationship. These ideas provided motivation for dynamic scaling and development of renormalization group ideas which is discussed later in this appendix.

Yang and Lee (1952) considered a system of N classical particles interacting via a two-body potential $u_{ij} = u(|r_i - r_j|)$, and showed that in the thermodynamic limit of $N \to \infty$, a phase transition is always associated with singular behavior of the partition function. Much work was done on studies of lattice models of systems undergoing phase transitions.

The Ising model, despite its simplicity, is crucial to the understanding of critical phenomena (Baxter, 1982). It is based on a regular lattice of N points in an n-dimensional space, with spin variables s_i placed at each lattice site and interacting with their nearest neighbors (denoted by $\langle i, j \rangle$) and an external magnetic field H parallel to the z axis. The Ising Hamiltonian is then:

$$H = - \sum_{\langle i,j \rangle} JS_i^z S_j^z - H \sum_i S_i^z. \qquad \text{(B.17)}$$

The number of nearest neighbors is labeled z_0 and can be 2 (for a one-dimensional lattice), 4 (for a two-dimensional square lattice), 6 (for a three-dimensional simple cubic or two-dimensional hexagonal lattice), 8 (for a three-dimensional body-centered cubic) or any other number, depending on the dimensionality and geometry of the lattice. The spin-spin interactions are assumed to be isotropic. If $J > 0$, they are of ferromagnetic-type and for $J < 0$ they are antiferromagnetic. The partition function:

$$Z = \sum_{S_1^z} \sum_{S_2^z} \cdots \sum_{S_N^z} e^{-\beta H(\{S_i\})}, \qquad \text{(B.18)}$$

has 2^N terms and has been calculated exactly in one and two dimensions. The result in one dimension leads to the free energy:

$$F = -Nk_B T \ln[e^{\beta J} \cosh \beta H + \sqrt{e^{2\beta J} \sinh^2 \beta H + e^{-2\beta J}}], \qquad (B.19)$$

and hence the mean magnetization per site is:

$$\frac{M}{N} = \frac{\sinh(\beta H)}{[\sinh^2(\beta H) + e^{-4\beta J}]^{\frac{1}{2}}}. \qquad (B.20)$$

Hence, it is easy to see that no spontaneous magnetization (and hence no phase transition) for $T > 0$ exists in one dimension. In two dimensions, Onsager (1944) obtained, using rather involved algebra, the Ising free energy in two dimensions demonstrated that the associated critical temperature is:

$$T_c = \frac{2.269 J}{k_B} \qquad (B.21)$$

The resultant critical exponents in 2D were found to be:

$$\alpha = 0(\log), \ \beta = \frac{1}{8}, \ \gamma = \frac{7}{4}, \ \delta = 15, \ \nu = 1, \eta = \frac{1}{4}. \qquad (B.22)$$

No exact results have been obtained in three dimensions, but a number of approximations have been worked.

Allowing the spins to interact via all three components coupled to form a scalar product leads to the Heisenberg model. However, the model exhibits no phase transitions in one- and two-dimensional spaces (Mermin and Wagner, 1966). A step in the direction of a three-dimensional Ising model is the *spherical model*. It uses the Ising Hamiltonian of Equation (B.17), and also accounts for the constraint (Berlin and Kac, 1952) that:

$$N = \sum_{i=1}^{N} S_i^2, \qquad (B.23)$$

which is a conserved quantity. Using the method of steepest descent, a three-dimensional lattice (and its higher dimensional extensions) possessed ferromagnetic properties in the low temperature regime. Following Joyce (1972), Table B.3 summarizes the values of critical exponents in the spherical model. Stanley (1968) proved that by generalizing the Ising model to q-component spins, its free energy approaches that of the spherical model when $q \to \infty$, i.e., at the classical limit.

The XY model is one of the simplest generalizations of the Ising model. It is assumed that spins have two components and interact through a bilinear scalar product, i.e., the Hamiltonian is taken as:

$$H = -J \sum_{\langle i,j \rangle} (S_i^x S_j^x + S_i^y S_j^y) = -JS^2 \sum_{\langle i,j \rangle} \cos(\phi_i - \phi_j), \qquad (B.24)$$

TABLE B.3
Critical Exponents in the Spherical Model

Dimensionality	β	δ	γ	η	ν	ν'	α	α'
$d < 3$			No phase transitions					
$d = 3$	$\dfrac{1}{2}$	5	2	0	1	Indet	0	0
$d = 4$	$\dfrac{1}{2}$	3	1	0	0	Indet	0	0
$d > 4$	$\dfrac{1}{2}$	3	1	0	1	Indet	0	0

Indet = indeterminate

where the spin magnitude is S and its angle with respect to the x axis is ϕ_i for the ith spin. This Hamiltonian, which is invariant with respect to uniform local spin rotation $\phi_i \rightarrow \phi_i + \phi_0$, has a ferromagnetic ground state with energy $E_0 = -z_o NJS^2/2$. A continuum approximation gives an estimate of Equation (B.24) as:

$$H \cong E_0 + \frac{JS^2}{2a^{d-2}} \int d^d r (\Delta\phi)^2, \tag{B.25}$$

where a is lattice spacing and d is system dimensionality. An important calculation in two-dimensional space shows that the correlation function exhibits an algebraic fall-off as:

$$\Gamma(r) \sim r^{-\eta(T)}, \tag{B.26}$$

with $\eta(T) = \frac{k_b T}{2JS^2\pi}$ demonstrating lack of long range order. Mermin and Wagner (1966) and Hohenberg (1967) proved a more general result that states that any two-dimensional system with short range interactions whose ordered phase has a continuous symmetry does not support long-range order. The reason is that at least one branch of collective excitations has energy that tends to zero continuously as its wavelength goes to zero. This type of asymptotic excitation is a Goldstone boson.

Kosterlitz and Thouless (1973) also showed that a phase transition exists in the xy model and it takes place at the transition temperature:

$$T_{KT} = \frac{\pi JS^2}{2k_B}, \tag{B.27}$$

and signifies a transformation between bound vortex-antivortex pairs (at low temperatures) and single vortices (at high temperatures). See Figure B.4. The energy of an isolated vortex is:

$$E_{\text{vortex}} = \frac{\pi JS^2}{2} n^2 \ln\left(\frac{L}{a}\right), \tag{B.28}$$

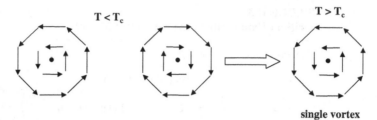

FIGURE B.4
Kosterlitz–Thouless transition in the XY model.

where L is the linear dimension of the system and n is the vortex strength. The energy of a vortex-antivortex pair is:

$$E_{\text{pair}} = -2\pi J S^2 n_1 n_2 \ln \left| \frac{\vec{r}_1 - \vec{r}_2}{a} \right|, \tag{B.29}$$

where n_1 and n_2 are respective vortex strengths and $\vec{r}_1$ and $\vec{r}_2$ are the position vectors of their centers.

A generalization of the Ising model to q components of the spin variable involves the Hamiltonian (Wu, 1982):

$$H = -J \sum_{\langle i, j \rangle} \sum_{\alpha=1}^{q} S_i^{\alpha} S_j^{\alpha}. \tag{B.30}$$

For a fixed value of q, there is always such dimensionality $d_c(q)$ that for $d > d_c(q)$, mean-field behavior prevails. Conversely, for a fixed dimensionality d, the $q_c(d)$ is such that for $q > q_c(d)$, the behavior is of mean-field type, e.g., $d_c(2) = 4$ and $q_c(2) = 4$. Table B.4 shows a set of associated critical exponents for the Potts model in two spatial dimensions following Wu (1982).

It is impossible to change the symmetry of a solid gradually (Landau and Lifshitz, 1959). For simple fluids, no symmetry breaking takes place because the vapor pressure curve ends on a critical point that can be circled around on the transition from liquid to gas phase. However, neither the fusion nor the sublimation curve possesses a critical point and transitions leading to the solid phase are of symmetry-breaking type. On this basis, Landau deduced that second order phase transitions are associated with symmetry breaking and can be qualitatively described by using an order parameter ψ which is zero in the symmetric (ordered) and nonzero in the dissymmetric (disordered) phases. Assuming that free energy depends on V, T, and ψ, the conditions on thermodynamic equilibrium are:

$$\left. \frac{\partial F(T, V, \psi)}{\partial \psi} \right|_{\psi=\psi_0} = 0 \text{ and } \left. \frac{\partial^2 F(T, V, \psi)}{\partial \psi^2} \right|_{\psi=\psi_0} > 0, \tag{B.31}$$

where ψ_0 is the equilibrium value of the order parameter.

TABLE B.4
Critical Exponents for the Potts Model
in Two Dimensions

q	$\alpha = \alpha'$	β	$\gamma = \gamma'$	δ	ν	η
0	$-\infty$	$\dfrac{1}{6}$	∞	∞	∞	0
1	$-\dfrac{2}{3}$	$\dfrac{5}{36}$	$2\dfrac{7}{18}$	$18\dfrac{1}{5}$	$\dfrac{4}{3}$	$\dfrac{5}{24}$
2	0	$\dfrac{1}{8}$	$\dfrac{7}{4}$	15	1	$\dfrac{1}{4}$
3	$\dfrac{1}{3}$	$\dfrac{1}{9}$	$\dfrac{13}{9}$	14	$\dfrac{5}{6}$	$\dfrac{4}{15}$
4	$\dfrac{2}{3}$	$\dfrac{1}{12}$	$\dfrac{7}{6}$	15	$\dfrac{2}{3}$	$\dfrac{1}{2}$

The cornerstone of the Landau theory of phase transitions is the assumption that sufficiently close to T_c, on both sides of it, F can be expanded in a Taylor series of ψ as:

$$F(T, V, \psi) \cong F_0(T, V) + F_1(T, V)\psi + F_2(T, V)\psi^2$$
$$+ F_3(T, V)\psi^3 + F_4(T, V)\psi^4 + \cdots \qquad (B.32)$$

Landau argued that unless a generalized force (e.g., magnetic field) is coupled to ψ, $F_1 = 0$, the coefficient F_3 may also vanish as a result of the invariance conditions: $F \to F$ as $\psi \to -\psi$ due to time reversal or parity invariance symmetries. In addition, introducing a control parameter for spontaneous second order phase transitions in the form of reduced temperature $\varepsilon \equiv (T - T_c)/T_c$, Landau postulated the simplest possible such expansion as:

$$F(T, V, \psi) \equiv F_0 + a\varepsilon\psi^2 + A_4\psi^4. \qquad (B.33)$$

where $a > 0$ and $A_4 > 0$ for stability reasons. As shown in Figure B.5, on changing the sign of ϵ, F transforms from a single- to a double-well shape. Simultaneously, solving the equilibrium conditions for ψ yields:

$$\psi = 0 \text{ for } \varepsilon > 0 \qquad \text{(a)}$$

$$\psi = \pm\left(-\frac{a\varepsilon}{2A_4}\right)^{\frac{1}{2}} \text{ for } \varepsilon < 0 \qquad \text{(b)}$$

$$\qquad (B.34)$$

This gives the value of the critical exponent β as $\beta = 0.5$. Calculating the entropy gives:

$$S = \frac{\partial F}{\partial T} = \begin{cases} S_0 & \varepsilon > 0 \\ S_0 + \frac{a^2}{2A_4}\frac{\varepsilon}{T_c} & \varepsilon \leq 0 \end{cases} \qquad (B.35)$$

(a)   (b)

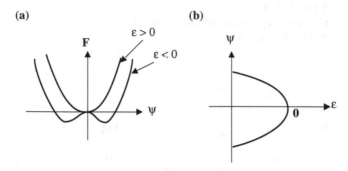

FIGURE B.5
Prototype of a second order phase transition according to Landau.

(a)   (b)

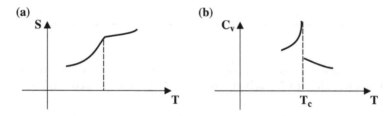

FIGURE B.6
Plots of $S(T)$ and $C_v(T)$ in the Landau model of second order phase transitions.

where S_0 is the entropy of the disordered phase. This allows us to evaluate the specific heat as:

$$C_v = T\frac{\partial S}{\partial T} = \begin{cases} C_0 & \varepsilon > 0 \\ C_0 + \frac{a^2}{2A_4 T_c}T & \varepsilon \leq 0 \end{cases} \qquad (B.36)$$

where C_0 is the specific heat of the disordered phase. Hence, a discontinuity occurs at T_c which amounts to:

$$\Delta C = \frac{a^2}{2A_4 T_c}. \qquad (B.37)$$

The plots of S and C_v are given in Figure B.6. Consequently, the second critical exponent α is also found to take on the classical value, i.e. $\alpha = 0$. To find the values of the remaining critical exponents, the free energy must now explicitly include an external field h coupled to ψ, so that:

$$F = F_0 + a\varepsilon\psi^2 + A_4\psi^4 - h\psi. \qquad (B.38)$$

Minimizing F with respect to ψ now yields an equation of state in the form:

$$h = 2\psi \lfloor a\varepsilon + 2A_4\psi^2 \rfloor \qquad (B.39)$$

This can be differentiated with respect to h on both sides, and remembering that $\chi \equiv \frac{\partial \psi}{\partial h}$, the following expression is arrived at:

$$\chi = [2a\varepsilon + 12A_4\psi^2]^{-1}. \tag{B.40}$$

Since $\psi \to 0$ as $T \to T_c$, we find that the third exponent is $\gamma = 1$. At $T = T_c$, the equation of state (B.39) simplifies to:

$$h \cong 4A_4\psi^3, \tag{B.41}$$

and hence $\psi \sim h^{1/3}$, giving $\delta = 3$. We conclude that the quartic expansion in the Landau model invariably leads to classical critical exponents.

The sixth power expansion has been widely used in the past to model first order phase transitions, but that requires $A_4 < 0$ (Binder, 1987). In the language of catastrophe theory, this expansion describes a "butterfly catastrophe" (Thompson and Stewart, 1986). A connection between catastrophe theory and structural stability can be found in de Alfaro and Rasetti (1978). The equation of state with a sixth power term in the free energy can be readily solved to yield:

$$\psi = \left\{ \frac{-A_4 \pm [A_4^2 - 3a\varepsilon A_6]^{\frac{1}{2}}}{3A_6} \right\}^{\frac{1}{2}}. \tag{B.42}$$

As $A_4 \to 0$, $\psi \sim \varepsilon^{1/4}$ and $\beta = 1/4$ for the tricritical point (TCP). The plus and minus signs correspond to $A_4 > 0$ and $A_4 < 0$, respectively. The transition temperature is T_c for $A_4 > 0$ (second order) and:

$$T_c^* = T_c + \frac{A_4^2}{4aA_6}, \tag{B.43}$$

for $A_4 < 0$ (first order). In the presence of an external field h, the corresponding susceptibility becomes:

$$\chi = [2a\varepsilon + 12A_4\psi^2 + 30A_6\psi^4]^{-1}. \tag{B.44}$$

and $\psi \sim \varepsilon^{1/4}$ for $A_4 \to 0$ and $\varepsilon \to 0$. Thus, the exponent γ should also remain 1 near the TCP. On the other hand, as $T \to T_c$:

$$h \sim 30A_6\psi^5 \tag{B.45}$$

and hence $\psi \sim h^{1/5}$ with $\delta = 5$ for the TCP. Specific heat, likewise, retains its discontinuous behavior and $\alpha = 0$ in both cases.

It is important to note a thermal hysteresis phenomenon associated with first order phase transitions. This is depicted in Figure B.7 for the order parameter. Figure B.8 shows free energy profiles. The disordered phase terminates its stability at $T = T_c$ and the ordered one does so at $T = T_s$ where:

$$T_s = T_c + \frac{A_4^2}{3aA_6}. \tag{B.46}$$

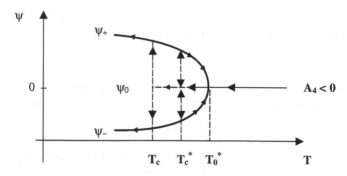

FIGURE B.7
Thermal hysteresis effect for first order phase transitions.

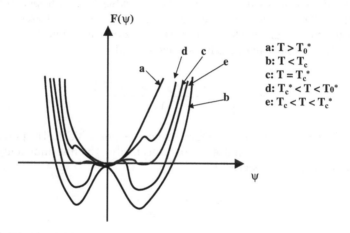

FIGURE B.8
Free energy forms for different temperatures close to a first order phase transition temperature.

In spite of very positive qualities, the Landau theory is not capable of describing spatial fluctuations that dominate the scene close to the transition point. This is why no predictions can be made for exponents ν and η within the Landau theory.

For the order parameter $\psi(r)$ treated as a function of spatial coordinates, the two-point correlation function was given in Equation B.11. Through the use of the fluctuation–dissipation theorem (Pathria, 1972), it can be shown that the correlation function is proportional to the generalized susceptibility function. A particular example useful in modeling the liquid–vapor phase transition describes the number of particles N in volume V:

$$\langle (N - \langle N \rangle)^2 \rangle = \langle N \rangle n k_B T \kappa_T, \tag{B.47}$$

where $n = \langle N \rangle / V$ and $\kappa_T = (-1/V)(\partial V/\partial P)_{T,N}$ is the isothermal compressibility. Ornstein and Zernike (Stanley, 1972) demonstrated that in the vicinity of the critical point, the particle number correlation function diverges in proportion to $(T - T_c)^{-1}$ so that κ_T must be a divergent function of $(T - T_c)$ with a critical exponent γ of 1.

To probe spatial dependence of the correlation function close to criticality, scattering experiments must be performed. If k_0 is the incident wave vector k_s, the scattered wave vector and $|q| = |k_0 - k_s| = 2k \sin(\theta/2)$ with $k = |k_0| \cong |k_s|$, then the intensity of scattered radiation taken as a fraction of the intensity of the incident radiation (Mermin and Wagner, 1966) is:

$$\frac{I(q)}{I_0(q)} = \frac{1}{n} \int d^3r \, e^{-i\vec{q}\cdot\vec{r}} \Gamma(r) \equiv \frac{1}{n} S(q), \tag{B.48}$$

where $S(q)$ is the form factor. Ornstein and Zernike provided a good fit to experimental results using a Lorentzian approximation for $S(q)$:

$$S(q) \sim (\xi^2 + q^2)^{-1}, \tag{B.49}$$

where ξ is the correlation length. Using Equation (B.48) and assuming a d-dimensional space yields for the correlation function:

$$\Gamma(r) = \int S(q) \left[\int_0^\pi e^{iq \cdot r \cos\theta} (\sin\theta)^{d-2} d\theta \right] d\Omega_{d-1} q^{d-1} dq, \tag{B.50a}$$

where $d\Omega_{d-1}$ is a solid angle element in $(d-1)$ dimensions. The correlation length can be evaluated as:

$$\xi(T) = \left[\frac{\int r^2 \Gamma(r, T) d^d r}{\int \Gamma(r, T) d^d r} \right]^{\frac{1}{2}}, \tag{B.50b}$$

The integral in the square bracket of Equation (B.50a) gives $\dfrac{J_{\frac{d}{2}-1}(q^r)}{(q^r)^{\frac{d}{2}-1}}$, where J is a Bessel function of the first kind (Abramowitz and Stegun, 1965). Using the Lorentzian form of Equation (B.49) yields:

$$\Gamma(r) \sim \begin{cases} e^{-r/\xi} & d = 1 \\ \ln r \, e^{-r/\xi} & d = 2 \\ e^{-r/\xi}/r & d = 3 \end{cases} \tag{B.51}$$

which fails to describe the situation in two-dimensional space. To repair this deficiency, Fisher (Stanley, 1972) proposed a correction:

$$S(q)\big|_{T=T_c} \sim q^{-2+\eta} \quad (q \approx 0), \tag{B.52}$$

that results in:

$$\Gamma(r)\big|_{T=T_c} \sim r^{-(d-2+\eta)} \quad (r \to \infty). \tag{B.53}$$

The Landau theory can be easily extended to account for spatial fluctuations. Ginzburg and Landau (1950) proposed a free energy functional rather than a simple function. The simplest form consistent with Landau's principles for expansions is:

$$F(\psi(r), T) = \int d^3r [A_2 \psi^2 + A_4 \psi^4 - h\psi + D(\nabla\psi)^2] \qquad (B.54)$$

where D is a phenomenological parameter describing the contribution due to spatial inhomogeneities. Applying a variational principle to F to find functional free energy minima gives:

$$h = 2A_2 \psi + 4A_4 \psi^3 - 2D\Delta^2 \psi. \qquad (B.55)$$

Perturbative solutions are obtained as small deviations from mean field results so that

$$\psi(r) = \psi_0(T) + \phi(r), \qquad (B.56)$$

Linearizing Equation (B.55), we obtain:

$$\nabla^2 \phi - \mu \frac{A_2}{D} \psi_0 = -\frac{h_0}{D} \delta(r), \qquad (B.57)$$

where we have assumed h to be a localized perturbation of the form $h_0 \delta(r)$ and $\mu = 1$ for $T > T_c$ while $\mu = -2$ for $T < T_c$. A solution of Equation (B.57) in three-dimensional spherical coordinates is:

$$\phi = \frac{h_0}{4\pi D} \frac{e^{-r/\xi}}{r}, \qquad (B.58)$$

where $\xi \sim A_2^{-1/2}$ is the correlation length. Note that Equation (B.58) agrees with the result of Ornstein and Zernike given in Equation (B.51).

Another insight into the role of inhomogeneities in the Landau model can be gained by Fourier transforming the order parameter according to Ma (1976):

$$\psi(r) \equiv L^{-d/2} \sum_{k<k_0} \psi_k e^{ik \cdot r} \qquad (B.59)$$

where k_0 is the cutoff wavelength $k_0 = \Lambda^{-1}$ corresponding to the smallest periodicity in the system (atomic spacing). The free energy then becomes:

$$F = \sum_{k<k_o} |\psi_k|^2 (A_2 + Dk^2) + L^{-d} \sum_{k,k',k''<k_o} A_4 \psi_k^* \psi_{k'}^* \psi_{k''} \psi_{k+k'-k''}, \qquad (B.60)$$

neglecting the external field h. Ignoring mode–mode coupling in the last term above provides the basis for the so-called Gaussian approximation where:

$$F \cong \sum_{k<k_o} |\psi_k|^2 (A_2 + Dk^2). \qquad (B.61)$$

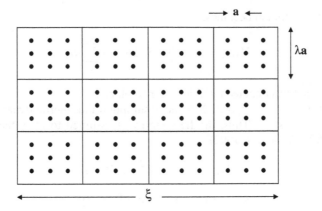

FIGURE B.9
The Kadanoff construction.

The Fourier transform of the correlation function is:

$$\Lambda(k) \equiv L^{-d} \int e^{-ik(r-r')} \langle \psi(r)\psi(r') \rangle \, dr dr'$$

$$= \langle |\psi_k|^2 \rangle \cong \frac{n}{2}(A_2 + Dk^2)^{-1} \tag{B.62}$$

Therefore, as $T \to T_c$, we find that $\psi = 0$ and consequently $\Gamma(k) \sim k^{\eta-2}$ with $\eta = 0$ in the Gaussian approximation. $\Gamma(r) \sim r^{-(d-2)}$. The Gaussian approximation introduces a new value of the specific heat critical exponent $\alpha = 2 - d/2$. The Gaussian approximation fails for d < 4 but gives a correct result for $d \geq 4$ (Reichl, 1979).

To extend the validity of the Landau–Ginzburg model further, we now examine dynamic scaling (or real space scaling) and renormalization group theory. Kadanoff proposed an extension of the concept of scaling at criticality to include size scaling understood as coarse graining. This means that all the essential features of the system should remain unchanged as we increase the length scale of the lattice by a factor λ such that:$1 \ll \lambda \ll \frac{\xi}{a}$ where ξ is the correlation length and a is the lattice spacing (see Figure B.9).

If N is the number of lattice sites and d the dimensionality of the system, $m = N/\lambda^d$ is the number of blocks obtained as a result of the first rescaling. Simultaneously, the lattice variables (spins) and the interaction parameters are redefined. The effective spins now refer to each block and interaction parameters refer to the interacting blocks. Thus, using the Ising model as an example, we have the cell Hamiltonian:

$$H_{\text{cell}} = -J \sum_{\langle i,j \rangle}^{N} S_i S_j - h \sum_i^{N} S_i \tag{B.63}$$

The block Hamiltonian becomes:

$$H_{\text{block}} = -\tilde{J} \sum_{\langle \alpha, \beta \rangle}^{m} \tilde{S}_\alpha \tilde{S}_\beta - \tilde{h} \sum_{\alpha}^{m} \tilde{S}_\alpha, \tag{B.64}$$

where the quantities denoted by tildes refer to blocks.

A crucial step in this approach is the assumption that thermodynamic potentials scale with block size. Thus, the free energy is assumed to obey the relationship (Stanley, 1972; Yeomans, 1992):

$$F_{\text{block}}(\tilde{\varepsilon}, \tilde{h}) = \lambda^d F_{\text{cell}}(\varepsilon, h), \tag{B.65}$$

where the tilde field $\tilde{h}$ and reduced temperature $\tilde{\varepsilon}$ satisfy:

$$\tilde{h} = \lambda^x h; \quad \tilde{\varepsilon} = \lambda^y \varepsilon, \tag{B.66}$$

and the exponents x and y are a *priori* unknown. It is also conjectured that F is a homogeneous function so that F_{block} and F_{cell} are of the same form so we can write:

$$F(\lambda^{y/d} \varepsilon, \lambda^{x/d} h) = \lambda F(\varepsilon, h). \tag{B.67}$$

However, earlier on when discussing static scaling, a statement was made regarding F:

$$F(\lambda^{a_\varepsilon} \varepsilon, \lambda^{a_H} H) = \lambda F(\varepsilon, H). \tag{B.68}$$

This allows us to identify the new exponents x and y with the already known exponents a_ε and a_H. Indeed, we find:

$$y = a_\varepsilon d; \, x = a_H d. \tag{B.69}$$

This result is known as the *static scaling hypothesis*. An application of the Kadanoff construction to the pair correlation function:

$$\Gamma(r, \varepsilon) \equiv \langle (S_i - \langle S \rangle)(S_j - \langle S \rangle) \rangle, \tag{B.70}$$

gives $y = \frac{1}{\nu}$ for temperature variable scaling. Recalling that, from static scaling, $y = da_\varepsilon$ and $a_\varepsilon = (2 - \alpha')^{-1}$:

$$\alpha = \alpha' \text{ and } d\nu = 2 - \alpha. \tag{B.71a}$$

Scaling the field variable yields $x = da_H$ which, with $a_H = \delta/(\delta + 1)$ results in:

$$(2 - \eta)\nu = \delta. \tag{B.71b}$$

We have found that the two scaling hypotheses simplify the theory of phase transitions so that only two independent exponents that were a *priori* unknown are required. We know that as $T \to T_c, \xi \to \infty$. Hence, in principle, size scaling should be allowed to continue ad *infinitum* as T approaches T_c.

TABLE B.5
Universality Classes

Universality Class		System	Order Parameter
$d = 2$	$n = 1$	Absorbed films	Surface density
$d = 2$	$n = 2$	Superfluid He^4 film	Superfluid wave function
$d = 3$	$n = 1$	Uniaxial ferromagnets	Magnetization
$d = 3$	$n = 1$	Fluids	Density difference
$d = 3$	$n = 1$	Mixtures, alloys	Concentration difference
$d = 3$	$n = 2$	Planar ferromagnets	Magnetization
$d = 3$	$n = 2$	Superfluids	Superfluid wave function
$d = 3$	$n = 3$	Isotropic ferromagnets	Magnetization

Denoting the original cell Hamiltonian by H_o and the Hamiltonian after the nth step of rescaling as H_n, we represent the chain of renormalization transformations R as:

$$R(H_0) = H_1, R(H_1) = H_2, \ldots, R(H_n) = H_{n+1}, \ldots, R(H^*) = H^*, \quad \text{(B.72a)}$$

where H^* denotes a fixed point Hamiltonian that is invariant with respect to scaling transformations. Each step in the RG transformation chain reduces the number of degrees of freedom by λ^d. The partition function is preserved so that:

$$Z_N(H_o) = Z_m(H_m) \quad \text{(B.72b)}$$

where $m = N/\lambda^d$. It is believed that the values of critical exponents are characteristics of groups of diverse physical models, not individual Hamiltonians.

Numerous models may lead to the same fixed point. Indeed, the so-called *universality hypothesis* states that for any two physical systems with the same dimensionality d and the same number of order parameter components n, all such systems belong to the same universality class and each fixed point corresponds to one universality class. Several important physical examples of the universality class concept are listed in Table B.5 based on Carreri (1984). A summary of the critical exponent values for the key theoretical models is given in Table B.6 following Stanley (1972), Reichl (1979), and Yeomans (1992).

For $d \geq 4$, the RG approach confirms earlier predictions that the Landau (mean field) model is exact. The founder of the RG theory, K.G. Wilson (1972), described his own work as a second stage of the Landau theory. He remarked that while the microscopic LG Hamiltonian:

$$H = D(\nabla \psi)^2 + A_2\psi^2 + A_4\psi^4, \quad \text{(B.73)}$$

TABLE B.6
Summary of Critical Exponent Values for the Most Important Theoretical Models

Model	α	β	γ	υ	δ	η
Ornstein–Zernike			2	1	5	0
Classical (MFT)	0 (disc.)	$\frac{1}{2}$	1	–	3	–
Classical (TCP)	0 (disc.)	$\frac{1}{4}$	1	–	5	–
Spherical $d = 3$		$\frac{1}{2}$	2	1	5	0
Spherical $\varepsilon > 0$	$-\dfrac{\varepsilon}{2-\varepsilon}$	$\frac{1}{2}$	$\dfrac{2}{2-\varepsilon}$	$\dfrac{1}{2-\varepsilon}$	$\dfrac{6-\varepsilon}{2-\varepsilon}$	0
Ising $d = 2$	0(log)	$\frac{1}{8}$	$\frac{7}{4}$	1	15	$\frac{1}{4}$
Ising (approx.) $d = 3$	0.12	0.33	1.25	0.64	4.8	0.04
Heisenberg (approx.) $d = 3$	-0.12	0.36	~ 1.39	$+0.71$	4.8	$+0.04$
S^4 model $d > 4$	$\dfrac{\varepsilon}{2}$	$\dfrac{1}{2} - \dfrac{\varepsilon}{4}$	1	$\dfrac{1}{2}$	$3 + \varepsilon$	0
S^4 model $d = 4$	0	$\dfrac{1}{2}$	1	$\dfrac{1}{2}$	3	0
S^4 model $d < 4$	$\dfrac{\varepsilon}{6}$	$(\dfrac{1}{2} - \dfrac{\varepsilon}{6})$	$(1 + \dfrac{\varepsilon}{6})$	$(\dfrac{1}{2} + \dfrac{\varepsilon}{12})$	$3 + \varepsilon$	0
S^4-model: $d = 3$	0.17	0.33	1.17	0.58	4	0
XY-Model: $d = 3$	0.01	0.34	1.30	0.66	4.8	0.04

$\varepsilon \equiv 4 - d$.
disc. = discontinuous

is essentially correct, the microscopic generalization in the form of the LG free energy:

$$F = \int d^3x [\tilde{A}_2 \psi^2 + \tilde{A}_4 \psi^4 + \tilde{D}(\nabla \psi)^2], \qquad (B.74)$$

is incorrect because it ignores the variations of its parameters (expansion coefficients) with the size of the region sampled.

References

Abramowitz, M. and Stegun, I.A., Eds., *Handbook of Mathematical Functions*, Dover, New York, 1965.

Anderson, P.W., *Basic Notions of Condensed Matter Physics*, Benjamin Cummings, Menlo Park, CA, 1984.

Baxter, R.J., *Exactly Solved Models in Statistical Mechanics*, Academic Press, New York, 1982.

Berlin, T.H. and Kac, M., *Phys. Rev.*, 86, 821, 1952.

Binder, K., *Rep. Progr. Phys.* 50, 783, 1987.

Bruce, A.D. and Cowley, R.A., *Structural Phase Transitions*, Taylor & Francis, London, 1981.

Carreri, G., *Order and Disorder in Matter*. Benjamin Cummings, Menlo Park, CA, 1984.

de Alfaro, V. and Rasetti, M. *Fort. Phys.*, 26, 143, 1978.

Dixon, J.M., Tuszynski, J.A., and Clarkson, P.A., *From Nonlinearity to Coherence*, Oxford University Press, Oxford, U.K., 1997.

Ginzburg, V.L. and Landau, L.D., *Zh. Eksp. Teor. Fiz.*, 20, 1064, 1950.

Hohenberg, P.C., *Phys. Rev.*, 158, 383, 1967.

Huang, K., *Statistical Mechanics*, John Wiley & Sons, New York, 1987.

Joyce, G.S., in *Phase Transitions and Critical Phenomena*, Vol. 2, Domb, C. and Green, M.S., Eds., Academic Press, New York, 1972.

Kittel, C., *Introduction to Solid State Physics*, John Wiley & Sons, New York, 1956.

Kosterlitz, J.M. and Thouless, D.J., *J. Phys.* C, 6, 1181, 1973.

Landau, L.D. and Lifshitz, E.M., *Statistical Physics*. Pergamon, London, 1959.

Ma, S.K., *Modern Theory of Critical Phenomena*, Benjamin, New York, 1976.

Mermin, N.D. and Wagner, H., *Phys. Rev. Lett.*, 17, 1133, 1966.

Onsager, L., *Phys. Rev.*, 65, 117, 1944.

Pathria, R.K., *Statistical Mechanics*, Pergamon, Oxford, 1972.

Phillips, J.C., *Physics of High Temperature Superconductors*, Academic Press, New York, 1989.

Reichl, L.E., *A Modern Course in Statistical Physics*, University of Texas Press, Austin, 1979.

Stanley, H.E., *Introduction to Phase Transitions and Critical Phenomena*, Oxford University Press, Oxford, U.K., 1972.

Stanley, H.E., *Phys. Rev.*, 176, 718, 1968.

Thompson, J.M.T. and Stewart, H.B., *Nonlinear Dynamics and Chaos*, John Wiley & Sons, New York, 1986.

White, R.H. and Geballe, T., *Long Range Order in Solids*. Academic Press, New York, 1979.

Wilson, K.G., *Rev. Mod. Phys.*, 55, 583, 1983.

Wilson, K.G., *Sci. Am.*, 241, 140, 1979.

Wilson, K.G., *Phys. Rev. Lett.*, 28, 248, 1972.

Wu, F.Y., *Rev. Mod. Phys.*, 54, 235, 1982.

Yang, C.N. and Lee, T.D., *Phys. Rev.*, 87, 404, 1952.

Yeomans, J.M., *Statistical Mechanics of Phase Transitions*, Oxford University Press, Oxford, U.K., 1992.

Appendix C

Foundations of Nonlinear Physics

Multistability and bifurcation

Properties of systems near instability points may be manifested by the creation of new ordered states that highlight *broken symmetries* in the system. As a result, *long-range order* may be established and the emergence of *nonlinear dynamical modes* of behavior takes place. We can measure *nonlinear responses* to external stimuli or perturbations such as hysteresis loops.

These responses are indications of the changes of a system's generalized rigidity. The average *fluctuations* in physical quantities may be more or less pronounced, depending on the structure of the system. In the critical state, unbounded growth of fluctuations takes place. Near criticality, we select a finite number of most important (driving) degrees of freedom, referred to, following Landau, as *order parameters* (Landau and Lifshitz, 1959).

The remaining "fast" degrees of freedom (slaved modes) can be incorporated through fluctuations, temperature-dependent coefficients, or dissipative terms in the equation of motion or ignored altogether. Landau and Lifshitz (1959) proposed a theory of systems undergoing phase transitions and concentrated on a single thermodynamic quantity called the order parameter. The effects of all the other degrees of freedom were incorporated in temperature-dependent coefficients of Landau's free energy expansions (see Appendix B).

The concept of reducing the number of degrees of freedom was fully justified for critical systems through development of scaling procedures, known as the renormalization group theory (RGT) of Kadanoff and Wilson (1983). A similar ideology can be found in chemical kinetics through the works of Nicolis and Prigogine on reaction–diffusion equations (1989) and in biological self-organization expressed via Haken's theory of synergetics (1983). Feigenbaum's discovery of chaotic maps (Cvitanovic, 1984) and Mandelbrot's seminal work on fractals (Jullien and Botet, 1987) are two other illustrations of the power of scale invariance. A much more comprehensive presentation of the ideas covered in this appendix can be found, for example, in Dixon, Tuszynski, and Clarkson (1997).

Suppose a system under consideration can be described by a macroscopic order parameter η. Under the influence of an external field σ coupled linearly to η, the state of the system is determined by an *equation of state* that takes the form:

$$P(\sigma, \eta) = 0, \qquad (C.1)$$

where P is usually a polynomial function of η representing the first derivative with respect to η of an associated thermodynamic potential $V(\eta)$. Thus, $P = \partial V / \partial \eta$

481

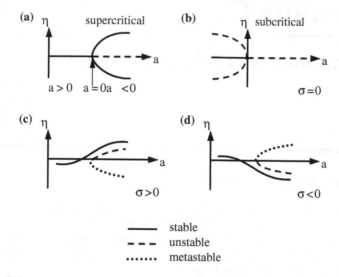

FIGURE C.1
Pitchfork bifurcation without an external field σ in the supercritical case (a), subcritical case (b), and for two possible signs of the external field (c) and (d).

can be viewed as a conservative nonlinear force. Assuming the presence of at least one control parameter acting through one of the variable coefficients in $P(\sigma, \eta)$ leads directly to the problem of *catastrophes,* an area investigated by René Thom (Thompson and Stewart, 1986). The cusp catastrophe is described by the potential:

$$V(\eta) = \frac{\varepsilon}{4}\eta^4 + \frac{1}{2}a\eta^2 - \sigma\eta \qquad (C.2)$$

where $\varepsilon = +1$ or -1 and a is a control parameter. This potential describes second order phase transitions both in the absence ($\sigma = 0$) and presence of external fields ($\sigma \neq 0$). The butterfly catastrophe, on the other hand, employs:

$$V(\eta) = \frac{1}{6}\eta^6 + \frac{a}{4}\eta^4 + \frac{b}{3}\eta^3 + \frac{c}{2}\eta^2 - \sigma\eta \qquad (C.3)$$

and has been used to model first order phase transitions, both field- and temperature-induced. The parabolic umbilic catastrophe potential is:

$$V(\eta_1, \eta_2) = \eta_1^2\eta_2 + \eta_2^4 + a\eta_1^2 + b\eta_2^2 - \sigma_1\eta_1 - \sigma_2\eta_2 \qquad (C.4)$$

and it has applicability in coupled critical systems such as ferroelastics or ferroelectrics (Tolédano and Tolédano, 1987).

To illustrate the related phenomenon of *bifurcation* we take Equation (C.2) which yields the equation of state:

$$\varepsilon\eta^3 + a\eta = \sigma \qquad (C.5)$$

where a is the only control parameter present. A plus sign in front of η^3 can be changed to a minus to designate a subcritical bifurcation. Figure C.1 shows bifurcation diagrams for the plots of η as a function of a when $\sigma = 0$ and $\sigma \neq 0$.

For $\sigma = 0$, the plot is commonly referred to as a *pitchfork bifurcation* and it illustrates a transition between a single stable solution for $a > 0$ and a bistable situation when $a < 0$. In the case of $\sigma \neq 0$, the bifurcation takes place at $a = 0$. However, new features are the phenomena of external field-induced hysteresis and metastability. Stability corresponds to a solution for which $\partial^2 V/\partial \eta^2 > 0$ and, if more than one solution of the equation of state (C.1) is stable, we call the higher energy solutions metastable. Unstable solutions correspond to $\partial^2 V/\partial \eta^2 < 0$ while the coincidence of $\partial V/\partial \eta = \partial^2 V/\partial \eta^2 = 0$ highlights an instability (or inflection) point. Figure C.2 shows the difference in the response of the order parameter to the application of an external field. In unistable situations, η as a function of σ is a smooth single-valued function. Multistability results in double valuedness (at least) in some ranges of the external fields. Figure C.2c demonstrates the regions of multistability in the parameter space. The dividing line is given by:

$$-4a^3 + 27\sigma^2 = 0 \qquad (C.6)$$

This indicates the parameter set for which a triple solution of the cubic expression in Equation (C.5) is obtained. Figure C.3 is an extension of these concepts to the butterfly catastrophe case exhibiting thermal hysteresis (if control parameter a involves temperature dependence) and double hysteresis under the influence of an external field.

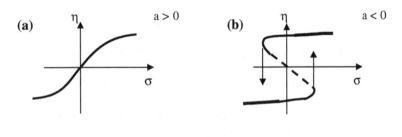

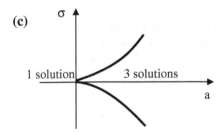

FIGURE C.2
Response of order parameter to an externally applied field (a) in a unistable state, (b) in a bistable state, and (c) in separation of the control parameter space into multiplicity regions.

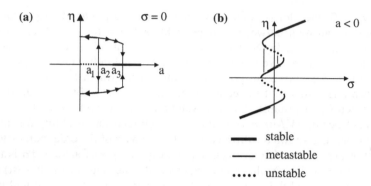

stable

metastable

..... unstable

FIGURE C.3
(a) Bifurcation plot and (b) double hysteresis in the butterfly catastrophe case.

Stochastic analogue of a bifurcation

If the order parameter η is not a constant and is a probabilistic variable linked to the potential function:

$$V(\eta) = a_2\eta^2 + a_4\eta^4, \tag{C.7}$$

the corresponding Boltzmann-weighted probability distribution is:

$$P(\eta) = P_0 e^{-\beta V} = P_0 \exp[-\beta(a_2\eta^2 + a_4\eta^4)] \tag{C.8}$$

where $\beta = (k_B T)^{-1}$, k_B is the Boltzmann constant, T the absolute temperature, and P_0 a normalization function. As the coefficient a_2 changes its sign from positive to negative, the potential function $V(\eta)$ changes from a single well to a double well and $P(\eta)$ transforms from a single-peaked to a double-peaked function of η (see Figure C.4).

The quartic non-Gaussian distribution function in Equation (C.8) can be used in exact calculations of all statistical quantities as follows. Writing $\lambda_2 = \beta a_2$ and $\lambda_4 = \beta a_4$, the partition function Z (Abramowitz and Stegun, 1965) is:

$$Z = (2\lambda_4\beta)^{-1/4}\,\Gamma\left(\frac{1}{2}\right)\exp\left(\frac{\lambda_2^2}{8\lambda_4}\right)D_{-1/2}\left[\frac{\lambda_2}{\sqrt{2\lambda_4/\beta}}\right] \tag{C.9}$$

and Z is analytic everywhere. Γ is the gamma function and D_v is a parabolic cylinder function (Tuszynski et al., 1986). In the limit $\lambda_4 \to 0$ or $\lambda_2 \to \infty$, we recover the Gaussian result (Ma, 1976):

$$Z_\infty = \Gamma\left(\frac{1}{2}\right)(\beta\lambda_2)^{-1/2} \tag{C.10}$$

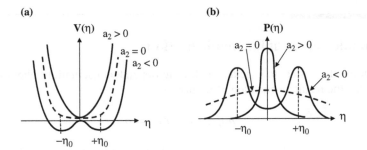

FIGURE C.4
Changes affecting the form of (a) $V(\eta)$ and (b) $P(\eta)$ resulting from a_2 changing its sign.

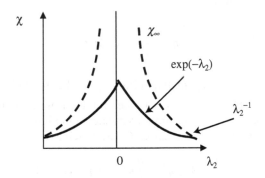

FIGURE C.5
Plot of χ as a function of λ_2 which is proportional to control parameter a_2.

The second moment of the quartic non-Gaussian distribution is the static susceptibility function χ and can be obtained as:

$$\chi = \langle \eta^2 \rangle = (8\lambda_4\beta)^{-1/2} \frac{D_{-3/2}[\frac{\lambda_2}{\sqrt{2\lambda_4/\beta}}]}{D_{-1/2}[\frac{\lambda_2}{\sqrt{2\lambda_4/\beta}}]} \qquad (C.11)$$

Again, it is an analytic function everywhere that tends to its Gaussian limit for $\lambda_4 \to 0$ or $\lambda_2 \to \infty$:

$$\chi_\infty = (2\lambda_2)^{-1}. \qquad (C.12)$$

Figure C.5 shows the plot of χ as a function of λ_2.

Harmonic versus anharmonic motion

If we allow the system to execute small time-dependent oscillations about its *stable* equilibria, the resultant equation of motion is:

$$m\ddot{\eta} + a\eta + b\eta^3 = \sigma \tag{C.13}$$

where $\ddot{\eta} = \partial^2 \eta / \partial t^2$. For $a > 0$ (and η small) this can be approximated as a shifted harmonic oscillator equation:

$$m\ddot{y} \cong -ay, \tag{C.14}$$

where $y = \eta - \dfrac{\sigma}{a}$, so that:

$$\eta = \frac{\sigma}{a} + \cos \Omega_0 t, \tag{C.15}$$

with $\Omega_0 = \sqrt{\frac{a}{m}}$ representing small amplitude periodic oscillations about the stable equilibrium $\eta = \frac{\sigma}{a}$. However, for $a < 0$, the problem is much more complicated and involves three types of oscillating motions as shown in Figure C.6. Figures C.6a and b represent the first integral of Equation (C.13) which is:

$$\frac{m}{2}\dot{\eta}^2 + \frac{a}{2}\eta^2 + \frac{b}{4}\eta^4 - \sigma\eta + c_0 = 0, \tag{C.16}$$

where c_0 is a constant of integration. This can be illustrated in terms of phase-space diagrams and their trajectories, as shown in Figures C.6c and d. An important special trajectory denoted by 2 in Figure C.6d is called a *separatrix* and it represents a boundary between oscillations around a single well and those around both wells.

A new situation arises when we consider periodic nonlinearities, as for example in a *simple pendulum*. The equation of motion is the Sine–Gordon (SG) equation (Campbell, 1987):

$$\ddot{\theta} + \frac{g}{\ell} \sin \theta = 0, \tag{C.17}$$

that can be integrated once to:

$$\frac{1}{2}\dot{\theta}^2 - \frac{g}{\ell} \cos \theta + c_0 = 0, \tag{C.18}$$

where c_o is an integration constant. Figure C.7 illustrates periodically located stable elliptic points around which closed-orbit vibrations are concentrated. Large amplitude motion exists as rotations around the point of suspension of the pendulum. These two types of trajectories are divided by a separatrix.

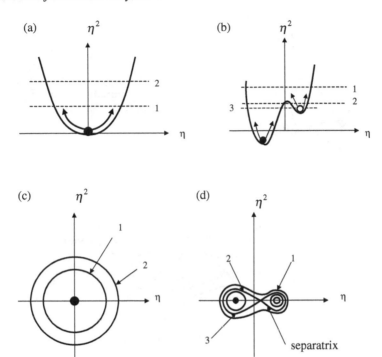

FIGURE C.6
Comparison of (a) a harmonic and (b) anharmonic potential and corresponding orbits in the $(\dot{\eta}, \eta)$ phase space (c) and (d).

External driving and dissipation

Another important feature is to include effects due to energy *dissipation* and external, time-dependent *driving*. For bistable systems, the equation of motion takes the form of the Duffing oscillator:

$$m\ddot{\eta} + a\eta + b\eta^3 + \gamma\dot{\eta} = \sigma(t), \qquad (C.19)$$

where $\gamma\dot{\eta}$ is a friction term and the external force $\sigma(t)$ is time-dependent. Since the system is unistable for $a > 0$, this equation can be linearized about $\eta = 0$ in this regime. Using $\sigma(t)$ in harmonic form, we obtain a damped-driven harmonic oscillator equation:

$$m\ddot{\eta} + a\eta + \gamma\dot{\eta} = \sigma_0\cos(\omega_0 t), \qquad (C.20)$$

with the well known periodic solution:

$$\eta = \sigma_0 A\cos(\omega_0 t + \theta), \qquad (C.21)$$

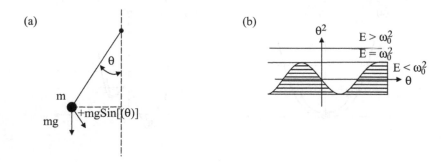

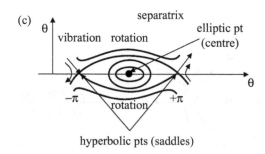

FIGURE C.7
(a) Simple pendulum of mass m; (b) first integral; (c) phase-space diagram.

where:

$$A = [(m\omega_0^2 - a)^2 + \gamma^2 \omega_0^2]^{-1/2} \tag{C.22a}$$

and:

$$\theta = \tan^{-1}\left[\frac{\gamma \omega_0}{a - m\omega_0^2}\right] \tag{C.22b}$$

signifying resonance at $\omega_0 = (a/m)^{1/2}$.

The situation changes dramatically if we reincorporate nonlinearity. Let us first look at nonlinearity and forcing without dissipation. Scaling the independent variable only in Equation (C.19) according to $s = \omega_0 t$, we obtain:

$$\omega^2 \ddot{\eta} + \eta + \beta \eta^3 = F \cos(s), \tag{C.23}$$

where $\beta = b/a$, $F = \dfrac{\sigma_0}{a}$, and $\omega^2 = \dfrac{\omega_0^2 m}{a}$. Taking as initial conditions $\eta(0) = A$ and $\dot{\eta}(0) = 0$ for small values of F, we find (Davis, 1962):

$$\omega \cong 1 + \beta\left(\frac{3}{8}A^2 - \frac{F}{2A}\right) + \dots \tag{C.24}$$

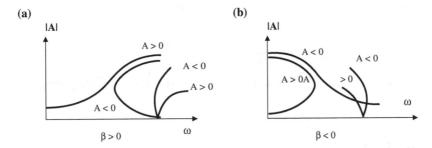

FIGURE C.8
Amplitude versus frequency behavior for the Duffing oscillator for (a) $\beta > 0$ and (b) $\beta < 0$.

Taking the forcing term to be large, compared to nonlinearity gives a different type of solution, namely (Davis, 1962):

$$\omega = 3 + \frac{9}{8}\beta \left(A^2 + \frac{AF}{8} + \frac{F^2}{32} \right) + \ldots \tag{C.25}$$

with bifurcation points of A as a function of ω signifying the emergence of so-called subharmonics. A global range of behavior of A as a function of ω is shown in Figure C.8.

In the other limiting case, namely nonlinearity and dissipation without forcing, the equation of motion becomes (Dixon et al., 1991):

$$\ddot{\eta} + \gamma\dot{\eta} + \eta + \beta\eta^3 = 0, \tag{C.26}$$

and is integrable only for $\gamma = \dfrac{3}{\sqrt{2}}$, in which case we find a so-called kink solution (using $\beta = -1$) in the form:

$$\eta = \pm\frac{1}{2}\left[1 - \tanh\left(\frac{1}{2\sqrt{2}}(s - s_0) \right) \right], \tag{C.27}$$

and a class of damped oscillatory solutions in terms of elliptic functions (see Figure C.9).

Finally, an analysis of the complete Duffing equation with nonlinearity, dissipation, and forcing for Equation (C.19) has also been carried out (Cvitanovic, 1984). What we see now is a *limit* cycle for $\gamma = \gamma_0$ (a particular value). Decreasing γ_0 leads to a period doubling cascade and a related bifurcation cascade (see Figure C.10), leading eventually to chaos.

Next, we briefly discuss the *damped-driven pendulum* equation:

$$m\ell\frac{d^2\theta}{dt^2} + \gamma\frac{d\theta}{dt} + W\sin\theta = A\cos(\omega_0 t), \tag{C.28}$$

(a)   **(b)**

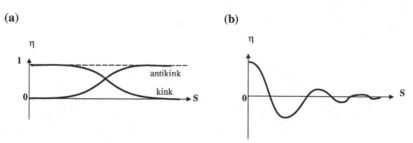

FIGURE C.9
(a) Kink and (b) damped oscillatory solutions of Equation (C.26) and $\beta = -1$.

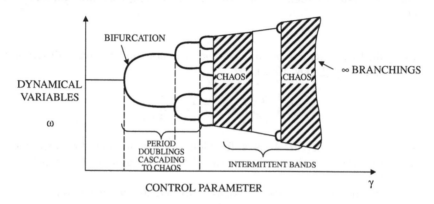

FIGURE C.10
Period doubling cascade and a transition to chaos for the frequency ω of the damped-driven Duffing oscillator.

where m is the pendulum mass, ℓ the pendulum length, γ the damping coefficient, W the restoring force (gravitation), A the forcing amplitude, and ω_0 the forcing frequency. In dimensionless form, this equation of motion becomes:

$$\frac{d^2\theta}{dt^2} + \frac{1}{q}\frac{d\theta}{dt} + \sin\theta = g\cos(\omega_0 t), \qquad (C.29)$$

where q is the damping parameter and g the forcing amplitude. First, ignoring both friction and forcing, we find that Equation (C.29) can be integrated once to yield an elliptic type of differential equation:

$$\frac{1}{2}\left(\frac{d\theta}{dt}\right)^2 = \cos\theta + c_0, \qquad (C.30)$$

where c_o is an integration constant. Figure C.11 shows a phase portrait of Equation (C.30). We have denoted the angular velocity as $\omega \equiv d\theta/dt$. Note the rotational and vibrational orbits, a separatrix orbit separating the two rotational and vibrational regimes, an elliptic focus point, and a hyperbolic instability saddle point.

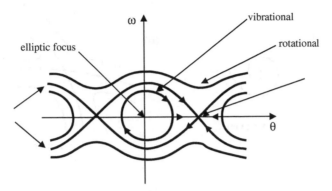

FIGURE C.11
Phase portrait of free pendulum equation.

The period of motion increases with increasing amplitude in contrast to the linearized pendulum. The inclusion of friction can be analyzed by assuming that $\omega \cong Be^{\lambda t}$ and linearizing the equation about $\theta = n\pi$ ($\sin\theta \cong \theta - n\pi$ for θ close to $n\pi$ with n even) to get a discriminant for the growth rates λ of each mode:

$$\lambda^2 + \lambda \pm 1 = 0, \tag{C.31}$$

where the plus sign corresponds to n even and the minus sign to n odd. Figure C.12 illustrates the emergence of focus points and saddle points and the new phase diagram (Baker and Gollub, 1990) when friction is included. It is important to note the presence of the *basins of attraction* associated with each focus point.

The other limiting case, i.e., periodic perturbation without friction, results in a distorted phase portrait with a *stochastic layer* around the separatrix where motion is virtually unpredictable (see Figure C.13). The separatrices divide oscillatory and rotational trajectories. Periodic perturbation destroys the separatrix, and a stochastic layer forms in its vicinity. The motion is chaotic within the layer. The region occupied by the stochastic layer has holes into which a stochastic trajectory cannot enter. An infinite number of stability islands within which an infinite number of thinner stochastic layers reside are present.

Relaxation dynamics and asymptotic stability

In many time-dependent processes in nature, the mechanism of evolution of the concentrations of the molecular species involved is dominated by relaxation dynamics rather than conservative oscillating motion. Examples of such phenomena include chemical kinetics of reacting species, exciton migration in molecular crystals, recombination processes in semiconductors, mixing of binary alloys and fluids,

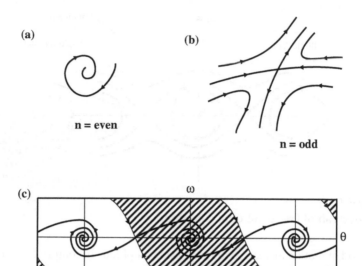

FIGURE C.12
**Critical points in phase space: (a) focal point and (b) saddle point. In (b), the
trajectories going to the saddle point are stable; the trajectories coming from the
saddle point are unstable. (c) Phase-space diagram of the damped pendulum.
Alternate shaded and unshaded regions are basins of attraction. All points within
a particular basin are attracted to the focal point within the basin.**

etc. Considering a single order parameter η relaxing to its equilibrium value with
relaxation constant Γ, the evolution equation in the absence of spatial dispersion is:

$$\eta_t = \frac{1}{\Gamma} P(\eta), \tag{C.32}$$

where $P(\eta)$ is the nonlinear force function appearing in the equation of state (C.1).

The main points are the number and type of bifurcation points of $P(\eta)$. Depending
on whether a given point η_i is a stable or unstable bifurcation point, the system will
tend asymptotically to it or away from it and toward the nearest asymptotically stable
value of η. In this respect, we now define *asymptotic stability* of a solution $\eta(t)$ of
a differential equation. For example, equilibrium $\eta = 0$ is asymptotically stable if
it has a neighborhood of initial conditions $0 < |\eta(0)| < \varepsilon$ and if for all $\eta(0)$ in this
neighborhood (1) the trajectory $\eta(t)$ satisfies $|\eta(t)| < \varepsilon$ for $t > 0$ and (2) $|\eta(t)| \to 0$
as $t \to \infty$.

Figure C.13 shows a number of typical behaviors for the most important types of
bifurcations of a scalar order parameter η (Hale and Kocak, 1991).

Saddle-node bifurcation:

$$\dot{\eta} = \eta^2 + c, \tag{C.33}$$

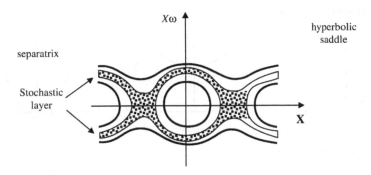

FIGURE C.13
Phase portrait of perturbed pendulum.

Transcritical bifurcation:

$$\dot{\eta} = c\eta + \eta^2, \tag{C.34}$$

Pitchfork bifurcation (where the plus sign corresponds to the subcritical and minus sign to the supercritical cases):

$$\dot{\eta} = d\eta \pm \eta^3, \tag{C.35}$$

Single hysteresis bifurcation:

$$\dot{\eta} = c + \eta - \eta^3. \tag{C.36}$$

A combination of pitchfork and single hysterisis bifurcations with two control parameters c and d:

$$\dot{\eta} = c + d\eta - \eta^3. \tag{C.37}$$

In Figure C.14, open circles indicate repulsive points (asymptotically unstable equilibria) and full circles denote attractive points (asymptotically stable equilibria). Half-filled circles correspond to attraction from one side and repulsion from the other. The curve in Figure C.14e is produced by the cubic discriminant equation $4d^3 = 27c^2$.

Coupled systems and limit cycles

Some situations may involve competing subsystems, orders, or sets of degrees of freedom requiring two or more dependent variables treated equally. This is described by the set of coupled equations below:

$$\dot{\eta}_i = f_i(\{n_j\}), 1 \le i, j \le n, \tag{C.38}$$

(a) Saddle – node bifurcation

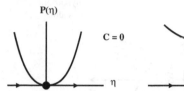

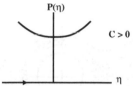

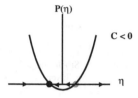

(b) Transcritical bifurcation

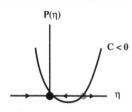

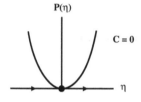

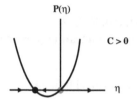

(c) Pitchfork bifurcation (supercritical)

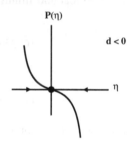

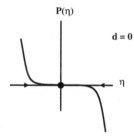

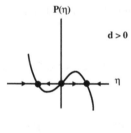

(d) The single hysteresis

(e) The combination

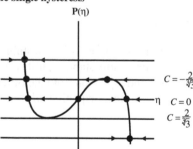

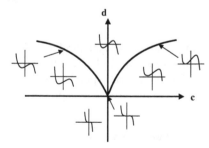

FIGURE C.14

Phase portraits of the main types of bifurcation dynamics: (a) saddle node bifurcation; (b) transcritical bifurcation; (c) pitchfork bifurcation (supercritical); (d) single hysteresis bifurcation; and (e) combination diagram for (c) and (d).

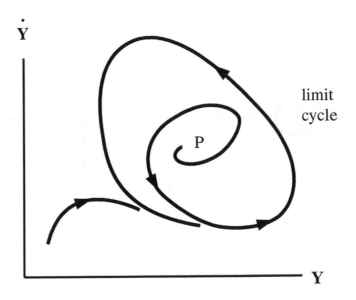

FIGURE C.15
Brusselator-type limit cycle.

where n is the number of components of the order parameter (or the number of competing order parameters).

An important example in the area of reaction–diffusion dynamics is Prigogine's Brusselator system (Nicolis and Prigogine, 1989) where:

$$\dot{X} = X^2 Y - BX + A - X, \qquad (C.39a)$$

and:

$$\dot{Y} = -X^2 Y + BX, \qquad (C.39b)$$

which leads to a *limit cycle* shown in Figure C.15.

In a *Hopf bifurcation*, the constituent Poincaré–Andronov–Hopf equations can be represented as:

$$\frac{dr}{dt} = \lambda r - r^3, \qquad (C.40a)$$

and:

$$\frac{d\varphi}{dt} = \omega = -1, \qquad (C.40b)$$

in polar coordinates. The phase-space picture changes from a stable focus point only to a stable limit cycle as a function of λ, the bifurcation occurring at $\lambda = 0$ as shown in Figure C.16.

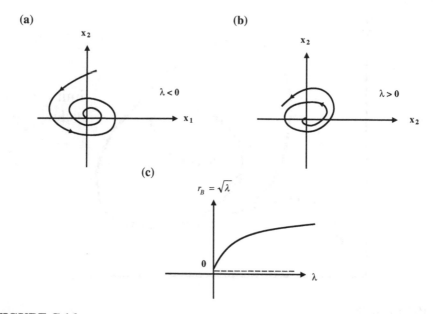

FIGURE C.16
Hopf bifurcation for (a) $\lambda < 0$ and (b) $\lambda > 0$; (c) shows the radius of the limit cycle.

The radius of the limit cycle has $\sqrt{\lambda}$ dependence for $\lambda > 0$. The question of structural stability of such solutions can be tackled using Lyapunov stability analysis, at least for two coupled equations. We can find in general attractive, repulsive, and mixed orbits (Crawford, 1991). A summary of the various possibilities for two-component systems is given in Figure C.17 and shows nodes (sinks and sources), centers, saddles, foci, and limit cycles (Hale and Kocak, 1991; Crawford, 1991).

Nonlinear waves and solitons

We may extend the number of dependent variables to infinity, as can be done for a lattice or a chain. Numerous physical applications of this approach can be readily found and embrace such phenomena as: (1) electric LC circuits, (2) chains of coupled pendula, (3) masses connected by elastic springs, (4) coupled acoustic resonators, and (5) interacting magnetic moments, among others.

Historically, the first such construction in terms of a numerical simulation appears to have been attempted by Fermi, Pasta, and Ulam (Scott et al., 1973) for a set of coupled nonlinear equations:

$$\ddot{x}_i = (x_{i+1} + x_{i-1} - 2x_i) + \alpha[(x_{i+1} - x_i)^2 - (x_i - x_{i-1})^2], \qquad \text{(C.41)}$$

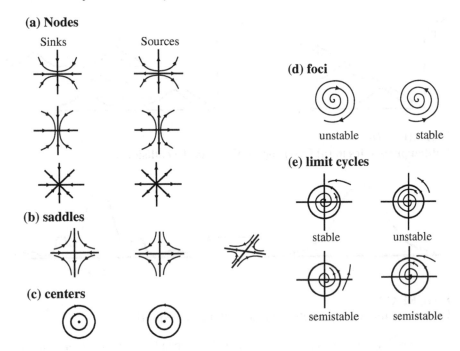

(a) Nodes

Sinks   Sources

(d) foci

unstable   stable

(e) limit cycles

(b) saddles

stable   unstable

(c) centers

semistable   semistable

FIGURE C.17
Topologically important points in phase portraits of two-parameter systems.

for $i = 1, 2, \ldots, n$ which represents n masses on a lattice that are anharmonically coupled through nonlinear springs. The astonishing result of the computer simulation was that the energy initially injected into a chosen mode was not thermalized and remained concentrated in a few modes. These localized, stable nonlinear modes of behavior (Scott et al., 1973; Zabusky and Kruskal, 1965; Drazin and Johnson, 1989) propagate at constant velocities without changes of profile and preserve their shapes and identities on collisions, as shown in Figure C.18. Only in special cases such as the so-called Toda lattice and Ablowitz–Ladik equation can exact solutions be found.

The behavior of nonlinear differential equations is in stark contrast to the properties of *linear wave equations*. In the latter case, *dispersion* due to space-dependent derivatives results in the spread of a wave packet, say:

$$\psi = \int A_k e^{i(\omega t - kx)}\, dk, \tag{C.42}$$

so that with the dispersion relation $\omega = \omega(k)$, the group velocity $v_g = d\omega/dk$ depends on the wave number k (see Figure C.19).

Nonlinearity, in many cases, provides a balance to dispersion and a steepening of the wave prevents it from spreading. See Figure C.20 (Scott et al., 1973). Nonlinear wave equations have four major types of nonsingular solutions: extended traveling waves, nontopological solitary waves, topological solitary waves, and breathers (shown in Figure C.21).

(a)   (b)

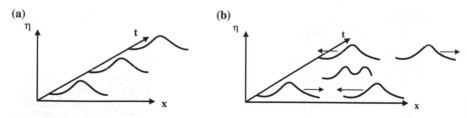

FIGURE C.18
Soliton properties in (a) free propagation and (b) collisions.

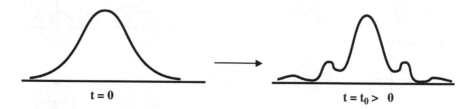

$t = 0$ $t = t_0 > \ 0$

FIGURE C.19
Dispersion for wave packets in linear differential equations.

A *traveling wave* $\phi_T(\xi)$, is a solution that depends on x and t only through $\xi = x - vt$ with v constant.

A *solitary wave* $\phi_s(\xi)$ is a localized traveling wave solution of a differential equation that exhibits a transition from one constant asymptotic state at $\xi = -\infty$ to another at $\xi = +\infty$. It is essentially localized in ξ over a region whose width is Δ.

A *topological solitary wave* (kink or antikink) connects two asymptotic and different values $\phi(+\infty) \neq \phi(-\infty)$ corresponding to the emergence of two asymptotic equilibrium plateaus. Their topological charge can be defined as $Q = \phi(+\infty) - \phi(-\infty)$ which can be a conserved quantity.

A *breather* can be represented as $\phi = f_1(t)f_2(x)$ where $f_1(t)$ is a periodic function of time and $f_2(x)$ is a localized function of space. Unlike kinks, breathers require little energy to be activated. They can be seen as bridging the gap between highly nonlinear modes such as topological solitons and linear (phonon) modes.

A particular type of a solitary wave is called a *soliton* when it asymptotically preserves its shape and velocity on collisions with other solitary waves.

Nonlinear Schrödinger (NLS) equation

The cubic nonlinear version reads:

$$\eta_{xx} + i\eta_t + k|\eta|^2\eta = 0, \tag{C.43}$$

where η is a complex field. It represents self-modulation of an almost monochromatic wave with linear dispersion and weak nonlinearity. Its soliton solution is:

$$\eta = \sqrt{2}ae^{i\phi}\text{sech}[a(x - bt)], \tag{C.44a}$$

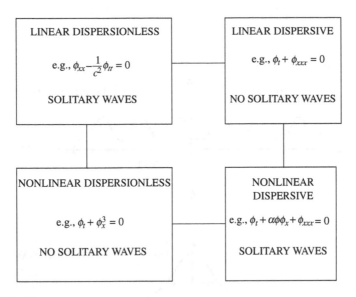

FIGURE C.20
Wave equations and their properties.

where a and b are arbitrary constants and ϕ defines the carrier wave with:

$$\phi = \frac{1}{2}bx - \left(\frac{1}{4}b^2 - a^2\right)t, \qquad \text{(C.44b)}$$

and the sech function provides a bell-shaped envelope.

A generalized nonlinear Schrödinger equation can be defined as:

$$\eta_{xx} + i\eta_t + W(|\eta|)\eta = 0, \qquad \text{(C.45)}$$

where W tends to infinity as $\eta \to \infty$ and is bounded from below everywhere. Although it frequently appears in physical applications, only its cubic version is integrable. A general result has been obtained with $W = |\eta|^{2n}$ which states that stable soliton solutions exist only if $n < 1 + 2/d$ where d is the dimensionality of the physical space (Rasmussen and Rypdal, 1986).

The *Sine–Gordon equation* is:

$$\eta_{tt} - c_0^2\eta_{xx} + m^2\sin\eta = 0. \qquad \text{(C.46)}$$

It possesses a wealth of exact solutions, including a kink and an antikink:

$$\eta_k = 4\arctan[\exp(\pm\gamma(\xi - v\tau))], \qquad \text{(C.47)}$$

where $\gamma = (1 - v^2)^{-1/2}$; $\xi = mx/c_o$; $\tau = mt$, a kink–antikink pair:

$$\eta_{k,\bar{k}} = 4\arctan\left[\frac{\sin h\frac{v\tau}{\sqrt{1-v^2}}}{v\cos h\frac{\tau}{\sqrt{1-v^2}}}\right], \qquad \text{(C.48)}$$

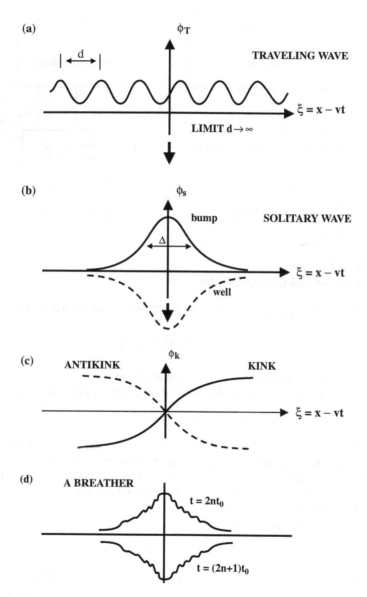

FIGURE C.21
Four types of solutions to nonlinear wave equations: (a) traveling waves;
(b) solitary waves (nontopological solitons); (c) solitary waves (topological
solitons); and (d) breather.

along with a stationary breather:

$$\eta_B = 4 \arctan\left[\frac{\varepsilon \sin(t/\sqrt{1+\varepsilon^2})}{\cos h(\varepsilon x/\sqrt{1+\varepsilon^2})}\right]. \tag{C.49a}$$

and a moving breather:

$$\eta_{MB} = 4 \arctan\left[\frac{\sqrt{1-\lambda^2}}{\lambda}\frac{\sin[\gamma\lambda\,(t-cx)]}{\cosh[\gamma\sqrt{1-\lambda^2}\,(x-ct)]}\right], \tag{C.49b}$$

Also, a multikink solution has been developed (Dodd et al., 1982) and it can be viewed as a nonlinear superposition of individual kinks or antikinks moving in their reference frames at their own velocities. Except for phase shifts, their collisions preserve their shapes and velocities. Its generalization, called the double Sine–Gordon equation:

$$\eta_{tt} - \eta_{xx} = \sin\eta + \sin 2\eta, \tag{C.50}$$

is not integrable.

The *nonlinear Klein–Gordon equation (NLKG)* is, in scaled variables, defined by:

$$\eta_{xx} - \eta_{tt} = F(\eta) \tag{C.51}$$

where η is a real field and $F(\eta) = \pm(\eta - \eta^3)$ is the most frequent choice of the nonlinearity that models bistable properties of the system. Although this equation has no soliton solutions it has two types of solitary waves, kinks (and antikinks) given by:

$$\eta = \pm\tanh\left(\frac{1}{\sqrt{2}}\frac{x-vt}{\sqrt{1-v^2}}\right) \tag{C.52}$$

and bumps (and wells) given by:

$$\eta = \pm\mathrm{sech}\left(\frac{1}{\sqrt{2}}\frac{x-vt}{\sqrt{1-v^2}}\right). \tag{C.53}$$

These solutions can bounce off each other, lock, annihilate each other on collision, and even emit oscillatory disturbances (radiation) as a result of collisions (Fokas and Ablowitz, 1984).

The concept of the soliton has been generalized to higher dimensional equations. The inverse scattering method has been extended to solve $(2+1)$–dimensional equations such as the Kadomtsev–Petviashvili (Ablowitz et al., 1983; Fokas and Ablowitz, 1983; Manakov, 1981) and Davey–Stewartson equations (Fokas and Ablowitz, 1984). Whereas inverse scattering formalism has been developed in $(n+1)$-dimensional space (Nachman and Ablowitz, 1984), at present no known nonlinear partial differential equations are solvable by these techniques (Ablowitz and Clarkson, 1991).

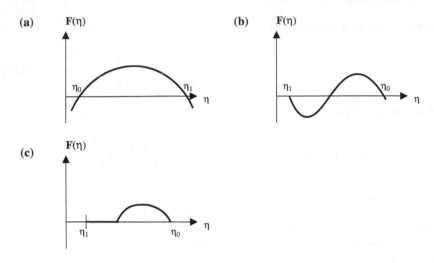

FIGURE C.22
Generalized nonlinear forces in reaction–diffusion systems: (a) Fisher case, (b) bistable case, and (c) ignition case.

Numerical studies (Spatschek et al., 1990; Bishop et al., 1986; Bishop et al., 1988; Taki et al., 1989) showed that the interplay of nonlinearity, dissipation, and forcing leads to a typical bifurcation diagram with a period doubling cascade resulting in chaos, as we have already seen for ordinary differential equations such as the Duffing oscillator. Analogous behavior has been seen for the damped driven Sine–Gordon equation:

$$\eta_{xx} - \eta_{tt} + \sin \eta = -\gamma \, \eta_t + \Gamma \sin \Omega t \qquad (C.54)$$

Obviously, only a small number of partial differential equations are integrable and thus support solitons. We do not find soliton solutions in reaction–diffusion type equations:

$$\eta_t = \eta_{xx} + F(\eta) \qquad (C.55)$$

However, traveling wave fronts exist in the form $\eta = \eta(x - vt)$ with velocity v that connect stable asymptotic states η_0 and η_1 such that $F(\eta_0) = F(\eta_1) = 0$. Figure C.22 illustrates three important cases (Fife, 1984): the Fisher case, the bistable case, and the ignition case.

The Fisher case involves a critical velocity $v^* > 0$ such that for every $v > v^*$ a front exists connecting η_1 and η_0. In the bistable case, however, with $F'(\eta_0) < 0$ and $F'(\eta_1) < 0$, a unique wave front connects η_0 and η_1. This can be seen as a possible mechanism of velocity selection. In the ignition case, a unique wave front connects the two asymptotically stable equilibria.

Pattern formation

This section covers selected physical systems and phenomena that exhibit pattern formation and are describable in terms of theoretical models, for example, the Landau–Ginzburg.

Pattern formation in fluid dynamics

Many types of fluid flows lead to pattern formation. For example, convective flows in binary fluid mixtures exhibit interesting transitions of flow patterns from rolls of linear oscillatory convection subjected to bending, compression, and pinching to cellular localized structures (Kolodner et al., 1987). Forced, unbounded shear flows possess (Browand and Ho, 1987) vortices, elliptic jets, and circular jets, and the flows can be of periodic or phase-dislocation type.

Rayleigh–Bénard convection cells are characterized by transitions from linear diffusive convection to convection roll patterns and they also develop chaotic behavior and eventually full turbulence (Newell and Whitehead, 1969). A fluid is placed between horizontal plates and heated from below. When the temperature difference ΔT exceeds a critical value ΔT_c, the heat can no longer be carried up by conduction alone and the fluid is set into motion with flow in the form of convective rolls whose characteristic spacing is of order d, the plate separation.

Another example of pattern formation in fluids is the Taylor–Couette flow. A fluid is placed between concentric cylinders and the inner cylinder is rotated. When the angular frequency Ω exceeds a critical value Ω_c, the flow is no longer purely azimuthal. Instability occurs and a pattern of Taylor vortices with an axial component of flow arises with a characteristic separation of order d, the distance between cylinders. Turbulence in Taylor–Couette flows (Hagseth et al., 1989) may also develop in the form of a spiral band rotating about the axis of symmetry with a constant velocity. Qualitative features of these phenomena are described using Landau phenomenology through Landau–Stuart expansions for the complex amplitude function $A(x, t)$ that satisfies:

$$\frac{\partial A}{\partial t} = D \nabla^2 A + v.\nabla A + f(A), \tag{C.56}$$

where D is the diffusion constant, the second term is the entrainment field, and f is a nonlinear function of A. This corresponds to the time-dependent Landau–Ginzburg (TDLG) equations when a transformation to a moving frame of reference is made. The control parameter is the Reynolds number and it is believed to critically affect the linear term's coefficient in $f(A)$.

Pattern formation in liquid crystals

Cholesteric liquid crystals have their symmetry axes shifted by a constant angle on going from plane to plane. This describes a helical pattern in the system.

Experiments on nematic liquid crystals (Joets and Ribotta, 1986) subjected to a variable potential difference revealed a series of pattern transformations with increasing voltage. The patterns changed from the rest state to normal rolls (which could develop defects in the form of dislocation edges), to undulated rolls, and to oblique rolls that produced period pinching (sheared varicose), and eventually showed a rectangular cell lattice that at high voltage disappeared, leading to chaos. Depending on whether the field was applied suddenly or gradually, the transitions were, respectively, accompanied by defects or were free of them. Joets and Ribotta (1986) proposed the Newell–Landau–Ginzburg equations (Newell and Whitehead, 1969) to describe dynamics of the system (characterized by two components of the order parameter A and B). These are identical to Equation C.56) with an added mutual coupling.

Pattern formation in polymers

Among the many possible transformations occurring in polymer systems, two deserve special attention. First, fluctuation-induced first-order phase transitions have been found (Bates et al., 1988) between a one-dimensional (lamellar) periodic structure and a disordered phase. The transition is accompanied by microphase separation occurring with a uniform speed.

The second case is a crumpling transition occurring at a finite temperature and characterized by a change from a stretched surface to a crumpled tethered one (Paczulski et al., 1988). Critical fluctuations lead to a weakly first-order transition. The order parameter is given by the coarse-grained vectors tangent to the surface. Figure C.23 illustrates the phases that arise in solutions of surfactant molecules (Corkhill and Goodman, 1969). In both cases, a phenomenological Landau–Ginzburg Hamiltonian in terms of the order parameter is postulated.

Crystal growth and structural transitions

The dynamics of crystal growth can be described starting from an Ising-like Hamiltonian given by:

$$H = -\sum_{(ij)} J_{ij}\, s_i\, s_j - \sum_i \Delta\mu . s_i + V(\{s\}), \qquad (C.57)$$

where $s_i = 2c_i - 1 = \pm 1$ is a generalized spin and c_i is the concentration variable for the ith lattice site such that $c_i = +1$ means that the site is filled (solid) while $c_i = -1$ that it is empty (vapor). $\Delta\mu$, is the chemical potential and $V(\{s\})$ is an effective on-site potential. The dynamics is then described using an Onsager relation:

$$\frac{\partial\phi}{\partial t} = -\frac{D}{kT}\frac{\delta F}{\delta\phi}, \qquad (C.58)$$

where ϕ is the trajectory normal to the surface (interface of an anisotropic medium) playing the role of a coarse-grained order parameter. In the continuum limit, the

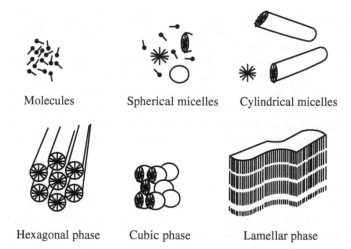

Molecules	Spherical micelles	Cylindrical micelles

Hexagonal phase	Cubic phase	Lamellar phase

FIGURE C.23
Idealized structures of surfactant molecules that can form in solution as the surfactant concentration increases.

relevant Hamiltonian becomes of the Landau–Ginzburg type and the Onsager equation takes the form of a TDLG equation again. Not surprisingly, typical crystal growth trajectories are spirals.

Once a crystal lattice is formed, it may undergo various changes and transitions. A typical model Hamiltonian for commensurate and incommensurate transitions is the Frank–van der Merve Hamiltonian (Eykhott et al., 1986), which predicts the formation of discommensuration lines and their interactions. The resultant patterns may change from striped to hexagonal and lattice melting is also possible. The continuum limit of this lattice Hamiltonian leads to the Sine–Gordon equation.

Chemical instabilities

Reaction–diffusion systems of two or more chemically active species are described using coupled TDLG equations. The results of computer simulations were confirmed experimentally and often lead to a variety of patterns, such as those of spiral type in the Belousov–Zhabotinskii system (Winfree, 1974).

Chaos and turbulence

In perturbed dynamical systems and in pattern-forming systems an often-encountered scenario is a period doubling cascade ending with fully developed chaos. Systems undergoing transitions to chaotic behavior typically exhibit universal features. First, plotting the given dynamical variable with respect to changes in the corresponding control parameter results in *bifurcations* or splitting into two distinct stable levels. Second, when the dynamical variable is related to the frequency of the system, such bifurcations represent *period doubling.*

Successive period doublings can lead to an infinite number of branchings representing one of the possible routes to chaos. As the control parameter is varied, the system may pass through alternate regions of stability and chaos. Such behavior is called *intermittency.* Two other features common to chaotic systems are evident in the phase-space diagram and can be treated analytically discretizing the equations of motion. These features are *sensitivity* to *initial conditions and stretching and folding* of phase space regions.

When a system rapidly branches or period doubles until the number of branches goes to infinity, it is said to *cascade to chaos.* The bifurcation diagram shows that chaotic behavior can be intermittent with respect to changes in system parameters, as shown by the interspersed bands of chaotic and periodic behavior (see Figure C.13).

The bifurcation diagram, regardless of the physical system that generates it, displays several characteristic features. The successive bifurcations divide the figure into self-similar shapes. Magnification or rescaling of the diagram tends to recover the initial shape. This self-similarity is equivalent to the fact that the geometry of the diagram is *fractal* in nature.

The currently accepted definition of chaos is based on initial conditions. If initial conditions are changed by an infinitesimal amount, the behavior of the system may be affected drastically at later times. We call a system chaotic if it has this property of very sensitive (exponential) dependence on initial conditions (Milloni et al., 1987).

Another indication of the onset of chaos is through the Fourier transform of any coordinate involved in chaotic motion. This is much different from quasiperiodic motion whose Fourier transforms consist of very sharp peaks. At the turn of the century, Poincaré noted that other bounded systems had continuous broad band spectra indicative of chaotic motion. During the past two decades, studies of chaotic motion in many systems indicate that the spectra typically evolve from quasiperiodic motion into chaos (Coullet, 1984). Chaos may appear in systems governed by very simple rules of behavior, such as first-order differential equations. Moreover, transition to chaos can take place only in a few universal ways. These transitions are called the routes to *chaos* and we describe them below.

Landau scenario

The oldest scenario for the transition from laminar to turbulent flow was proposed by Landau in 1944 (Ter Haar, 1965). He assumed that chaos would develop via an infinite sequence of Hopf bifurcations. When a control parameter characterizing the flow is increased, it causes the number of spikes in the Fourier spectrum to grow. This would constitute a transition to chaos by frequency generation. In general, this is not a realistic approach.

The number of spikes does not necessarily indicate chaos. The behavior in the Landau scenario may look more complicated, but is still of deterministic type. No sensitive dependence on initial conditions is required by the definition of chaos used here.

Ruelle–Takens–Newhouse scenario

The approach accepted as the most "generic"was proposed in 1971 by Ruelle and Takens (Cvitanovic, 1984). Here, a system will transform into chaos after only two Hopf bifurcations, not an infinite number advocated in Landau's scenario. Very small perturbations can result in this sudden shift after only two bifurcations, thus indicating a sensitive dependence on initial conditions. This route to chaos, often called the two-frequency route, has been seen in many experiments.

In order to analyze the transition to chaos, a control parameter must be defined to direct the evolution of the system. To quantitatively describe the various regions in the phase space, the Lyapunov characteristic exponent (LCE) is used. For the simplest case of the map, $x_{n+1} = f(x_n)$, the associated Lyapunov exponent is defined as:

$$\lambda \equiv \lim_{N \to \infty} \sum_{n=0}^{N} \ell n \left| f'(x_n) \right|, \tag{C.59}$$

and, in general, stable paths are characterized by $\lambda < 0$, superstable by $\lambda = -\infty$, bifurcation points by $\lambda = 0$, and unstable paths by $\lambda > 0$. This exponent can be interpreted in many ways depending on the system. For example, the LCE associated with a particle trajectory gives the average rate at which nearby trajectories diverge. The best way to identify chaotic behavior is to compute the LCE or compute the largest LCE for the system.

Feigenbaum picture

Feigenbaum's work (Feigenbaum, 1979; Feigenbaum, 1978) with discrete mappings of the form:

$$x_{n+1} = \Delta f(x_n) \tag{C.60}$$

led to his scenario for the development of chaos. According to Feigenbaum, chaos develops as a result of an infinite sequence of period doubling or pitchfork bifurcations. As Δ is varied, a transition to chaotic behavior occurs in a universal period doubling route to chaos.

The sensitivity to initial conditions that determines chaotic behavior lies in the location of attractors in the phase space. Points further from the attractors take different paths to approach the stable points. The paths depend significantly on the location of the initial state in phase space. Attractors in the phase space of the system are determined by the criterion Feigenbaum discussed in 1978. Equation (C.60) has highly bifurcated attractors (stable points) identified by:

$$X^* = \Delta f(X^*) \tag{C.61}$$

where X^* is the position of the attractor. Points nearby X^* converge to X^*, and it is considered to be a global attractor when most points in the system converge toward

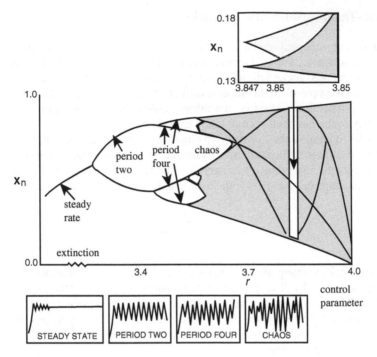

FIGURE C.24
Logistic map and its properties.

X^*. The most important map used in the analysis of chaos is the *logistic* map given by the iterative formula:

$$X_{n+1} = r\, X_n(1 - X_n) \tag{C.62}$$

where the control parameter r is contained within $0 < r \leq 4$ and the variable X_n within $0 \leq X_n \leq 1$. The behavior of its asymmetric solutions is shown in Figure C.23 following Corkhill and Goodman (1969).

In the logistic map (Equation C.62) for values of $r < 3$, the results eventually converge to a steady state called an attractor. See Figure C.24. However, for values of $r > 3$, the resultant oscillation does not settle down and remains stable, i.e., the behavior is periodic. The two possible values of $X_n(r)$ never converge and the curve $X_n(r)$ shows a bifurcation. Higher values of the control parameter r produce further splitting and doubling of the periodicities. Each period doubling is a bifurcation. Abruptly, at $r_c = 3.58$, the result for X_n no longer oscillates periodically but changes in a chaotic fashion. Although for the values of r above r_c chaos and randomness seem to prevail, a further increase of the control parameter introduces windows of regularity (intermittency) among chaos.

Computer simulations of the logistic map demonstrate that the structure is infinitely deep and self-similar. Parts of it, when magnified, show identical patterns ad *infinitum*.

The patterns are characteristic of the dynamic behavior of nonlinear systems leading to chaos and complexity. The Feigenbaum exponent δ is then found from the location of period doubling bifurcation points r_k as:

$$\delta = \lim_{k \to \infty} \frac{r_k - r_{k-1}}{r_{k+1} - r_k} = 4.669\ldots \tag{C.63}$$

Pomeau–Manneville scenario

This scenario has a subcritical pitchfork bifurcation as opposed to a supercritical one in the Feigenbaum picture. A periodic equilibrium state is achieved by control parameter changes at unequal intervals so that quasiperiodic phases occur irregularly and intermittently with chaotic impulses. Increasing the control parameter increases intermittency, eventually removing the quasiperiodic phases altogether and developing chaos.

The main surprise in the area of nonlinear dynamics was that chaos emerged as a *highly hierarchical* form of dynamic ordering. Another important discovery was the *strange attractor*. Strange attractors are equilibrium states in a chaotic region where convergence depends strongly on initial conditions. The Lorenz equations (see Figure C.25) exhibit strange attractors:

$$\left. \begin{aligned} \dot{X} &= \sigma(X - Y) \\ \dot{Y} &= -rX - Y - XZ \\ \dot{Z} &= -bZ + XY \end{aligned} \right\} \tag{C.64}$$

Fractals

Many pattern-forming systems, especially those far from thermodynamic equilibrium, exhibit growth of forms of fractal nature. The basic property of all fractal objects is their self-similarity, i.e., when we cut a part of an object and magnify it, the resulting object appears the same as the original. Another property of fractals is that their dimensionality is generally a real number, not an integer. In the simplest form, the fractal dimension D is given by the relationship:

$$V(R) \sim R^D \tag{C.65}$$

where $V(R)$ is the volume of the region bounded by the interface whose radius is R. Thus, the fractal dimension determines the extent to which an object fills space. A fractal object fills space unevenly because its parts are related or correlated. The fractal dimension quantifies the rate at which the object clumps together in space. The notion of fractal was principally introduced by B. Mandelbrot (1977).

In a physical context, it is useful to distinguish two general classes of fractals (Vicsek, 1989): (1) deterministic (see Figure C.26a) where a simple iterative rule is

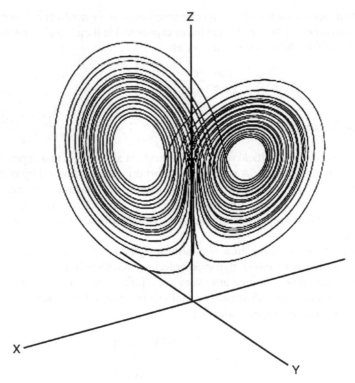

FIGURE C.25
Strange attractor for Lorenz equation.

present, e.g., involving a procedure to cut part of an object at each stage and replace it with a fixed element and (2) random (see Figure C.26b) where a stochastic approach is used so that a given operation, e.g., aggregation event, is predicted with a preselected probability level (Feder, 1988).

Self-organized criticality

In 1987, Bak et al., introduced a new type of critical phenomenon they designated self-organized criticality (SOC). The spatially extended dynamical systems to which SOC applies exhibit spatial and temporal self-similar scale invariance (Bak et al., 1988; Bak and Chen, 1991).

Spatially this is manifested in the form of fractal properties of the system, while temporally the presence of *flicker* (also called pink or 1/f) noise (Dutta and Horn, 1981), signals event lifetimes over all time scales. Physically, such dynamical systems are characterized by power law spatial and temporal correlations over many orders

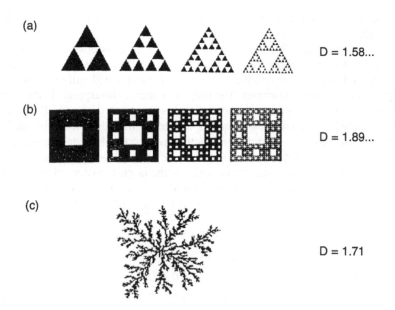

(a)   D = 1.58...

(b)   D = 1.89...

(c)   D = 1.71

FIGURE C.26
Two deterministic fractals: (a) triangular Sierpinski gasket; (b) Sierpinski carpet and random fractal; (c) diffusion-limited aggregation cluster.

of magnitude. A consequence of this power law behavior is that systems exhibiting SOC are weakly chaotic.

The concept of self-organization used in the context of SOC refers to the evolution of a system to the critical state independently of the initial conditions; the critical state is an attractor for its dynamics (Tang and Bak, 1988a,b).

Cellular automata are mathematical realizations of SOC in which a grid of separate cellular states is defined by a field (such as pressure) and a simple set of rules sets the system in motion (in discrete time steps). The only influence on the evolution of a square is the value of the field at all its neighboring sites in the previous time step. In the next time step, a site may grow or die, depending on the number of live neighbours and other parameters (Bak et al., 1988). The physical system to which the cellular automaton approach is most often employed in SOC is the sandpile.

Not all sites have critical slopes at their critical points. A system may evolve in such a way that the minimally stable clusters, defined to be areas accessible by avalanches starting from a single site perturbation, become intermixed with regions of less-than-minimal stability that impede noise. When these noise filters become numerous enough to stop propagation of perturbations over infinite length scales, the system has reached its critical point.

At a critical slope of the sandpile, no length or time scale is present and stationary conditions are reached through self-organization. We conclude that the combination of scale-invariant sand grain clusters and their minimal stability lead to a power spectrum resembling the $1/f$ type. The fact that these systems scale both in space and time in this manner suggests that the physics principles behind avalanches and small slides are the same.

In analogy with equilibrium phase transitions, the behavior of crucial parameters of a SOC system near its critical point may be characterized using critical exponents (Tang and Bak, 1987; Tang and Bak, 1988). To this end, we identify the slope θ of the sandpile as the control parameter. The order parameter corresponds to the current of sand flow j or equivalently the number of slides in a given time period. The external field h coupled to the order parameter is identified with the current of sand grains dropped on the pile. The correlation length ξ represents the cutoff in linear cluster size just below criticality. The susceptibility χ of the order parameter to a change in the external field is also introduced and s is the largest cluster volume for a given value of the control parameter. Similarly with equilibrium-critical phenomena, we postulate that the above system parameters follow power laws of the forms:

$$j \propto (\theta - \theta_c)^\beta \tag{C.66}$$

$$\chi \propto (\theta - \theta_c)^{-\gamma} \tag{C.67}$$

$$\xi \propto (\theta_c - \theta)^{-\nu} \tag{C.68}$$

$$s \propto (\theta_c - \theta)^{-1/\sigma} \tag{C.69}$$

and:

$$j(\theta = \theta_c) \propto h^{1/\delta} \tag{C.70}$$

Several physical observations about the behavior of the system at criticality can be readily deduced from these relations. Firstly, if $\gamma > 0$, χ is singular at the critical point; hence a small external field h (i.e., a single grain of sand) may cause sand avalanches of all sizes. Also, for $\nu > 0$, the correlation length tends to infinity at the critical point as in the theory of phase transitions. Numerical simulations support the conjecture about power law scaling. The mean field theory for SOC (Tang and Bak, 1988) leads to the exponents given by $\beta = 1, \gamma = 1, \delta = 2, \sigma = 1/2$, and $\nu = 1/2$. From simulations with cellular automata, values for the exponents in two dimensions are (Tang and Bak, 1988a) $\beta = 0.7, \gamma = 1.35, \nu = 0.74$, and $\sigma = 0.72$. As the spatial dimension of the simulation is increased, the values converge toward their mean field predictions, suggesting that the latter may be exact in a sufficiently high dimension.

The applicability of SOC theory to natural phenomena has received a great deal of attention in recent years (Bak and Chen, 1989). The use of cellular automata to simulate interacting ecosystems has shown that activity settles down to a critical state that may be indicative of SOC. This might explain the long-term stability (and resilience) of life to perturbations, while providing a natural explanation for large biomass extinctions as simply avalanches. The concept that turbulence can be represented as energy dissipation on a fractal has also been suggested in this context (Mandelbrot, 1974).

References

Ablowitz, M.J. and Clarkson, P.A., *Solitons, Nonlinear Evolution Equations and Inverse Scattering,* Cambridge University Press, Cambridge, U.K., 1991.

Ablowitz, M.J., Bar Yaccov, D., and Fokas, A.S., *Stud. Appl. Math.,* 69, 35, 1983.

Abramowitz, M. and Stegun, I.A., Eds., *Handbook of Mathematical Functions,* Dover, New York, 1965.

Bak, P. and Chen, K., *Physica D,* 38, 5, 1989.

Bak, P. and Chen, K., *Sci. Amer.* 264, 1, 1991.

Bak, P., Tang, C., and Wiesenfeld K., *Phys. Rev. Lett.,* 59, 381, 1987.

Bak, P., Tang, C. and Wiesenfeld, K. *Phys. Rev. A.* 38, 364, 1988.

Baker, G.L. and Gollub, J.P. *Chaotic Dynamics,* Cambridge University Press, Cambridge, U.K., 1990.

Bates, F.S. et al., *Phys. Rev. Lett.,* 61, 2229, 1988.

Bishop, A.R. et al., *Phys. Lett.,* 127A, 335, 1988.

Bishop, A.R. et al., *Physica,* 23D, 293, 1986.

Browand, F.K. and Ho, C.M., *Nucl. Phys.* B, Proc. Suppl., 2, 139, 1987.

Campbell, D.K., *Nucl. Phys.* B, Proc. Suppl., 2, 541, 1987.

Corkhill, J.M. and Goodman, J.F., Adv. *Colloid Interface Sci.,* 2, 297, 1969.

Coullet, P., in Kuramoto, Y., Ed., *Chaos and Statistical Methods,* Springer, Berlin, 1984.

Crawford, J.D., *Rev. Mod. Phys.,* 63, 991, 1991.

Cvitanovic, P., *Universality in Chaos,* Adam Hilger, Ltd., Bristol, 1984.

Davis, H.T., *Introduction to Nonlinear Differential and Integral Equations,* Dover, New York, 1962.

Dixon, J.M., Tuszynski, J.A., and Otwinowski, M., *Phys. Rev.,* A 44, 3484, 1991.

Dodd, R.K. et al., *Solitons and Nonlinear Wave Equations,* Academic Press, New York, 1982.

Drazin, P.G. and Johnson, R.J., *Solitons: An Introduction,* Cambridge University Press, Cambridge, U.K., 1989.

Dutta, P. and Horn, P.M., *Rev. Mod. Phys.,* 53, 3, 1981.

Eykhott, R. et al., *Physica,* D 23, 102, 1986.

Feder, J., *Fractals,* Plenum Press, New York, 1988.

Feigenbaum, M., *J. Stat. Phys.,* 21, 669, 1979.

Feigenbaum, M., *J. Stat. Phys.,* 20, 25, 1978.

Fife, P.C., in *Nonequilibrium Cooperative Phenomena in Physics and Related Fields,* Velarde, M.G., Ed., Plenum Press, New York, 1984.

Fokas, A.S. and Ablowitz, M.J., J. *Math. Phys.,* 25, 2494, 1984.

Fokas, A.S. and Ablowitz, M.J., *Stud. Appl. Math.,* 69, 211, 1983.

Hagseth, J.J. et al., *Phys. Rev. Lett.,* 62, 1257, 1989.

Haken, H., *Synergetics: An Introduction,* Springer, Berlin, 1983.

Hale, J. and Kocak, H., *Dynamics and Bifurcations,* Springer, Berlin, 1991.

Joets, A. and Ribotta, R., *J. Phys.,* 47, 595, 1986.

Jullien, R. and Botet, R., *Aggregation and Fractal Aggregates,* World Scientific, Singapore, 1987.

Kolodner, P. et al., *Nucl. Phys.* B, Proc. Suppl., 2, 97, 1987.

Landau, L.D. and Lifshitz, E.M., *Statistical Physics,* Pergamon Press, London, 1959.

Ma, S.K. *Modern Theory of Critical Phenomena,* Benjamin, Reading, MA, 1976.

Manakov, S.V., *Physica,* 3D, 420, 1981.

Mandelbrot, B., *J. Fluid. Mech.,* 62, 331, 1974.

Mandelbrot, B.B., *The Fractal Geometry of Nature,* W.H. Freeman, San Francisco, 1977.

Milloni, P.W., Shih, M.L., and Ackerhalt, J.R., *Chaos in Laser-Matter Interactions,* World Scientific, Singapore, 1987.

Nachman, A.I. and Ablowitz, M.J., *Stud. Appl. Math.,* 71, 243, 1984.

Newell, A.C. and Whitehead, J., *J. Fluid Mech.,* 38, 203, 1969.

Nicolis, G. and Prigogine, I., *Exploring Complexity,* W.H. Freeman, San Francisco, 1989.

Paczulski, M., Kardar, M., and Nelson, D.R., *Phys. Rev. Lett.,* 60, 2638, 1988.

Rasmussen, J.J. and Rypdal, K., *Phys. Scr.,* 33, 481, 1986.

Scott, A.C., Chu, F.Y.F., and McLaughlin, D.W., *Proc. IEEE,* 61, 1443, 1973.

Spatschek, K.H., Taki, M., and Eickermann, T., in *Nonlinear Coherent Structures,* Barthes, M. and Léon, J., Eds., Springer, Berlin, 1990.

Taki, M. et al., *Physica,* 40D, 65, 1989.

Tang, C. and Bak, P., *J. Stat. Phys.,* 51, 797, 1988a.

Tang, C. and Bak, P., *Phys. Rev. Lett.,* 60, 2347, 1988b.

Ter Haar, D., Ed., *Collected Papers of L.D. Landau,* Pergamon Press, Oxford, U.K., 1965.

Thompson, J.M.T. and Stewart, H.B., *Nonlinear Dynamics and Chaos,* John Wiley & Sons, New York, 1986.

Tolédano, J.C. and Tolédano, P., *The Landau Theory of Phase Transitions,* World Scientific, Singapore, 1987.

Tuszynski, J.A., Clouter, M.J., and Kiefte, H., *Phys. Rev.* B, 33, 3423, 1986.

Vicsek, T., *Fractal Growth Phenomena,* World Scientific, Singapore, 1989.

Wilson, K.G., *Rev. Mod. Phys.,* 55, 583, 1983.

Winfree, A.T., *Sci. Amer.,* 230, 83, 1974.

Zabusky, N.J. and Kruskal, M.D., *Phys. Rev. Lett.,* 15, 240, 1965.

Appendix D

Master Equations

The brief description of derivation of a master equation for a stochastic process is based on a presentation by Goel and Richter-Dyn (1974) where a much more comprehensive treatment can be found.

Let us designate the discrete states of a random variable by the integers ... $-2, -1, 0, 1, 2, \ldots$. We further assume that in the time interval $(t, t + \Delta t)$, the probability of transition from state n to $n + 1$ is $\lambda_n \Delta t + O(\Delta t)$ and from state n to $n - 1$ is $\mu_n \Delta t + O(\Delta t)$. Hence, the probability of staying in state n is $1 - (\lambda_n + \mu_n)\Delta t + O(\Delta t)$ where $\lim_{\Delta t \to 0} O(\Delta t)/\Delta t = 0$. Under these conditions, if the process was initially $t = 0$ in state m, the probability that $P_{n,m}(t)$ of the process will be in state n at time t satisfies two types of difference equations:

$$P_{n,m}(t + \Delta t) = [\lambda_{n-1}\Delta t + O(\Delta t)]P_{n-1,m}(t) + [1 - (\mu_n + \lambda_n)\Delta t$$
$$+ O(\Delta t)]P_{n,m}(t) + [\mu_{n+1}\Delta t + O(\Delta t)]P_{n+1,m}(t) \qquad \text{(D.1)}$$

$$P_{n,m}(t + \Delta t) = [\mu_m \Delta t + O(\Delta t)]P_{n,m-1}(t) + [1 - (\mu_m + \lambda_m)\Delta t$$
$$+ O(\Delta t)]P_{n,m}(t) + [\lambda_m \Delta t + O(\Delta t)]P_{n,m+1}(t) \qquad \text{(D.2)}$$

Dividing both sides of the equation above by Δt and using the limit $\Delta t \to 0$, we obtain:

$$\frac{dP_{n,m}(t)}{dt} = \lambda_{n-1}P_{n-1,m}(t) - (\mu_n + \lambda_n)P_{n,m}(t) + \mu_{n+1}P_{n+1,m}(t) \qquad \text{(D.3)}$$

$$\frac{dP_{n,m}(t)}{dt} = \mu_m P_{n,m-1}(t) - (\mu_m + \lambda_m)P_{n,m}(t) + \lambda_m P_{n,m+1}(t) \qquad \text{(D.4)}$$

Equations (D.3) and (D.4) belong to a general class known as *master equations*. The general form of a master equation results when, in time interval $(t, t + \Delta t)$, transitions allowed are from state n to states $n - 1$ and $n + 1$ and to any other state. Equation (D.3) is called a forward master equation and Equation (D.4) is a backward master equation.

These equations, each of which is a set of difference-differential equations, are to be solved subject to an initial condition and some boundary conditions. It is sufficient to solve them for the initial condition:

$$P_{n,m}(0) = \delta_{n,m} \qquad \text{(D.5)}$$

i.e., for the case when the process is initially in a definite state m, because the solution for an initially distributed process is related to the solution for the definite initial state. If $\wp_m$ denotes the probability that the process is initially in state m (with $\sum_m \wp_m = 1$), the solution $P_n(t)$ of the master equation is:

$$P_n(t) = \sum_{\substack{state \\ space}} P_{n,m}(t)\wp_m \tag{D.6}$$

Boundary conditions depend on the allowed set of states. Basically, the random processes may be of two types: one that imposes no restrictions on the allowed set of states (unrestricted) and the second that imposes restrictions in the sense that some states have special properties (restricted). The calculation of $P_{n,m}(t)$ is somewhat difficult, and if one is interested only in the first few moments of n, one must solve a simpler set of equations. For example, the differential equations for the first moment:

$$\langle n \rangle = \sum n P_{n,m}(t) \tag{D.7}$$

where the summation is on the allowed states is:

$$\frac{d\langle n \rangle}{dt} = \langle \lambda_n \rangle - \langle \mu_n \rangle \qquad \langle \lambda_n \rangle \equiv \sum \lambda_n P_{n,m} \tag{D.8}$$

This equation results if we multiply Equation (D.3) by n and sum both sides over n to obtain:

$$\frac{d\langle n \rangle}{dt} = \left[\sum n\lambda_{n-1} P_{n-1,m} - \sum n\lambda_n P_{n,m} \right]$$
$$+ \left[\sum n\mu_{n+1} P_{n+1,m} - \sum n\mu_n P_{n,m} \right] \tag{D.9}$$

Note that since $\sum n\lambda_{n-1} P_{n-1,m} = \sum (n+1)\lambda_n P_{n,m}$, the expression in the first bracket in (D.9) is $\lambda_n P_{n,m}$, with an analogous expression for the second bracket. In general, the kth moment:

$$\langle n^k \rangle = \sum n^k P_{n,m}(t) \tag{D.10}$$

satisfies the differential equation:

$$\frac{d\langle n^k \rangle}{dt} = \langle \{(n+1)^k - n^k\}\lambda_n \rangle - \langle \{n^k - (n-1)^k\}\mu_n \rangle$$
$$= \sum_{i=0}^{k-1} \binom{k}{i} \{\langle n^i \lambda_n \rangle + (-1)^{k-i} \langle n^i \mu_n \rangle\} \tag{D.11}$$

where:

$$\binom{k}{i} = \frac{k!}{i!(k-i)!} \tag{D.12}$$

In particular, $\text{var}(n) \equiv \langle n^2 \rangle - \langle n \rangle^2$ satisfies the differential equation:

$$\frac{d}{dt}\text{var}(n) = 2\left\{\langle n\lambda_n \rangle - \langle n\mu_n \rangle\right\} + (1 - 2\langle n \rangle)\langle \lambda_n \rangle$$
$$+ (1 - 2\langle n \rangle)\langle \mu_n \rangle \tag{D.13}$$

The initial condition for $\langle n^k \rangle$ follows from Equation (D.5):

$$\langle n^k \rangle(t = 0) = m^k \tag{D.14a}$$
$$\langle n \rangle(t = 0) = m, \qquad \text{var}(n)(t = 0) = 0 \tag{D.14b}$$

When λ_n and μ_n are at most linear in n, the right side of Equation (D.8) is devoid of a higher than first moment of n. It can be solved for $\langle n \rangle$. Likewise, Equation (D.11) can be solved for $\langle n^k \rangle$. If λ_n and μ_n have terms in n of degrees higher than 1, in general Equations (D.8) and (D.11) cannot be solved. For this case, Equation (D.8) for $\langle n \rangle$ involves $\langle n^2 \rangle$, the equation for which involves $\langle n^3 \rangle$, and so on. This hierarchy of equations can only be solved by a variety of approximations.

References

Goel, N.S. and Richter-Dyn, N., *Stochastic Models in Biology*, Academic Press, New York, 1974.

9 780367 578541